Mechanical Behaviour of Engineering Materials

Mechanical Behaviour of Engineering Materials

Volume 2: Dynamic Loading and Intelligent Material Systems

by

Yehia M. Haddad

Faculty of Engineering,
Mechanical Engineering,
University of Ottawa,
Canada

SPRINGER-SCIENCE+BUSINESS MEDIA, B.V.

A C.I.P. Catalogue record for this book is available from the Library of Congress.

ISBN 978-1-4020-0350-9 ISBN 978-94-010-0436-7 (eBook)
DOI 10.1007/978-94-010-0436-7
Transferred to Digital Print 2001

Originally published by Kluwer Academic Publishers in 2000

This Volume is dedicated
to the memory
of
Professor D. R. Axelrad.

Table of Contents

Preface

I wish to express my full indebtedness to all researchers in the field. Without their outstanding contribution to knowledge, this book would not have been written.

The author wishes to express his sincere thanks and gratitude to Professors M. F. Ashby (University of Cambridge), N. D. Cristescu (University of Florida), N. Davids (The Pennsylvania State University), H. F. Frost (Dartmouth College), A. W. Hendry (University of Edinburgh), F. A. Leckie (University of California, Santa Barbara), A. K. Mukherjee (University of California, Davis), T. Nojima (Kyoto University), J. T. Pindera (University of Waterloo), J. W. Provan (University of Victoria), K. Tanaka (Kyoto University), Y. Tomita (Kobe University) and G. A. Webster (Imperial College), and to Dr. H. J. Sutherland (Sandia National Laboratories).

Permission granted to the author for the reproduction of figures and/or data by the following scientific societies, publishers and journals is gratefully acknowledged: ASME International, ASTM, Academic Press, Inc., Addison Wesley Longman (Pearson Education), American Chemical Society, American Institute of Physics, Archives of Mechanics / Engineering Transactions (*archiwum mechaniki stosowanej / rozprawy inzynierskie, Warsaw, Poland*), British Textile Technology Group, Butterworth-Heinemann Ltd. (USA), Chapman & Hall Ltd. (International Thomson Publishing Services Ltd.), Elsevier Science-NL (The Netherlands), Elsevier Science Limited (U.K.), Elsevier Sequoia S. A. (Switzerland), John Wiley & Sons, Inc., IOP Publishing Limited (UK), Kluwer Academic Publishers (The Netherlands), Les Editions de Physique Les Ulis (France), Pergamon Press Ltd. (U.S.A), Society for Experimental Mechanics, Inc. (USA), Springer-Verlag (Heidelberg, Germany), Steinkopff Verlag Darmstadt, Tappi, Technomic Publishing Co., Inc. (U.S.A.) and The Institute of Physics (UK).

I wish to express my deep appreciation to Drs. Karel Nederveen and Sabine Freisem (Publishing Editors), the sub-editorial staff at *Kluwer academic publishers*, and to *Kluwer* for the reviewing, editing, and the efficient production and distribution of the book.

I like to extend my thanks to past and present graduate students: Mrs. Y. Ping, and Messrs J. Feng, S. Iyer, P. Mirfakhraei and G. Molina, for their conscientious assistance during the preparation of the text. I am, also, grateful to my family for their understanding, patience and support.

I hope that the work presented in this book will provide guidance to science and engineering students, educators and researchers who are working in the field. Also, it is hoped that the book will be of value to scientists and engineers who are involved in the production, processing, and application of engineering materials.

Y. M. Haddad
University of Ottawa, Ottawa, Canada

List of Symbols

A	Energy per unit mass
A_0	Mean free energy
$\mathbf{A}$	Displacement vector of a particle along the plane of the wave
c_1, c_2	Magnitude of wave velocity (dilatational, rotational)
$C(t)$	Creep function
ds	Line element
$D(\cdot)/Dt$	Material derivative
$\mathbf{e}_k (k=1,2,3)$	Unit vectors associated with an external Cartesian frame of Reference
E	Elastic modulus
$E_1 (\omega)$	Storage modulus (frequency-dependent)
$E_2 (\omega)$	Loss modulus (frequency-dependent)
$E^*(i\omega)$	Dynamic complex modulus
$\mathbf{E}(t)$	Nonlinear strain measure
$f(\cdot)$	Constitutive functional
$\mathbf{F}$	Deformation measure
$\mathbf{g}$	Deformation gradient
$h(t)$	Characteristic function of a continuous system
$h_d(k);\ k=0,1,2,\ldots$	Characteristic series of a discrete-time system
$H(t)$	Heaviside sep function
$H_d(z)$	Transfer function of a discrete-time system
$\mathbf{I}$	Identity matrix
I_1, I_2, I_3	Stress invariants
II_1, II_2, II_3	Strain invariants

$J^*(i\omega)$	Dynamic creep compliance
K	Bulk modulus
L	Velocity gradient
L(·)	Constitutive function
n	Unit normal vector
$\overset{1}{\boldsymbol{n}}, \overset{2}{\boldsymbol{n}}, \overset{3}{\boldsymbol{n}}$	Unit vectors associated with the stress tensor principal axes
N(s), N'(s)	Frequency distribution (creep, relaxation)
r(t), **R**	Position vector (deformed, undeformed state)
$s = 1/\lambda$	Frequency
s	Laplace transform parameter
t, τ	Time parameter
Δt, ΔT	Time interval of sampling
T	Absolute temperature
T	Stress-traction vector
u	Displacement vector
$\mathbf{u}_I$	Irrotational displacement field
$\mathbf{u}_R$	Rotational displacement field
v	Velocity
v_{ij}	Rate of deformation tensor
W	Elastic energy per stress cycle
x, **X**	Position vector (deformed, undeformed state)
x(t)	Input to a dynamic system
$\{x_i\}$ (i=0,1,2,...)	Discrete-time series of an input
y(t)	Output of a dynamic system
$\{y_i\}$ (i=0,1,2,...)	Discrete-time series of an output

$Y(s)$	Laplace transform of an output signal
$(\cdot)^T$	Designates a transpose
	Indicates a transform
$\Vert \cdot \Vert$	Denotes a norm
α, β	Normalization factors (Also, microstructural elements)
$\alpha_m, \beta_m (m=1,2,\ldots,p)$	Parameters of a discrete-time system
α_m $(m=1,2,\ldots,p)$	Estimated values of parameter α_m
β_m $(m=1,2,\ldots,p)$	Estimated values of parameter β_m
γ	Specific rate (per unit mass) of entropy production
$\gamma_m (m=1,2,\ldots)$	Exponent factors
Γ_1^2, Γ_2^2	Wave operator (dilatational, rotational)
Δ	Dilatation
δ_{ij}, $\delta_{\alpha\beta}$	Kronecker delta (three-dimensional, two-dimensional)
∇^2	Laplace operator
ϵ_{ijk}, $\epsilon_{\alpha\beta}$	Alternating tensor (three-dimensional, two-dimensional)
ϵ_{ij}	Infinitesimal strain tensor
ϵ_{ij}'	Deviatoric strain tensor
η	Viscosity modulus
θ	Empirical temperature
λ_i $(i=1,2,\ldots,p)$	Roots of the characteristic equation of a discrete-time system
υ	Poisson's ratio
ξ_i $(i=1,2,\ldots,p)$	Roots of the characteristic equation of a continuous system
ρ, ρ_0	Mass density (current, reference configuration)
$\boldsymbol{\sigma}(\mathbf{t})$, $\sigma_{ij}(t)$	Cauchy stress tensor (time-dependent)

σ_{ij}'	Stress deviator
$\bar{\sigma}$	Mean stress
$\sigma_1, \sigma_2, \sigma_3$	Principal components of stress
Σ	Piola-Kirchhoff stress tensor
χ	Body force

INTRODUCTION

During the past four decades, the research efforts of investigating the mechanical response behaviour of engineering materials, under various types of loading, have been ultimately significant. The interpretations and applications of mechanical response data have simulated powerful advances in research interest and in engineering practice. In this context, widespread research work on the subject has established well-profound concepts, principles and results.

The purpose of this monograph is to introduce the principles of the mechanical response of various classes of engineering materials, the identification and interpretation of the mechanical response data, properties evaluation, and, whenever possible, application of the data to structure-properties relationships. The monograph deals with the subject matter in two volumes. Volume I, contains eight chapters and three appendices, concerns itself with the basic concepts as pertain to the entire monograph, together with the response behaviour of engineering materials under static and quasi-static loading, Thus, Volume I is dedicated to the introduction, the basic concepts and principles of the mechanical response of engineering materials, together with the pertaining analysis of elastic, elastic-plastic, and viscoelastic behaviour. Volume II, consists of ten chapters and one appendix, concerns itself with the mechanical behaviour of various classes of materials under dynamic loading, together with the effects of local and microstructural phenomena on the response behaviour of the material. Volume II contains also selected topics concerning intelligent material systems and pattern recognition and classification methodology for the characterization of material response states. In the majority of the presentation, the two volumes of the monograph treat the considered subjects in a generalized three-dimensional fashion.

Static loading?

In the case of static loading, one has, at any particular instant of time, a condition of static equilibrium. A conventional static tensile test of a material specimen within its linear elastic range would be a typical example of this situation.

Quasi-static Loading?

A quasi-static deformation process, although it is, in general, time-dependent, is in reality a sequence of states of static equilibrium. Typical illustrations of a quasi-static deformation process are the quasi-static creep and relaxation processes of engineering materials.

Dynamic Loading?

The deformation process that occurs in the material under dynamic loading differs to a large extent from those due to static or quasi-static loading. When the material is subjected to dynamic loading, e.g., a very high rate impulse, the portion of the body that contains the point of impact is stressed instantaneously, while the other portions may not have yet experienced the effect of the imposed impact. This is due to the fact that the imposed dynamic effect will require time to travel, i.e., to propagate, through the body. Such propagation of the dynamic effect through the body occurs with a particular velocity of propagation which would depend on the specific characteristics of the material and the boundary conditions at the instant of time considered. This phenomenon is referred to as "*wave propagation*". That is, the dynamic deformation of materials, under dynamic loading, involves stress wave propagation, whereby the inertia and inner kinetics of the material specimen play important roles.

At strain rates of the order 10^{-6} to 10^{-5} s^{-1}, the creep behaviour of the material is the primary consideration and creep laws are used to describe the mechanical behaviour. At higher rates, e.g., in the range of 10^{-4} to 10^{-3} s^{-1}, a uniaxial test, or a quasi-static stress-strain curve obtained from a constant strain-rate test is used to describe the material behaviour. Although the quasi-static stress-strain curve is often treated as an inherent property of the material, it is a valid description of the material only at the strain rate at which the test was conducted. At higher strain rates, the mechanical response of the material may change, and alternate testing techniques have to be used. The range of strain rates from 10^{-1} to 10^{2} s^{-1} is generally referred to as the intermediate or medium strain-rate regime. Within this regime, strain-rate effects become a consideration in most materials (e.g., metals), although the magnitude of such effects may be quite small. Strain rates of 10^{3} s^{-1} or higher are generally treated as the range of high strain-rate response. It is within the high strain-rate range (10^{3} s^{-1} or higher) that inertia and wave-propagation effects become important in interpreting experimental data. At these high rates of strain, care must be taken to distinguish between average and local values of stress and that may be the result of one, or more, high-intensity stress wave propagating through the material. At strain rates of 10^{5} s^{-1} or higher, we are generally dealing with "*shock waves*" propagating through the material. At these high rates, there exists a transition from nominally isothermal conditions to adiabatic conditions.

In the mechanics of deformable media, we deal with physical events, e.g. deformation and flow, that occur and evolve, in both space and time, independent of any particular coordinate system that may be used to observe them. In a proper mathematical description, such events and their governing laws are expressed in terms of tensorial quantities. The invariance of tensors under coordinate transformation highlights a principal reason for employing tensor calculus in the study of the mechanics of deformable media. When transformation is carried out from one homogeneous (rectangular) coordinate system to another, the resulting tensors are identified as "*Cartesian tensors*". However, in dealing with tensor transformation between general "*curvilinear*" coordinate systems, the pertaining tensorial quantities are referred to as "*Curvilinear" or "General"* tensors. In *Chapter 1*, the reader is introduced to Cartesian tensors. Curvilinear tensors, however, are considered in *Appendix A (Volume 1)*.

Two mechanical approaches are generally considered in the study of phenomena and problems concerning the mechanics of deformable media, i.e., the "*microstructural*" approach and the "*continuum mechanics*" approach.

In the "*microstructural*" approach, the macroscopic medium is considered to consist of a large number of structural elements. Such elements are assumed to be in continuous interaction with each other, and, hence their individual responses are seen to be mutually inter-dependent. The behavior of a statistical ensemble of such elements may be studied using, for instance, statistical or stochastic mechanics.

Conventionally, however, the description of material behavior is based on "*continuum mechanics*" models that generally refer to homogeneous media. In the "*continuum mechanics*" approach, the actual microstructure of the medium is disregarded and the medium is pictured as a "*continuum*" without gaps or empty spaces. Hence, the configuration of the assumed continuous medium would be described by a continuous mathematical model whose geometrical points are identified with material particles of the actual physical medium. The aim of *Chapter 2* is to provide the reader with a concise introduction of the basic assumptions and principles of *Continuum Mechanics* with an emphasis on those specifically used in the remainder of the book.

As mentioned earlier, engineering materials, when subjected to external loading, experience deformation and flow that evolve in space and time. Thus, in *Chapter 3*, we first consider the kinematics of involved deformation in the continuous material body and the determination of the pertaining strain by adopting a number of conventional measures. Second, we analyze the relationships between the sequential configurations that the parts of a "*continuous*" material body may acquire with the passage of time. Subsequently, in *Chapter 4*, we attempt to study the restrictions that classical thermodynamics impose on the theory of deformation of solids, and to seek information concerning the thermodynamics of continuous media.

Different materials of the same geometry may respond differently under identical external effects. Such difference in response is often attributed to the inherent constitution of the material. Consequently, the response behaviour of a particular material, or of a class of such material, is described mathematically by so-called "*constitutive relations*". These constitutive equations define the response behaviour of idealized media within a specific range of external effects. Accordingly, they only approximate the response characteristics of real materials, within a specified domain of actual service conditions. Constitutive relations establish, under certain physical and thermodynamical restrictions, the connection between the stimuli acting on the material specimen and the evolution of the occurring response. Thus, *Chapter 5* attempts to guide the reader throughout a transition between the general concepts and principles, which are presented in Chapters 1 to 4, and the task of establishing the response behaviour of engineering materials, as presented in Chapters 6 to 15. In this, the elastic response behaviour of the material is dealt with first.

Elastic behaviour of an engineering material depends only upon the stress level in the material, meanwhile, it is not strain- or time-history dependent. Further, an elastic deformation process is described, from a thermodynamical point of view, as dealt with in *Chapter 5*, to be a *reversible* process. Thus, upon the removal of the load, a complete recovery to the undeformed configuration would take place. An elastic response of an engineering material is formulated within the realm of "*classical elasticity*". Such an elastic response could be linear or nonlinear pending on the form of the constitutive law that is used in its description. In this context, *Chapter 6* deals first with the general nonlinear elastic behavior, then it introduces the required assumptions and postulates in order to reduce such response to the idealized case of linear (perfect) elastic behavior.

Two ways in which the behaviour of real solids deviates from a perfect elastic one:

First, the stress-strain relationship may be nonlinear and may also depend on the loading path. Further, the pertaining stress-strain curve may show hysteresis loops. Thus, the resulting stress-strain relationships may not be "*uni-valued*".

Second, the stress-strain relationship may be time-dependent. Thus, phenomena such as creep and stress-relaxation could become of importance, in determining the mechanical response of the real solid.

In general, "*inelastic*" solids show the above mentioned two types of deviation from a perfect (linear) elastic behavior. That is, the stress-strain relation is both time-dependent and nonlinear. Thus, inelastic deformation depends, in general, as dealt with in *Chapter 7*, on the stress level and both the strain- and time-history of the material. A transition to the important subject of creep and stress-relaxation of metallic systems is dealt with at the end of Chapter 7.

With the recent advances in material science and the parallel extensive industrial demands on advanced industrial materials such as high polymers and polymeric base composite systems, the identification of the viscoelastic response of engineering materials has gained recently a strong momentum in the realms of industrial techniques and applications.

High polymeric materials are organic substances of high molecular weight, the technical importance of which depends on their particular microstructure. This class of materials may include, for example, rubber in its various forms, synthetic rubber-like materials, commercial plastics, and natural and synthetic textile fibres. Other few examples of a viscoelastic material would include a wide range of inorganic polymeric systems such as silicones and glass resins, constituents of polymeric base systems, natural fibres such as wood and the by-products of such fibres as, for instance, paper and board, building materials such as concrete, and a large class of biomaterials, among others. These materials are "*time-dependent*" in response and possess a "*time-memory*". Attempts to characterize the behaviour of such materials under the action of external loading, consequently, gave rise to the science of *"rheology"* within which the phenomena now labelled "*viscoelasticity*" is well defined and intended to convey mechanical behaviour combining response characteristics of both an elastic solid and a viscous fluid. A viscoelastic material is, thus, characterized by a certain level of rigidity of an elastic solid body, but, at the same time, it flows and dissipates energy by frictional losses as a viscous fluid. *Chapter 8* treats the subject of viscoelasticity of engineering materials in a quite comprehensive manner. The important subject of thermoviscoelasticity is also dealt with in Chapter 8.

The significant importance of the subject of the dynamic response of engineering materials has, also, gained in recent years a strong momentum in a wide scope of engineering practice. Dynamic properties of materials appear to be receiving more attention at present as a result of such current applications as space structures, machine components, advanced aircraft, and nondestructive evaluation of engineering materials and structures. Familiar applications of the study of dynamic deformation of engineering materials may include, for instance,

- identification, modelling, and prediction of the response behaviour of different classes of engineering materials under the effect of rapidly changing loads.
- development of new materials that can perform favourably from a design point of view when subjected to dynamic loading.
- study of the dynamic response of engineering members and structures with the inclusion of the dynamic behaviour of the pertaining materials.
- identification of the response of materials during dynamic fabrication processes, e.g. metal forming operations under rapidly changing loads, explosive welding and compaction operations.
- development of nondestructive evaluation techniques that are based on dynamic-effect phenomena, e.g., acoustic emission, ultrasonics and acousto-ultrasonics.
- shock synthesis to produce new elements or compounds.
- study of crash worthiness.
- development of anti-collision shielding for space vehicles.
- traditional and novel armour and anti-armour concepts for military applications.

In *Chapter 9*, we introduce the subject of the response of metallic materials to dynamic loading. In this, the distinction of higher rates from lower rates is made not on the basis of time-dependence of the material behavior, as we dealt with, for instance, in Chapter 8, but rather on the necessity of including inertia forces in the pertaining dynamic analysis.

Chapter 10 deals with the subject of plastic instability and localization effects in engineering materials. In this context, a decrease in stiffness due to geometrical change and/or material softening caused by deformation is responsible for the occurrence of instability phenomena in engineering materials within the plastic range; i.e., beyond the yield point. Such phenomena manifest themselves in various ways; e.g., buckling, bulging, necking and shear banding. Once such instabilities are started, they tend to persist and the stiffness of the specific cross-sectional area of the specimen decreases; therefore deformation intensifies locally and eventually leads to final collapse and/or failure. Because the occurrence of such instabilities is an important precursor to collapse or failure, computational prediction of the onset and of the augmentation of these instabilities is essential and indispensable in understanding the ultimate strength of engineering materials, and in predicting and improving, for instance, the formability of ductile solids.

In rigid body dynamics, it is assumed that, when an external force is applied to any one point of the body, the resulting effect sets every other point of the body instantaneously in motion, and the applied force can be considered as producing a linear acceleration of the whole body, together with an angular acceleration about its center of gravity. In the theory of deformable media, however, the body is considered to be in equilibrium under the action of the external applied forces, and the occurring deformations are assumed to have reached their equilibrium static values. This assumption could be sufficiently accurate for problems in which the time between the application of the force and the setting up of effective equilibrium is short compared with the time in which the observation is made. Meanwhile, If the external force is applied for only a short period of time, or it is changing rapidly, the resulting effect must be considered from the point of view of "*stress wave*" motion. Thus, when a localized disturbance is applied suddenly into a medium, it will propagate to other parts of this medium. The local excitation is not detected at other positions of the medium instantaneously, as some time would be necessary for the disturbance to propagate from its source to other parts of the medium. This simple fact constitutes a general basis for the interesting subject of "*wave propagation*". In the particular case, when the suddenly applied disturbance is mechanical, e.g., an impact force, the resulting waves in the medium are due to mechanical stress effects and, thus, these waves are referred to as "*mechanical stress waves*", or simply "*stress waves*".

The propagation of stress waves in solids can be divided into two categories, "*elastic*" and "*inelastic*". When loading conditions result in stresses below the yield point, solids behave elastically and obey Hook's Law, and consequently stress waves are "*elastic*". As the intensity of applied loading is increased, the response of the material is driven out of the elastic range to a possible inelastic behavior. The behavior here may involve large deformation, internal heat generation, and often failure of the solid through a variety of mechanisms. In this context, "*plastic*" waves can be propagated in a material, such as a metal, which exhibits the phenomenon of yielding, when stressed beyond its proportional limit. The subject of elastic wave propagation in engineering materials is dealt with in *Chapter 11*. Meanwhile, in *Chapter 12*, we consider the plastic response of engineering materials under dynamic loading, whereby a rate-effect phenomenon might be occurring in the material and the inertia forces would be included in the equation of motion.

Chapter 13 deals with the identification problem of the linear viscoelastic response behaviour of an engineering material using dynamic experimental measurements. In this context, a linear viscoelastic material is considered as a dynamic system, whereby, a dynamic system identification method is developed for the determination of the relaxation or creep function of the material.

In most of viscoelastic material components, the presence of mechanical dissipation can effectively change the nature of wave motion in them. In addition to the significant mechanical dissipation that can occur in viscoelastic materials, it is well-recognized that these materials are "*dispersive*". In view of the latter property, phase velocity of a wave propagating in a viscoelastic material will depend on wave frequency. More specifically, waves of high frequency will propagate in viscoelastic materials with a greater phase velocity than if these waves have a low frequency. Consequently, a mechanical disturbance would continually change in shape during its motion in a viscoelastic medium. Further, the attenuation of high frequency waves in viscoelastic materials is greater than that of waves of low frequency. In this context, *Chapter 14* concerns itself with the phenomenon of wave propagation in a viscoelastic solid and the associated with boundary value problem.

The current technology of the design and manufacturing of laminated and fibre-reinforced composites is faced with problems essentially related to the inherent nature of the mechanical response of the different constituents of the microstructure, the formation of interfaces between such constituents and the evolution of the associated deformation processes under loading. Optimal design of such material systems is becoming a very progressive and challenging domain in both applied mechanics and material science. Thus, the increasing use of such materials is inciting new developments to be made within the context of macro- and micro-mechanical constitutive modelling, applications of such materials under variable boundary conditions, experimental testing methods, computational methods of analysis, and optimization. A new dimension of optimal design is being realized by building new composite systems through direct tailoring of the microstructure, e.g., by judicious reinforcement and mixing (hybridization) of the constituents of the microstructure within a specific topological frame of reference and to satisfy the boundary conditions involved. In this context, theoretical and experimental studies of the dynamic stress-strain relations of hybrid composites have become significantly important. The increased interest in the subject matter has been motivated recently by the increasing number of engineering applications and, as well, by the contributions provided by such studies to a better understanding of the mechanisms of deformation of such material systems when subjected to a dynamic loading environment. In this, *Chapter 15* reviews recent research efforts pertaining to the micromechanics of polymeric fibre-composite systems, in general, and the optimization of the microstructure in the case of short-fibre composite systems.

Chapter 16 deals with the microstructural or microscopic effects on the response behaviour of structured material systems. In this, the material system is considered as a heterogenous medium of actual microstructural elements. These elements are seen to exhibit random geometric and physical characteristics. Due to the discrete nature of the microstructure , the pertaining deformation process and its space- and time-evolutions are considered to be stochastic in character. Thus, the overall response behaviour of the material is formulated by the use of probabilistic concepts and statistical theory. An important feature of the presented analysis is the introduction of a so called "*Material Operator*" of the structured material system that contains in its argument the significant response characteristics of the microstructure. These concepts are, first, utilized to formulate the outlines of a stochastic micromechanical model for the deformation of a heterogenous elastoplastic system. Then, the presented approach is extended to include the analysis of probable internal damage mechanisms in this class of material.

Engineering materials are used either for their inherent structural strength or for their functional properties. Often a feed back control loop is designed so that the mechanical response of the material is monitored and the environment that is causing such a response can be controlled. The evolution of a new kind of material termed "*Intelligent*", "*Smart*", or "*Adaptive*" witnesses a significant development in materials science whereby the referred-to smart material adapts itself to suit the environment rather than necessitating to control the same. In this context, development in the area of materials research aims at incorporating intelligence into engineering materials, enabling them to sense the external stimuli and alter their own properties to adapt to the changes in the environment.
Chapter 17 presents *"an overview"* of possible forms of intelligence that may be incorporated in these materials. Three basic mechanisms of intelligent materials, namely, the sensor, processor and actuator functions are described. Implementation of these in the microstructure of various materials, as well as associated algorithms and techniques are illustrated. Different models, control algorithms and analyses are reviewed and their potential applications in engineering materials are presented.

Chapter 18 deals with the design procedure of a computer-based expert system, in conjunction with a non-destructive quantitative examination technique, e.g., acousto-ultrasonics, for the identification of material response states.

Acousto-ultrasonics (AU) is a relatively new quantitative non-destructive examination technique that combines aspects of conventional "*Ultrasonic*" and "*Acoustic Emission*" practices. It has been proven to be a suitable approach to quantify microstructural and morphological states of materials and the related mechanical properties.

In the AU practice, the multi-interactions of the ultrasonic-wave with the material microstructure usually result in complicated waveforms that are quite difficult to analyse. A relatively new approach to the analysis of AU signals is the use of "*Pattern-recognition and Classification Methodologies*". In this approach, acousto-ultrasonic waveforms are identified as belonging to a number of classes, where each class represents one of different states of the tested material-property. For this purpose, each waveform is mathematically treated as a multi-parametric entity, which is called a "*pattern vector*". Each component of such a pattern vector represents a value of a parameter, called "*feature*", which is used for the identification of the AU signal. In the pattern-recognition practice, a computer-based pattern-recognition system, labelled "*Pattern-recognition Classifier*", is designed on the basis of AU signals pertaining to known material states of a particular tested response or material property. Two case studies are being dealt with in Chapter 18, i.e., characterization of the stress-relaxation response of a class of polymeric system, and the identification of residual impact properties of such a system.

Throughout the text, generalized tensorial notations are used. For simplification, however, the presentation has been limited, as much as possible, to Cartesian tensors. *Appendix A* (Volume I), however, introduces to the reader the basics of "*Curvilinear or General tensors*". This will prove to be particularly useful when reading Chapter 10. *Appendix B* (Volume I) presents the definition and a summary of the properties of both the delta and step functions. These functions are used frequently throughout the text. *Meantime,* the important subject of integral transformation is dealt with in *Appendix C* (Volume I). *Appendix D* (Volume II) deals with the definition and basic properties of z-transform. The latter is employed throughout Chapter 13.

In the presentation, vectors and unindexed tensor quantities are indicated in general by bold. The author has used majuscules to identify the undeformed configuration or the original state of the material and minuscules to designate the corresponding deformed state. Equations, figures and tables are numbered within the chapter; for example, Fig. 2.1 identifies Fig. 1 of Chapter 2.

Further Reading

Atkins, A. G. (1969) *The Rheology of Solids: High Speed Testing*, Vol. II, Wiley Interscience, New York.

Campbell, J. D. (1970) *Dynamic Plasticity of Metals*, Course held at the Department of Mechanics of Deformable Bodies, Udine, July 1970, Springer-Verlag, New York.

Cristescu, N. (1967) *Dynamic Plasticity*, North Holland, Amsterdam.

Davison, L. and Graham, R. A. (1979) Shock compression of solids, *Phys. Rep.* **55** (4), 255-379.

Duvall, G. F. (1962) Shock waves in the study of solids, *Appl. Mech. Rev.* **15**, 849-54.

Edington, J. W. (1968) Effect of strain rate on the dislocation substructure in deformed Niobium single crystals, In *Mechanical Behavior of Materials under Dynamic Loads*, edited by U. S. Lindholm, Symposium held in San Antonio, Texas, Sept. 6-8, 1967, Springer-Verlag, New York, pp. 191-240.

Greszczuk, L. B. (1982) Damage in composite materials due to low velocity impact, in *Impact Dynamics*, Zukas, J. A., Nicholas, T., Swift, H. F., Greszczuk, L.B. and Curran (eds.), John Wiley & Sons, New York, pp. 55-94.

Haddad, Y. M. (1995) *Viscoelasticity of Engineering Materials*, Kluwer, Dordrecht, The Netherlands.

Haddad, Y. M., editor (1998) *Advanced Multilayered and Fibre-Reinforced Composites*, Kluwer, Dordrecht, The Netherlands.

Hahn, G. T., Kanninen, M. F., Mukherjee, A. K. and Rosenfield, A. R. (1968) The speed of ductile-crack propagation and the dynamics flow in metals. In: *Mechanical Behavior of Materials under Dynamic Loads*, edited by U. S. Lindholm, Symposium held in San Antonio, Texas, Sept. 6-8, 1967, Springer-Verlag, New York, pp. 96-133.

Hopkins, H. G. (1968) *Engineering Plasticity*, University Press, Cambridge,

Huffington, N. J., Jr. , ed. (1965) *Behavior of Materials Under Dynamic Loading*, ASME, New York.

Johnson, W. (1979) Application: Processes involving high strain rates. In *Mechanical Properties at High Rates of Strain*, 1979, Proceedings of the Second Conference on the Mechanical Properties of Materials at High Rates of Strain, J. Harding (editor), Oxford, 1979, The Institute of Physics, Conference Series Number 47, pp. 337-59.

Kolsky, H.(1963) *Stress Waves in Solids*, Dover, New York.

Kornhauser, M. (1964) *Structural Effects of Impact*, Cleaver Hume, London.

Kumar, S. (1968) Introduction- Mechanics/Materials, Aspects of Dynamic Loading. In: *Mechanical Behavior of Materials under Dynamic Loads*, edited by U. S. Lindholm, Symposium held in San Antonio, Texas, Sept. 6-8, 1967, Springer-Verlag Inc., New York, pp. ix-xv.

Lindholm, U. S. (1965) Dynamic deformation of metals, in *Behavior of Materials Under Dynamic Loading*, Huffington, N. J., Jr. , (ed.), ASME, New York, pp. 42-61..

Lindholm, U. S. (1968) Some experiments in dynamic plasticity under combined stress, In: *Mechanical Behavior of Materials under Dynamic Loads*, edited by U. S. Lindholm, Symposium held in San Antonio, Texas, Sept. 6-8, 1967, Springer-Verlag, New York, pp. 77-95.

Meyers, M. A. (1994) *Dynamic Behaviour of Materials*, John Wiley & Sons, New York.

Nicholas, T. (1982) Material behavior at high strain rates, in *Impact Dynamics*, Zukas, J. A., Nicholas, T., Swift, H. F., Greszczuk, L.B. and Curran (eds.), John Wiley & Sons, New York, pp. 277-332.

Perzyna, P. (1968) On thermodynamic foundations of viscoplasticity, In *Mechanical Behavior of Materials under Dynamic Loads*, edited by U. S. Lindholm, Symposium held in San Antonio, Texas, Sept. 6-8, 1967, Springer-Verlag, New York, pp. 61-76.

Ripperger, E A. and Watson, H., Jr. (1968) The relationship between the constitutive equation and one-dimensional wave propagation, In: *Mechanical Behavior of Materials under Dynamic Loads*, edited by U. S. Lindholm, Symposium held in San Antonio, Texas, Sept. 6-8, 1967, Springer-Verlag, New York, pp. 294-313.

CHAPTER 9

TRANSITION TO THE DYNAMIC BEHAVIOUR OF ENGINEERING MATERIALS

9.1. Introduction

In this chapter, we introduce the subject of the response of metallic materials to dynamic loading. In this context, if we consider the term "*dynamic*" to be solely characterized as "*time-dependent*", then we are in fact, as Lindholm (1962, 1964 and 1978) pointed out, including the entire range of material performance. In other words, the commonly called "***static***" or "***quasi-static***" deformation, e.g., creep and relaxation, is, in effect, "***dynamic***" or "***time-dependent***". The majority of us, however, may be more accustomed to thinking of dynamic loading as associated with high loading rates or high deformation rates, with the adjective "high" referring to rates above those achieved on a standard testing machine. *In this, the distinction of higher rates from lower rates is often made, however, not on the basis of time-dependence of the material behaviour, but rather on the necessity of including inertia forces in the pertaining dynamic analysis.*

However, according to the laws of mechanics, *see Chapter 2*, the inertia forces are generally included in the equations of motion, not in the constitutive relations of the material. Further, one may often argue that the mechanisms which lead to time-dependency in the constitutive equations for plastically deforming metals are not significantly different, if they are not basically the same for low and high rates of loading, so that on the basis of the constitutive relations alone, the distinction between static and dynamic loading may not be easily made (e.g., Lindholm, 1962, 1964 and 1978).

9.1.1. LOADING REGIMES

Following Lindholm (1962, 1964 and 1978), and the presentation in the introduction to this book, the following loading regimes of engineering materials may be identified:

- ***Sub-static regime.*** The lowest strain rate regime is that generally associated with creep, where, as dealt with in Chapters 7 and 8, the specimen is deformed under constant load or stress and the strain *vs.* time or creep rate is recorded. The counter part of creep is the so-called "***stress -relaxation***" where the material specimen is instantaneously strained to a specific strain level, which is maintained constant for the entire duration of the experiment, and the stress vs. time or relaxation-rate is recorded.

- ***Static regime.*** The next regime is generally referred-to as static testing. The term "***static***" is emphasized as it refers, in general, to stationary or time-independent behavior. In this regime, the strain rate level should be maintained constant, and be identified in conjunction with the test results. Here, standard hydraulic, or preferably screw machines are used to apply the load.

- ***Dynamic regime.*** In this regime, inertia forces become important, and mechanical resonance in the machine and the specimen must be considered. High strain rates from approximately 50 s^{-1} to 10^4 s^{-1} can be obtained with mechanical impact from a moving mass, or by explosively generated pulses. At very high or "***hypervelocity***" impact, the impacting projectile is usually accelerated by means of a light gas gun or explosive generator. The analysis of the impact results must include the propagation of elastic and plastic waves. At the highest impact velocities, "***material compressibility***" becomes dominant and shock waves are developed. In this range, strain is not the appropriate deformation indicator, but rather the time-dependent particle and wave velocities are measured. Thus, the response of the material is considered within the realm of wave propagation theory. In this chapter, we introduce some results pertaining to the behavior of various metallic materials to high strain-rate loading, meantime the subject of wave propagation in such materials are dealt with within the scope of elastic wave propagation in Chapter 11, and in the context of dynamic plasticity in Chapter 12.

The mechanical response of metals under high rates of loading may differ significantly, from the corresponding response within the static regime. In this context, one may refer to the early experiments by Hopkinson (1905) when he conducted a series of dynamic experiments on steel and concluded that the dynamic strength was at least twice as high as its low-strain-rate strength. In this context, significant increases in flow stress were reported (e.g., Clifton, 1979) as strain rates of the order of 10^5 s^{-1} are attained. Such significant rise in flow stress leads some researchers to believe that there might be indeed a "***limiting strain-rate***" at which the strength of the material might approach infinity. On the other hand, however, it is known that steels undergo a ductile-to-brittle transition when the strain rate is significantly increased.

Kolsky (1960) devised a method for measuring the stress-strain behavior at very high rates of loading, without setting up stress waves in the material specimen. Applying his method, Kolsky used specimens in the form of thin circular disks which were placed between two steel bars of the same diameter as the discs along which stress pulses were propagated. With this arrangement, the inertia of the specimen and plastic wave propagation in it could be neglected, and since the wave propagation in the steel bars was elastic and amenable to calculation, the stress-strain relation of the specimen could be determined; see, also, Davies and Hunter (1963) and Kolsky (1965). An apparatus was subsequently developed by Kolsky and Douch (1962) to carry out such measurements. In this method, short cylindrical metal rods were fired from an air gun to impinge axially on a steel bar of the same diameter as the rods. The stress was measured by observing the amplitude of the elastic wave propagated in the steel bar. Meanwhile, the plastic strains were measured by examining the specimens after impact. By using a series of specimens fired over a range of velocities, dynamic stress-strain curves were obtained. The "***Kolsky bar***" often referred-to as the "***Split Hopkinson Pressure Bar***" is shown schematically in Figure 9.1.

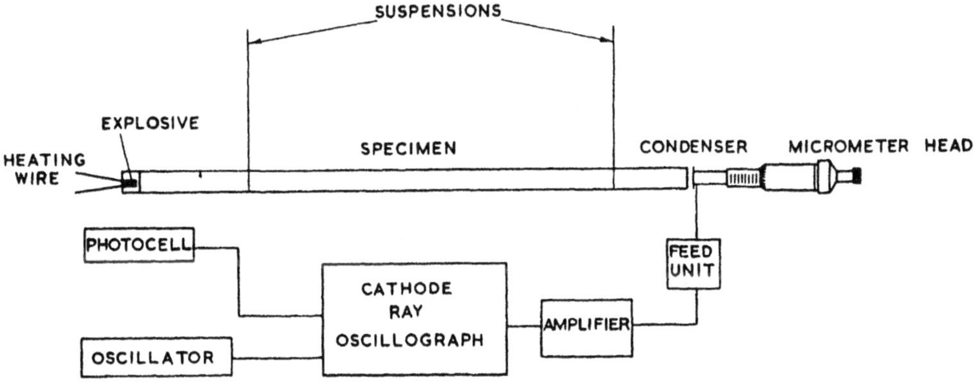

Figure 9.1. The "Kolsky bar". (Source: Kolsky, H. (1960) Viscoelastic waves, in Int. Symp. on Stress Wave Propagation in Materials, Ed. N. Davids, Interscience Publishers, London, pp. 59-90. Reprinted with permission).

Figure 9.2 is due to Kolsky (1965). It shows a comparison between the "dynamic" stress-strain curve for annealed bars of ***aluminum*** [better than 99.5 per cent pure] obtained according to the above described method, and the "*static*" stress-strain curve for similar bars measured on a conventional testing machine. It is seen from the figure that there is a definite strain-rate effect. Meanwhile, Figure 9.3 shows a correlation between the velocity of impact and the magnitude of the permanent strain in the material.

As we discussed in Chapter 7, materials such as metals exhibit, in general, nonlinear stress-strain relations and plastic yielding, and the stress-strain curve for unloading is generally different from that for loading (*see, also, Chapter 12*). When the strain rate increases, the deformation process changes gradually from fully isothermal to fully adiabatic, as there is not enough time for the heat generated during the deformation process to escape out of the body. This gives rise, in some cases, to adiabatic shear instabilities that have a profound effect on the mechanical response of the material. This phenomenon is discussed in the following chapter (*Chapter 10*) within the scope of plastic instability and localization effects.

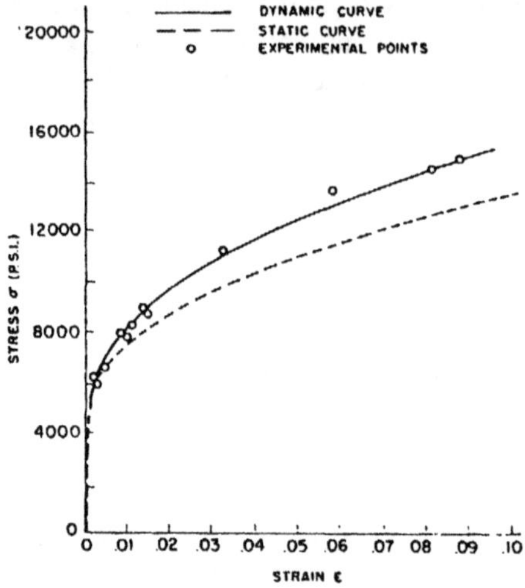

Figure 9.2. Dynamic and static stress-strain curves for ***annealed aluminum***. Reprinted from "Kolsky, H., The propagation of mechanical pulses in anelastic solids, in: *Behaviour of Materials under Dynamic Loading*, edited by N. J. Huffington, Jr., The American Society of Mechanical Engineers, New York, 1965, 1-18", with kind permission of The American Society of Mechanical Engineers.

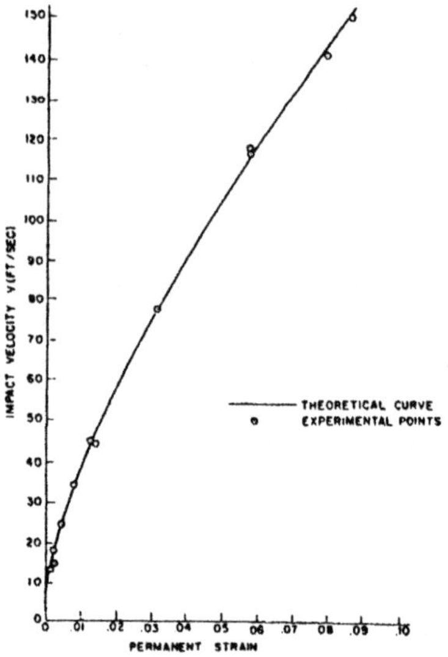

Figure 9.3. Variation of strain with impact velocity for ***aluminum***. Reprinted from "Kolsky, H., The propagation of mechanical pulses in anelastic solids, in: *Behaviour of Materials under Dynamic Loading*, edited by N. J. Huffington, Jr., The American Society of Mechanical Engineers, New York, 1965, 1-18", with kind permission of The American Society of Mechanical Engineers.

The rate dependency of the mechanical behavior of materials in general is a key factor in understanding the fundamental mechanisms involved in the deformation process (Campbell, 1968). According to Campbell (1968), the realization of this fact for the cases of metals and non-metals was delayed for two reasons: First, many common alloys are relatively rate-insensitive at normal rates of strain, so that it is necessary to use experimental techniques capable of measuring accurately small increments of stress and strain over a very wide range of strain rate. Second, the fundamental processes of plastic flow in metals could not be investigated thoroughly until adequate techniques were developed for observing dislocations and their properties. In this context, the possibility of direct measurement of dislocation velocities was a major step towards the goal of relating macroscopically observable quantities, e.g., stress, strain and strain-rate, to basic microstructural deformation mechanisms in metals, such as dislocations and other microstructural rate-controlling processes.

9.2. Response Behaviour of Metals under Dynamic Loading

9.2.1. STRAIN-RATE SENSITIVITY / STRAIN-RATE HISTORY

Some metals, e.g., aluminum (FCC) and copper (FCC), may show sensitivity to both strain-rate and strain-rate history. Other metals may show sensitivity to strain-rate only, e. g., steels (BCC) and titanium (HCP).

9.2.2. THE JUMP TEST "INCREMENTAL STRAIN-RATE TEST"

A well-known experiment to study the history effects in metals is the so-called "***jump test***", often referred-to as the "***incremental strain-rate test***". The main objective of performing the referred-to jump tests is to obtain information concerning the dynamic response of the material that may be used in the development of the pertaining constitutive equations.

A jump test is effected by combining two types of loading: a quasi-static loading is applied first and, without unloading, it is followed by a dynamic loading. Typical of the apparatus used to perform a jump test is the stored-torque "*Split Hopkinson Bar*"; e.g., Campbell *et al.* 1977 and Duffy, 1979. In this experimental set-up, the specimen is a thin-walled tube placed near the center of the bar and loaded in torsion. Quasi-static loading is first applied by, e.g., an electric motor at one end of the specimen, which turns the bar against a clamping mechanism. Dynamic loading is then applied from the other end by the sudden release of a stored torque. In Campbell's apparatus (Campbell *et al.*, 1977), the clamp is the crucial part of the apparatus; its release is effected by the fracture of a brittle bolt. The referred-to clamp must meet two requirements: First, it must provide as short time as possible between first arrival of the pulse and the establishment of a constant strain rate. In referred-to Campbell's bar, this rise-time is 25 or 30 μs which corresponds to about 1.5% strain accumulation in the specimen before a constant strain-rate of 10^3 s^{-1} is attained. Second, the pulse must be pure torsional, i.e., not be accompanied by pulses in other directions, e. g., an axial pulse or a bending pulse. This functional requirement of the apparatus can be incorporated in the design of the clamp itself (Duffy, 1979).

Figure 9.4 shows schematically an example of the stress-strain curve (Senseny, Duffy and Hawley, 1978) which was obtained during a jump test. The lowest curve in the Figure is for loading at a constant low strain rate, $\dot{\gamma}_i$, e.g., of the order of 10^{-4} or 10^{-3} s^{-1}. The highest curve was obtained at a constant high strain rate, $\dot{\gamma}_r$, e.g. 10^3 s^{-1}. During a jump test, the stress-strain curve follows the path *ABCD*. As shown in the figure, at point *B*, the strain rate changes abruptly from $\dot{\gamma}_i$ to $\dot{\gamma}_s$, with a resulting increase in stress of $\Delta\tau_s$.

For many purposes, the "***strain-rate sensitivity***" of a material is evaluated by comparing values along constant strain rate curves, e.g., the curves corresponding to constant strain rates $\dot{\gamma}_i$ and $\dot{\gamma}_r$ in Fig.9.4. However, a comparison of conditions at points C and F shows that the strain and strain rate remain the same at theses points, while the stress is different. Thus, one may consider the possibility that the behaviour of the material may depend on some other factor, one involving, say the "***deformation history***". Hence, with reference to Fig. 9.4, $\Delta\tau_h$ is often viewed as a measure of the "***deformation history***" dependency, while $\Delta\tau_s$ is considered to be related to the "***true strain rate sensitivity***".

Figures 9.5 and 9.6 show the results, due to Lindholm (1964), for cyclic loading of ***aluminum*** (FCC). In this context, Figure 9.5 shows the true stress *vs.* the true strain from cyclic static-dynamic-static loading, while Figure 9.6 demonstrates the corresponding results in the case of cyclic dynamic-static-dynamic loading. It is evident, from Figure 9.5, that the stress in a dynamic test following static pre-loading is not equal to the stress found at the same strain in all-dynamic loading (the dotted line). This difference is apparently due to "***strain rate history***" (Duffy, 1979).

Bodner and Partoum (1975) and Bodner and Merzer (1978) define an "*internal variable*", based ultimately on a relation between dislocation velocity and stress. With this internal variable, these researchers evaluate the parameters in their constitutive equations from the results of tests carried out, using jump tests, at constant strain rates. The results from the jump tests are then used further to establish the validity of their constitutive equations. Reference, in this context, is made Klepaczko (1968, 1975), Campbell *et al.* (1978) and Ponter (1978).

Does the strain-rate history effect is influenced by the "dwell time"?

- For ***aluminum*** **(FCC)**, for instance, Lindholm (1964) considered the effect of dwell time at zero load. In this context, Lindholm loaded a specimen dynamically at 8% strain, unloaded, and then reloaded dynamically. In this context, the results of Fig. 9.7 show a "*history effect*" for a dwell-time of three minutes and greater, while for a dwell-time of 450 μs none is observed.

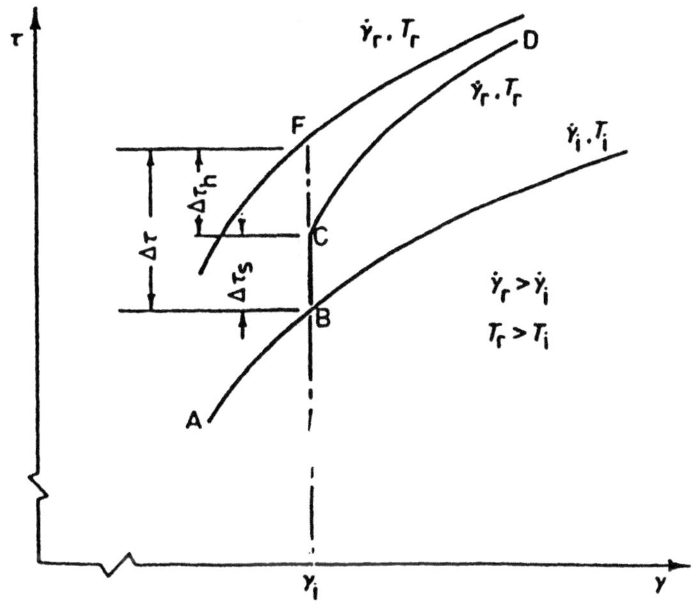

Figure 9.4. Schematic stress-strain curves showing effect of strain rate and strain-rate history on flow stress. From: Senseny, P. E., Duffy, J. and Hawley, R. H. (1978) Experiments on strain rate history and temperature effects during plastic deformation of close-packed metals, *J. Appl. Mech.* **45**, March 1978, 60-6. Reprinted with permission from ASME International.

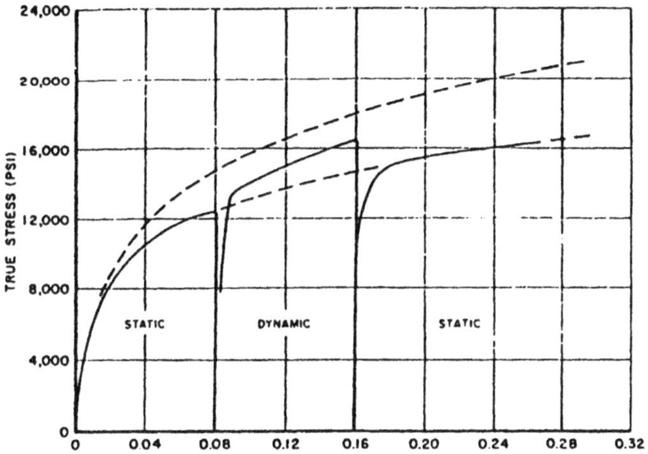

Figure 9.5. Cyclic static-dynamic-static loading for ***aluminum***. "Reprinted from *J. Mech. Phys. Solids* **12**, Lindholm, U.S., Some experiments with the Split Hopkinson Pressure Bar, 317-35, (1964), with permission from Elsevier Science".

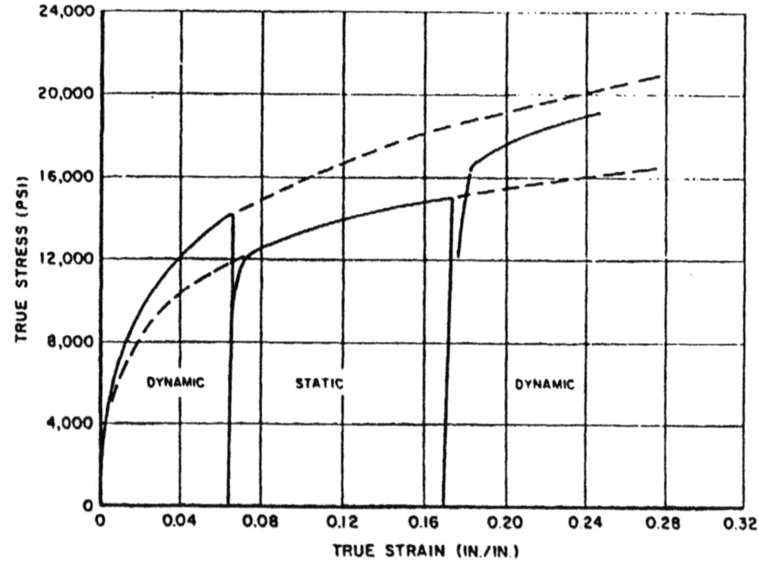

Figure 9.6. Cyclic dynamic-static-dynamic loading for ***aluminum***. "Reprinted from *J. Mech. Phys. Solids* **12**, Lindholm, U.S., Some experiments with the Split Hopkinson Pressure Bar, 317-35, (1964), with permission from Elsevier Science"

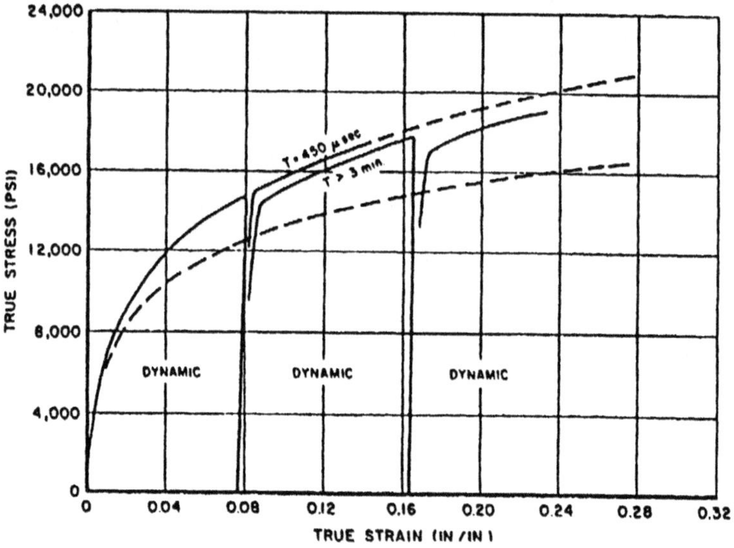

Figure 9.7. Cyclic dynamic loading for ***aluminum***. "Reprinted from *J. Mech. Phys. Solids* **12**, Lindholm, U.S., Some experiments with the Split Hopkinson Pressure Bar, 317-35, (1964), with permission from Elsevier Science".

- For ***steels* (BCC),** jump tests were performed by Barraclough and Sellars (1974), Both ***stainless steel*** and ***low alloy steel*** were considered. Rods of the material were loaded in torsion at a temperature of about 1000 °C. Their results indicate that steel is strongly influenced by strain rate (at least at this temperature). On the other hand, steel appeared almost insensitive to strain-rate history, as far as the jump test is concerned. Jump tests to higher strain rates were performed by Wilson *et al* (1979). Again the results show a strong strain rate sensitivity, but an insensitivity to strain rate history (see Duffy, 1979).

A peculiar aspect of the behavior of steels was also noticed by examining the results of Eleiche and Campbell (1976a) and pointed out by Duffy (1979). In this, it was shown that all high strain curves reach a maximum and then turn down. The latter effect was suspected by Duffy (1979) to be due to a material instability effect; see, e.g., Costin et al (1979). In general, it appeared that steel (BBC) is not as sensitive to strain rate history effects as are FCC metals, but that strain rate effects are high.

Within the strain rate range of 10^{-4} to 10^{3} s^{-1}, constant and incremental strain rate tests were performed by Tanaka and Nojima (1979) on 0.02% and 0.45% C. steels: 1) A 0.02% C. steel (0.02% C; 0.01% Si; 0.31% Mn; 0.008% P; 0.012% S; balance is Fe), and 2) A 0.45% C steel (0.45% C; 0.24% Si; 0.64% Mn; 0.002% P; 0.13% S; balance. Fe). In this context, after machining, the material specimens (5 mm in diameter, and 5 mm in length) were annealed for 1 hr. at 800°C and cooled in a vacuum furnace.
A split Hopkinson pressure bar apparatus and an Instron testing machine were used for the high strain rate (10^{2} to 10^{3} s^{-1}) and low strain rates (10^{-4} to 10^{-2} s^{-1}) tests, respectively. In the incremental strain rate tests, a stepped striker bar was used in the Hopkinson bar apparatus. In the tests in which deformation was rapidly stopped at high strain rates, a device (a stopper) was installed between the input and output bars in the apparatus (*see* Tanaka and Nojime, 1979).
Constant strain rate tests were performed at plastic strain rates of ϵ_p = 10^{-4} to 10^{3} s^{-1}, and temperatures of 78 to 290 °K. The relations between flow stress σ and the plastic strain rate (log $\dot{\epsilon}_p$), at the plastic strain ϵ_p = 5% are shown in Figures 9.8 a&b for 0.02%C and 0.45% C steels, respectively. As shown in these figures, the stresses are considerably affected by the plastic strain rate, especially at high temperatures.
Incremental strain rate change tests were performed by Tanaka and Nojima (1979) at both high and low strain rates, and the values of the strain rate sensitivity of the stress as defined by equation (9.1) below were determined.

Strain rate sensitivity of the stress: $$h_\tau = \Delta\sigma / \Delta\log \dot{\epsilon}_p \qquad (9.1)$$

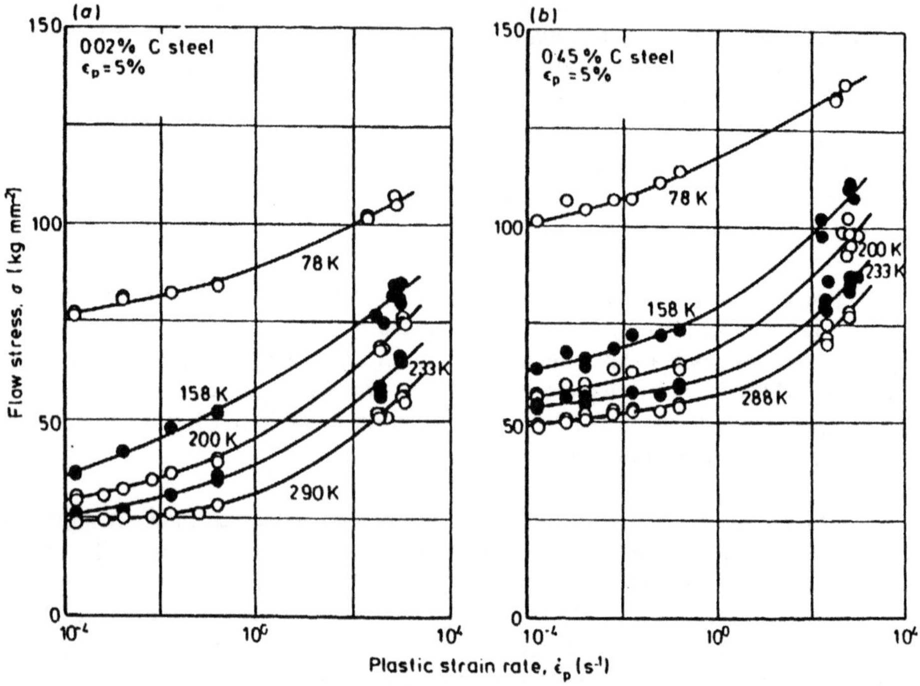

Figure 9.8. Flow stress σ- plastic strain rate $\dot{\epsilon}_p$ relations. (a) ***0.02% C. Steel***; (b) ***0.45% C. steel***. From: Tanaka, K. and Nojima, T. (1979) Dynamic and static strength of steels, In: Mechanical Properties at High Rates of Strain, *Proceedings of the Second Conference on the Mechanical Properties of Materials at High Rates of Strain*, J. Harding (editor), Oxford, 28-30, 1979, The Institute of Physics, Conference Series Number 47, pp. 166-73. Reprinted with kind permission of the Institute of Physics.

The obtained values of h_τ are shown in Figures 9.9 a&b. Values of the strain rate sensitivity, $k_\tau = \sigma / \log \dot{\epsilon}_p$, were also determined from the slopes of the σ - $\log \dot{\epsilon}_p$ relations which were obtained from constant strain rate tests. In both types of steel, it was found by the authors that the values of h_τ are larger than those of k_τ especially at low strain rates.

More typical of the behavior of ***copper*** (FCC) are the results of Klepaczko *et al* (1977). These results were obtained by means of a torsional "Kolsky bar", Fig. 9.10, in which the pulse is initiated explosively rather than by means of a stored torque. This technique (for explosive loading) was developed by Duffy *et al* (1971). The technique has the advantage of producing a pulse with a much shorter rise-time (8 μs), but of shorter duration (about 100 μs). An extensive series of jump tests were performed with this bar by Senseny *et al* (1978) whereby four metals were tested; namely, ***aluminum*** (FCC), ***copper*** (FCC), ***magnesium*** (HCP) and ***zinc*** (HCP). The results are shown in Fig. 9.11.

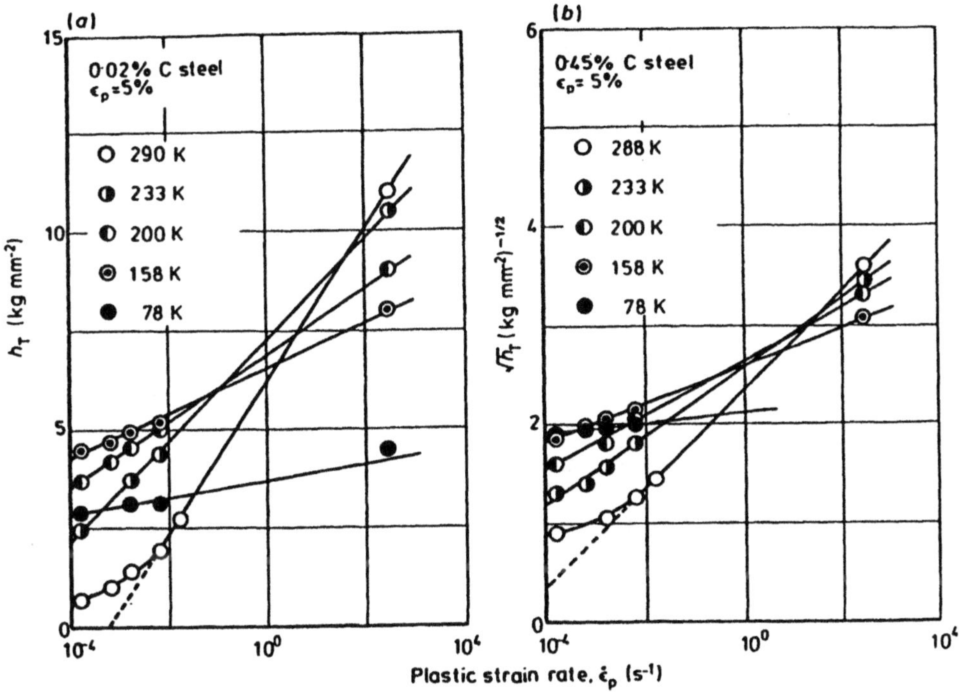

Figure 9.9. (a) h_T - $log\ \dot{\epsilon}_p$ relations (***0.02%C Steel***); (b) $\sqrt{h_T}$ - $log\ \dot{\epsilon}_p$ relations (***0.45% C steel***). From: Tanaka, K. and Nojima, T. (1979) Dynamic and static strength of steels, In: Mechanical Properties at High Rates of Strain, *Proceedings of the Second Conference on the Mechanical Properties of Materials at High Rates of Strain*, J. Harding (editor), Oxford, 28-30, 1979, The Institute of Physics, Conference Series Number 47, pp. 166-73. Reprinted with kind permission of the Institute of Physics.

Extensive series of jump tests was performed by Eleiche and Campbell (1976a) and Campbell *et al.* (1977), whereby ***copper*** (FCC), ***titanium*** (HCP) and ***mild steel*** (BCC) were tested. The tests were performed over a range of temperatures and to up to 60% of shear strain. The results of these tests confirmed that ***copper*** is sensitive to strain rate history, while ***titanium*** and ***mild steel*** are less sensitive to strain-rate history, but more sensitive to direct effects of strain rate. Stelly and Dormeval (1978) performed experiments of the cyclic type, involving complete unloading before reloading at a new strain rate, with specimens of ***copper*** (FCC). In these tests, loading was in compression, using a Kolsky bar.

Other strain rate histories can be imposed besides that characterizing the jump test. Eleiche and Campbell (1976b), for instance, performed tests, on a moderately sensitive magnesium alloy, in which the strain rate is reversed in sign while being changed in magnitude from 10^3 to 10^{-3} s^{-1}; Figures 9.12 to 9.14.

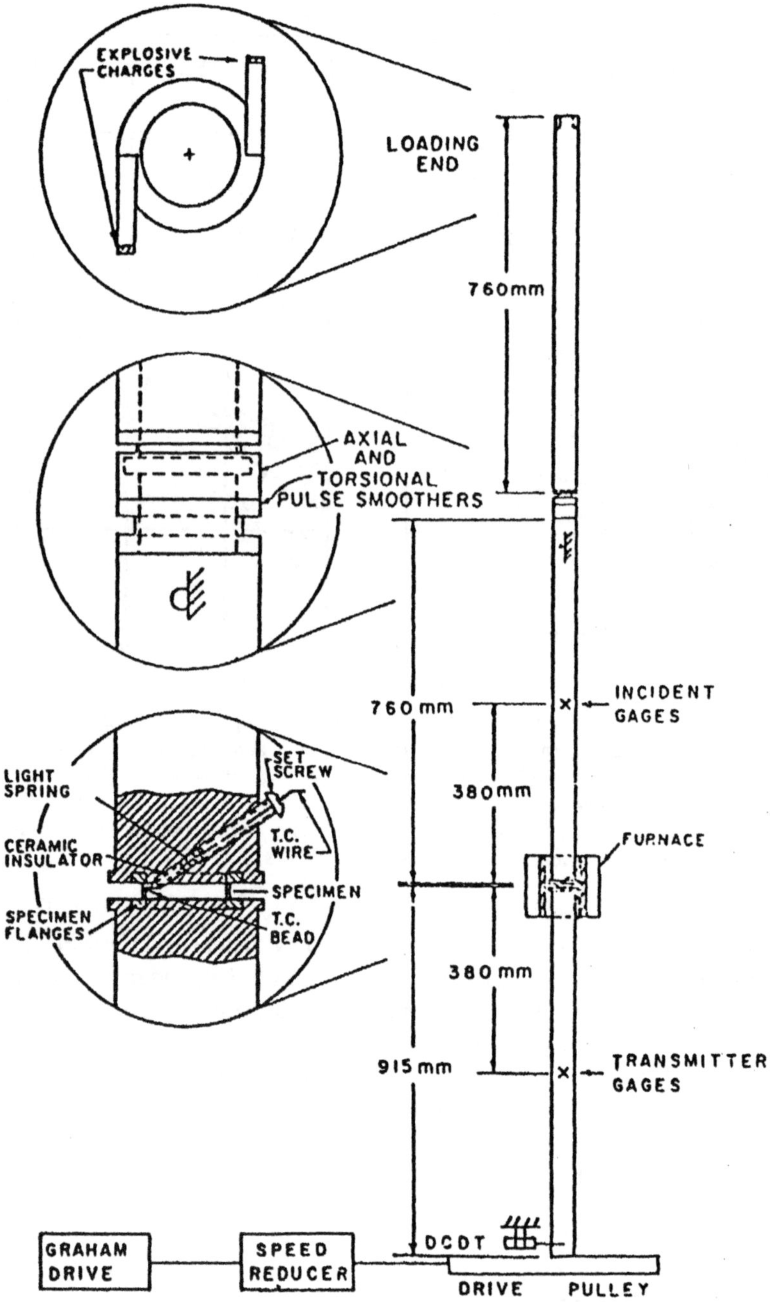

Figure 9.10. Schematic of the explosively loaded "***Kolsky bar***" for incremental testing. From: Senseny, P. E., Duffy, J. and Hawley, R. H. (1978) Experiments on strain rate history and temperature effects during plastic deformation of close-packed metals, *J. Appl. Mech.* **45**, March 1978, 60-6. Reprinted with permission from ASME International.

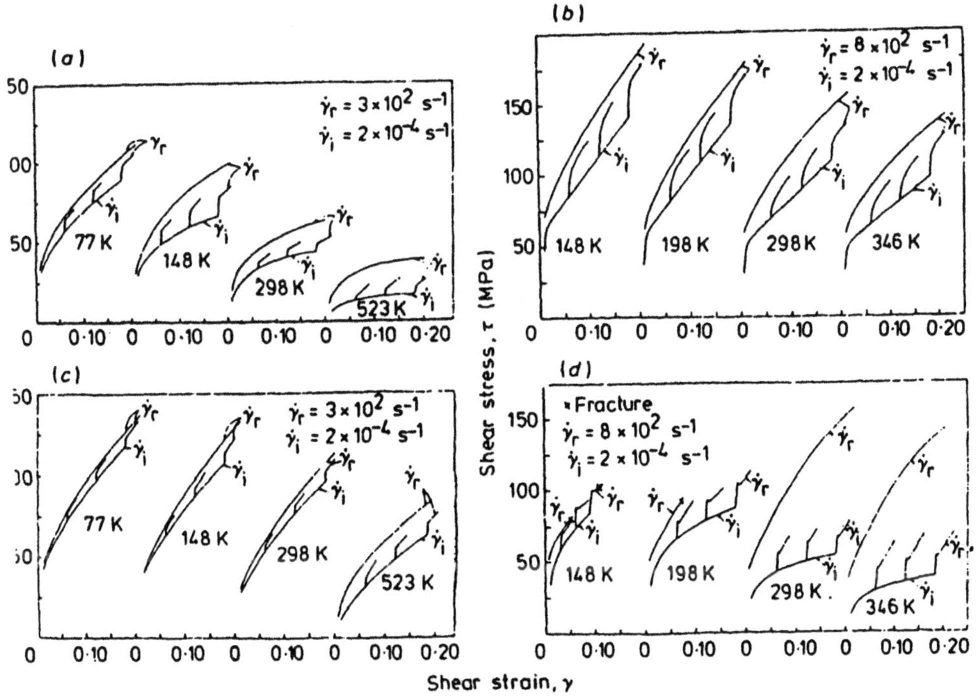

Figure 9.11. Stress-strain curves from incremental and constant strain-rate tests. (a) ***1100-0 aluminum***, (b) ***OFHC copper***, (c) ***AZ31 B magnesium***, and (d) Commercially ***pure zinc***. From: Senseny, P. E., Duffy, J. and Hawley, R. H. (1978) Experiments on strain rate history and temperature effects during plastic deformation of close-packed metals, *J. Appl. Mech.* **45**, March 1978, 60-6. Reprinted with permission from ASME International.

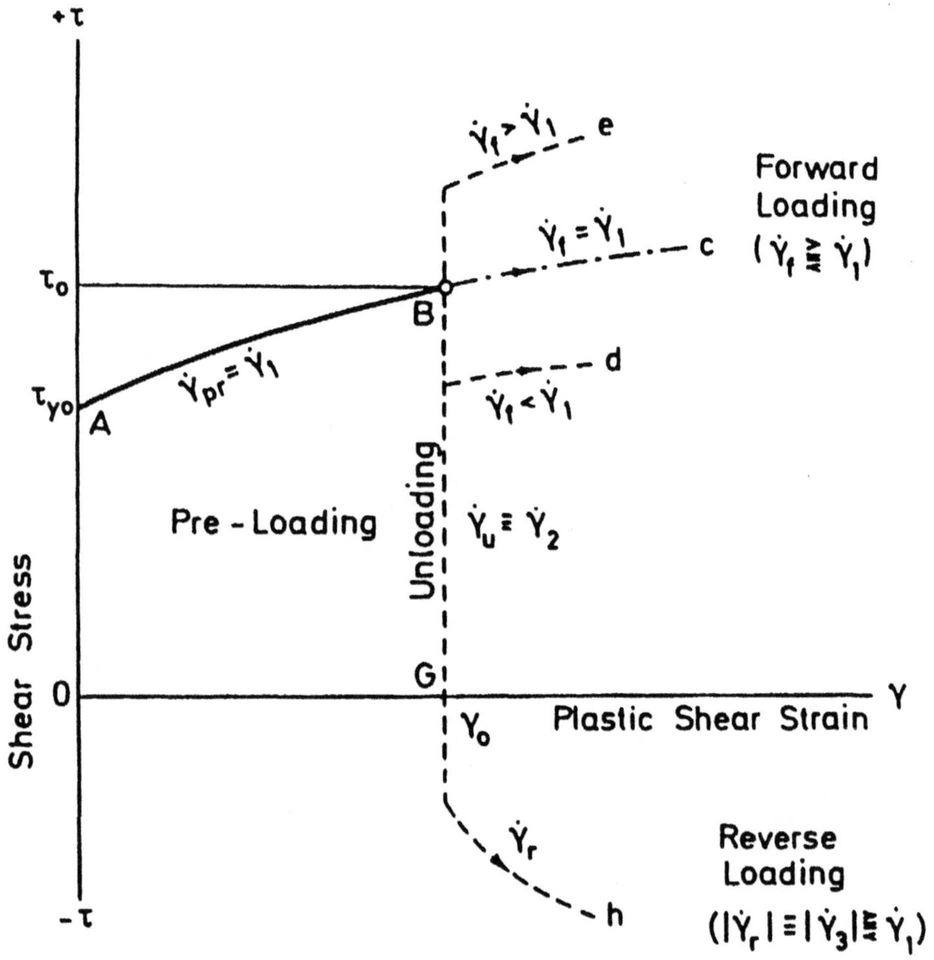

Figure 9.12. Schematic showing stress-plastic strain path, BGh, during reverse-shear testing. Material: Moderately rate-sensitive ***magnesium alloy***. From: Eleiche, M. A. and Campbell, J. D. (1976b) Strain-rate effects during reverse torsional shear, Experimental Mechanics 16 (8), 281-290. Reprinted with kind permission of the Society for Experimental Mechanics.

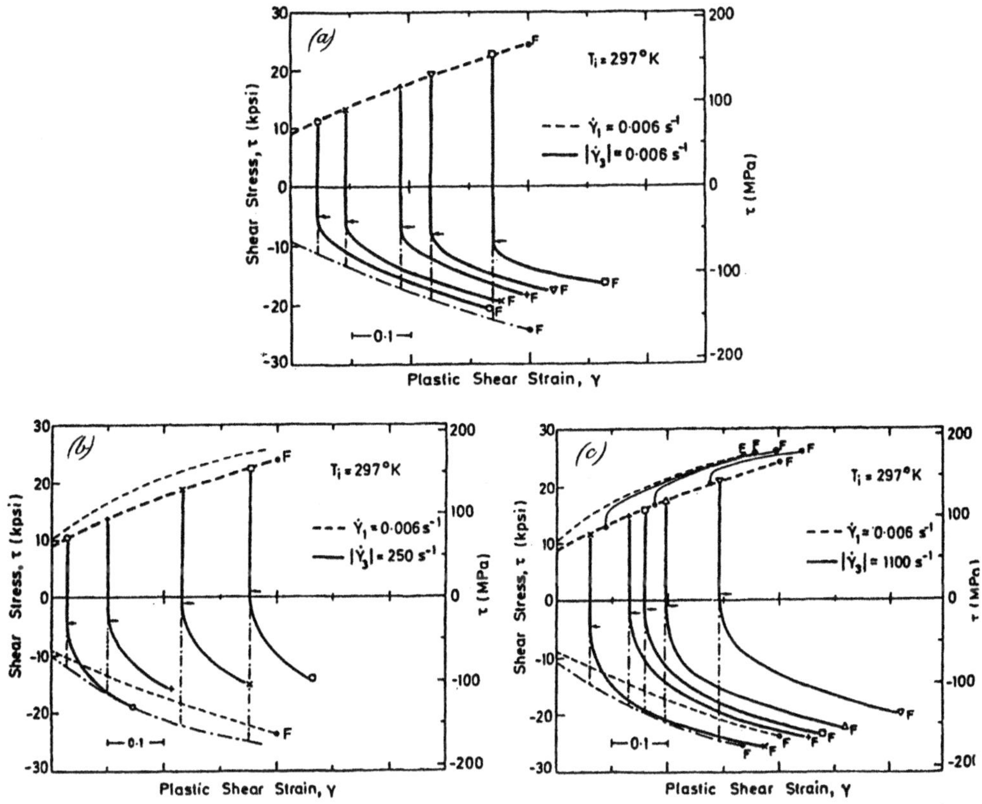

Figure 9.13. Flow-stress-strain curves in reverse shear. Material: Moderately rate-sensitive ***magnesium alloy***. (a) $|\dot{\gamma}_s| = \dot{\gamma}_1 = 0.006\ s^{-1}$; (b) $|\dot{\gamma}_s| = 250\ s^{-1}$ and $\dot{\gamma}_1 = 0.006\ s^{-1}$; (c) $|\dot{\gamma}_s| = 1100\ s^{-1}$ and $\dot{\gamma}_1 = 0.006\ s^{1}$. From: Eleiche, M. A. and Campbell, J. D. (1976b) Strain-rate effects during reverse torsional shear, *Experimental Mechanics* **16 (8)**, 281-290. Reprinted with kind permission of the Society for Experimental Mechanics.

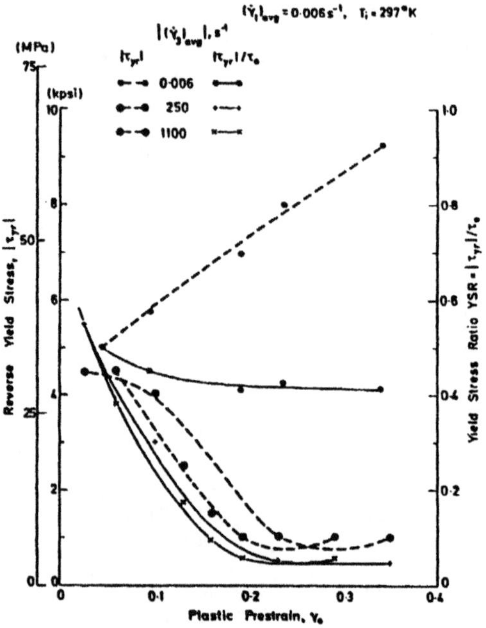

Figure 9.14. Variation of reverse yield stress, $|\tau_{yr}|$, and yield-stress ratio, YSR= $|\tau_{yr}|/\tau_o$, with prestrain and reverse-strain rate. Material: Moderately rate-sensitive magnesium alloy. From: Eleiche, M. A. and Campbell, J. D. (1976b) Strain-rate effects during reverse torsional shear, *Experimental Mechanics* **16** **(8)**, 281-290. Reprinted with kind permission of the Society for Experimental Mechanics.

The work described above refers entirely to polycrystalline metals. Recently, some progress has been made within the domain of establishing the history effects during the deformation of single crystals in the dynamic range of strain rates. Chiem and Duffy (1979), for instance, carried out jump tests in shear with single crystals of LiF. Their tests were performed on small cuboid specimens, four of which were mounted symmetrically in a torsional Kolsky bar.

Summary

- The FCC metals, e.g. *aluminum* and *copper*, are not strongly sensitive to strain rate. However, history effects appear to be important.
- *Steel* (BCC) and *titanium (HCP)*, on the other hand, show a greater strain rate sensitivity but only a small history effect.
- For the HCP metals, in general, *e.g., magnesium* and *zinc*, it appears that insufficient data are available.

Some results obtained by Lindholm (1965), using the split Hopkinson pressure bar are presented in Figures 9.15, 9.16 and 9.17 for three commercially pure, annealed, face centered cubic metals: ***copper***, ***aluminum*** and ***lead***, respectively, and in Figure 9.18 for ***iron***. In these figures, stress and strain are the true or instantaneous values. In Fig. 9.18, the negative slope of the stress *vs.* strain rate curve at the higher strains is generally associated with "***strain ageing***". Strain ageing occurs by diffusion of *interstitials*, e.g., carbon and nitrogen, into the active dislocation sites. The stress required to maintain flow is dependent upon the number of dislocations which are either free or bound by this atmosphere and, thus, on the effective interstitial diffusion rates, the temperature and deformation rate during the test (Lindholm, 1965).

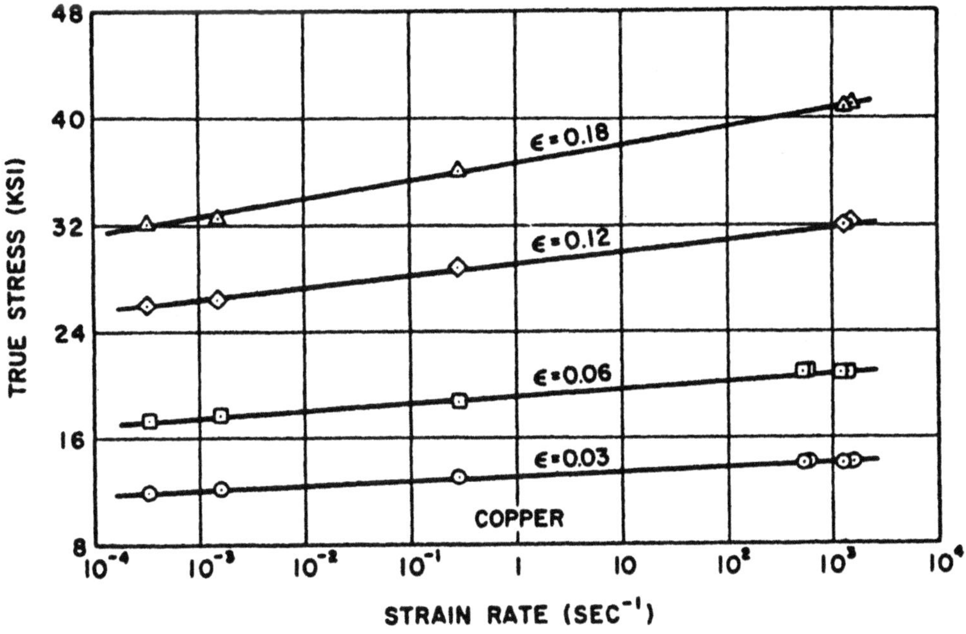

Figure 9.15. Flow stress as a function of strain rate for ***copper***. Reprinted from "Lindholm, U. S., Dynamic deformation of metals, in: *Behavior of Materials under Dynamic Loading*, edited by N. J. Huffington, Jr., The American Society of Mechanical Engineers, New York, 1965, 42-61", with kind permission of The American society of Mechanical Engineers.

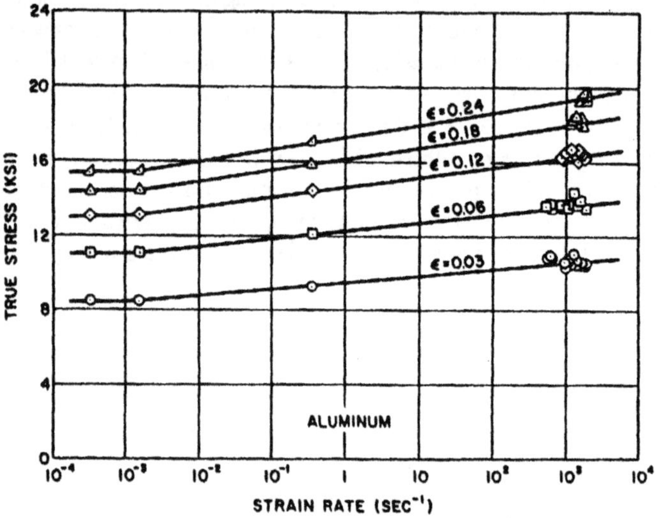

Figure 9.16. Flow stress as a function of strain rate for ***aluminum***. Reprinted from "Lindholm, U. S., Dynamic deformation of metals, in: *Behavior of Materials Under Dynamic Loading*, edited by N. J. Huffington, Jr., The American Society of Mechanical Engineers, New York, 1965, 42-61", with kind permission of The American Society of Mechanical Engineers.

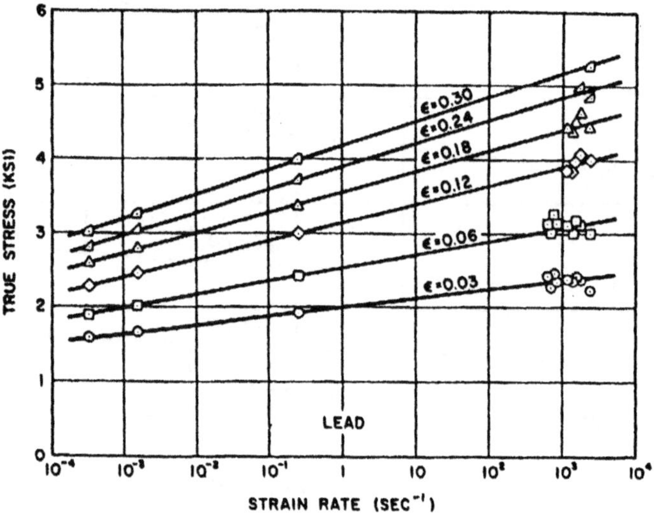

Figure 9.17. Flow stress as a function of strain rate for ***lead***. Reprinted from "Lindholm, U. S., Dynamic deformation of metals, in: *Behavior of Materials under Dynamic Loading*, edited by N. J. Huffington, Jr., The American Society of Mechanical Engineers, New York, 1965, 42-61", with kind permission of The American Society of Mechanical Engineers.

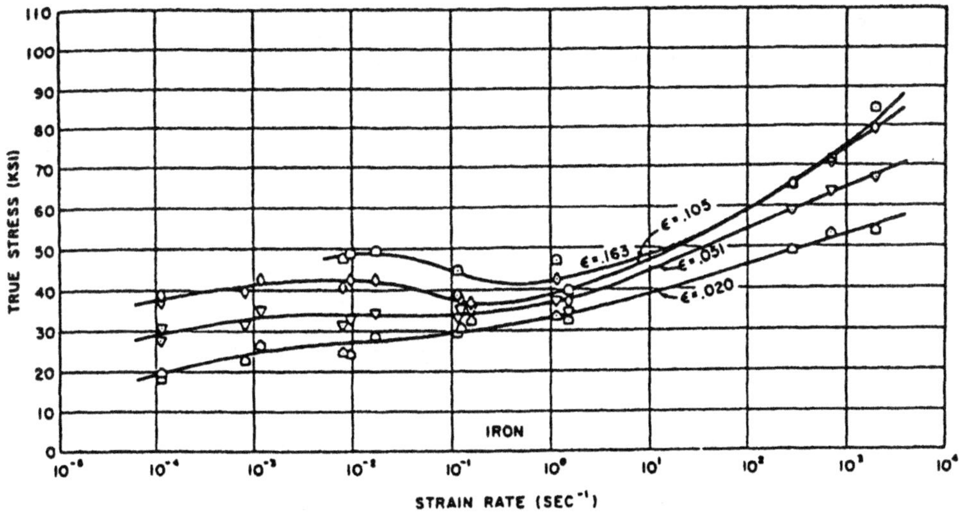

Figure 9.18. Flow stress as a function of strain rate for **iron**. Reprinted from "Lindholm, U. S., Dynamic deformation of metals, in: *Behavior of Materials under Dynamic Loading*, edited by N. J. Huffington, Jr., The American Society of Mechanical Engineers, New York, 1965, 42-61", with kind permission of The American Society of Mechanical Engineers.

9.2.3. DYNAMIC BIAXIAL LOADING

Figure 9.19 is due to Lindholm (1965). It shows a plot of stress and strain as functions of time in the case of biaxial loading of mild steel. In this figure, σ and τ are the tensile and shear stresses, respectively, and ϵ and γ are the corresponding tensile and shear strains. For this record, as discussed by Lindholm (1965), yield occurred about 6 milliseconds after initial application of the load. There is strong instability in torsion, whereas yield is hardly noticeable on the tensile stress trace. This may be due to both the material instability in mild steel, associated with the upper and lower yield stresses, and the region of zero work hardening during the lower yield point elongation. While the strain increments remain roughly proportional, the stress increment vector assumes a direction tangent to the yield surface and therefore normal to the strain increment vector during the period of zero work hardening.

Fig. 9.20 (Lindholm, 1965) demonstrates the results from 20 tests on mild steel at varying loading rates and stress ratios; from pure tension to pure torsion. The two curves correspond to two different measures of stress. The abscissa is the square root of the second invariant of the elastic strain tensor. This measure of the strain rate is nearly proportional to the reciprocal of the time to yield or delay time. For the lower curve, of Fig. 9.20, correlation is made with the square root of the second invariant of the deviatoric stress tensor, which is equivalent to a distortion energy or octahedral shearing stress criteria. For this curve, there appears a tendency for the tensile stress points to be consistently high (Linholm, 1965).

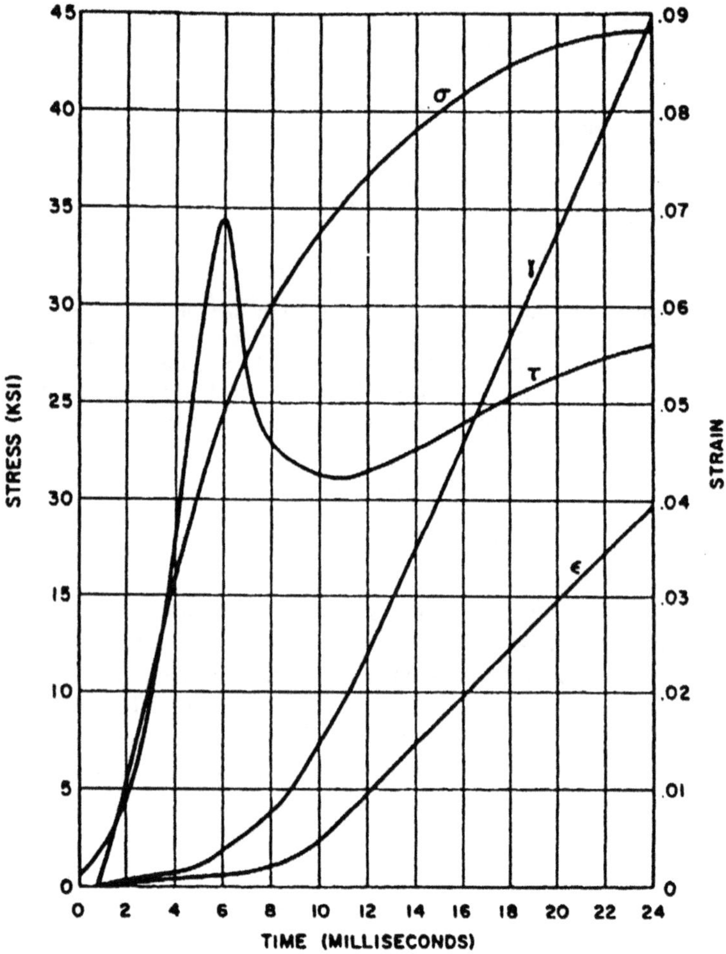

Figure 9.19. Typical dynamic record for combined loading of ***mild steel***. Reprinted from "Lindholm, U. S., Dynamic deformation of metals, in: *Behavior of Materials under Dynamic Loading*, edited by N. J. Huffington, Jr., The American Society of Mechanical Engineers, New York, 1965, 42-61", with kind permission of The American Society of Mechanical Engineers.

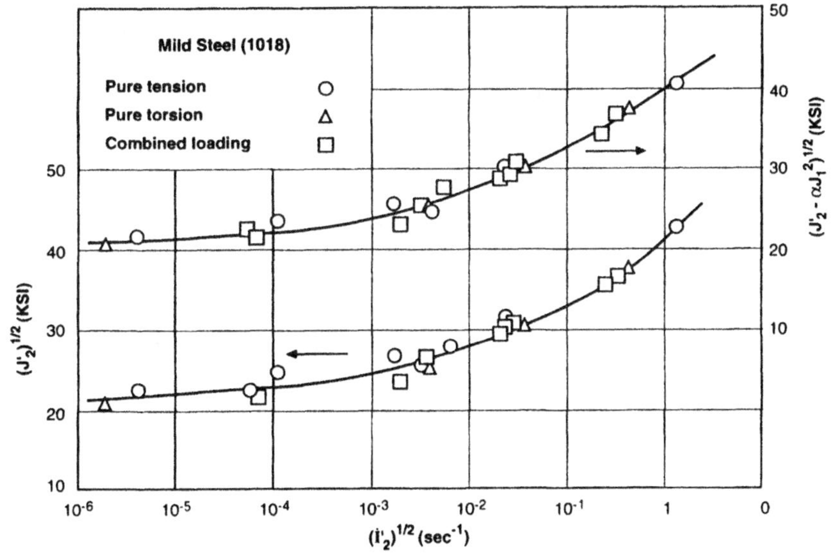

Figure 9.20. Correlation of results for the upper yield stress in ***mild steel*** under combined dynamic loading. Reprinted from "Lindholm, U. S., Dynamic deformation of metals, in: *Behavior of Materials Under Dynamic Loading*, edited by N. J. Huffington, Jr., The American Society of Mechanical Engineers, New York, 1965, 42-61", with kind permission of The American Society of Mechanical Engineers.

9.3. Metallurgical Effects

9.3.1. STRAIN-RATE EFFECTS

A large number of investigations, carried out in the last four decade, or about, in the area of dynamic behavior of materials, have shown that effects due to high strain-rate could be quite significant, e.g., the flow stress and the ductility of materials, the deformation and fracture mechanisms are often quite different from those exhibited under static or quasi-static loading. At very high strains and strain rates, there can be abrupt changes in deformation mode leading to noticeably different microstructures. These lead to noticeable metallurgical effects, e.g., microstructurally related flow stress, ductility, hardness and other related mechanical property changes.

Deformation induced metallurgical effects are now generally well documented to be the result of stress or strain-induced microstructures, or microstructural changes in crystalline (polycrystalline) metals and alloys. In many cases, strain hardening, work hardening, or other controlling deformation mechanisms can be described by the generation, movement and interactions of dislocations. These dislocations can produce drag or a range of impedances, including obstacles to further motion.

While dislocations may be involved in a range of metallurgical effects which are evident in the response behavior of metals and alloys, there are of course the controlling effects of temperature, strain, strain-rate and the associated mechanical state.

Metallurgical effects, characterized mainly by the relationships between deformation induced microstructures and residual mechanical properties, are therefore the result of the complex interrelations between stress, stress state, strain, strain rate and temperature.

For instance, changes in plastic stress in a uniaxial tensile stress state may be expressed by the following expression (Murr, 1987):

$$d\sigma = \left(\frac{d\sigma}{d\varepsilon}\right)_{\dot{\varepsilon},T} d\varepsilon + \left(\frac{d\sigma}{d\dot{\varepsilon}}\right)_{\varepsilon,T} d\dot{\varepsilon} + \left(\frac{d\sigma}{dT}\right)_{\varepsilon,\dot{\varepsilon}} dT \qquad \textbf{(9.2)}$$

where σ, ε, $\dot{\varepsilon}$ and T are the stress, strain, strain rate, and absolute temperature, respectively.

The above expression indicates that even if the loading or deformation parameters are controlled externally to the deforming material, there can be functional relationships which could override that control. For instance, temperature in a deforming material can be raised by increasing the strain, and by adiabatic heating at high strain rates. In addition, very high pressures in the shock loading regime can create both transient and residual heating.

Figure 9.21 is due to Murr (1987). It illustrates a range of microstructures which include planar dislocation arrays at relatively low levels of strain which evolve into more dense and microstructurally different arrays at higher strain levels. These different arrays (microstructures) are composed of twin-faults and martensite, whereby the martensite forms at the intersections of twin-fault bundles.

The dislocation density changes may be related to changes in stress (or strain) through expressions of the form (Murr, 1987):

$$\sigma = \sigma_0 + K\sqrt{\rho} \qquad \textbf{(9.3)}$$

$$\rho = \rho_0 + A\varepsilon \qquad \textbf{(9.4)}$$

where σ is the flow stress, ρ is the current dislocation density (ρ_0 is the initial dislocation density) and σ_0, K and A are constants. In the context of flow stress or residual yield stress (at constant strain), Eqn. (9.3) also expresses the fact that the residual yield or flow stress will be increased by the creation of dislocations. Since hardness and yield stress are inter-related, the former is also expected to increase by the creation of such dislocations (Murr, 1987).

9.3.2. SHOCK LOADING AND RESULTING SHOCK WAVES

In describing the metallurgical effects of shock and/or high strain-rate loading, one may consider the effects of such loading on the evolution of the microstructure.

Shock Waves

Shock loading represents a regime at the extreme end of the high-strain rate deformation range (~ 10^6 to 10^7 s^{-1}) as opposed to tensile or compression loading at strain rates of ~ 10^{-4} s^{-1}). The pulse duration is very short, usually never exceeding 10 μs. Peak pressure is the dominant shock loading parameter in residual microstructure production where strain is small or negligible.

Shock waves are characterized by an **abrupt** pressure front and a state of uniaxial strain. This characterization includes a hydrostatic component of stress which, when greater by several factors than the ***dynamic flow stress*** in the material, allows for the assumption that the solid has no shear resistance (G = 0), i.e., ***a "hydrodynamic"*** *behavior* (e.g., Eichelberger, 1965).

In reality, however, this pressure front may not be abrupt. Thus, a shock wave propagating into or through a material might be illustrated in the context of time and pressure as shown schematically on Fig. 9.22 (Murr, 1987).

As shown in Fig. 9.22, the shock front is shown as a region where the material is subjected to increasing pressure (stress) up to the peak shock pressure (P). The time of application of P (the peak shock pressure) is referred to as the "***shock pulse duration***", Δt. As the pressure of the wave declines or is attenuated, the shock wave is called a "***rarefaction wave***" or "***wave portion***".

Both the peak shock pressure and its duration can be expected to have some effect on the shock dynamic behavior and the residual properties of the material. This is due to the fact that the disturbance created within the shock front and the dynamic behavior of the material will be altered by these two parameters. Fig. 9.23 shows examples of shock-induced microstructures in face-centered cubic metals having a range of stacking fault free energies (e.g., Murr and Meyers, 1983 and Murr, 1987).

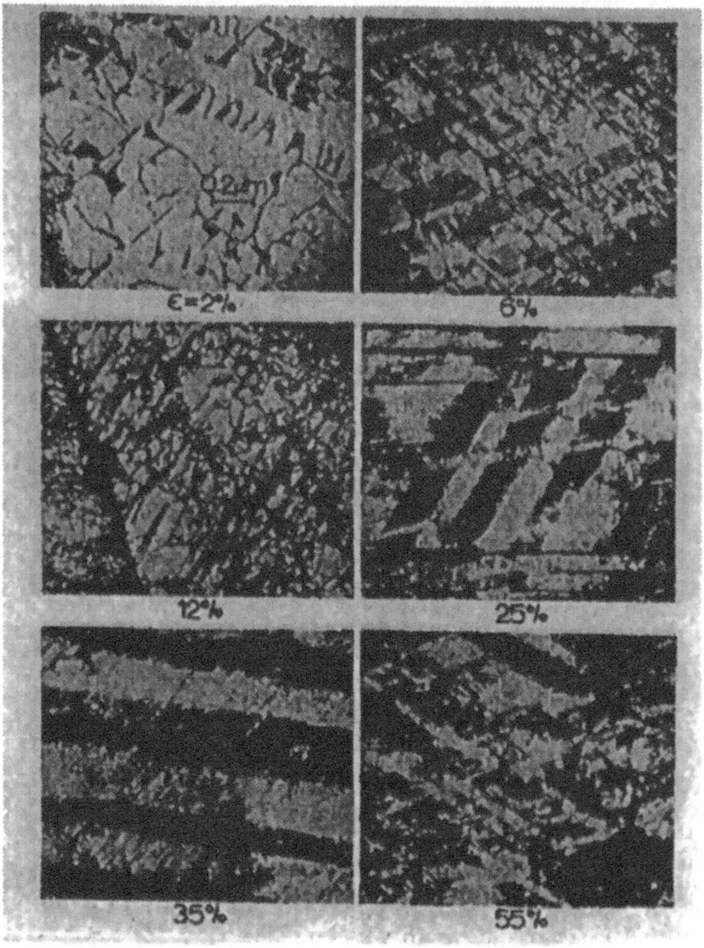

Figure 9.21. Examples of residual microstructures in type ***304 stainless steel*** after deformation in uniaxial tension to an average total strain ε as indicated (room temperature; $\dot{\epsilon}=10^{-3}\ s^{-1}$). The microstructures are characterized by increasing densities of dislocation arrays, stacking faults and twin faults, with α'-martensite forming at twin-fault intersections and constituting a prominent volume fraction nearly equivalent, as a volume fraction percent, to the total strain value from about 25% strain "Reprinted from Murr, L. E., Metallurgical Effects of Shock and High-Strain-Rate Loading, in Blazynski, T. Z. (editor), *Materials at High Strain Rates* (1987), 1-46, with kind permission from Chapman & Hall".

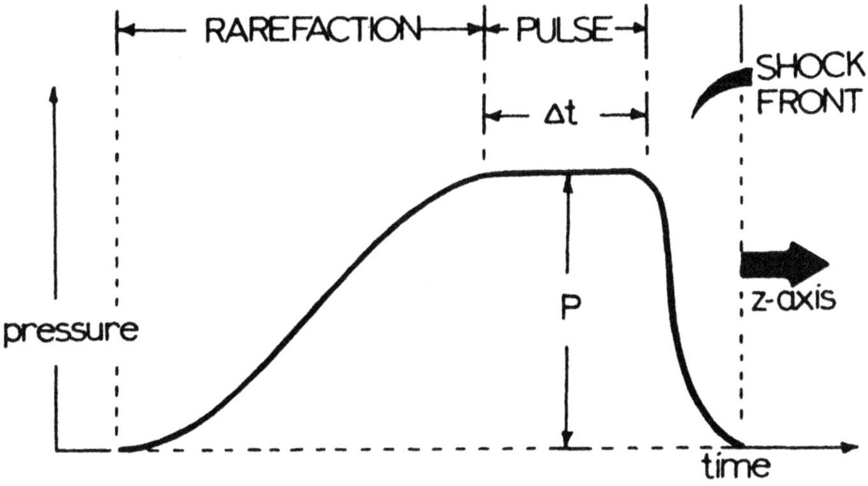

Figure 9.22. Idealized (schematic) view of a shock pulse traveling through a solid material. The z-direction is assumed to be normal to the plane shock wave front and to the specimen surface "Reprinted from Murr, L. E., Metallurgical Effects of Shock and High-Strain-Rate Loading, in: Blazynski, T. Z. (editor), *Materials at High Strain Rates* (1987), 1-46, with kind permission from Chapman & Hall".

9.3.3. SHOCK-INDUCED MICROSTRUCTURE AND MECHANICAL PROPERTY CHANGES

The peak pressure of the shock wave characterizes the shock front. This, by consequence, influences the stress-induced generation of dislocations and other defects in metals and alloys. In view of the very high strain rates involved in shock loading, the peak pressure may result in some of the following unique deformation phenomena:

- Pressures (shock stresses) of two orders of magnitude greater than the yield or flow stress of metals and alloys are common, and, in most controlled plane wave shock loading, strains are minimal (<5%). But because the rapid movement of the shock wave, dislocations interact within the shock front forming jogs which favor high vacancy production (Kressel and Brown, 1967, and Murr, Inal and Morales, 1967). In many shock-loaded metals and alloys, vacancies and vacancy clusters can contribute to residual metallurgical effects such as hardness, ductility and thermal recovery.

- At very high pressures, the heating associated with the high-pressure state (within the shock front) can become very significant, and dislocations or other defects created by the propagating shock front can be partially or completely annihilated. Residual heating after shock front may also create recovery, recrystallization and related microstructure which may be characterized, for instance, by short twin segments, sub-grains, etc..

- Some of the above mentioned features are illustrated in Fig. 9.24, which shows residual hardness reduction as a result of shock-thermal recovery in ***nickel*** and ***type 304 stainless steel*** at plane shock pressures above about 60 Gpa.

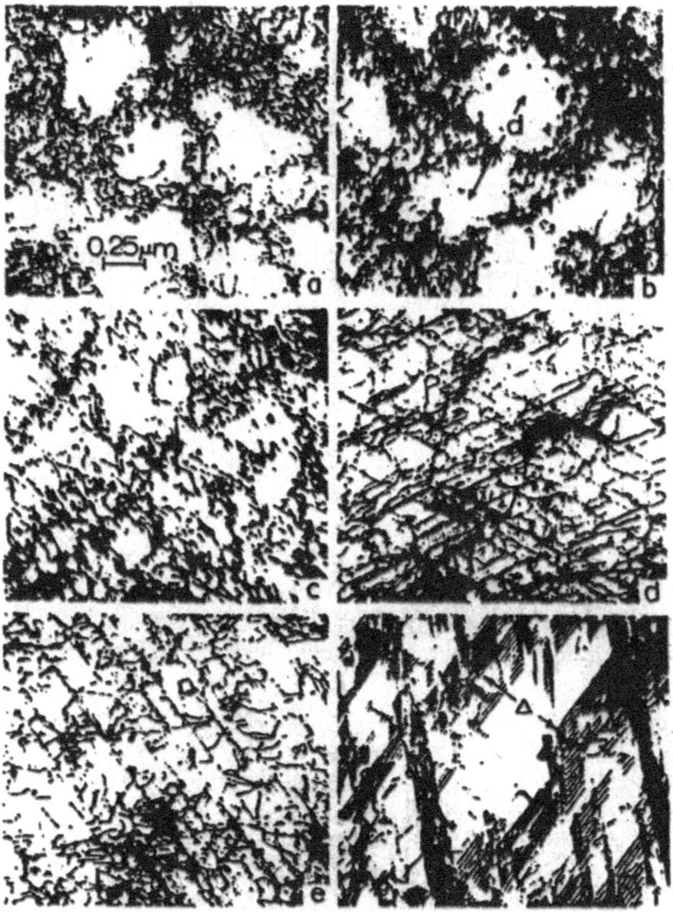

Figure 9.23. Examples of shock-induced microstructures in face-centered cubic metals having a range of stacking fault free energies. (a) ***Ni*** (15 GPa); (b) ***Cu*** (15 Gpa); (c) ***Fe***-34% Ni (10 Gpa); (d) ***Ni Cr*** (8 Gpa); (e) ***Inconel*** 600 (8GPa); (f) ***304 stainless steel*** (15 Gpa). "Reprinted from Murr, L. E., Metallurgical Effects of Shock and High-Strain-Rate Loading, in: Blazynski, T. Z. (editor), *Materials at High Strain Rates* (1987), 1-46, with kind permission from Chapman & Hall"; After Murr and Meyers (1983).

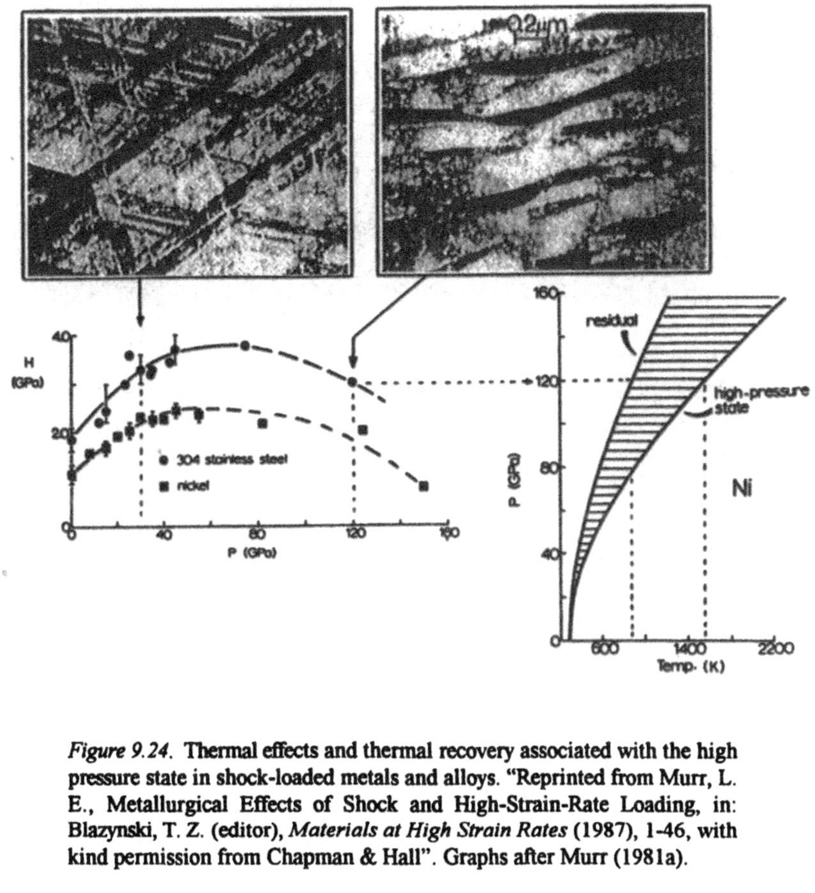

Figure 9.24. Thermal effects and thermal recovery associated with the high pressure state in shock-loaded metals and alloys. "Reprinted from Murr, L. E., Metallurgical Effects of Shock and High-Strain-Rate Loading, in: Blazynski, T. Z. (editor), *Materials at High Strain Rates* (1987), 1-46, with kind permission from Chapman & Hall". Graphs after Murr (1981a).

- During the production of dislocations (and other defects) in the shock front, heating occurs. The latter, combined with the actual defect production, contributes to an internal energy change across the shock front. On the basis of the work done on a solid during rarefaction (stored energy calculations), Murr (1987) advanced (see, also, Murr and Meyers, 1983) the following expression for the residual yield or flow stress of a metal or alloy subjected to a 'planar' shock:

$$(\sigma - \sigma_0) = 2\,\alpha\,G\,|b|\,\{f(P)\}^{1/2} \qquad \textbf{(9.5)}$$

Where σ_o and α are constants for any particular material, G is the shear modulus, **b** is Burgers vector and P is the peak shock pressure.
In the above expression (9.5), it is readily apparent that the left hand side of this expression is dependent on the changes in residual mechanical properties of the material under consideration, e.g., yield strength, ultimate tensile strength, hardness, following the passage of a shock wave. These changes will be shock-pressure dependent. As mentioned earlier, such dependence is the result of shock-wave-induced defects, for instance, dislocations as illustrated schematically in Fig. 9.24.

– As the peak pressure, in the plane-wave compressive shock loading, is increased, the dislocation density increases and as a consequence in high-stacking-fault free energy materials (such as nickel) the dislocation cell size (or cell center spacing) decreases. In low-stacking-fault free energy materials where twin -faults form, the density or volume fraction of twin-faults will increase. If the twin-fault bundle thickness does not change much, the consequence of this increased volume fraction is a corresponding decrease in the twin-fault spacing. These parametric changes with peak shock pressure, at constant shock pulse duration, are illustrated in the experimental data graphs in Fig. 9.24 (Murr,1987). It is particularly important to observe in this figure that:

- The residual hardness is functionally related to the square root of the peak pressure, for a great variety of shock loaded metals and alloys.

- The two obvious deviations in the slopes of the straight lines occur for metals dispersed with fine particles of thoria (ThO_2), or thoria dispersed particles. The thoria dispersion not only hardens the material, but also locks up dislocations created by the shock front. This feature is apparent on comparing the annealing responses for NiCr (chromel A) and Thoria dispersed-NiCr, Fig. 9.25. The latter figure attests not only to the hardness difference for dispersion-hardened metals shown in Fig. 9.26, but also to the unique locking ability of dispersed particles in shock-loaded materials: While dislocations can be created by the shock front passage in spite of the presence of the dispersed particles, the particles could effectively prevent the dislocations created by the shock front from annealing out, thereby maintaining the shock-induced high hardness to very high temperatures.

- A summary review of the effects of peak shock pressure for plane-wave, shock-loaded polycrystalline metals and alloys is shown in Figure 9.26 and, also, as a microstructure-property map in Fig. 9.27. Both figures are due to Murr (1987).

9.3.4. TWINNING IN SHOCK-LOADED METALS AND ALLOYS

One of the unique metallurgical effects of planar shock loading is the occurrence of twins in crystalline metals and alloys at some critical pressure. This is especially unique because some metals such as nickel and molybdenum do not normally twin when subjected to other modes of loading.

In FCC-metals and alloys, twining is expected to occur initially in (001) orientations. Twinning also occurs preferentially in low-stacking fault free energy metals and alloys as the preponderance of stacking faults provides opportunities for thin twins to form. This process is, however, irregular, leading to the formation of bundles of intermixed stacking faults (intrinsic, extrinsic and other irregular faults) and thin twins, often referred to as "*twin faults*". Thus, the critical pressure at which twinning occurs in FCC- metals and alloys appears to be dependent on stacking fault free energy. This is illustrated in Fig. 9.28 (due to Murr, 1987).

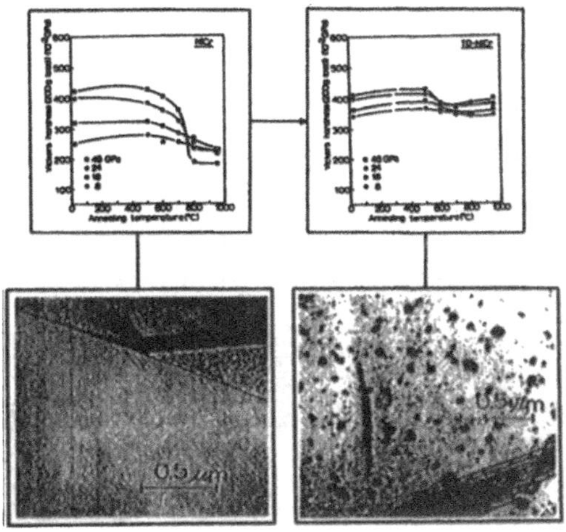

Figure 9.25. Comparison of hardness and hardness recovery in shock-loaded Ni_{80} Cr_{20} and ***TD-NiCr*** (the same alloy with 2 vol % ThO_2 included as a dispersed phase) for a constant annealing time of 1 h. The corresponding unshocked microstructures are also shown for comparison. "Reprinted from Murr, L. E., Metallurgical Effects of Shock and High-Strain-Rate Loading, in: Blazynski, T. Z. (editor), *Materials at High Strain Rates* (1987), 1-46, with kind permission from Chapman & Hall". Graphs after Murr (1981a).

Twinning in aluminum (where the stacking fault free energy is approximately 160 mJ/m^2), while estimated from Fig. 9.28 to occur at about 40 Gpa, should not likely to occur because of the low melting point for ***aluminum*** (660 °C) and the shock heating which would occur at that pressure (Fig. 9.27) leading to complete recovery (annealing) at pressures below the critical twinning pressure. This has yet to be demonstrated experimentally (Murr, 1987). The data in Fig. 9.28 correspond generally to ambient temperatures or above (Fig. 9.27) and very low (or zero) strain. Consequently, changing the shock temperature or altering the strain should have a significant effect on the critical twinning conditions implicit in Fig. 9.28.

In low-stacking fault free energy alloys such as ***brass*** and ***stainless steel***, planar dislocation arrays and stacking faults at very low peak shock pressures (<10 GPA) lead to increasing densities of twin-faults at increasing pressures above the critical twinning pressures (10-20 GPA). In high-stacking fault free energy metals, such as ***nickel*** and ***copper***, dislocation cells densify with increasing peak shock pressure, resulting in a reduction in the average dislocation cell size ***d*** and a saturation of cell size at the critical twinning pressure. Twins and twin-faults develop with increasing density and in orientations other than (001) above the critical twinning pressure. There is, therefore, a microstructural transition in metals like ***copper*** and ***nickel***, e.g., dislocation cells decreasing in size up to the critical twinning pressure where twins and twin-faults increasing in density occurs. These features are illustrated in Fig. 9.29 (due to Murr, 1987).

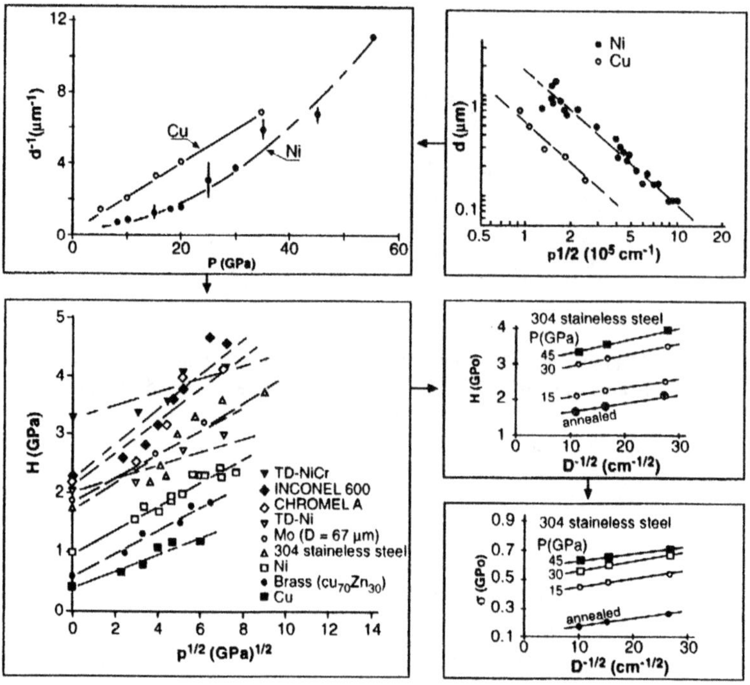

Figure 9.26. Parametric variations (metallurgical effects and variations of residual mechanical properties) in shock-loaded metals and alloys "Reprinted from Murr, L. E., Metallurgical Effects of Shock and High-Strain-Rate Loading, in: Blazynski, T. Z. (editor), *Materials at High Strain Rates* (1987), 1-46, with kind permission from Chapman & Hall".

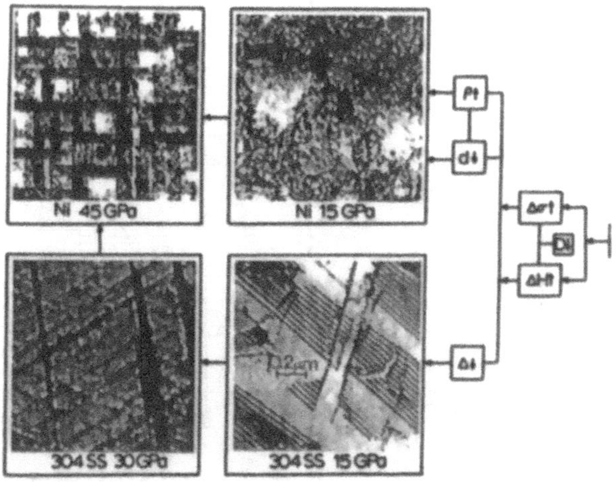

Figure 9.27. Microstructure-property hypermap for crystalline shock-loaded metals and alloys. Arrows indicate parametric increase (↑) or decrease (↓) "Reprinted from Murr, L. E., Metallurgical Effects of Shock and High-Strain-Rate Loading, in: Blazynski, T. Z. (editor), *Materials at High Strain Rates* (1987), 1-46, with kind permission from Chapman & Hall".

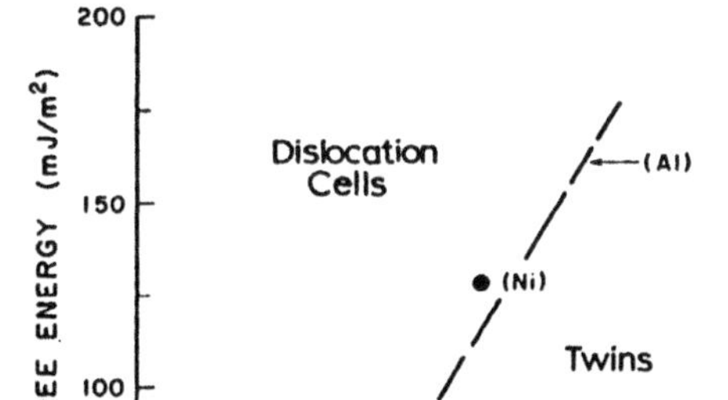

Figure 9.28. Critical twinning pressure versus stacking-fault free energy for a number of fcc metals and alloys. (Critical pressure values are estimated from shock-loading data of Murr (1981); stacking-fault free energy values are from Murr (1975)) "Reprinted from Murr, L. E., Metallurgical Effects of Shock and High-Strain-Rate Loading, in: Blazynski, T. Z. (editor), *Materials at High Strain Rates* (1987), 1-46, with kind permission from Chapman & Hall".

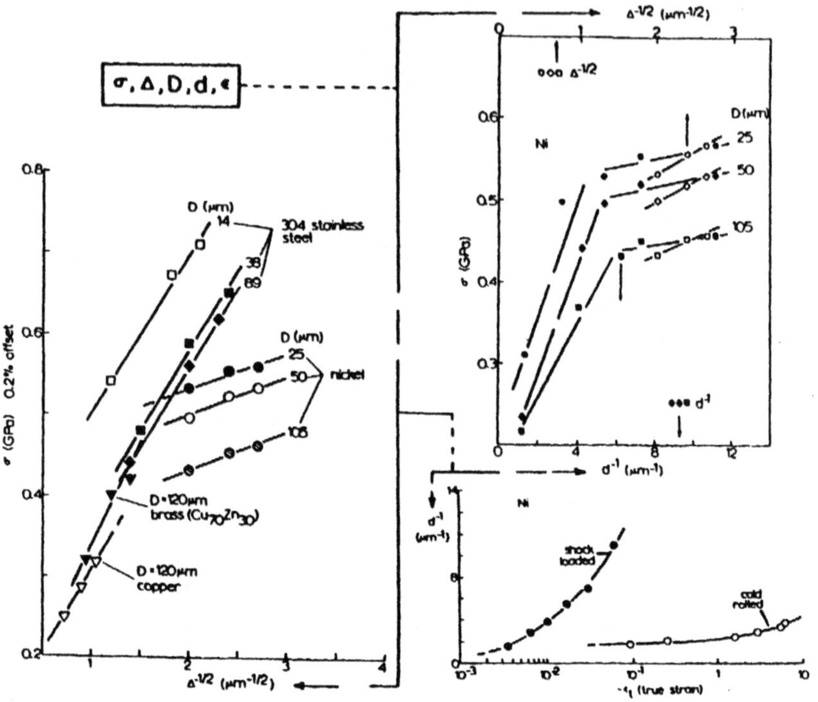

Figure 9.29. A comparison of dislocation cell size changes and twin-fault spacing changes for shock-loaded and cold-rolled metals and alloys "Reprinted from Murr, L. E., Metallurgical Effects of Shock and High-Strain-Rate Loading, in: Blazynski, T. Z. (editor), *Materials at High Strain Rates* (1987), 1-46, with kind permission from Chapman & Hall".

9.3.5. METALLURGICAL EFFECTS OF SHOCK PULSE DURATION

The pulse duration in shock loading serves to equilibrate defects generated in the shock front by maintaining the applied peak pressure for some interval of time. As mentioned earlier, the pulse duration is very short, usually never exceeding 10 μs. Thus, a pulse duration range of 0.1 to 10 μs might represent a strain rate of approximately 10^7 to 10^5 s^{-1}.

Murr (1981a&b) summarized the effects of pulse duration on the residual structure and properties of shock-deformed metals and alloys:

- While longer pulse durations seem to allow for larger twin or twin-fault volume fractions in metals and alloys which twin at sufficiently high peak shock pressures, there is no significant effect on the residual hardness and related mechanical properties.

- In high stacking fault free energy metals such as ***nickel,*** where dislocation cells are formed, larger pulse durations do not alter the cell sizes but simply allow the cells to be more well defined or better developed, Fig. 9.30.

- While twin volume and martensite volume fractions have been observed to increase with increasing shock pulse duration over the range of about 1 to 10 μs (Fig. 9.30), the corresponding hardness does not change because the deformation gradient wavelength is not altered significantly.

- In low-stacking fault free energy alloys, at very short pulse durations (<0.50 μs), irregular behaviour may occur due to peak pressure instabilities and uncertainties which result from the use of flyer plates to create a planar shock wave. The result can result in, for instance, variations in residual hardness (see Murr, 1981).

9.3.6. STRAIN RATE EFFECTS OF UNIAXIAL STRESSES

Most metals and alloys exhibit effects of varying strain rate on deformation mechanisms. Plastic strain rate is commonly expressed, with the inclusion of the microstructure, by the so-called Orowan expression:

$$\dot{\varepsilon}^{p} = b\,\rho_m(\varepsilon^p, \sigma_s)\,\bar{v}(\varepsilon^p, \sigma_s) \tag{9.6}$$

where

b	is Burgers vector
ρ_m	is the mobile dislocation density
$\bar{v}$	is the average dislocation density

and both ρ_m and $\bar{v}$ are considered to be functions of the stress σ_s and plastic strain ε^p.

9.3.7. STRAIN-RATE SENSITIVITY

Strain-rate sensitivity, at constant strain, is often expressed by

$$\sigma = K(\dot{\varepsilon})^{\beta_\epsilon} \tag{9.7}$$

where K is a constant.

Strain-rate sensitivity has been measured experimentally to vary significantly when defined as a function of flow stress as

$$\beta_\epsilon = \left(\frac{\partial\sigma}{\partial \log \dot{\varepsilon}}\right)_\epsilon \tag{9.8}$$

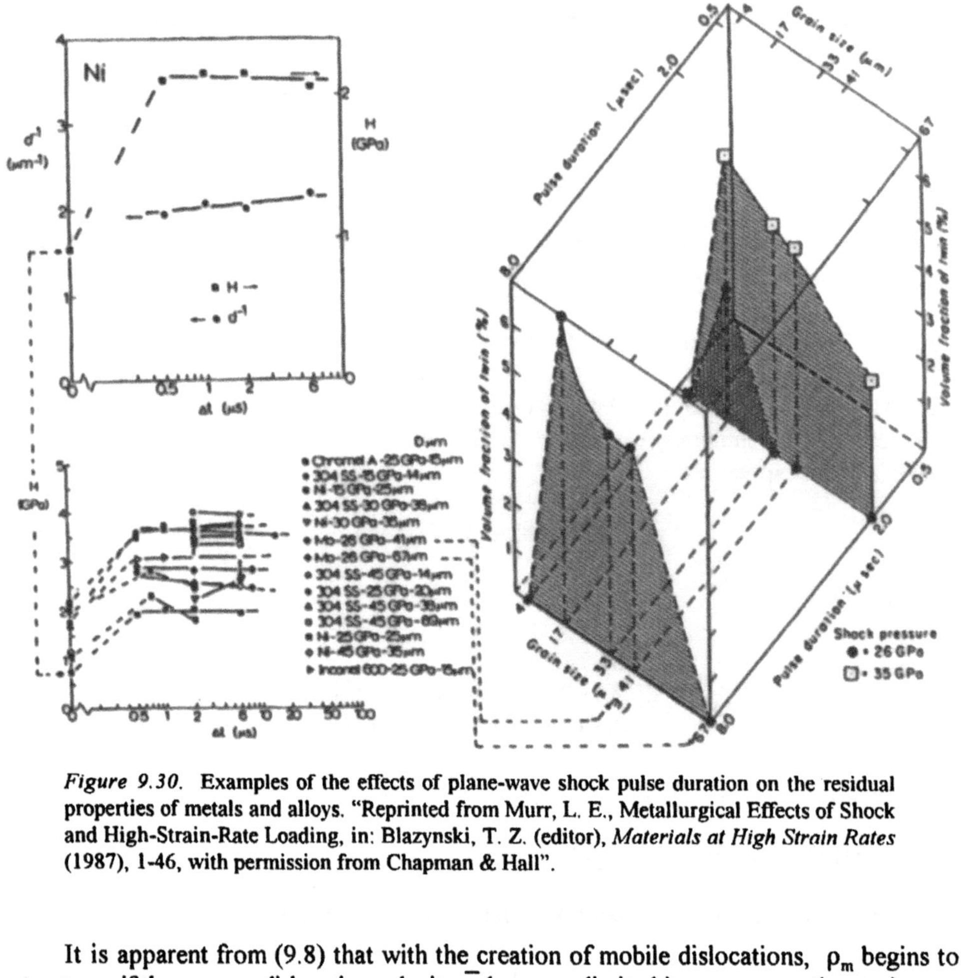

Figure 9.30. Examples of the effects of plane-wave shock pulse duration on the residual properties of metals and alloys. "Reprinted from Murr, L. E., Metallurgical Effects of Shock and High-Strain-Rate Loading, in: Blazynski, T. Z. (editor), *Materials at High Strain Rates* (1987), 1-46, with permission from Chapman & Hall".

It is apparent from (9.8) that with the creation of mobile dislocations, ρ_m begins to saturate, or if the average dislocation velocity $\bar{v}$ becomes limited in some way, the strain-rate sensitivity, β_{ε}, will change noticeably: In this context, β_{ε} is found to increase when $\dot{\varepsilon}$ exceeds roughly 10^3 s^{-1} (e. g., Campell, 1970 and Lindholm, 1978), but below that range, β either does not increase or the change is irregular.

Figure 9.31 illustrates some stress/strain/strain-rate curves for ***copper*** at various strains and over a range of strain-rates, along with similar, smoothed curve data for ***Nitronic 40*** and ***type 316 stainless steels*** based upon some of the experimental results of Follansbee (1986).

The increased rate sensitivity above 10^3 s^{-1} (denoted as the high strain rate region) is quite apparent for copper (Fig. 9.31a), while, for the **stainless steels** (Fig. 9.31b), the increased rate sensitivity appears to begin at strain rates as low as 10^2 s^{-1}.

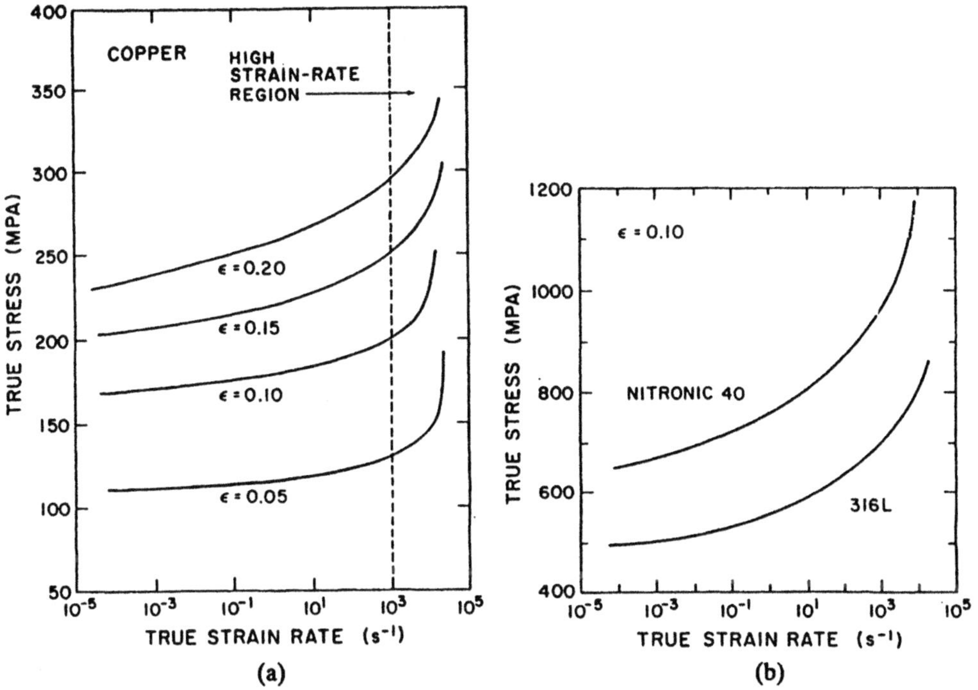

Figure. 9.31. Flow stress versus strain rate curves for some face-centered cubic metals and alloys. (a) Copper measured at various strains; (b) Nitronic 40 and 316L stainless steels at a constant strain. "Reprinted from Murr, L. E., Metallurgical Effects of Shock and High-Strain-Rate Loading, in: Blazynski, T. Z. (editor), *Materials at High Strain Rates* (1987), 1-46, with kind permission from Chapman & Hall".

Figure 9.31 illustrates the connection between uniaxial shock loading and uniaxial high-strain rate deformation. The figure shows a comparison of the mechanical threshold stress (measured at a constant strain of 0.0825) as a function of strain rate for ***copper*** from the experiments of Follansbee (1986). The estimated strain-rate range for a corresponding shock loading experiment 10^5 to 10^7 s^{-1} is indicative of the fact that the increased strain-rate sensitivity of the threshold stress noted at strain rates exceeding ~ 10^3 s^{-1} continues into the shock loading regime.

Measurements of the mechanical threshold stress (Fig. 9.31) coupled with an analysis of the dislocation-obstacle interactions led Follansbee (1986) to the conclusion that the increased strain-rate sensitivity arises from the rate sensitivity of the microstructure evolution rather than from any change in the involved deformation mechanism. Murr (1987) supported the latter conclusion by referring to the fact, revealed from examinations of residual microstructures over a range of strain rates for ***copper*** and ***stainless steel***, that the microstructures may change neither abruptly nor characteristically. In ***copper***, for instance, changes in dislocation cell size and density are consistent with the mentioned conclusion. In ***type 304 stainless steel***, the microstructure and its evolution support this conclusion (Murr, 1987).

9.3.8. ADIABATIC SHEAR PHENOMENA

When metals and alloys are deformed at very large strains and at very high strain rates such as in ballistic impact and penetration, forging and machining, localized shearing can occur, leading to localized deformation and a localization of heat generation. At high strain rates, heat generated in the localized bands provides some self-acceleration to the localization, and even melting. This concentration of deformation leads to two categories of adiabatic shear bands; namely, "*deformation*" and "*transformation*" bands. The microstructure associated with these bands includes dynamic re-crystallized microstructures, dislocations, microtwins and twin-faults as a result of the shear deformation in the localized bands. These fine and intermixed microstructures lead to very small deformation gradient wavelengths, and dramatic increases in localized hardness or residual flow stress (Murr, 1987). The reader is referred, in this context, to the work of Bedford *et al.* (1974), Blicharski and Gorczyca (1978), Malin and Hatherly (1979), Rogers (1979, 1983), Aghan and Nutting (1980), Murr *et al.* (1986), Stelly and Dormeval (1986) and Dormeval (1987).

'***Adiabatic shearing***' is one aspect of high strain-rate deformation that has received much attention for some years due to the large number of applications in which it appears to play a significant role. Although this phenomenon was discovered and studied over five decades ago; e. g. Zener and Holloman (1944) and Zener (1948), the phenomenon was not considered for a long time, and it was only in early 1970s that researchers began to take a new interest in its study.
In metals, it has been determined that at room temperature about 90% of the work of deformation energy goes into heat. Adiabatic shearing is a particular situation in which the heat generated in localized bands cannot be dissipated because of the high level of strain rate in conjunction with the thermal properties of the material. An idealized adiabatic deformation does not exist, some part of the heat being always lost to the surrounding metal and the environment. However, the term '*adiabatic*' is taken to refer to the fact that a large portion of the heat is retained in the band.

"***Shear bands***" form as a result of a thermo-mechanical instability due to the presence of a local inhomogeneity, inducing local deformation and heating. If the thermal properties of the material are not sufficient to conduct the generated heat away, the deformation becomes unstable and is localized on surfaces of very small thickness (~ 10 to 50 microns).

On microscopic observation, these surfaces appear as narrow bands in which cracks can propagate (Fig. 9.32), inducing catastrophic failure of the material.

Adiabatic shearing is involved in a large number of processes where high strain rates occur, e.g., impact, penetration, fragmentation, machining, metal forming.

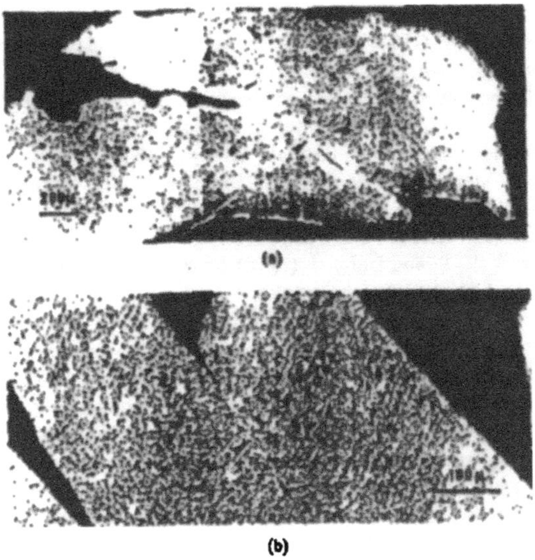

Figure 9.32. Evidence of adiabatic shear bands in TA6V titanium alloy: (a) fragment from an explosively expanded cylinder; (b) chip (machine-turned). "Reprinted from Dormeval, R., The Adiabatic Shear Phenomenon, in: Blazynski, T. Z. (editor), *Materials at High Strain Rates* (1987), 47-70, with kind permission from Chapman & Hall".

It is traditional to distinguish two types of adiabatic shear band:

- *Deformed bands.* They are characterized by a very high shear strain (up to 100) in a very thin zone of deformation. Inside the band, the grains are highly distorted, but there is no evidence of a change in the microstructure of the material.

- *Transformed bands.* In these bands, a crystallographic phase change occurs. In steels, for instance, they are often called '*white bands*', Fig. 9.32, as their appearance after itching is quite different from that of the matrix.

9.4. References

Aghan, R. L. and Nutting, J. (1980) Structure and properties of low-carbon steel after deformation to high strains, *Metal Sci.* **14**, 233-7.

Barraclough, D. R. and Sellars, C. M. (1974) in: *Mechanical Properties at High Rates of Strain,* Inst. Phys. Conf. Ser. **21**, pp. 111-23.

Bedford, A. J., Wingrove, A. L. and Thompson, K. R. L. (1974) The Phenomenon of adiabatic shear deformation, *J. Aust. Inst. Metals* **19 (1)**, 61-73.

Blicharski, M. and Gorczyka, J. (1978) Structural inhomogeneity of deformed austentic stainless steel, *Metal Sci.* **12**, 303-12.

Bodner, S. R. and Merzer, A. (1978) *MML Rep. No.* **55**, Technion, Israel.

Bodner, S. R. and Partom, Y. (1975) Constitutive equations for elastic-viscoelastic strain-hardening materials, *J. Appl. Mech.* **42**, June 1975, 385-9.

Campbell, J. D. (1968) Closing comments by session chairmen, in *Mechanical Behavior of Materials under Dynamic Loads*, Ed. U. S. Lindholm, Springer, New York. pp. 410-14.

Campbell, J. D. (1970) *Dynamic Plasticity of Metals,* Springer-Verlag, Vienna.

Campbell, J. D., Eleiche, A. M. and Tsao, M. C. C. (1977) Strength of metals and alloys at high strains and strain rates, in: *Battelle Colloq. on Fundamental Aspects of Structural Alloy Design*, R. I. Jaffe and B. A. Wilcox (editors), Plenum, New York, pp. 545-63.

Campbell, J. D., Ponter, A. R. S. and Duffy, J. (1978) *Euromech-Symp. III on Constitutive Modelling in Inelastisity*, Marianske Lazne, Czechoslovakia.

Chiem, C. Y. and Duffy, J. (1979) Brown University Rep. NSF ENG 75-18532/9

Clifton, R. J. (1979) Plastic wave theory-supported by experiments, in: *Proceedings of the Second Conference on the Mechanical Properties of Materials at High Rates of Strain*, J. Harding (editor), Oxford, March 28-30, 1979, The Institute of Physics, Conference Series Number **47**, pp. 174-86.

Costin, L. S., Crisman, E. E., Hawley, R. H. and Duffy, J. (1979) On the localization of plastic flow in mild steel tubes under dynamic torsional loading, in: *Proceedings of the Second Conference on the Mechanical Properties of Materials at High Rates of Strain*, J. Harding (editor), Oxford, 28-30, 1979, The Institute of Physics, Conference Series Number **47**, pp. 90 -100.

Davies, E. D. H. and Hunter, S. C. (1963) The dynamic compression testing of solids by the method of the Split Hopkinson Bar, *J. Mech. And Phys. of Solids* **11**, 155-79.

Dormeval, R. (1987) The adiabatic shear phenomenon, in: *Materials at High Strain Rates*, T. Z. Blazynski (Ed.), Elsevier, London, 47-70.

Duffy, J. (1979) The J. D. Campbell memorial lecture: Testing techniques and material behaviour at high rates of strain, in: Proceedings of the Second Conference on the Mechanical Properties of Materials at High Rates of Strain, J. Harding (editor), Oxford, March 28-30, 1979, The Institute of Physics, Conference Series Number **47**, pp. 1-15.

Duffy, J. Campbell, J. D. and Hawley, R. H. (1971) On the use of a torsional split Hopkinson bar to study rate effects in 1100-0 aluminum, *J. Appl. Mech.* **38**, March 1971, 83-91.

Eichelberger, R. J. (1965) Hypervelocity impact, in *Behavior of Materials Under Dynamic Loading*, N.J. Huffington, Jr. (Ed.), ASME, New York, pp.155-87.

Eleiche, A. M. and Campbell, J. D. (1976a) Tech. Rep. ARML-TR-76-90, Air Force Materials Laboratory, Wright-Patterson Air Force Base, USA.

Eleiche, A. M. and Campbell, J. D. (1976b) Strain-rate effects during reverse torsional shear, *Exptl Mech.* **16 (8)**, 281-90.

Follansbee, P. S. (1986) In. *Metallurgical Applications of Shock-Wave and High-Strain-Rate Phenomena*, L. E. Murr, K. P. Staudhammer and M. A. Meyers (editors), Marcel Dekker, New York, Chap. 24.

Hopkinson, B. (1905) *Proc. Roy. Soc. A* **74**, 498.

Klepaczko, J. (1968) Strain rate history effects for polycrystalline aluminum and theory of intersections, *J. Mech. Phys. Solids* **16**, 255-66.

Klepaczko, J. (1975) *Mater. Sci. Engng.* **18,** 121-36.

Klepaczko, J., Frantz, R. A. and Duffy, J. (1977) Polska Akademia Nauk, Instytut Podstawowych Problemow Techniki, Engng. Trans. **25**, 3-22.

Kolsky, H. (1960) Viscoelastic waves, *Proc. International Symp. on Stress Wave Propagation*, N. Davids, Ed., Interscience, New York, pp. 693-711.

Kolsky, H. (1965) The propagation of stress pulses in linear viscoelastic solids, *Philosophical Magazine* **1**, pp. 693-711.

Kolsky, H. and Douch, L. S. (1962) Experimental studies in plastic wave propagation, *J. Mechanics and Physics of Solids* **10**, 195-223.

Kressel, H. and Brown, N. (1967) Lattice defects in shock-deformed and cold-worked nickel, *J. Appl. Phys.* **38**(4), 1618-25.

Lindholm, U.S. (1962) Some problems involved in testing materials at high strain rates, *Proc. Army Conf. on Dynamic Behavior of Materials and Structures,* Springfield, Mass., September 1962, pp. 14-41.

Lindholm, U. S. (1964) Some experiments with the Split Hopkinson Pressure Bar, *J. Mech.and Phys. of Solids* **12**, 317-35.

Lindholm, U. S. (1965) Dynamic deformation of metals, in *Behavior of Materials under Dynamic Loading*, N.J. Huffington, Jr. (Editor), ASME, New York, pp. 42-61.

Lindholm, U. S. (1978) Deformation maps in the region of high dislocation velocity, in: *High Velocuty Deformation of Solids*, K. Kawata and J. Shioiri (editors), Springer-Verlag, New York.

Malin, A. S. and Hatherly, Y. M. (1979) Microstructure of cold-rolled copper, *Metal Sci.* 13, 463-72.

Murr, L. E. (1975) *Interfacial Phenomena in Metals and Alloys*, Addison-Wesley, Reading, MA.

Murr, L. E. (1981a) Residual microstructure - mechanical property relationships in shock loaded metals and alloys, in: *Shock-Wave and High-Strain-Rate Phenomena in Metals*, edited by M. A. Meyers and L. E. Murr, Plenum Press, New York, Chap. 37, pp. 607-73.

Murr, L. E. (1981b) Effects of peak pressure, pulse duration, and repeated loading on the residual structure and properties of shock deformed metals and alloys, in: *Shock-Wave and High-Strain-Rate Phenomena in Metals*, edited by M. A. Meyers and L. E. Murr, Plenum Press, New York, Chap. 42, pp. 753-77.

Murr, L. E. (1987) Metallurgical effects of shock and high-strain-rate loading, in: *Materials at High Strain Rates*, T. Z. Blazynski (Editor), Elsevier, London, 1-46.

Murr, L. E., Inal, O. T. and Morales, A. A. (1967) Direct observations of vacancies and vacancy-type defects in molybdenum following uniaxial shock-wave propagation, *Acta Met.* **24**, 261-270.

Murr, L. E. and Meyers, M. A. (1983) Metallurgical effects of shock and pressure waves in metals, in: *Explosive Welding, Forming and Compaction*, T. A. Blazynski (Editor), Applied Science Publishers, London, Chapter 3, 83-121.

Murr, L. E., Staudhammer, K. P. and Meyers, M. A. (editors), (1986) *Metallurgical Applications of Shock-Wave and High-Strain-Rate Phenomena,* Marcel Dekker, New York.

Ponter, A., R. S. (1978) Rep. No. 78-12, Dept. of Engineering, University of Leicester.

Rogers, H. C. (1979) *Ann. Rev. Mat. Sci.* **9**, 283.

Rogers, H. C. (1983) Adiabatic shearing - General nature and material aspects, in: *Material Behavior under High Stress and Ultra High Loading* Rates, edited by J. Mescall and V. Weiss, Plenum Press, New York, pp. 101-18.

Senseny, P. E., Duffy, J. and Hawley, R. H. (1978) Experiments on strain rate history and temperature effects during plastic deformation of close-packed metals, *J. Appl. Mech.* **45**, March 1978, 60-6.

Stelly, M. and Dormeval, R. (1978) Some results on the dynamic deformation of copper, *IUTAM Symp. on High Velocity Deformation of Solids*, Tokyo, Japan, Aug. 24-27, 1977, K. Kawata and J. Shioiri (Eds.), Springer-Verlag, New York, pp. 82-97.

Stelly, M. and Dormeval, R. (1986) Adiabatic shearing, In: *Metallurgical Applications of Shock-wave and High-strain-rate Phenomena*, Murr, L. E., Staudhammer, K. P. and Meyers, M. A. (editors), Marcel Dekker, New York., Chap. 32, pp. 607-32.

Tanaka, K. and Nojima, T. (1979) Dynamic and static strength of steels, in: *Proceedings of the Second*

Conference on the Mechanical Properties of Materials at High Rates of Strain, J. Harding (editor), Oxford, 28-30, 1979, The Institute of Physics, Conference Series Number **47**, pp. 166-73.
Wilson, M. L., Hawley, R. H. and Duffy, J. (1979) Brown University, Rep. NSF ENG 75-18532/8.
Zener, C. and Hollomon, J. H. (1944) Effect of strain rate upon plastic flow of steel, *J. Appl. Phys.* **15**, 22-32.
Zener, C. (1948) The micro-mechanism of fracture, in: *Fracturing of Metals*, ASM, Cleveland, Ohio, pp.3-31.

9.5. Further Reading

ASM Handbook (1985) *Mechanical Testing*, ASM Int. 1, Vol. 8, Melno Park, Ohio 1985, pp.187-297.
Bell, J. F. (1959) Propagation of plastic waves in solids. J. *Appl. Phys.* **30 (2)**, 196-201.
Blazynski, T. Z. (Ed.) (1983) *Explosive Welding, Forming and Compaction*, Applied Science Publishers, London.
Crossland, B. (1954) The effect of fluid pressure on the shear properties of metals, *Proc. Inst. Mech. Engineers* **68**, 935-46.
El-Sobky, H. (1983) Mechanics of explosive welding, in: *Explosive Welding, Forming and Compaction*, Blazynski, T. Z. (Ed.), Applied Science Publishers, London, pp. 189-217.
Harding, J. (Editor) (1979) Mechanical Properties at High Rates of Strain, 1979, *Proceedings of the Second Conference on the Mechanical Properties of Materials at High Rates of Strain*, Oxford, March 28-30, 1979, The Institute of Physics, Conference Series **47**.
Harding, J. (1983) High-rate straining and mechanical properties of materials, in: *Explosive Welding, Forming and Compaction*, Blazynski, T. Z. (Ed.), Applied Science Publishers, London, pp. 123-58.
Hauser, F. E. and Simmons, J. A. and Dorn, J. E. (1961) Strain rate effects in plastic wave propagation." in *Response of Metals to High Velocity Deformation, Proc. Metallurgical Soc. Conf.*, P. G. Shewmon and V. F. Zackay, Eds., Vol. 9, Interscience, N.Y , pp. 93-114.
Hill, R. (1968) On constitutive inequalities for simple materials - I, *J. Mech. Phys. Solids* **16** , 229-42.
Hopkins, H. G. (1966) Dynamic non-elastic deformations of metals, in: *Applied Mechanics Surveys*, Abramson, H. N., Liebowitz, H., Crowley, J. M. and Juhasz, S. (Eds.), Spartan Books, Inc., Washington, D. C., pp. 847-67.
Jahsman, W. E. (1971) Re-examination of the Kolsky technique for measuring dynamic material behavior, *Trans. ASME, J. Appl. Mech.*, March 1971, 75-82..
Kolsky, H. (1949) An investigation of the mechanical properties of materials at very high rates of loading, *Proc. Phys. Soc.*, Sec. B, Vol. 62, 676--700.
Lindholm, U. S. (Editor) (1968) *Mechanical Behavior of Materials Under Dynamic Loads*, Springer, New York.
Lindholm, U. S. and Yeakley, L. M. (1965) Dynamic deformation of single and polycrystalline aluminum, *J. Mech. Phys. of Solids* **13**, 41-53.
Malvern, L. E. and Efron, L. (1964) *Stress Wave Propagation and Dynamic Testing; Longitudinal Plastic Wave Propagation in Annealed Aluminum Bars*, Tech. Rept. No.1, Grant G-24898, National Science Foundation, Michigan State University, East Lansing, Mich., September 1964.
Manjoine, M. J. (1944) Influence of rate of strain and temperature on yield stresses of mild steel, *J. Appl. Mech.* **11**, Trans. ASME **66**, Ser. A, 211-18.
Meyers, M. A. (1978) A mechanism for dislocation generation in shock-wave deformation, *Scripta Metallurgica* **12**, 21-6.
Meyers, M. A. and Murr, L. E. (1983) Propagation of stress waves in metals, in: *Explosive Welding, Forming and Compaction*, Blazynski, T. Z. (Ed.), Applied Science Publishers, London, pp. 17-82.
Murr, L. E. and Meyers, M. A. (1983) Metallurgical effects of shock and pressure waves in metals, in *Explosive Welding, Forming and Compaction*, T. Z. Blazynski (Editor), Applied Science Publishers, New York, Chapter 3, pp. 83-121.

Pearson, J. (1983) Introduction to high-energy rate metal working, in: *Explosive Welding, Forming and Compaction*, Blazynski, T. Z. (Ed.), Applied Science Publishers, London, pp. 1-15.

Perrone, N. (1965) On a simplified method for solving impulsively loaded structures of rate-sensitive materials, *J. Appl. Mech.*, Paper No. 65-APM-10.

Prümmer, R. (1983) Powder Compaction, in: *Explosive Welding, Forming and Compaction*, Blazynski, T. Z. (Ed.), Applied Science Publishers, London, pp. 369-95.

Tardif, H. P. and Marquis, H. (1963) Some dynamic properties of plastics, *Canadian Aeronautics and Space* **I** (9), 205-13.

Wongwiwat, K. and Murr, L. E. (1978) Effect of shock pressure, pulse duration, and grain size on shock-deformation-twinning in molybdenum, *Materials Science and Engineering* **35**, 273-85.

CHAPTER 10

PLASTIC INSTABILITY AND LOCALIZATION EFFECTS

10.1. Introduction

A decrease in stiffness due to geometrical change and/or material softening caused by deformation is responsible for the occurrence of instability phenomena in engineering materials within the plastic range; i.e., beyond the yield point. Such phenomena manifest themselves in various ways; e.g., buckling, bulging, necking and shear banding. Once such instabilities are started, they tend to persist and the stiffness of the specific cross-sectional area of the specimen decreases; therefore deformation intensifies locally and eventually leads to final collapse and/or failure.

Because the occurrence of much instabilities is an important precursor to collapse or failure, computational prediction of the onset and of the augmentation of these instabilities is essential and indispensable in understanding the ultimate strength of the structures and materials, and in predicting and improving plastic solids formability.

The onset of plastic instability is likely to be related to the point where "***bifurcation***" from the fundamental path becomes possible

The point of bifurcation maybe obtained by applying "*Hill's bifurcation theorem (1958)*" for "***associative materials***" under "***conservative loading***"; Hill (1958).

A more elaborate theorem must be employed for "***non associative***" and "***nonlinear***" materials; see, e.g., Tomita (1994).

10.2. Onset of Shear Banding

The onset of shear banding can be analysed within the framework given by Hill (1962a), and Rice (1976).

The necessary conditions for the earliest possible localization of instabilities may be determined by the linear instability theory (Leroy and Ortiz, 1989).

Under specific conditions, post bifurcation behaviour, immediately after the bifurcation point, may be expressed by the sum of the fundamental solution at the bifurcation

point and a suitably
normalized bifurcation mode (Hutchinson 1973a).

Regardless of the problem associated with the material characteristics and loading conditions, the so-called "***Growth of Plastic Instability***" may be predicted with proper computational conditions.

10.2.1. BASIC EQUATIONS

Virtual work principle:

In this section, the governing equations for an elastic-plastic boundary value problem are given within the context of large-strain theory. An updated Lagrangian formulation of the field and constitutive equations is employed

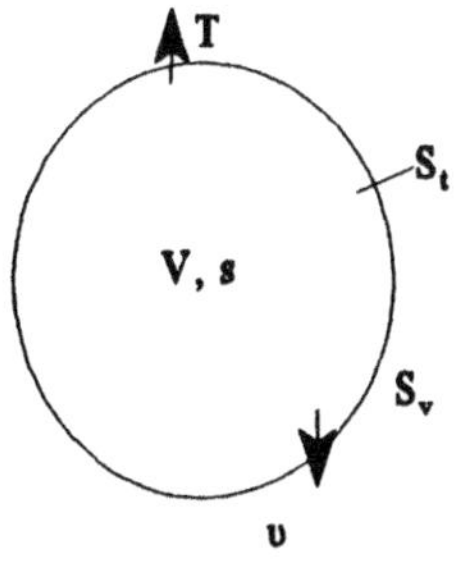

An equilibrium state of the body

Consider an equilibrium state for a body, with volume V and surface S, subjected to a velocity constraint on S_v and traction on the remaining part of S, i.e., S_t. Each particle is labelled by a set of curvilinear coordinates x^i (see Appendix D). The latter are embedded in the body in the current state and serve as independent variables. In the deformed configuration, the covariant component of metric tensor are denoted by G_{ij}. The weak form of the equation governing the rate of stress and traction yields the virtual work principle (Hill, 1958, Seguchi et al., 1971, Kitagawa et al., 1972).

$$\int_V \left(\Sigma^{ij} + \sigma^{lj} \upsilon^i,_\ell\right) \delta\, \upsilon_{i,j}\, d V = \int_{S_t} \dot{T}^i\, \delta\, \upsilon_i\, d S \tag{10.1}$$

where:

- Σ^{ij} is the Kirchhoft stress (*Chapter2*). It is identical to Cauchy stress σ^{ij} in the current configuration.
- $\dot{T}^i$ is the nominal traction rate.
- $\delta\upsilon_i$ is the virtual velocity satisfying the homogeneous boundary condition over surface
- . (an over dot) denotes a material derivative.
- $(.),_\ell$ denotes the covariant derivative with respect to the current coordinates (*Appendix A*).

For the body with configuration dependent loading, the nominal traction rate $\dot{P}^i$ in (10.1) is given as

$$\dot{T} = \dot{T}_0^i + \dot{T}_\upsilon^i \tag{10.2}$$

where $\dot{T}_0^i$: **Configuration independent traction rate**; $\dot{T}_\upsilon^i$: **Configuration dependent traction rate**

Following Hill (1962a), Sewell (1967) and Dubey (1970), the configuration-dependent traction rate $\dot{T}_\upsilon^i$ may be expressed as a function in the velocity υ_i and velocity gradient $\upsilon_{i,j}$ as

$$\dot{T}_\upsilon^i = A^{ji}\upsilon_j + n_j R^{ijk\ell}\upsilon_{\ell,k} \tag{10.3}$$

n_j: **the surface unit normal**

The tensors A^{ji} and $R^{ijk\ell}$ are independent of the velocity. The first term in (10.3) may represent the traction induced by elastic formulation and the second term may represent the follower force; see, e.g., Timoshenko and Gere (1961). Under certain conditions, the configuration dependent traction rate may have potential and the variational principle can be established (Tomita 1994).

For a body with pressure p on the portion of the surface S_t, $\dot{T}_0^i$ and $\dot{T}_\upsilon^i$ are given by

$$\begin{aligned} \dot{T}_0^i &= -\dot{p}\, n_j G^{ij} \\ \dot{T}_\upsilon^i &= n_j R^{jik\ell}\upsilon_{\ell,k} \\ R_{jik\ell} &= -p\,(G^{ij}G^{k\ell} - G^{ki}G^{j\ell}) \end{aligned} \tag{10.4}$$

Meanwhile, the weak form of the energy balance equation for the same body subjected to heat flux $q = -n_i q_i = Q$ on S_q and temperature constraint on S_T can be established by multiplying the local form of the energy balance equation by δT which satisfies the homogeneous boundary condition on S_T as expressed by specific heat — Thermal conductivity tensor

$$\int_V \delta T \, \rho \, c \, \dot{T} \, dV + \int_V \delta T_{,i} \, \kappa^{ij} \, T_{,j} \, dV = \int_V \delta T \, w^{P} \, dV + \int_{S_q} \delta T \, Q \, dS \qquad (10.5)$$

mass density — fraction of irreversible work

In (10.5), the fraction of invertible work $w^{P} = \sigma'^{ij} D_{ij}^{P}$ which is converted to heat is α, where α is in the range of 0.85 - 0.95 for many metals; Taylor and Quinney (1934). Meantime, the specific form of heat flux Q depends on the respective boundary conditions.

Constitutive Equations

Predictions of instability behaviour strongly depend on material response.

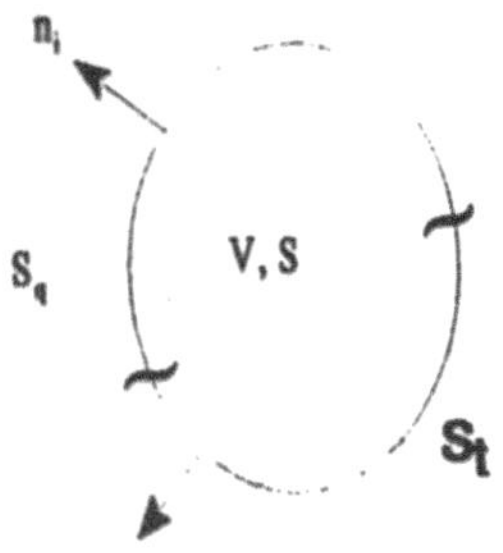

In elastic-plastic material response, the plastic part $D_{k\ell}^{P}$ of the strain rate $D_{k\ell}$, $D_{k\ell} = (\upsilon_{k,\ell} + \upsilon_{\ell,k})/2$ is usually specified through various classes of constitutive equations.

Elastic material response takes the form of a linear relationship between the elastic strain rate, $D_{k\ell}^{e} = D_{k\ell} - D_{k,\ell}^{P}$, and a suitable objective stress rate as

$$\Sigma^{ij} = D_e^{ijk\ell} \left(D_{k\ell} - D_{k\ell}^{P} \right) \qquad (10.6)$$

An elastic constitutive tensor

Meantime, the constitutive equation for elastic-plastic response can be expressed as

$$\Sigma^{ij} = L^{ijk\ell} D_{k\ell} \tag{10.7}$$

where L^{ijkl} is an elastic-plastic constitutive tensor which depends on the current stress, the deformation history and the choice of an objective stress rate. Reference in this context is made to Dafalias (1983), Loret (1983), and Dafalias and Aifantis (1990), among others.

For the material obeying the flow rule, plastic strain rate D_{ij}^{P} can be expressed by

$$D_{ij}^{P} = \frac{1}{h} n_{mn} \Sigma^{mn} m_{ij} \tag{10.8}$$

h: the hardening modulus; n_{mn}: normal to the yield surface; m_{ij}: directions of plastic strain components

For $n_{ij} = m_{ij}$, the constitutive equations derived for associated flow rule are recovered. The constitutive equations following "***J_2 flow theory***" (Hill 1958, Hutchinson, 1973b), "***J_2 Kinematic hardening theory***" (Tvergaard 1978) and many anisotropic theories (see, e.g., Neale 1980, Tomita 1994) fall into the special case of eqn. (10.8).

Rudnicki and Rice (1975) expressed in their model n_{ij} and m_{ij} by the following relationship

$$n_{ij} = \sqrt{3}\, \sigma'_{ij} / 2 \bar{\sigma} + B\, G_{ij} / 3$$

and

$$m_{ij} = \sqrt{3}\, \sigma'_{ij} / 2 \bar{\sigma} + M\, G_{ij} / 3 \tag{10.9}$$

Specific values of B and M can be determined, for instance, by a Gurson-type yield function (Gurson 1977, Tvergaard, 1981).

A kinematic hardening version of the material was suggested by Mear and Hutchinson (1985) and Tvergaard (1987). Tomita (1994) extended the model to account for the change in elasticity modulus due to the void volume fraction.

The deformation type constitutive equation, originally proposed by Budiansky (1959) has been generalized to account for finite strain (Stören and Rice, 1975, Needleman and Tvergaard, 1977, Hutchinson and Neale, 1973) and anisotropy (Tomita and Shindo, 1985). The plastic strain constitutive equation is

$$D_{ij}^{n} = \frac{1}{h} n_{nm} \Sigma^{nm} n_{ij} + \frac{1}{h_s} (\Sigma_{ij}^{'} - n_{nm} \Sigma^{nm} n_{ij}) \tag{10.10}$$

a new harderning modulus

The constitutive eqn. (10.10) is valid for the deformation satisfying the total loading condition (Budiansky 1959).

For the strongly non-proportional stress histories, Christofersen and Hutchinson (1979) proposed the "***corner theory***" in which an angular measure ϕ of the stress rate direction with respect to the corner direction of the yield surface is defined. The instantaneous moduli for nearly proportional loading, $\phi \leq \phi_o$, are chosen equal to those of deformation theory, and for increasing derivation from proportional loading $\phi_o \leq \phi \leq \phi_c$, the moduli stiffen monotonously until they coincide with the linear elastic moduli for elastic deformation or unloading.

The plastic strain rate constitutive equation can be expressed by

$$D_{ij}^{n} = \frac{\partial^2 E^p}{\partial \Sigma^{ij} \partial \Sigma^{k\ell}} \Sigma^{k\ell}$$

$$E_c^p = \frac{1}{2} f(\theta) C_{ijk\ell} \Sigma^{ij} \Sigma^{k\ell} \tag{10.11}$$

$f(\theta)$: A transition function; it is unity throughout the total loading range, $0 \leq \theta \leq \theta_0$, and is identically zero for $\theta_c \leq \theta \leq \pi$. Here, $f(\theta)$ decreases monotonically from unity to zero as θ increases form θ_0 to θ_c.

$C_{ijk\ell}$: Plastic compliance tensor

The Bauschinger effect (Tomita et al., 1986) and anisotropy (Tomita et Shindo 1990) have been concretely introduced in the constitutive eqn. (10.11); see, also, Gotoh (1985).

10.3. Strain-Rate and Temperature Effects

Engineering materials generally do possess strain rate and temperature sensitivities to various extents. These have an important effect on instability behaviour. Here we restrict our attention to an isotropic material and assume that the relation of representative stress $\bar{\sigma}$, representative viscoplastic strain $\bar{\varepsilon}^{\upsilon}$, representative viscoplastic strain rate $\dot{\bar{\varepsilon}}^{v}$ and absolute temperature T has the form

$$\bar{\sigma} = F\left(\bar{\varepsilon}^{v}, \dot{\bar{\varepsilon}}^{v}, T\right)$$

or

$$\dot{\bar{\varepsilon}}^{v} = G\left(\bar{\sigma}, \bar{\varepsilon}^{v}, T\right) \qquad \textbf{(10.12)}$$

Tomita (1994) advanced that the constitutive equation for plastic strain rate, eqn. (10.10) may be modified to include the effect of strain rate and temperature sensitivity. The viscoplastic strain rate D_{ij}^{v} is proposed to be given by

$$D_{ij}^{v} = \frac{1}{h}\left(n_{nm}\,\Sigma_{mn} + \beta\,\dot{T}\right)n_{ij} + \frac{1}{h_{s}}\left(\Sigma_{ij}' - n_{nm}\,\Sigma^{nm}\,n_{ij}\right)$$

$$n_{ij} = \frac{\sigma_{ij}'}{\sqrt{2/3}\,\bar{\sigma}}, \quad h = \frac{2\dot{\bar{\sigma}}}{3\dot{\bar{\varepsilon}}^{v}}, \quad h_{s} = \frac{2\,\bar{\sigma}}{3\,\bar{\varepsilon}^{v}\omega} \qquad \textbf{(10.13)}$$

In (2.13), ω accounts for the degree of non-coaxiality of the viscoplastic strain rate to the stress tensor (Tomita and Shindo 1985), and β stands for the temperature sensitivity of the flow stress. Naturally the situation $\omega \to 0$ provides a generalized constitutive eqn. following the "***J_2 flow theory***"; Tomita (1994).

When the total strain rate D_{ij} is assumed to be the sum of an elastic strain rate D_{ij}^{e}, accounting for the temperature dependent elastic response, and a viscoplastic strain rate D_{ij}^{v}, Eqn. (10.13), the constitutive equation for stress rate Σ_{ij}, strain rate $D_{k\ell}$ and the rate of charge in temps $\dot{T}$ is then established (Tomita et al, 1990).

A concrete form of (10.12) which is often used has the form

$$\bar{\sigma}=\sigma_0(T)\left(1+\frac{\bar{\varepsilon}^v}{\varepsilon^y}\right)^n\left(1+\frac{\dot{\bar{\varepsilon}}^v}{\dot{\varepsilon}^y}\right)^m \tag{10.14}$$

where σ_o (T) is temperature dependent stress characterizing the thermal softening affect, and n and m are strain and strain rate sensitivity components, respectively. Here ε_y and $\dot{\varepsilon}_y$ are reference strain and strain rate, respectively.

According to the experimental observation of the response of the material under a multi-axial stress condition and high rate of deformation, the constitutive equation is quite complicated, and the strain- rate sensitivity exponent m, generally depends on the strain rate applied and increases as the deformation rate increases.

Furthermore, for specific materials, an abrupt increase in the strain rate during the deformation process causes a substantial increase/decrease in flow stress as seen in steel/copper (Campbell et al. 1977, Mimura and Tomita, 1991, Tomita and Higo 1993).

Such substantial increase/decrease in the flow stress will be referred to as "positive" and "negative" strain-rate history dependence, respectively. Such an effect may be accounted for (see Tomita, 1994) by

$$\bar{\sigma}=\sigma_0(T)\left(1+\frac{\bar{\varepsilon}^v}{\varepsilon_y}\right)^n\left(1+\frac{\dot{\bar{\varepsilon}}^v}{\dot{\varepsilon}^y}\right)^m \times\left\{1+A\left(B\,\bar{\varepsilon}_1^v\right)^n \ell n\left(\frac{1+\dot{\bar{\varepsilon}}^v}{1+\dot{\bar{\varepsilon}}_1^v}\right)\right\} \tag{10.15}$$

In (10.15), A and B account for the material strain rate and strain history dependence of the flow stress, respectively, and ε_1^v and $\dot{\bar{\varepsilon}}_1^v$ are viscoplastic representative strain and its rate before abrupt charge in strain rate, respectively (Tomita and Higo, 1993).

In order to avoid numerical instability and maintain the required accuracy, suitable integration schemes for the rate-type constitutive equation must be employed. In this context, for the temperature-independent case, in the Euler method, the size and the time steps must be determined such that the stress exactly satisfies the yield condition in the course of yielding (Yamada et al 1968), and the magnitude of the increment of the displacement as well as the rotation is restricted to avoid numerical instability (Nagtegaarl and Jong, 1981).

These methods have been extended to the strain rate and temperature-dependent constitutive equations (Peirce et al. 1984, Rashid and Nemat-Nasser 1992, and Nemat-Nasser and Li, 1992).

Iterative methods such as the radial return method (Krieg and Krieg, 1977) and mean normal method (Rice and Tracey 1973) have been developed and extended to different types of materials.

Return mapping algorithms capable of accommodating the general yield condition and arbitrary flow hardening rules; nonlinear elastic response for general rate-independent and rate-dependent behaviour (Ortiz and Simo, 1986); versatile integration algorithms including their application to the treatment of nonsmooth yield surfaces (de Borst, 1987, Simo et al 1988, Runesson et al 1988), their accuracy (Ortiz and Popov 1985, and Ortiz and Simo 1986) and consistent tangent operators (Simo and Taylor 1985, Runesson et al. 1986, Simo et al. 1988) have been extensively investigated to obtain a converged and accurate solution.

Runneson et al. (1988) also provides an excellent brief review of the development of integration schemes. Furthermore, the treatment of large increments of strain (Hughes and Winget, 1980, Pierce et al, 1984, Simo and Ortiz, 1985, Runesson et al, 1986, Rashid and Nemat Nasser 1992, Nemat-Nasser and Li, 1992) is indispensable for large strain and displacement analysis.

As long as the deformation is sufficiently small, the elastic-plastic boundary value problem has a unique solution which is referred to as the <u>fundamental solution</u>.

When the deformation reaches a certain value, bifurcation from the fundamental solution becomes possible. <u>The point of bifurcation can be found through the use of Hill's general theory of bifurcation and uniquesness (Hill,1958) for elastic-plastic solids.</u> This theory states that the solution is not unique when a nontrivial solution can be found for the eigenvalue problem given by the following variational equation

Bifurcation functional

$$\delta I = 0$$

$$\textcircled{I} = \int_V (\Sigma^{\cdot ij} + \sigma^{ij} v^{\cdot}_{i,j} \, dV - \int_{S_t} \dot{T}_v^{\cdot i} v^{\cdot}_i \, dS \qquad (10.16)$$

Bifurcation condition (10.16) is valid only when the material follows an associative flow law, whereby the superscripted asterisk denotes the difference between the fundamental solutions and the second one. The surface integral in (10.16) arises from the configuration dependence of the loading. Meantime stress rate Σ^{*ij} is related to strain rate $D^{*}_{k\ell}$ by 1

$$\Sigma^{*ij} = L^{ijk\ell} D^{*}_{k\ell} \qquad \textbf{(10.17)}$$

Is assumed to be symmetric when the material follows an associative flow law.

Is assumed to be a constitutive tensor for a linear comparison solid, in which the plastic parrt of the constitutive tensor is employed for the current plastic zone. (see Tomita,1994).

When the bifurcation functional I (Eqn.10.16) is approximated in terms of finite elements, one can arrive at an approximate functional of the following form

$$I = \{\delta^{*}\}^{T} \{K\} \{\delta^{*}\} \qquad \textbf{(10.18)}$$

where $\{\delta^{*}\}$ denotes the values of v_i^{*} at the nodal points. The stationary condition of the approximate functional with respect to $\{\delta^{*}\}$ yields the following homogeneous algebraic equation

$$\{K\} \{\delta^{*}\} = 0 \qquad \textbf{(10.19)}$$

When the equation (10.19) has a nontrivial solution, bifurcation may occur.

At every computational step, the vanishing point of the determinant of the coefficient matrix of (10.19), i.e.

$$\det [K] = 0 \qquad \textbf{(10.20)}$$

is checked. Usually, when the sign of the det [K] changes at a specific incremental step, an iterative method is used to determine the accurate vanishing point of the determinant.

The bifurcation mode is obtained as the eigen-mode of the homogeneous equation (10.19).

In this context, reference is made to Kitagawa et al (1980, 1982), de Borst (1989), Bardet (1990), and Leroy and Chapuis (1991). The special case where two or more eigenvalues of [K] may simultaneously change sign, has been treated by de Borst (1989).

With reference to condition (10.20), at every computational step of the analysis, the vanishing of the determinant of the coefficient matrix is checked against several modes. The first bifurcation point is referred to as the "***Critical Bifurcation Point***", and its mode is referred as the "***Critical Bifurcation Mode***"

10. 4. Bifurcation Analysis for Specific Constitutive Equations

When the material follows the nonassociative flow law, the bifurcation condition (10.16) becomes invalid because of the nonsymmetry of the tensor L^{ijkl} in (10.17).

EXAMPLE :

I. Linear Constitutive Equation

Raniecki (1979) and Raniecki and Brunhs (1981) introduced two comparison solids to determine the bifurcation point for the material obeying the non-asssociative flow law.

The first comparison-solid is from the "*one-parameter family*" of "*linear comparison solids*" with the following strain rate constitutive form

$$D^{ijk\ell} = D_E^{ijk\ell} - \frac{1}{G} p^{ij} q^{k\ell}$$

$$p^{ij} = q^{ij} = D_E^{ijk\ell} n_{k\ell} + \xi D_E^{ijk\ell} m_{k\ell}$$
$$G = 4\xi (H + m_{ij} D_E^{ijk\ell} n_{k\ell}) \tag{10.21}$$

where H is the hardening modulus and m_{ij} and n_{ij} are the directions of plastic strain rate and the normal to the yield surface, respectively. Here ξ is a positive parameter.

The solid obeying the constitutive eqn. (10.21) is referred to as an "***Alternative Comparison Solid***". Raniecki and Brunhs (1981) proved that if uniqueness is certain for these comparison solids, then bifurcation is precluded for the underlying materials. The bifurcation point for these comparison solids provides the lower bound to a solid with the "*non-associative flow law*". However, the still undetermined positive parameter ξ in the constitutive tensor is a function of the particle position and should be optimized to give the closest lower bound. For the homogeneous fundamental deformation, the optimal lower bound can be determined as (e.g., Raniecki, 1979)

$$\xi = \sqrt{\frac{m_{ij} \, D_E^{ijk\ell} \, m_{k\ell}}{n_{nm} \, D_E^{mnrs} \, n_{rs}}} \qquad \mathbf{(10.22)}$$

The lower bound to the bifurcation point is not ordinarily the bifurcation point of the underlying material. Consequently, the search for the genuine bifurcation state is replaced by a search for upper and lower bounds:

i) the lower bound: Here the occurrence of bifurcation is checked against the vanishing point of the ξ-value determinant matrix, det [K (ξ)], expressed in (10.20), at every step of the analysing of the fundamental solution. With note of the positiveness of these determinants up to the bifurcation point, the maximization of the determinant with respect to the positive parameter ξ at each step of fundamental analysis substantially improves the accuracy of the lower bond.

ii) the upper bound: Here the concept of a second-comparison solid, a "***nonloading solid***" is introduced and shown with the nonassociative flow law such that the first eigenstate of such a comparison solid identifies an upper bound to the bifurcation point of the underlying solid.

II. Nonlinear Constitutive Equation

For a material obeying a nonlinear constitutive equation such as Christoffersen and Hutchinson's corner theory (1979), the bifurcation theory should be generalized (Triantafyllidis 1985).

At same stage of deformation, stress σ^{ij}, displacement u_i and any state variables in the constitutive equation, as well as their corresponding rates, are known and unique. Then, the following bifurcation functional, quadratic in $\dot{v}_i^*$, and bifurcation condition for displacement prescribed loading are defined as (see Tomita, 1994)

$$\begin{aligned} &\delta I = 0 \\ &I = \int_V \left(\Sigma^{*ij} + \sigma^{ij} \; \dot{v}^{*i}_{,\ell} \; \dot{v}^{*}_{i,j} \right) A V \end{aligned} \qquad \mathbf{(10.23)}$$

where $\Sigma^{\cdot ij}$ is related to $\dot{D}^{*}_{k\ell}$ by

$$\Sigma^{\cdot ij} = L^{ijk\ell}\dot{D}^{*}_{k\ell} \tag{10.24}$$

In (10.24), $L^{ijk\ell}$ is the constitutive moduli tensor of an actual solid in the fundamental state.

The bifurcation functional (10.23) has been used to determine a lower bound for the first bifurcation. On the other hand, Tvergaard (1982) assumed the uniqueness of stress σ^{ij}, displacement u_i and any state variable in the constitutive equation and employed the bifurcation functional (10.16) with the actual constitutive moduli tensor in the fundamental state to obtain the upper bound for the critical load.

10.5. Post Bifurcation Analysis

The solution of the boundary value problem at the bifurcation point can be expressed by the sum of the fundamental solution and a suitably normalized eigen mode for the variational equation.

The specific amplitude of the eigen mode is determined so that loading occurs everywhere in the current plastic zone, except at one point where neutral loading takes place (Hutchinson, 1973a). This solution reveals the postbifurcation behaviour just after the bifurcation point.

Due to the highly nonlinear nature of the postbifurcation behaviour, numerical analysis appears indispensable, e.g., by employing the virtual work principle with finite element approximation.

When the materials obeys the constitutive equation derived by the nonassociative flow law or expressed by the nonlinear relation between the stress and strain rates, the bifurcation point obtained does not necessary provide the real bifurcation point. Thus, the post-bifurcation behaviour must be traced approximately by employing bodies with initial imperfections, through, again, the virtual work principle.

The proper magnitude of imperfection which depends on the problems and the significant features of the computational facility, must be introduced to simulate approximate bifurcation and post bifurcation behaviour (Tomita et al. 1984).

10.6. Plastic Instabilities in Specific Problems

10.6.1. INSTABILITY BEHAVIOUR OF CIRCULAR TUBES

The problem of predicting the deformation behaviour of an elastic-plastic tube subjected to a combined load is an important one in mechanics and in engineering applications. Thus, significant research effort has been undertaken on this topic so far.

Following Tomita (1994), the presentation below concentrates on axisymmetric and non-axisymmetric bifurcation and post-bifurcation behaviour of elasto-plastic circular tubes under lateral pressure and axial load.

For axially plane strain problems, the bifurcation functional has the form (Chu 1979, Tomita et al., 1981).

$$\begin{aligned} I_m &= \pi \int_b^a \hat{v}_m^T B_m \hat{v}_m \, r \, dr - p_a \pi a^2 \left(\hat{v}_m^T C_m \hat{v}_m \right)_{r=a} \\ &\quad + p_b \pi b^2 \left(\hat{v}_m^T C_m \hat{v}_m \right)_{r=b} \end{aligned} \tag{10.25}$$

$$\hat{v}_m = \left(\frac{\hat{v}_{rm}}{r} \quad \frac{\hat{v}_{\theta m}}{r} \quad \hat{v}_{m,r} \quad \hat{v}_{\theta m,r} \right)$$

where p_a and p_b are internal and external pressure, respectively. $\mathbf{B}_m$ and $\mathbf{C}_n$ are matrices depending on the physical components of the constitutive tensor and a bifurcation mode m.

The bifurcation functional (10.25) and its slightly extended version have been extensively employed in the prediction of the onset of bifurcation for internal pressure (Chu 1979, Tomita et al 1981, Reddy 1982) and external pressure (Tomita and Shindo 1982) under plane strain conditions, and for the combined loading condition of internal pressure and axial force (see Tomita, 1994). These are considered frequent collapse problems for design purposes.

On the other hand, although the number of investigations is rather restricted, the initial to intermediate post-bifurcation behaviour (Tomita et al 1981, Tomita and Shindo 1982) and localization of the deformation accompanied by shear bands (Larrson et al. 1982) have been clarified. However, these studies are restricted to the deformation under axially plane strain conditions (Tomita, 1994)..

Tomita et al. (1984, 1986) investigated the loading path-dependent bifurcation and post-bifurcation behaviour of tubes subjected to axial tension and internal pressure, and the localization behaviour of tubes under axial load and external pressure.

Fig. 10.1 shows a few results obtained by Tomita et al. (1986). The positive axial loads substantially lower the maximum pressure, however, these effects diminish as deformation proceeds. The Bauschinger effect is quite noticeable for a tube with positive axial load. In Fig. 10.1b, the critical displacements at which the stress system first satisfies Hill and Hutchinson's surface instability and shear band formation condition (Hill and Hutchinson 1975) are shown. The influence of axial bond, the Bauschinger effect and corner formation, including the corner angle, and the mobility of the yield surface on the formation of uneveness and shear band, and their growth were investigated (Tomita, 1994).

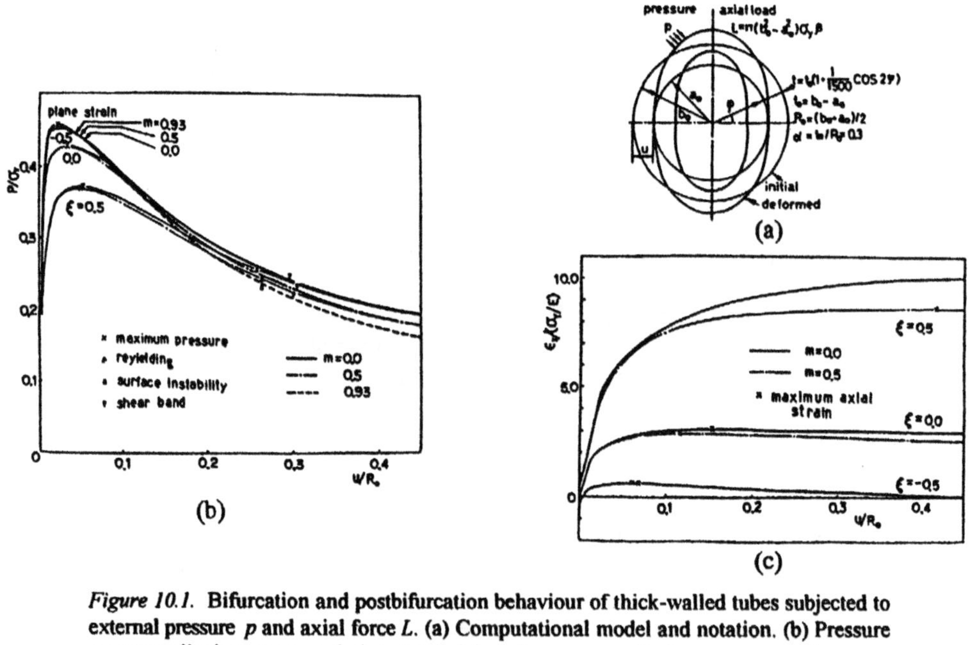

Figure 10.1. Bifurcation and postbifurcation behaviour of thick-walled tubes subjected to external pressure p and axial force L. (a) Computational model and notation. (b) Pressure p versus displacement u relation. (c) Axial strain ε_z versus displacement u relation. m: parameter defining partial translation of yield surface. $m = 0$: no rotation, σ_y: initial yield stress. "Reprinted from *Int. J. Mech. Sci.* **28 (5)**, Tomita, Y., Shindo, A., Kim, Y. S. and Michiura, A., Deformation behaviour of elastic-plastic tubes under external pressure and axial load, pp. 263-74, 1986, with permission from Elsevier Science".

10.7. Instability Propagation (Metallic and Polymeric Materials)

For many ductile materials, once inhomogeneous deformation has started, further flow localization is accompanied by an increase in deformation, which in turn leads to final fracture.

On the other hand, the necking of many polymers initially develops in the specimen in a manner similar to that observed for ductile materials, and this subsequently propagates along the specimen under an essentially steady-state condition (Hutchinson and Neale 1983). In case of polymers, however, the "re-stiffening effect", which is observation in the high-strain region in polymer and is generally caused by the alignment of material chains, randomly oriented in the undeformed state, is seen to be responsible for neck development and propagation.

The mechanical aspects of instability propagation in polymeric material have recently received much attention:

- Hutchinson and Neale (1983) and Chater and Hutchinson (1984) investigated the neck propagation of tension blocks, bulge propagation in long cylindrical balloons and the buckle propagation of tubes under lateral pressure in terms of simple one-dimensional anlaysis or approximate steady-state analysis.

- Fig. 10.2 shows the uniaxial stress-strain relation, corresponding elongation curves and deformed shape of the specimens (Tomita and Hayashi, 1991& 1993). After the maximum load point, necking starts and it develops until the load attains the load minimum. Then it propagates with an almost constant load. The propagation may not appear when the strain at the re-stiffening point is smaller than strain at the maximum load point.

- Except under conditions of very slow deformation, the propagation behaviour of instability manifests different features associated with frictional heating of the polymer undergoing large deformation.

In a subsequent study, the effects of strain rate sensitivity (Tugcu and Neale, 1987&1988) and the temperature dependency (Tugcu and Neale, 1990, Tugcu et al, 1991, and Tomita and Hayashi, 1991&1993) on the neck propagation behaviour have been investigated with a constitutive equation similar to (10.13).

Figure 10.3 shows the thermo-elastoviscoplastic neck propagation behaviour. In this Figure, $\dot{U}$ is the normalized end displacement rate and AD stands for a locally adiabatic process, otherwise thermocoupled analysis is performed.

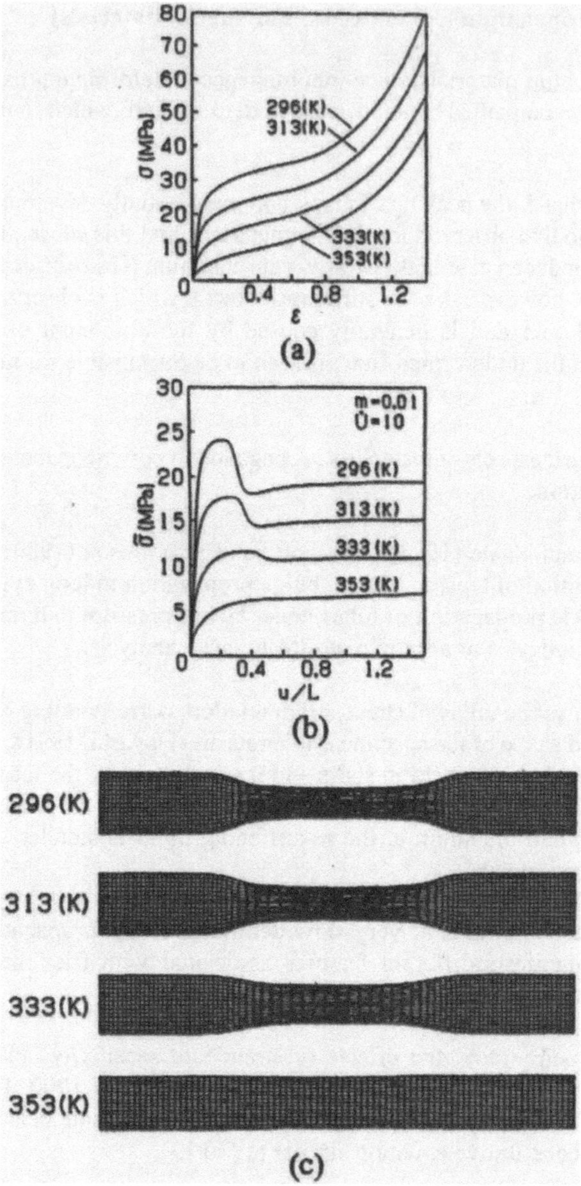

Figure 10.2. Deformation behaviour of polymeric material under tension. (a) Uniaxial true stress σ-natural strain ε relations for different temperatures, u: end displacement, $2L$: initial length of the specimen. (b) Load σ- elongation u/L curves under quasi-static deformation rate. (c) Deformed specimen profiles at u/L = 0.8 for quasi-static and isothermal deformation. "Reprinted from *Int. J. Solids Structures* 30(2), Tomita, Y., and Hayashi, K., Thermo-elasto-viscoplastic deformation of polymeric bars under tension, pp. 225-35, 1993, with permission from Elsevier Science". See, also Tomita and Hayashi (1991).

As shown in Figure 10.3, at a low rate of deformation, the stabilization effect by re-stiffening overcomes the destabilization effect due to thermal softening, and the neck propagates along the tensile direction with a heat source which can be seen in the temperature distribution along the tensile axis.

Thus, the deformation-induced heating and its conduction strongly affect the neck propagation behaviour for a relatively low rate of deformation.

As a result, predictions based on steady-state analysis with the adiabatic assumption will provide an improper estimation because the deformation-induced heating tends to cause nonsteady-state deformation, which increases as the material strain rate sensitivity increases (Tomita and Hayashi, 1991&1993).

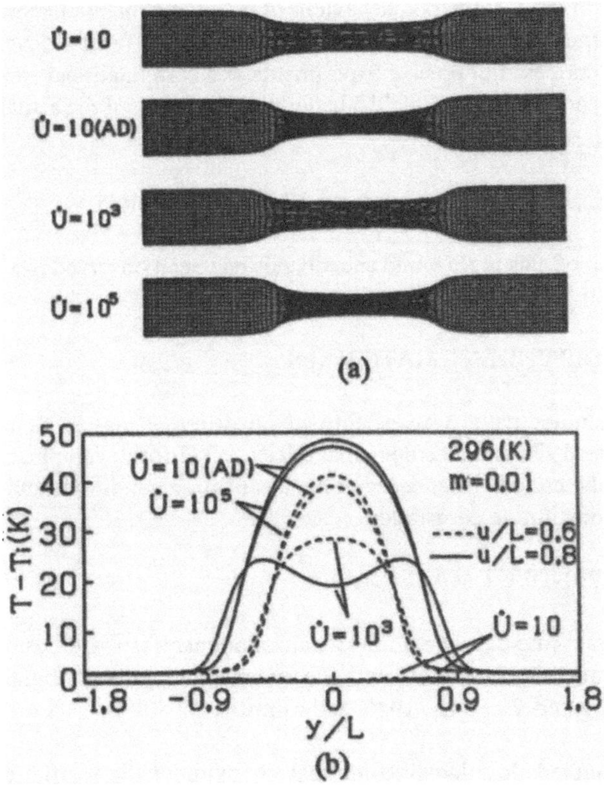

Figure 10.3. Thermo-elasto-viscoplastic neck propagation behaviour. (a) Deformed profiles for different strain rates of $2\text{x}10^{-5}$ U, AD: with assumption of locally adiabatic process. (b) Temperature distribution along tension axis *y*. *u*: end displacement, 2*L*: initial length, T_i : initial temperature, 296 °K, m: strain-rate sensitivity exponent. "Reprinted from *Int. J. Solids Structures* **30(2)**, Tomita, Y., and Hayashi, K., Thermo-elasto-viscoplastic deformation of polymeric bars under tension, pp. 225-35, 1993, with permission from Elsevier Science". See, also Tomita and Hayashi (1991).

Furthermore, anisotropy caused by microscopic mechanisms of the molecular chains and the distribution of their orientation due to excessive deformation is quite important:

Boyce et al (1988) developed a three-dimensional constitutive model describing the inelastic response, including the strain rate, temperature and strain softening/hardening of glassy polymers, based on the micromolecular structures of materials and corresponding micromechanisms of plastic response.

Boyce and Arruda (1990) verified that the constitutive equation can predict the major aspects of the material well.

Since the identification of the constitution response is quite difficult due to the complicated nature of the deformation behaviour of polymeric material, including the special attention to experimental methods (G'Sell and Jonas, 1979), further developments in the hybrid strategy in cooperating precise experiments and computational simulation (Tomita and Hayashi, 1991) are expected to yield a better understanding of the actual response of the polymeric material.

10.8. Flow Localization of Thermo-Elasto-Viscoplastic Solids

Localization of plastic flow into shear bands has been observed in various materials and is recognized to be a very important precursor to failure.

10.8.1. RATE-INDEPENDENT MATERIALS

Intensive studies have been performed on different classes of rate independent materials (e.g., Rice 1976 and Needleman and Rice, 1978 for development of localization and have clarified the critical dependence of the localization conditions and the localization processes on the constitutive description.

10.8.2 RATE-DEPENDENT MATERIALS

The real strain-rate-dependent flow localization manifests itself as different features depending on the rate of deformation, with the understanding that, thermocoupled analysis is inevitable (Chung and Wagoner, 1986, and Tomita et al, 1990)

Thermo-coupled flow localization analyses without the inertial effect have been carried out by Lemonds and Needleman (1986 a, b), Kim and Anand (1987), Nemat-Nasser (1988), Nemat-Nasser et al (1989), Tomita et al (1990), Tomita and Nakao (1991, 1992) and Zbib and Jubran (1992) for plane strain tension.

Kim and Anand (1987), Nemat-Nasser (1988), Nemat-Nasser et al (1989) and Zbib and Jubran (1992) assumed the adiabatic process which represents an upper bound on the temperature, whereas Lemonds and Needleman (1986a,b), Tomita et al. (1990) and Tomita and Nakao (1991,1992) accounted for the heat conduction. Plane strain quadrilateral elements with hourglass control (Nemat-Nasser et al 1989, Zbib and Jubran 1992) and crossed trangular elements (Lemonds and Needleman, 1986 a, Tomita et al, 1990, and Tomita and Nakao, 1991, 1992) are employed. An intensification of shear localization has been observed for large specimens, and the adiabatic assemption may provide suitable information on the specific order of the strain rate, which increases as the specimen size decreases (Tomita et al, 1990). Further, it has been clarified that the localization of the deformation is delayed by the strain-rate effect, strain-gradient dependent of the yield stress, as seen in equation (10.26) below

$$\sigma = \sigma(\varepsilon, \dot{\varepsilon}, T) - c\nabla^2 \varepsilon \qquad \textbf{(10.26)}$$

where σ is the local flow stress.

The results also illustrate that the interactions of material properties and thermal softening and the growth of voids are two competing and interacting softening mechanisms in porous materials.

Dynamic flow localization analyses have been carried out by Needleman (1989) and Batra and Liu (1989, 1990) for plane strain compression. Needleman (1989) employed the softening constitutive equation as a simple model for a thermally softening solid. Then the problem is treated from a purely mechanical point of view with initial homogeneity of the flow stress near the centre of the block.

Batra and Liu (1989, 1990) investigated a similar problem by introducing a temperature bump at the center of the block obeying thermally softening viscoplastic solids. Thermocoupled analyses have been performed. The results are in qualitative agreement with those of Needleman (1989). Except for a significant delay in shear band development due to the inertial effect, the main features of shear band development are the same as under the quasi-static loading condition.

Wright and Walter (1987) studied the problem of dynamic simple shear of a finite slab of incompressible material and showed that in the late localization stages, the conductivity and strain rate set the width of the shear band.

Batra and Zhang (1990) and Batra and Zhu (1991) investigated shear band development in a viscoplastic cylinder and bimetallic body containing two voids under dynamic loading.

Figure 10.3 shows the results of the plane strain tension blocks under the average deformation rate $\dot{u}/L = 2000/s$ and with both ends free and fixed under conditions without and with inertial force (Tomita and Higo 1993). A locally adiabatic condition is assumed.

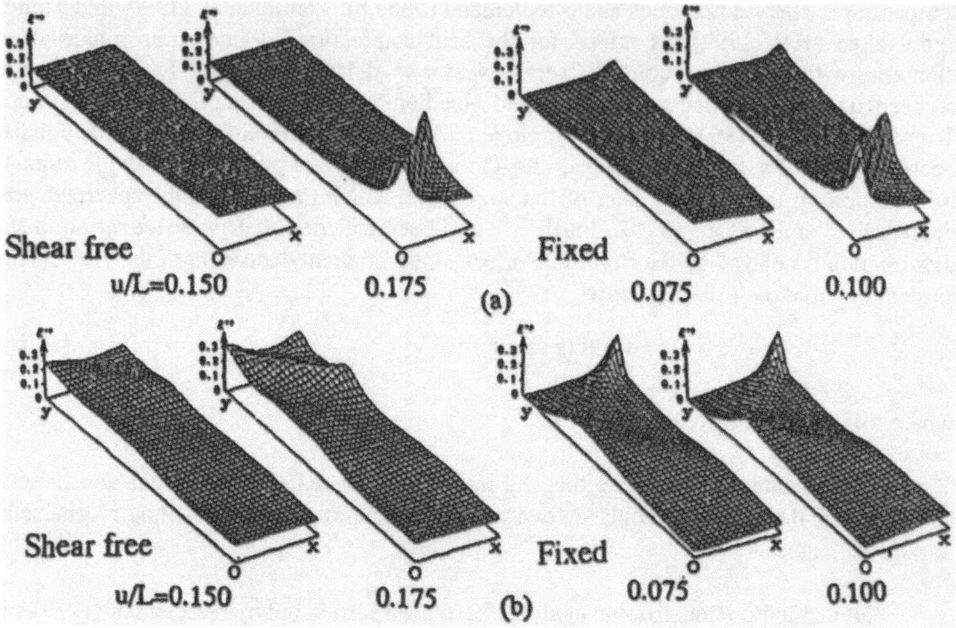

Figure 10.4. Representative strain distribution in plane strain blocks, L=24 mm, m=0.01. (a) Without inertial effect. (b) With inertial effect. "Reprinted from Int. *J. Mech. Sci.* **35(12)**, Tomita, Y. and Higo, T., Plane-strain flow localization in tension and compression of thermo-elasto-viscoplastic blocks under high rates of deformation, pp. 985-94, 1993, with permission from Elsevier Science".

With reference to Fig. 10.4, in the case without inertial force (a), regardless of the boundary condition, localization predominantly develops near the center of the specimen, whereas irregular flow localization is observed in the case with inertial force (b). The propagation of the dynamic force and boundary constraints play an important role in the onset and development of flow localization.

Needleman and Ortiz (1991) gives a complete mechanistic explanation concerning the interaction between shear bands, free surfaces and interfaces.

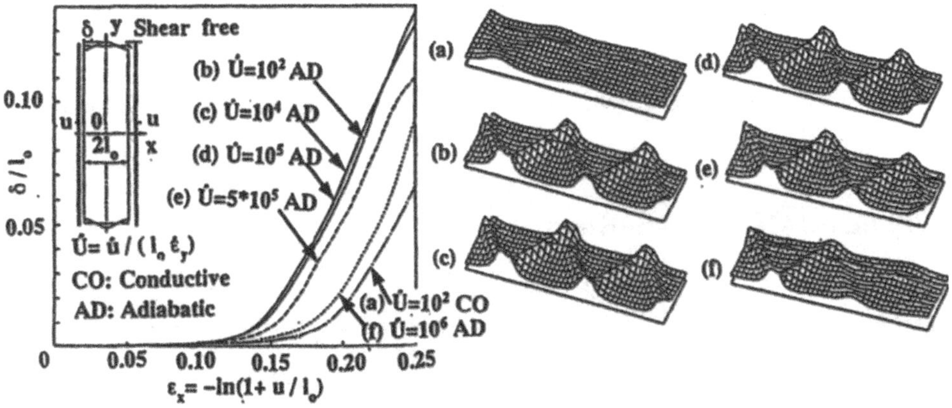

Figure 10.5. Growth of undulation for dynamic compression of blocks near the stress-free surface. "Reprinted from *Int. J. Mech. Sci.* 35(12), Tomita, Y. and Higo, T., Plane-strain flow localization in tension and compression of thermo-elasto-viscoplastic blocks under high rates of deformation, pp. 985-94, 1993, with permission from Elsevier Science".

Fig. 10.5 (Tomita and Higo, 1993) shows the compression of the strip under a wide range of deformation rates $\dot{U} = \dot{u}/(\ell_0\, \dot{\varepsilon}_y) = 10^2 - 10^6$:

- A deformation process with heat conduction (CO) and a locally adiabatic condition (AD) have been assumed for a low and high rates of deformation, respectively (see, also Tomita et al. 1993).

- Fig. 10.4 (left) shows the evolution of the undulation of the stress force surface.

- Fig. 10.4 (right) depicts the representative strain distribution at a specific stage of compression $\varepsilon_x = -0.25$.

- The general feature of the flow localization is essentially the same as that seen in the quasi-static case (Tvergaard, 1982, Kitagawa and Matsushita, 1987).

- The surface undulation abruptly starts to increase at a specific point (which can be approximately obtained by linear perturbation analysis), and leads to the development of a shear band connecting the highly strained regions beneath the highly strained region in a zigzag faction.

- At a low rate of deformation, thermal softening is substantially suppressed by the heat conduction and causes significant delay in the evolution of the undulation and strain distribution.

- Competing effects of thermal softening and inertia are observed in the evolution of undulation and representative strain at a relatively high rate of deformation

At very high rates of strain, over $10^3\ s^{-1}$, the inertial effect overcomes the thermal softening and causes significant delay in flow localization and greater thickness of the shear localization zone.

10.9. Effect of Material Rate History

Fig. 10.6 (Tomita and Higo 1993) shows the effect of material strain rate history dependence on flow localization behaviour. Five different types of computations with the end velocity shown in Fig. 10.6a have been performed:

The difference observed between cases I and II is attributed to the dynamic effect. As discussed above, the inertial force again stabilizes the deformation.

Comparison of cases III and IV clarifies the effect of the material strain-rate history dependence on the flow localization.

The comparison for cases II and III clarifies that the dynamic deformation subsequently applied to the quasi-static deformation stabilizes the deformation.

Since the stabilization effect of strain rate history dependence suppresses the development of flow localization, the ductility of the material is clearly increased by subsequent dynamic loading after pre-straining. The efficiency of increasing the ductility depends on the magnitude of pre-straining, as seen in case V.

10.10. Three-Dimensional Effects

The three dimensional aspects of localized deformation without inertial effect (Leroy and Ortiz 1990; Zbib and Jubran (1992) have been investigated:

- Zbib and Jubran (1992) assumed adiabatic deformation and clarified the smooth transition of plane stress to plane strain deformation by employing very thin to thick specimens.

- Fig. 10.7 (Zbib and Jubran, 1992) shows the deformed meshes with shear bands.
- A very strong three-dimensional geometric effect on the shear banding is observed. The orientation of the shear bands are 35.25° and 45°, respectively, and they are consistent with the theoretical predictions. Again, a softening mechanism and an initial imperfection are among the many cause of shear banding. The multiaxial effect stabilizes the deformation and yields a delay in localization (Zbib and Jubran 1992).

10.11. Problems

1. Explain briefly the following terms:

   - Associative material
   - Conservative loading
   - Bifurcation state
   - Critical bifurcation point *vs.* Critical bifurcation point
   - Upper and lower bounds of a bifurcation state.

2. Comment briefly on the aim of "*Hill's bifurcation theorem*".

3. What is configuration-dependent loading ?

4. What constitute uniqueness criteria in mechanics of solids ?

5. What are conservative and non-conservative problems ?

6. Discuss briefly the three-dimensional effect on shear banding in a metallic material.

10.12. References

Bardet, P. (1990) Finite element analysis of plane strain bifurcation within compressible solids, *Computers & Structures* **36**, 993-1007

Batra, R. C. and Liu De-Shin (1989) Adiabatic shear banding in plane strain problems, *J. appl.Mech.* **56**, 527-34.

Batra, R. C. and Liu De-Shin (1990) Adiabatic shear banding in dynamic plane strain compression of a viscoplastic material, *Int. J. Plasticity* **6**, 231-46.

Batra, R. C. and Zhang, X. T. (1990) Shear band development in dynamic loading of a viscoplastic cylinder containing two voids, *Acta Mech.* **85**, 221-34.

Batra, R. C. and Zhu, Z. G. (1991) Dynamic shear band development in a thermally softening bimetallic body containing two voids, *Acta Mech.* **86**, 31-52.

Borst, R. de (1987) Integration of plasticity equation for singular yield functions, *Computers & Structures* **26**, p. 823-29.

Borst, R. de (1989) Numerical methods for bifurcation analysis in geomechanics, *Ing. Arch.* **59**, 160-74.

Boyce, M. C., Parks, D. M. and Argon, A. S. (1988) Large inelastic deformation of glassy polymers, Part I: Rate dependent constitutive model, *Mech. Mat.* **7**, 15-33.

Boyce, M. C. and Arruda, M. (1990) An experimental and analytical investigation of the large strain compressive and tensile response of glassy polymers, *Pol. Engng. Sci.* **30**, 1288-98.

Budiansky, B. (1959) Assessment of deformation theory, *J. Appl. Mech.* **26**, 259-64.

Campbell, J.D., Eleiche, A.M. and Tsao, M.C.C. (1977) *Fundamental Aspect of Structural Alloy Design*, Plenum, New York, pp. 545.

Chater, E. and Hutchinson, J. W. (1984) On the propagation of bulges and buckles, *J. appl. Mech.* **51**, 269-77.

Christoffersen, J. and Hutchinson, J. W. (1979) A class of phenomenological corner theories of plasticity, *J. Mech.Phys. Solids* **27**, 465-87.

Chu, C. C. (1979) Bifurcation of elastic-plastic circular cylindrical shells under internal pressure, *J. Appl. Mech.* **46**, 889-94.

Chung, K. and Wagoner, R. (1986) Invariance of neck formation to material strength and strain rate for power-law -materials, *Metall Trans.* **A17**, 1632-3.

Dafalias, J.F. (1983) Co-rotational rates for kinematic hardening at large plastic of formations, *J. Appl. Mech.* **50**, 561-5.

Dafalias, J.F. and Aifantis, E.C. (1990) On the macroscopic origin of the plastic spin, *Acta Mech.* **82**, 31-48.

Dubey, R. N. (1970) Variational method for non conservation problems, *J. Appl. Mech.* **37**, 133-6.

Gotoh, M. (1985) A class of plastic constitutive equation with vertex effect - I, *Int J. Solids & Structures* **21**, 1101-16.

Gurson, A.L. (1977) Continuum theory of ductile rupture by void nucleation and growth, Part 1 - Yield criteria and flow rules for porous ductile media, *J. Eng. Mat. Tech.* **99**, p. 2-15.

G'Sell, G. and Jonas, C. (1979) Determination of the plastic behaviour of solid polymers at constant true strain rate, *J. Mater. Sci.* **14**, 583-91.

Hill, R. (1958) A general theory of uniqueness and stability in elastic-plastic solids, *J. Mech. Phys. Solids* **8**, 236-49.

Hill, R. (1962a) Acceleration waves in solids, *J. Mech. Phys. Solids* **10**, 1-16.

Hill, R. (1962b) Uniqueness criteria and extremum principles in self-adjoin problems of continuum-mechanics, *J. Mech. Phys. Solids* **10**, 185-94.

Hill, R. and Hutchinson, J.W.(1975) Bifurcation phenomena in the plane tension test, *J. Mech. Phys. Solids* **23**, 236-64.

Hughes, T. J. R. and Winget, J. (1980) Finite rotation effects in numerical integration of rate constitutive equations arising in large deformation analysis, *Int. J. Num. Meth. Engng.* **15**, 1862-7.

Hutchinson, J.W. (1973a) Post bifurcation behaviour in the plastic range, *J. Mech. Phys. Solids* **21**,163-90.
Hutchinson, J.W. (1973b) Finite strain analysis of elasto-plastic solids and structures, in *Numerical Solution of Nonlinear Structural Problems*, Hartung, R.F. (Ed.), ASME, pp. 17-29.
Hutchinson, J. W. and Neale, K.W. (1973) Sheet necking. II. Time independent behaviour, in *Mechanics and Sheet Metal Forming*, Koistinen, D.P. and Wang, N.M. (Eds), Plennum Press, pp. 127-53.
Hutchinson, J. W. and Neale, K. W. (1983) Neck propagation, *J. Mech. Phys. Solids* **31**, 405-26.
Kim, K. H. and Anand, L. (1987) A note on adiabatic flow localization in visco-plastic solids, in: *Computational Methods for Predicting Material Processing Defects*, M. Predeanu (ed.), Elsevier, London, pp.181-92.
Kitagawa, H. and Matsushita, H. (1987) Flow localization in elastic-plastic material developing from stress-free surface, *Int. J. Solids Structures* **23**, 351-68.
Kitagawa, H., Seguchi, Y. and Tomita, Y. (1972) An incremental theory of large strain and large displacement problems and its finite element application, *Ing. Arch.* **41**, 213-24.
Krieg, R. D. and Krieg, D. B. (1977) Accuracies of numerical solution method for the elastic-perfectly plastic model, *Trans ASME, J. Pressure Vessel Tech.* **99**, 510-15.
Larrson, M., Needleman, A., Tvergaard, V. and Storakers, B. (1982) Instability and failure of internally pressurized ductile metal cylinder, *J. Mech. Phys. Solids* **30**, 121-54.
Lemonds, J. and Needleman, A. (1986a) Finite element analyses of shear localization in rate and temperature dependent solids, *Mech. Materials* **5**, 339-61.
Lemonds, J. and Needleman, A. (1986b) An analysis of shear band development incorporating heat conduction, *Mech. Materials* **5**, 363-73.
Leroy, Y.M. and Ortiz, M. (1989) Finite element analyses of strain localization in frictional materials, Int. J. Num. Anal. Mech. Geomech. 13, 53-74
Leroy, Y.M. and Ortiz, M. (1990) Finite element analysis of transient strain localization phenomena in frictional solids, *Int. J. Num. Anal. Meth. Geomech.* **13**, 53-74.
Leroy, Y.M. and Chapuis, O. (1991) Localization in strain-rate-dependent solids, *Comp. Meth. Appl. Mech. Engng.* **90**, 969-86.
Loret, B. (1983) On the effects of plastic rotation in finite deformation of anisotropic elastplastic materials, *Mech. Materials* **2**, 287-304
Minura, K. and Tomita, Y. (1991) Constitutive relations of mold steel and alpha-titanium at high rates under multiaxial loading condition, *Journal de Physique* **IV**, *DYMAT* 91, Les editions de Physique, France, 813-20.
Nagtegaarl, J.C. and de Jong, J.E. (1981) Some computational aspects of elastic-plastic large strain analysis, *Int. J. Num Meth. Eng.* **17**, 15.
Neale, K.W. (1980) Phenomenotogical constitutive laws in finite plasticity, *SM arch* **6**, 79-128.
Needleman, A. (1989) Dynamic shear band development in plane strain, *J. appl. Mech.* **56**, 1-9.
Needleman, A. and Ortiz, M. (1991) Effects of boundaries and interfaces on shear band localization, *Int. J. Solid Structures* **28**, 859-77.
Needleman, A. and Rice, J. R. (1978) Limits to ductility set by plastic flow localization, in: *Mechanics of Sheet Metal Forming*, Koistinen, D. P. and Wang, N -M. (Eds.), Plenum Press, New York, pp. 237-67.
Needleman, A. and Tvergaard, V. (1977) Necking of biaxially stretched elastic plastic circular plates, *J. Mech. Phys. Solids* **25**, 159.
Nemat-Nasser, S. (1988) Micromechanics of failure at high strain rate: Theory, experiments, and computations, *Computer & Structures* **30**, 95-104.
Nemat-Nasser, S., Chung, D. T. and Taylor, L.M. (1989). Phenomonological modelling of rate-dependent plasticity for high strain rate problems, *Mech. Materials* **7**, 319-44.
Nemat-Nasser, S. and Li, Y.F. (1992) A new explicit algorithm for finite-deformation elastoplasticity and elastoviscoplasticity: Performance evaluation, *Computers and Structures* **44**, 937-63.

Ortiz, M. and Popov, E.P. (1985) Accuracy and stability of integration algorithms for elastoplastic constitutive relations, *Int. J. Num. Mech. Eng.* **21**, pp. 1561-76.

Ortiz, M. and Simo, J. C. (1986) An analysis of a new class of integration algorithms for elasto plastic constitutive relation, *Int. J. Numerical Meth Eng.* **23**, p.353-66.

Peirce, D., Shih, C.F. and Needleman, A (1984) A tangent modulus method for rate dependent solids, *Computers & Structures* **18**, 875.

Raniecki, B. (1979) Uniqueness criteria in solids with non-associated plastic flow laws at finite deformations, *Bull. Acad. Pol. Ser. Sci. Tech.* **27**, 7219.

Raniecki, B. and Brunhs, O. (1981) Bounds to bifurcation stresses in solids with non-associated plastic flow law at finite strain, *J. Mech. Phys. Solids* **29**, 153-72.

Rashid, M.M. and Nemat-Nasser, S. (1992). A constitutive algorithm for rate dependent plasticity, *Comp. Mech Appl. Mech. Eng.* **94**, 201-28.

Reddy, B. D. (1982) A deformation-theory analysis of the bifurcation of pressurized thick-walled cylinders, *Q. J. Mech. Appl. Math.* **35**, 183-96.

Rice, J.R. (1976) The localization of plastic deformation, *Proc. 14th IUTAM congress*, Koiter, W. (ed.), North Holland, Amsterdam, 207-20.

Rice, J.R. and Tracey, D.M. (1973) *Numerical and Computational Methods in Structural Mechanics*, Academic press, pp. 585.

Rudnicki, J.W. and Rice, J.R. (1975) Conditions for the localization of deformation in pressure, sensitive dilutant materials, *J. Mech. Phys. Solids* **23**, p. 371-94.

Runneson, K., Samuelsson, A. and Bernspang, L. (1986) Numerical Technique in plasticity including solution advancement control, *Int. J. Num. Math. Engng* **22**, 769-88.

Runneson, K., Sture, S. and William, K. (1988) Integration in computational plasticity, *Computers & Structures* **30**, 119-30.

Sewell, M.J. (1967) On configuration dependent loading, *Arch. Rate Mech. Anal.* **23**, 327-51.

Simo, J.C., Kennedy, J.G. and Govindjee, S. (1988) Nonsmooth multi-surface plasticity and viscoplasticity. Loading /unloading conditions and numerical algorithms, *Int. J. Num Mech. Eng.* **26**, 2161-85.

Simo, J.C. and Ortiz, M. (1985) A unified approach to finite deformation elastoplastic analysis based on the use of hyper elastic constitutive equations, *Com. Meth. Appl. Mech. Engng* **49**, pp. 221-45.

Simo, J.C. and Taylor, R.L. (1985) Consistant tangent operators for rate -independent elastoplasticity, *Comp. Meth. Appl. Mech. Eng.* **48**, 101-18.

Stören, S. and Rice, J.R. (1975) Localized necking in thin sheets, *J. Mech. Phys. Solids* **23**, 221-41.

Timoshenko, S. P. and Gere, J. M. (1961) *Theory of Elastic Stability*, 2nd Edition, McGraw-Hill, New York.

Taylor, G.I. and Quinney, H. (1934) The latent energy remaining in a metal after cold working, *Proc. Roy Soc. London* **A 143**, p. 307.

Tomita, Y. (1994) Simulations of plastic instabilities in solid mechanics, *Appl. Mech. Rev.* **47(6)**, Part 1, 171-205.

Tomita, Y. and Hayashi, K. (1991) Deformation behaviour in elasto-viscoplastic polymeric bars under tension, *Proc. Int. Symp. on Plasticity and its Current Applications, Grenoble, France*, Boehler, J. P. and Khan, A.S. (Eds), Elsevier, pp. 524-27.

Tomita, Y. and Hayashi, K. (1993) Thermo-elasto-viscoplastic deformation of polymeric bars under tension, *Int. J. Solids Structures* **30(2)**, 225-35.

Tomita, Y. and Higo, T. (1993) Plane-strain flow localization in tension and compression of thermo-elasto-viscoplatic blocks under high rates of deformation, *Int. J. Mech. Sci.* **35**, 985-94.

Tomita, Y. and Nakao, T. (1991) Flow localization of elasto-viscoplastic tension blocks, Proc. ICM6, Jono, M. and Inoue, T. (Eds.), Pergamon Press, New York, pp. 1997-2002.

Tomita, Y., Mimura, K. and Sasayama, T. (1993) Effects of intermediate process annealing on surface roughening and forming limit in polycrystalline thin sheet metal subjected to tension, MECA-MAT'91, *Large plastic deformations*, Teodosiu, C., Raphanel, J. L. and Sidoroff, F. (Eds.),

Balkema, Rotterdam, pp. 177-83.

Tomita, Y. and Nakao, T. (1992) Shear localization in thermo-elasto-viscoplastic plane strain block, Proc. IUTAM Symposium, Finite Inelastic Deformations, Theory and Application, Besdo, D. and Stein, E. (Eds), Springer-Verlag, 179-88.

Tomita, Y. and Shindo, A. (1982) On the bifurcation and post-bifurcation behaviour of thick circular tubes under lateral pressure, *Comp. Meth. Appl. Mech. Engng.* **35**, 207-19.

Tomita, Y. and Shindo, A. (1985) Bifurcation behaviour of thin square elastic-plastic orthotropic plates subjected to diagonal tension, Proc. Considere Memorial Symposium, Salencon, J. (ed.), Ecole Nationale des Ponts et Chaussés, pp. 203-13.

Tomita, Y. and Shindo, A (1990) Wrinkling behaviour in thin-walled bodies during plastic forming, computational plasticity, Inoue, T., Kitagawa, H. and Shima, S. (Eds), CJMR7, Elsevier, pp. 165-78.

Tomita, Y., Shindo, A., Kim, Y. S. and Michiura, K. (1986). Deformation behaviour of elastic plastic tubes under external pressure and axial load, *Int. J. Mech. Sci.* **28**, 263-74.

Tomita, Y., Shindo, A., and Kitagawa, H. (1981) Bifurcation and post bifurcation behaviour of internally pressurized elastic-plastic circular tubes under plane strain conditions, *Int. J. Mech. Sci.* **23**, 723-32.

Tomita, Y., Shindo, A. and Nagai, M. (1984). Axisymmetric deformation of circular elastic-plastic tubes under axial tension and internal pressure, *Int. J. Mech. Sci.* **26**, 437-44.

Tomita, Y., Shindo, A. and Sasayama, T. (1990) Plane strain tension of thermoelastic viscoplastic blocks, *Int. J. Mech. Sci* **32**, 613-22.

Triantafyllidis, N. (1985) Puckering instability phenomena in the hemispherical cup test, *J. Mech. Phys. Solids* **33**, 117-39.

Tugcu, P. and Neale, K. W. (1987) Analysis of plane-strain neck propagation in viscoelastic polymeric films, *Int. J. Mech. Sci.* **29**, 793-805.

Tugcu, P. and Neale, K. W. (1988) Analysis of neck propagation in polymeric fibers including the effects of viscoplasticity, *J. Engng. Mat. Tech.* **110**, 395-400.

Tugcu, P. and Neale, K. W. (1990) Cold drawing of polymers with rate and temperature dependent properties, *Int. J. Mech. Sci.* **32**, 405-16.

Tugcu, P., Neale, K. W., Marques-Lucerno, A. (1991) Effect of deformation induced heating on the cold drawing of polymeric films, *J. Engng. Mat. Tech.* **113**, 104-11.

Tvergaard, V. (1978) Effects of kinematic hardening on localization necking in biaxially stretched sheets, *Int. J. Mech. Sci.* **20**, 651-58.

Tvergaard, V. (1981) Influence of voids on shear band instabilities under plane strain localization, *Int. J. Fracture* **171**, p. 389-437.

Tvergaard, V. (1982) Influence of void nucleation on ductile shear fracture at a free surface, *J. Mech. Phys. Solids* **30**, 399-425.

Tvergaard, V. (1987) Effect of yield surface curvature and void nucleation on plan flow localization, *J. Mech. Phys. Solids* **35**, p. 43-60.

Wright, T. W. and Walter, J. W. (1987) On stress collapse in adiabatic shear bands, *J. Mech. Phys. Solids* **35**, 701-20.

Yamada, Y., Yoshimura, N. and Sakurai, T. (1968) Plastic stress stain matrix and its application for the solution of elastic-plastic problems by finite element method, *Int. J. Mech. Sci.* **10**, 343-54.

Zbib, H. M. and Jubran, J. S. (1992) Dynamic Shearbanding: A three-element analysis, *Int. J. Platicity* 8, 619-41.

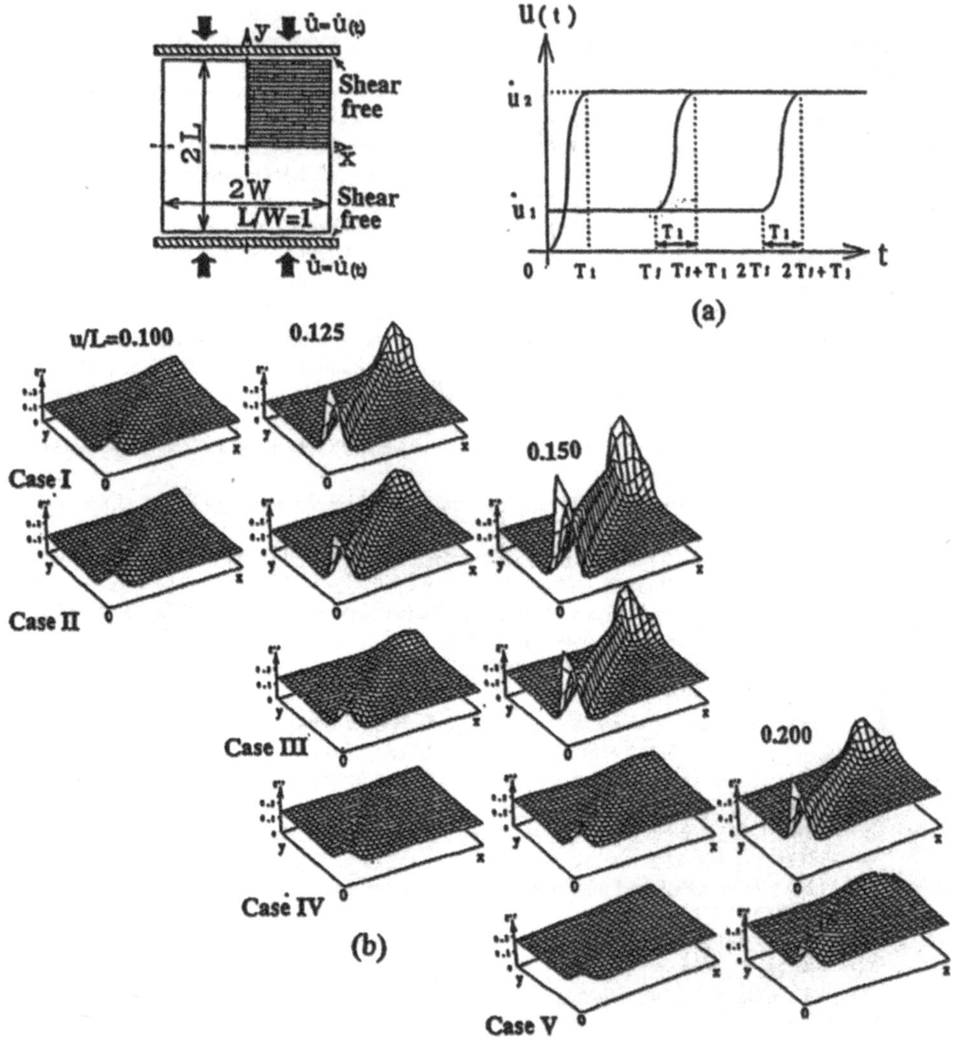

Figure 10.6. (a) Computational model for compression of blocks. Cases I, II: Constant velocity and without and with inertial effect, respectively. $u_1 = 0$, $T_j = 0$, $u_2/L=10^4\ s^{-1}$. Cases III, IV and V: with velocity jump at $t = T_j$ and flow stresses exhibiting negative dependence, no dependence and positive dependence on strain rate history, respectively. $u_1/L = 0.002\ s^{-1}$, $u_2/L = 10^4\ s^{-1}$. Case VI: Case V with velocity jump at $t = 2T_j$, $L = 1.0$ mm. (b) Representative strain distribution at different stages of deformation during the dynamic compression of plane-strain blocks. Case I-VI correspond to those in (a). "Reprinted from Int. *J. Mech. Sci.* 35(12), Tomita, Y. and Higo, T., Plane-strain flow localization in tension and compression of thermo-elasto-viscoplastic blocks under high rates of deformation, pp. 985-94, 1993, with permission from Elsevier Science".

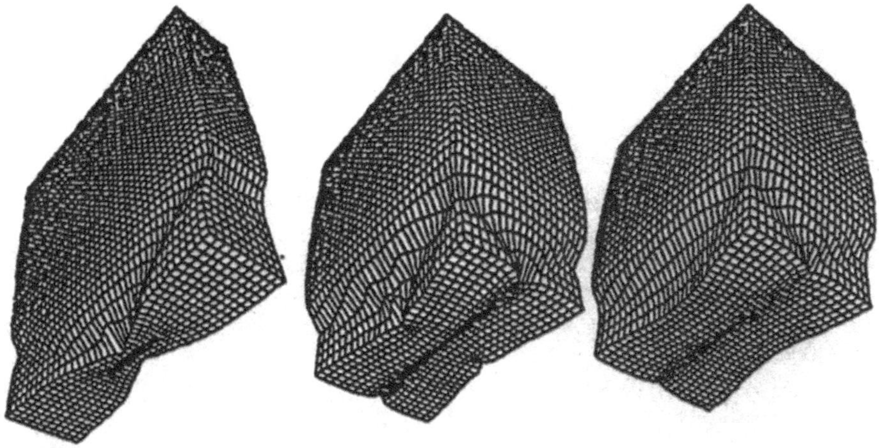

Figure 10.7. Dynamic tension of blocks with different thicknesses. "Reprinted from *Int. J. of Plasticity* **8**, Zbib, H. M. and Jubran, J. S., Dynamic shear banding: A three-dimensional analysis, pp. 619-41, 1992, with permission from Elsevier Science".

CHAPTER 11

ELASTIC WAVE PROPAGATION

11.1. Introduction

When a localized disturbance is applied suddenly into a medium, it will propagate to other parts of this medium. The local excitation is not detected at other positions of the medium instantaneously, as some time would be necessary for the disturbance to propagate from its source to other parts of the medium. This simple fact constitutes a general basis for the interesting subject of "***wave propagation***". Well-cited examples of wave propagation in different media include, for instance, the transmission of sound in air, the propagation of a seismic disturbance in the earth, the transmission of radio waves, among others. In the particular case, when the suddenly applied disturbance is mechanical, e.g., an impact force, the resulting waves in the medium are due to mechanical stress effects and, thus, these waves are referred to as "***mechanical stress waves***", or simply "***stress waves***". Our attention in this text is restricted to the study of the propagation of stress waves in engineering materials.

In rigid body dynamics it is assumed that, when an external force is applied to any one point of the body, the resulting effect sets every other point of the body instantaneously in motion, and the applied force can be considered as producing a linear acceleration of the whole body, together with an angular acceleration about its center of gravity. In the theory of deformable media, on the other hand, the body is considered to be in equilibrium under the action of the external applied forces, and the occurring deformations are assumed to have reached their equilibrium static values. This assumption could be sufficiently accurate for problems in which the time between the application of the force and the setting up of effective equilibrium is short compared with the time in which the observation is made. Meanwhile, If the external force is applied for only a short period of time, or it is changing rapidly, the resulting effect must be considered from the point of view stress wave motion.

Mechanical stress waves originate due to a forced motion of a portion of a deformable medium. As the other parts of the medium are deformed, as a result of such motion, the disturbance is transmitted from one point, of the medium, to the next and the disturbance, or wave, progresses through the medium. In this process, the resistance offered to deformation by the consistency of the medium, as well as to the resistance to motion due to the inertia, must be overcome. As the disturbance propagates through the medium it carries along various amounts of of kinetic and potential energies. Energy can be transmitted over considerable distances by wave motion. The transmission of energy is effected because motion is passed on from one particle to the next and not by any sustained bulk motion of the entire medium. Mechanical waves are characterized by the transport of energy through motions of particles about an equilibrium position. Thus, bulk types of motion of a medium such as those occur, for instance, in the turbulence of a fluid are not classified as wave motion.

As mentioned above, ***deformability*** and ***inertia*** are essential properties of a medium for the transmission of mechanical waves. If the medium were not deformable, any part of the medium would immediately experience a disturbance in the form of a rigid body acceleration upon the application of the localized excitation. Similarly, if a hypothetical medium were without inertia there would be no delay in the displacement of particles and the transmission of the disturbance from particle to particle would be affected instantaneously to the most distant particle.

In our presentation of the subject of wave propagation, we consider the solid medium to be a continuum. Hence, the mechanics of wave motion in the medium is dealt with from a continuum mechanics point of view. The basic concepts of continuum mechanics are briefly introduced in Chapter 2. In a continuum, the disturbance is generally considered to spread outward, from the source (the original disturbance), in a three-dimensional fashion. During their motion, waves propagating in a solid may encounter or interact with boundaries of the medium. On striking a boundary, a part or whole of an incident wave may be reflected and the mode of propagation of the wave may change.

In recent years, there has been considerable interest in the subject of wave propagation from both theoretical and experimental points of view. Such interest was motivated primarily by the advancements in the area of testing and measurement techniques. With the recent progress in fields such as electronics and laser optics, stress waves of high frequency can be now produced and detected easily. This has been particularly pronounced in the important domains of ultrasonics and acoustic emission. Another equally important reason for the ensuing interest in the subject of wave propagation is the continuous emerging of newly developed industrial materials. In this, the study of the phenomenon of wave motion has been able to identify microstructural problems and assist in the development of homogeneous and inhomogeneous material systems.

For a historical background of the subject of wave propagation, the reader is referred to Kolsky (1963), Tolosty (1973), Graff (1975) and Davis (1988), among others. For a review of the experimental methods that are commonly employed in producing and detecting stress waves in solids, reference is made, for instance, to the books by Hetenyi (1950), Dove and Adams (1964), Dally and Riley (1965), Keast (1967), and Magrab and Blomquist (1971). Comprehensive review articles in this area are due to Hillier (1960), Worely (1962), among others.

11.2. Elastic *vs.* Inelastic Waves

The propagation of stress waves in solids can be divided into two categories, ***"elastic"*** and ***"inelastic"*** waves. When loading conditions result in stresses below the yield point, solids behave elastically and obey Hook's Laws, and consequently stress waves are "***elastic***". As the intensity of applied loading is increased, the response of the material is driven out of the elastic range to a possible inelastic behavior. The behavior here may involve large deformation, internal heat generation, and often failure of the solid through a variety of mechanisms. In this context, ***"plastic"*** waves, for instance, can be propagated in a material, such as a metal, which exhibits the phenomenon of yielding, when stressed beyond its proportional limit. The theory of the propagation of such waves was first considered by Donnell (1930). The theory, as originally conceived, was based on a non-linear stress-strain relationship which was independent of the rate of loading. Subsequent experimental studies have shown that the time-rate dependence of the stress-strain relation has a considerable influence on the nature of wave propagation. Although Malvern (1969) has made an important first step in this direction, a theoretical approach which takes such time dependence into account leads to rather involved mathematical analysis. The subject of plastic wave propagation is dealt with in Chapter 12.

The mechanical properties of viscoelastic solids, such as plastics and rubber, have been studied extensively during recent years, and the subject of rheology is, to a large extent, devoted to the description of such viscoelastic behavior. An important development of these studies has been a consideration of the propagation of stress waves through such materials. The problems involved here are of particular interest, in that one is here dealing with media which are ***"dispersive"*** with respect to both velocity and attenuation. The study of the propagation, reflection and refraction of stress waves under these conditions leads to a number of problems which are not only of mathematical and physical interest, but also of practical importance in their bearing on the use of high polymers as vibration and shock absorbers, and the response of complete viscoelastic structures to rapid mechanical loading.

The third type of inelastic waves which have been studied are termed ***"shock waves"***. This class of waves arises, when an instantaneous, very large load is applied to the solid medium and lateral movement is restrained. Such conditions are normally encountered in "explosive" loading, or during the impact of high speed projectiles. Such shock waves may arise due to the fact that the effective bulk modulus of the material increases with increasing pressure. The importance of these shock waves lies, on the theoretical side, in obtaining the equation of state of solids at pressures which may not otherwise achieved , and, on the practical side, in military and mining applications.

In the case of metals, for instance, as the intensity of the applied load increases, the material is driven beyond its elastic limit and becomes plastic. In this state, Two waves propagate in the solid: an elastic wave (or *precursor*) followed by a much slower but more intense plastic wave. If the characteristics of the medium are such that the velocity of propagation of large disturbances is greater than the propagation velocity of smaller ones, the stress pulse develops a steeper and steeper front on passing through the medium, and the thickness of this front is ultimately determined by the constitution of the medium. The shock wave (or steep pressure pulse) thus formed differs from the high pressures generated by conventional methods in that it relies on the inertial response of material to the developed wave accelerations rather than on the structural constraints.

There are variety of applications of wave phenomena is engineering. First is the area of structures where the response of the structures to impact or blast loads are of significant importance. Although under transient loads of moderate strength completely elastic conditions may prevail throughout the structure and elastic wave theory may very well predict the response, the behavior of structures under high intensity loads, severe enough to cause permanent damage, would require the application of inelastic wave theories.

Another domain in the study of materials and structures involving wave phenomena is that of crack propagation or the interaction of dynamic stress fields with existing cracks, voids or inclusions in a material. Problems in this area are analogous to those pertaining to scattering and diffraction problems arising in acoustic and electromagnetic fields.

The field of ultrasonics represents another major area of application of wave phenomena. The general aspects of this area involve introducing a very low energy-level, high-frequency stress pulse of '***wave packet***' into a material and observing the subsequent propagation and reflection of this energy. In the majority of applications in this field, the means for introducing and detecting the stress waves are based on the piezoelectric effect in certain crystals and ceramics, whereby an electrical field applied to the material causes a mechanical strain or the inverse effect where a strain produces an electric field. Thus an electrical pulse is capable of launching a mechanical pulse. Detection is accomplished when a mechanical pulse strikes a piezoelectric crystal and generates an electrical signal. Many applications in ultrasonics are based on this reciprocal effect. For example, by studying propagation, reflection, and attenuation of ultrasonic pulses, it is possible to determine many fundamental properties of materials such as elastic constants and damping characteristics. The field of non-destructive testing makes wide use of ultrasonics to detect defects in materials. Meanwhile, the phenomenon of acoustic emission is a producer of stress waves and therefore of potential application.

11.3. Elastic Wave Propagation

In considering wave propagation in three dimensions we can, at a certain instant of time, draw a surface through all points undergoing an identical disturbance. As time goes on, such a surface, which is called a "***wavefront***", moves along showing how the disturbance propagates. The *wavefront* is a moving surface which separates the disturbed from the undisturbed part of the body. Consequently, particles of the medium that are located ahead of the *wavefront* are assumed to have experienced no motion, meantime, particles that are located behind the front are visualized to have experienced motion and may continue to vibrate for some time. In this context, a *wavefront* is considered to be associated with the outward propagating disturbance. The direction of propagation is always at tight angles to the *wavefront*. ***The field quantities and/or their derivatives are discontinuous at the wavefront.***

The normals defining the direction of wave propagation are called ***"rays"***. For an h isotropic medium, the rays are straight lines. If the wave propagation is limited to a single direction, the disturbance at a given instant will be the same at all points in a plane perpendicular to the direction of wave propagation. This situation is referred to as ***"plane wave"***. Other cases are ***"spherical waves"*** and ***"cylindrical waves"***, whereby the wave fronts are spherical and cylindrical surfaces, respectively.

Among the most important aspects of wave motion are the reflection and transmission of waves. When a wave encounters a boundary separating two media with different properties, part of the disturbance is reflected and part is transmitted into the second medium. If a body has finite cross-sectional dimensions, waves may bounce back and forth between the bounding surfaces. Although it is difficult to trace the actual occurring reflections, it can be noted that the general direction of energy transmission is in a direction parallel to the bounding surfaces. In such case, it is conventionally said that the waves are propagating in a ***"waveguide"***. The analysis of harmonic waves in waveguides leads to the notions of ***"nodes"*** of wave propagation, ***"frequency spectrum"***, ***"dispersion"***, and ***"group velocity"***.

When a pulse propagating through an elastic medium encounters an irregularity such as a void or an inclusion, the pulse is diffracted. As the wave strikes a crack, for instance, a stress singularity is generated at the crack tip which may give rise to the propagation of the crack and, thus, to the fracture of the body. The reader is referred, in this context, to Achenbach (1973), Gaff (1975), and Miklowitz (1978), among others.

The challenge in most of these problems stems from the complicated wave reflection, refraction and diffraction processes that occur at a boundary or interface in the continuum. This complexity evidences itself in the partial mode conversion of an elastic wave upon reflection from a traction-free or rigid boundary which converts, for example, compression into compression and shear. When there is a neighboring parallel boundary (forming then a waveguide), the so-created waves undergo multiple reflections between the two boundaries. This leads to dispersion, a further complicating geometric effect, which is characterized by the presence of a characteristic length (like the thickness of a plate). ***In the case of time-harmonic waves, dispersion leads to a frequency or phase velocity dependence on wavelength, and is responsible for the change in shape of a pulse as it travels along a waveguide.***

If we begin the analysis of wave propagation by considering the real case of finite or bounded solids, more likely we overlook the main concepts of the wave propagation. Therefore, we begin with an idealized, simplified case of having disturbance in an infinite elastic solid. Such a disturbance is necessarily simpler, because it is free of boundary effects such as reflection, refraction, diffraction and dispersion. Hence, the waves comprising this disturbance are referred to as ***"body waves"*** to distinguish them from ***"surface" or "interface waves"*** generated at, and propagating along a boundary. It is clear, however, since the displacement equations of motion underlie all elastodynamic problems, body waves play a role in all solutions. In an unbounded or infinite solid, which is idealized as an isotropic, elastic continuum, only <u>two</u> types of body waves can propagate. This is dealt with in the following subsection.

11.3.1. WAVE PROPAGATION IN UNBOUNDED ELASTIC SOLIDS

An unbounded solid is considered to extend indefinitely in the three dimensions of space so that the complications which might arise from reflections of waves at the boundaries of the medium might be disregarded.

The equations of motion of a continuum have been derived in Chapter 2, Section 2.4.3. These equations, (2.22), were presented in terms of the stress components acting on a small parallelepiped of the continuum without the inclusion of the response behaviour of the medium. However, in order to employ these equations in the study of wave propagation, one may substitute the stress components by the corresponding components of strain through the use of the constitutive relationships of the particular medium under consideration.

Following our presentation in Chapter 6, the stress-strain relations, for an isotropic elastic solid, can be expressed in component form as

$$
\begin{aligned}
{}_1 &= \lambda\Delta + 2\mu\varepsilon_{11},\ \sigma_{22} = \lambda\Delta + 2\mu\varepsilon_{22},\ \varepsilon_{33} = \lambda\Delta + 2\mu\varepsilon \\
{}_3 &= \mu\varepsilon_{23},\ \sigma_{31} = \mu\varepsilon_{31},\ \sigma_{12} = \mu\varepsilon_{12}
\end{aligned}
\tag{11.1}
$$

In the above relations, $\Delta = \epsilon_{kk} = \epsilon_{11} + \epsilon_{22} + \epsilon_{33}$, is the ***"dilatation"*** which represents the change in volume of unit cube of the solid and λ, μ are the ***Lamé's elastic constants***. In the theory of elasticity, four elastic (material) constants, not independent, are usually used. These are ***Young's modulus*** E, ***Poisson's ratio*** υ, ***Bulk modulus*** K and the ***rigidity (shear) modulus*** which is the ***Lame's constant*** μ. From the definitions of these constants and using equations (11.1) the following relations between the constants, in the case of an isotropic elastic solid, can be determined as

$$
E = \frac{\mu(3\lambda + 2\mu)}{\lambda + \mu},\quad \nu = \frac{\lambda}{2(\lambda + \mu)},\quad K = \lambda + 2\mu/3 \tag{11.2}
$$

Substituting from the constitutive relations (11.1) for the stress components in the equations of motion (2.22), the equation of motion for an isotropic elastic solid, in the absence of body forces, can be written in the x_1-direction in terms of the strain as

$$\rho \frac{\partial^2 u_1}{\partial t^2} = \frac{\partial}{\partial x_1}\left(\lambda\Delta + 2\mu\varepsilon_{11}\right) + \frac{\partial}{\partial x_2}\left(\mu\varepsilon_{12}\right) + \frac{\partial}{\partial x_3}\left(\mu\varepsilon_{13}\right) \tag{11.3}$$

where u_1 is the displacement component in the x_1-direction.

Replacing the strain components in (11.3) by the corresponding displacement components

$$\rho \frac{\partial^2 u_1}{\partial t^2} = (\lambda + \mu)\frac{\partial \Delta}{\partial x_1} + \mu\nabla^2 u_1 \tag{11.4a}$$

from equation (3.21), *Chapter 3*, it follows that

$$\nabla^2 = \left(\frac{\partial^2}{\partial x_1^{\,2}} + \frac{\partial^2}{\partial x_2^{\,2}} + \frac{\partial^2}{\partial x_3^{\,2}}\right)$$

where ∇^2 is the *Laplace operator* defined by

Similar relations to (11.4a) can be established for the other two components of the displacement vector, namely,

$$\rho \frac{\partial^2 u_2}{\partial t^2} = (\lambda + \mu)\frac{\partial \Delta}{\partial x_2} + \mu\nabla^2 u_2 \tag{11.4b}$$

and

$$\rho \frac{\partial^2 u_3}{\partial t^2} = (\lambda + \mu)\frac{\partial \Delta}{\partial x_3} + \mu\nabla^2 u_3 \tag{11.4c}$$

Equations (11.4) above are the equations of motion, in term of the displacement, for an isotropic elastic solid in the absence of body forces. These equations may be expressed conveniently in a vector form as

$$\rho \frac{\partial^2 \mathbf{u}}{\partial t^2} = (\lambda + \mu)\,\nabla\nabla\cdot\mathbf{u} + \mu\,\nabla^2 u \tag{11.5}$$

which is the form of the well-known *"**Navier's equation of motion**". The latter is conventionally adopted as the governing equation for the motion of an isotropic, elastic solid.* Equation (11.5) corresponds to the propagation of <u>two</u> types of waves through an unbounded isotropic, elastic solid; namely, "***dilatational***" and "***rotational***" waves.

Differentiating (11.4a) with respect to x_1, (11.4b) with respect to x_2 and (11.4c) with respect to x_3 and adding the resulting expressions, one obtains the following "***wave equation***" for an unbounded isotropic, elastic medium.

$$\rho \frac{\partial^2 \Delta}{\partial t^2} = (\lambda + 2\mu)\nabla^2 \Delta \qquad \textbf{(11.6)}$$

The above wave equation indicates that the dilatation Δ propagates through the medium with a velocity of magnitude $[(\lambda+2\mu)/\rho]^{1/2}$. Denoting the latter by c_1, then, $c_1 = [\lambda+2\mu)/\rho]^{1/2}$

In view of equations (11.2), the magnitude of the dilatational wave velocity c_1 may be expressed further by

$${}_1 = [(\lambda + 2\mu)/\rho]^{1/2} = \left[\frac{E(1-\nu)}{\rho(1+\nu)(1-2\nu)}\right]^{1/2} = \left[\frac{K + 4\mu/3}{\rho}\right]^{1/} \qquad \textbf{(11.7)}$$

It is noticed from (11.7) that the velocity c_1, of a dilatational wave, is dependent only on the elastic constants of the isotropic elastic material as well as its density. In an operational form, the wave equation (11.6) can be written as

$$\Gamma_1^2 \Delta = 0 \qquad \textbf{(11.8)}$$

where Γ_1^2 is a "***dilatational wave operator***" expressed (see, e.g., Chou, 1968) by

$$\Gamma_1^2 = \left[\nabla^2 - \frac{1}{c_1^2}\frac{\partial^2}{\partial t^2}\right] \qquad \textbf{(11.9)}$$

and $\Delta = \nabla \cdot u =$ dilatation

A "***dilatational***" wave, corresponding to the wave equation (11.8), is also referred to as "***irrotational***", since the propagation of such a wave involves no rotation of an elemental volume of the solid. A "***dilatational***" wave is also known as "***bulk wave***" or "***primary(P) wave***".

On the other hand, if we eliminate the dilatation Δ between (11.4b) and (11.4c), that is by differentiating (11.4b) with respect to x_3 and (11.4c) with respect to x_2 and subtract, it follows that

$$\rho \frac{\partial^2}{\partial t^2}\left(\frac{\partial u_3}{\partial x_2} - \frac{\partial u_2}{\partial x_3}\right) = \mu\nabla^2\left(\frac{\partial u_3}{\partial x_2} - \frac{\partial u_2}{\partial x_3}\right)$$

This equation can be written as

$$\rho \frac{\partial^2 \omega_1}{\partial t^2} = \mu\nabla^2 \omega_1$$

where ω_1 is the rotation about the x_1-axis (see Chapter 3). Similar relations can be obtained for ω_2 and ω_3 (the rotations about the x_2- and x_3-axis, respectively).
Thus, in generalized notation, one can write

$$\rho \frac{\partial^2}{\partial t^2} = \mu \nabla^2\omega \tag{11.10}$$

where $\omega = \nabla \times \omega/2$ is the rotation vector.

It follows from (11.10) that the rotational wave propagates in an isotropic, elastic solid with a velocity magnitude $(\mu/\rho)^{1/2}$. We denote the magnitude of the rotational wave velocity by c_2, then,

$$c_2 = (\mu/\rho)^{1/2} \tag{11.11}$$

It is noticed, from the above expression, that the rotational wave velocity c_2 is, similar to the dilatational velocity c_1,, dependent only on the elastic constants as well as the density of the material.

With reference to expressions (11.7) and (11.11), it is evident, in the case of an isotropic elastic solid, the two velocities of body waves are independent of the frequency. In other words, there is no dispersion (change of form) of these waves, i.e., body waves travel, in an isotropic elastic solid without change in form.

Applying the vector curl operator to (11.5), it can be shown that the vector form of the wave equation (11.10) can be expressed as

$$\Gamma_2^2 \omega = 0 \tag{11.12}$$

where Γ_2^2 is a rotational wave operator of the form (e. g., Chou,1968)

$$\Gamma_2^2 = \left[\nabla^2 - \frac{1}{c_2^2}\frac{\partial^2}{\partial t^2}\right] \tag{11.13}$$

and $\omega = \nabla \times u$ = rotation.

A "***rotational***" wave is also called "***equivoluminal***" wave, since there is no volume change would occur during the wave motion. A rotational wave is also known as "***distortional***" wave or "***secondary (S) wave***".

Equation (11.8), or (11.12), is *a necessary, but not a sufficient, condition* for the satisfaction of the *Navier's governing equation of motion* (11.5). Thus, for every displacement field that satisfies (5.5), the corresponding Δ and ω will satisfy (11.8) and (11.12), respectively. On the other hand, a displacement field with a dilatation satisfying (11.8), or a rotation satisfying (11.12) would not necessarily be a solution of the *Navier's governing equation* (11.5).

The particle motion in a dilatational wave is longitudinal, i.e., along the direction of wave propagation. In case of a rotational wave, the particle motion is transverse, that is perpendicular to the direction of propagation of the wave. Experimentally, one would generally attempt to generate one type of wave with the exclusion of the other. However, it should be emphasized that in the propagation of dilatational waves in an unbounded solid, the medium would not be simply subjected to pure compression, but to a combination of compression and shear. This is supported by the physical situation and mathematically by the appearance of both the bulk modulus and the shear modulus in the expression (11.7) of the dilatational velocity; see, for instance, Kolsky (1963).

11.3.2. IRROTATIONAL AND ROTATIONAL DISPLACEMENT FIELDS

Consider the displacement vector field u. In dynamic elasticity, u may be decomposed into an ***"irrotational field"***, say u_{IR}, associated with a *scalar potential* ϕ and a ***"rotational field"***, u_R, associated with a *vector potential* ψ. Thus, according to *Helmholtz theorem* (see Morse and Feshbach, 1953), for any displacement field, subject to mild continuity and boundary conditions, one may find at least one set of functions ϕ and ψ such that

$$\mathbf{u} = \nabla\phi + \nabla \times \omega, \qquad \nabla \cdot \psi = 0 \tag{11.14}$$

The condition $\nabla.\psi=0$ is necessary to uniquely determine the three components of the displacement vector u from the four components of (ϕ and ψ). Substituting (11.14) into Navier's equation (11.5) yields

$$c_1^2 \nabla\nabla^2\phi + c_2^2 \nabla x (\nabla^2\psi) = (\nabla\phi + \nabla x\psi)\, u \tag{11.15}$$

Every solution of (11.14), or (11.15), is always a solution of (11.5). Accordingly, equations (11.14) and (11.15) are also governing equations to the induced motion in an isotropic, elastic solid and each constitutes an exact equivalence to (11.5); see Chou (1968). A particular class of solutions of (11.15) is

$$\nabla \Gamma_1^{\,2}\phi = 0; \quad \nabla x\, \Gamma_2^{\,2}\psi = 0 \tag{11.16}$$

with a particular solution

$$\Gamma_1^{\,2}\phi = 0; \quad \Gamma_2^{\,2}\psi = 0 \tag{11.17}$$

This is with the understanding that the class of solutions presented by (11.16) and (11.17) is sufficient, but not necessary, for the satisfaction of (11.5). In equation (11.17), $\Gamma_1^{\,2}$ and $\Gamma_2^{\,2}$ are the dilatational and rotational wave operators introduced earlier by equations (11.9) and (11.13), respectively.

An Irrotational Field

A displacement field, u, is referred to as "***irrotational***" if

$$\nabla x\, \mathbf{u} = 0; \quad \mathbf{u} = \mathbf{u}_{IR} \tag{11.18}$$

For an irrotational wave, one has, following Eqn. (11.5),

$$\Gamma_1^{\,2}\mathbf{u}_{IR} = 0 \tag{11.19}$$

or, alternatively, according to *Potential theory*,

$$\mathbf{u}_{IR} = \nabla\phi \tag{11.20}$$

where ϕ is a scalar potential function. Equation (11.20) implies that, for an irrotational wave, the rotational vector ω is equal to zero in magnitude. Following (11.17), then, for an irrotational field

$$\Delta = \nabla^2\phi \quad \text{and} \quad \frac{\partial \Delta}{\partial x_i} = \nabla^2 u_i \qquad (i=1,2,3) \tag{11.21}$$

Accordingly, the scalar potential ϕ is seen to be associated with the dilatational (irrotational) part of the disturbance.

Substituting (11.21) into (11.4), one has, for an irrotational field,

$$\rho \frac{\partial^2 u_i}{\partial t^2} = (\lambda + 2\mu)\nabla^2 u_i \qquad (i=1,2,3) \tag{11.22}$$

Rotational Field
A displacement field, **u**, is called "***rotational***" if

$$\nabla \cdot \mathbf{u} = 0; \quad \mathbf{u} = \mathbf{u}_R \tag{11.23}$$

For a rotational field, the Navier's governing equation (11.5) results in

$$\Gamma_2^2 \mathbf{u} = 0 \tag{11.24}$$

$$\mathbf{u} = \nabla x \psi \tag{11.25}$$

i.e., the vector potential ψ is associated with the rotational part of the disturbance.

The above conditions for a rotational wave translates into that, in this case, the dilatation Δ=0. Hence, the set of equations (11.4) reduces, for a rotational wave, to

$$\rho \frac{\partial^2 u_i}{\partial t^2} = \mu \nabla^2 u_i \qquad (i=1,2,3) \tag{11.26}$$

$$\mathbf{u} = \mathbf{u}_{IR} + \mathbf{u}_R \tag{11.27}$$

Combining equations (11.14), (11.20) and (11.25), it follows that in an isotropic, elastic solid, a displacement field u is decomposed vectorially into an "irrotational field u_{IR}" and a "rotational one u_R". Further, in view of (11.17), (11.20) and (11.25), it may be concluded that for every displacement field that satisfies (11.5), there exists a set of functions $\mathbf{u}_{IR}$ and $\mathbf{u}_R$ such that [see, equations (11.19) and (11.24)]

$$\Gamma_1^2 \mathbf{u}_{IR} = 0 \quad \text{and} \quad \Gamma_2^2 \mathbf{u}_R = 0 \qquad \textbf{(11.28)}$$

This translates, physically, into the following:

A disturbance in an isotropic, elastic solid would generate two waves, one dilatational, involving no rotation, with velocity c_1 and the other is rotational, involving no volume change, that propagates at velocity c_2. The ratio of the two speeds may be expressed, with reference to (11.7) and (11.11), as

$$\frac{c_1}{c_2} = \kappa = \left(\frac{\lambda + 2\mu}{\mu}\right)^{\frac{1}{2}} = \left[\frac{2(1-\nu)}{1-2\nu}\right]^{\frac{1}{2}},$$

where ν = Poisson's ratio. Since $0 \le \nu \le 1$, it follows that $c_1 > c_2$.

In view of (11.28), the dilatational and rotational waves are not coupled within the continuous solid (except perhaps on the boundary where the prescribed boundary conditions must be satisfied).

Table 11.1 summarizes the relationships given in the foregoing, in terms of displacements, while Table 11.2 gives such relationships in terms of potentials (see, also, Chou,1968).

11.3.3. PLANE WAVES IN UNBOUNDED ELASTIC MEDIA

Plane waves are propagating disturbances in two- or three-dimensions where the motion of every particle in planes perpendicular to the direction of propagation is the same. An example of a propagating (three-dimensional) plane disturbance is given in Figure 11.1. As shown in this figure, the magnitude of the propagation velocity of the plane is denoted by c while the normal to the plane is designated by **n**. The position of an arbitrary point P on the plane is indicated by **r**.

For the plane wave illustrated in Figure 11.1, the motion of every particle along the plane is defined by

$$\mathbf{u}\cdot\mathbf{r} - ct = \text{constant} \qquad \textbf{(11.29)}$$

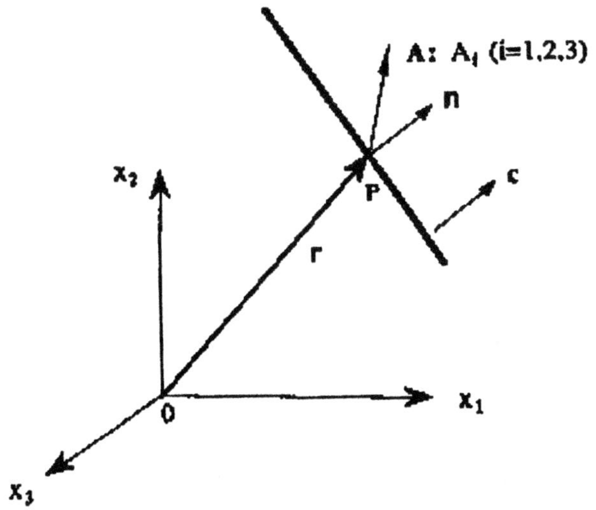

Figure 11.1. Plane wave motion in an unbounded elastic medium.

Consider now the plane wave

$$\mathbf{u} = \mathbf{A}f(\mathbf{n}\cdot\mathbf{r} - ct) \quad \textbf{(11.30)}$$

where **A** is the displacement vector of the particle along the plane of the wave and $f(\cdot)$ indicates an appropriate function of the shown argument. Substituting (11.30) in the **Navier's governing equation of motion,** (11.5), it can be shown that

$$(\lambda + \mu)A_j n_j n_i + \mu A_i = \rho c^2 A_i \quad \textbf{(11.31)}$$

Relation (11.31) above represents three homogenous equations in the amplitude components A_1, A_2, A_3. This leads, upon expanding the determinant of coefficients, to

$$(\lambda + 2\mu - \rho c^2)(\mu - \rho c^2)^2 = 0 \quad \textbf{(11.32)}$$

This equation gives the two roots

$$c_1 = \left[\frac{\lambda + 2\mu}{\rho}\right]^{1/2} \qquad \text{(a)}$$

$$c_2 = (\mu/\rho)^{1/2} \qquad \text{(b)} \qquad \textbf{(11.33)}$$

which again are, respectively, the magnitudes of the velocities of dilatational and rotational waves.

Accordingly, plane waves may propagate, without dispersion, at one or the other velocity (i.e., c_l or c_2) in the unbounded, isotropic elastic medium. Reference is made to Table 11.1 for representative values of these velocities, as calculated for various engineering materials.

11.3.4. WAVE PROPAGATION IN SEMI-INFINITE ELASTIC MEDIA

When a stress wave encounters a boundary between two media, energy is reflected and transmitted from and across the boundary. On the other hand, if the boundary is a free surface, reflection of the waves will be much more pronounced. It is well recognized that a characteristic phenomenon of the elastic wave-boundary interaction in solids is that of mode conversion. In this, an incident wave, either pressure or shear, on the boundary will be converted into two waves on reflection. Such mode-conversion phenomenon along with the fact that two types of waves may exist in an elastic solid, as discussed earlier, accounts for the relative complexity of wave propagation in solids in general as compared to equivalent problems in acoustics and electro-magnetics (e.g., Graff, 1975).

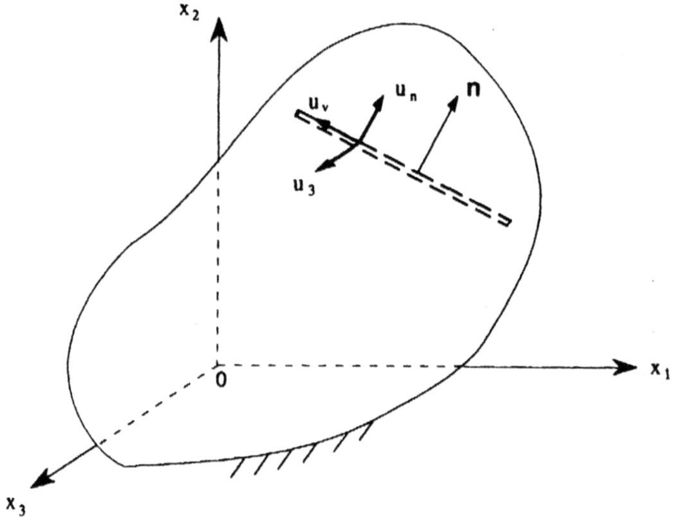

Figure 11.2. Wave motion in a semi-infinite elastic medium.

With reference to Fig. 11.2, we consider, following Graff (1975), plane harmonic waves propagating in the half-space $x_2 > 0$. It is assumed that the wave normal n lies in the x_1 x_2-plane. This plane will be referred to as the *vertical* plane while the x_1 x_3-plane, the surface of the half-space, will be referred to as the *horizontal* plane. Recalling the previous discussion concerning the propagation of plane waves in infinite media, Section 11.2.2, it is recognized that the particle motion due to dilatation will be in the direction of the wave normal and will, thus, be in the vertical plane only. The transverse particle motion, however, is due to shear and will have components both in the vertical plane and parallel to the horizontal plane. In Fig. 11.2, the normal displacement component is designated by u_n and the transverse components are denoted by u_v and u_3 which are, respectively, in the vertical and horizontal planes. As every particle along the plane of the wave is acquiring the same motion, the motion will be invariant with respect to x_3 if the wave normal is in the vertical plane. In terms of the potentials ϕ and ψ, the governing equations can be expressed as

$$u_1 = \frac{\partial \phi}{\partial x_1} + \frac{\partial \psi_3}{\partial x_2}$$

$$u_2 = \frac{\partial \phi}{\partial x_2} - \frac{\partial \psi_3}{\partial x_1}$$

$$u_3 = \frac{-\partial \psi_1}{\partial x_2} + \frac{\partial \psi_2}{\partial x_1}$$

$$\frac{\partial \psi_1}{\partial x_1} + \frac{\partial \psi_2}{\partial x_2} = 0 \tag{11.34}$$

$$\nabla^2 \phi = \frac{1}{c_1^2} \frac{\partial^2 \phi}{\partial t^2}, \quad \nabla^2 \psi_i = \frac{1}{c_2^2} \frac{\partial^2 \psi_i}{\partial t^2}$$

where ψ_i (i=1,2,3) are the components of the vector function ψ. In deriving the above governing equations both the postulate $\nabla \cdot \psi = 0$ and the x_3-independence of all quantities have been used.

Combining the displacement expressions in (11.34) with the stress-displacement constitutive relations for the isotropic elastic solid, the stress components can be established in terms of the potentials ϕ and ψ, i.e.,

$$\sigma_{11} = (\lambda + 2\mu)\left(\frac{\partial^2\phi}{\partial x_1^2} + \frac{\partial^2\phi}{\partial x_2^2}\right) - 2\mu\left(\frac{\partial^2\phi}{\partial x_2^2} - \frac{\partial^2\psi_3}{\partial x_1\,\partial x_2}\right)$$

$$\sigma_{22} = (\lambda + 2\mu)\left(\frac{\partial^2\phi}{\partial x_1^2} + \frac{\partial^2\phi}{\partial x_2^2}\right) - 2\mu\left(\frac{\partial^2\phi}{\partial x_1^2} + \frac{\partial^2\psi_3}{\partial x_1\,\partial x_2}\right)$$

$$\sigma_{12} = \mu\left(\frac{2\partial^2\phi}{\partial x_1\,\partial x_2} + \frac{\partial^2\psi_3}{\partial x_2^2} - \frac{\partial^2\psi_3}{\partial x_1^2}\right) \tag{11.35}$$

$$\sigma_{23} = \mu\left(-\frac{\partial^2\psi_1}{\partial x_2^2} + \frac{\partial^2\psi_2}{\partial x_1\,\partial x_2}\right), \qquad \sigma_{13} = 0$$

with boundary conditions

$$\sigma_{22} = \sigma_{21} = \sigma_{23} = 0, \qquad x_2 = 0 \tag{11.36}$$

Experimental studies on wave propagation in semi-infinite media may vary considerably in scope. Ultrasonic excitation is often used as an impulsive surface force, meantime, photo-elasticity has been conventionally adopted as a recording technique for the patterns of wave motion in elastic materials. Dally, Durelli and Riley (1960), for instance, used small explosive charges of lead azide (PbN_6) to dynamically load a low-modulus urethane rubber plate and the dynamic fringe propagation patterns were recorded by a high-speed camera (see, also, Dally, 1968 and Graff, 1975). Dally and Riley (1967) used an embedded polariscope technique to experimentally study the three-dimensional problem of a point load on half-space using a photo-elastic method (see, e.g., Pindera, 1986).

11.3.5 SURFACE WAVES

As per our earlier discussion concerning elastic wave propagation in an infinite elastic medium, only two types of waves can be propagated, i.e , ***dilatational** (primary, P -)* and ***rotational** (secondary, S -)* waves. In the case of a semi-infinite medium, however, a third type of wave may exist. The existence of the three types of waves in a semi-infinite medium was first encountered in seismology where it was observed that in an earthquake there were two early, rather minor, disturbances as a result of P- and S- waves, but the main damaging effect was done by the third shaking. Such a disturbance was not consistent with the elastic wave phenomenon in infinite media. This led to the realization of existence of a surface wave in semi-infinite media. In case of an earthquake, the relative significance of P- and S- waves is considered to be a consequence of volumetric dispersion of energy into the earth's interior, but, the significant amount of energy corresponding to the third wave suggested that this wave dissipated its energy less rapidly then the *P*- and *S*- waves. This could be rationalized by assuming it, the third wave, was basically limited to the surface. The other characteristics of the surface wave, other then it is confined to the surface zone, is that the velocity of surface wave is less than that of body waves, see, e. g., Kolsky (1963). We introduce below two types of surface waves; namely, *Rayleigh waves* and *Love waves*.

(A) Rayleigh Waves

When the solid has a free surface, "Rayleigh" surface waves can also exist. These waves were first introduced by Rayleigh (1887), see, also, Lamb (1904), who showed that their effect decays rapidly with depth and that their velocity is less than that of body waves c_1 and c_2. It is shown by Kolsky (1963) that Rayleigh waves do, in fact, travel with a fraction ξ of the velocity c_2 of distortional waves where ξ is obtained from the equation

$$\xi^6 - 8\xi^4 + (24 - 16b^2)\xi^2 + 16b^2 - 16 = 0 \tag{11.37}$$

In the above equation, b is an elastic constant of the material expressed by

$$b = [(1 - 2\nu)/(2 - 2\nu)]^{1/2} \tag{11.38}$$

where υ is Poisson's ratio.

In Rayleigh waves, the particle motion is parallel to the direction of wave propagation and it is in a plane perpendicular to the surface containing the waves during travel.

In case of an elastic solid, the velocity of a "Rayleigh" surface wave is independent of the frequency and depends, similar to the body waves, on the elastic constants of the material. In other words, there is no dispersion (change of form) of these waves.

It was Lord Rayleigh the first to investigate this type of surface wave in which the amplitude of the wave decays exponentially with depth, from the surface to the medium interior. Rayleigh waves spread only in two dimensions (see, e.g., Davis, 1988). It was anticipated by Rayleigh that waves of this type might approximate the behavior of seismic waves observed during earthquakes. We follow, below, the model of Achenbach (1973) to determine the displacement and velocity of Rayleigh waves.

The criterion for Rayleigh surface waves is that the displacement decays exponentially with distance from the free surface. Thus, we consider components of the form

$$u_1 = A\,e^{-bx_2} \exp[\,ik(x_1 - ct)\,] \tag{11.39a}$$

$$u_2 = B\,e^{-bx_2} \exp[\,ik(x_1 - ct)\,] \tag{11.39b}$$

$$u_3 = 0. \tag{11.39c}$$

The real part of b is supposed to be positive, so that the displacements decrease with increasing x_2 and tend to zero as x_2 increases beyond bounds.

Combining Equations (11.39) with the equation of motion (11.5) yields two homogeneous equations for the constants A and B. A non-trivial solution of this system of equations exists if the determinant of the coefficients vanishes, which leads to the equation

$$[c_L^2 b^2 - (c_L^2 - c^2)k^2]\,[c_T^2 b^2 - (c_T^2 - c^2)k^2] = 0 \qquad (11.40)$$

The roots of (11.40) are

$$b_1 = k\left(1 - \frac{c^2}{c_L^2}\right)^{\frac{1}{2}}, \qquad b_2 = k\left(1 - \frac{c^2}{c_T^2}\right)^{\frac{1}{2}}.$$

It is noted that b_1 and b_2 are real and positive if $c<c_T<c_L$, and if positive roots are considered.

The ratios (B/A) corresponding to b_1 and b_2 can now be computed as

$$\left(\frac{B}{A}\right)_1 = -\frac{b_1}{ik}, \qquad \left(\frac{B}{A}\right)_2 = \frac{ik}{b_2}.$$

Returning to Equations (11.39), a general solution of the displacement equations of motion may, thus, be written in the form

$$u_1 = [A_1 e^{-b_1 x_2} + A_2 e^{-b_2 x_2}]\exp[ik(x_1 - ct)] \qquad (11.41)$$

$$u_2 = \left[-\frac{b_1}{ik} A_1 e^{-b_1 x_2} + \frac{ik}{b_2} A_2 e^{-b_2 x_2}\right]\exp[ik(x_1 - ct)]. \qquad (11.42)$$

$$2\,b_1 A_1 + \left(2 - \frac{c^2}{c_T^2}\right) k^2 \frac{A_2}{b_2} = 0$$

$$\left(2 - \frac{c^2}{c_T^2}\right) A_1 + 2\,b_2 \frac{A_2}{b_2} = 0.$$

The constants A_1 and A_2 and the phase velocity c have to be chosen such that the stress tensor components σ_{22} and σ_{21} vanish at x_2=0. By substituting Equations (11.39), (11.41) and (11.42) into the expressions for σ_{22} and σ_{21} at $x_2 = 0$, we obtain after some manipulation

For a non trivial solution the determinant of the coefficients of A_1, A_2 must vanish, which yields the following well-known equation for the phase velocity of Rayleigh waves:

$$\left(2 - \frac{c^2}{c_T^2}\right)^2 - 4\left(1 - \frac{c^2}{c_L^2}\right)^{\frac{1}{2}}\left(1 - \frac{c^2}{c_T^2}\right)^{\frac{1}{2}} = 0 . \tag{11.43}$$

It is noted that the wave number does not enter in (11.43). Thus, surface waves at a free surface of an elastic half-space are thus nondispersive.

Since (11.43) is an equation for c^2, the two roots are each other's opposite. As noted earlier, Eq. (11.43) shows that the roots may be expected along the real axis for $-c_T < c < c_T$. Obviously, only the positive real root is of interest. The roots for c^2 are usually computed by rationalizing (11.43).

Denoting the phase velocity of Rayleigh waves by c_R, Eqn.(11.43) can be considered as an equation for c_R / c_T, with Poisson's ratio ν ($0 \le \nu \le 0.5$) as independent parameter.

A good approximation of c_R can be written as

$$c_R = \frac{0.862 + 1.14\nu}{1 + \nu}\, c_T . \tag{11.44}$$

As ν varies from 0 to 0.5, for most metals, the Rayleigh wave phase velocity increases monotonically from 0.862 c_T to 0.955 c_T.

Given suitable generating conditions, surface waves as well as body waves are generated at a bounding surface. For a two-dimensional geometry the surface waves are essentially one dimensional, but the body waves are cylindrical and undergo geometrical attenuation. Thus, at some distance from the source the disturbance due to the surface wave becomes predominant.

Rayleigh waves have been studied in great detail and they have found several applications. The attractive features are the absence of dispersion and the localization of the motion in the vicinity of the surface. For further study on the subject matter, the reader is referred to Viktorov (1967) and Graff (1975), among others.

(B) Love Waves

The ***"Love"*** wave is a shear surface wave confined to a relatively shallow zone. Recalling our earlier discussion, for Rayleigh waves, the material particles move in the plane of propagation. Thus, for propagation in the x_1-direction along the surface of the half-space $x_2 \geq 0$, the displacement u_3 vanishes for classical Rayleigh waves.

The question may now be raised whether surface waves with displacements perpendicular to the plane of propagation, the plane x_1, x_2, are possible in a homogeneous isotropic linearly elastic half-space. We recall that the S-waves are governed by the equation:

$$\frac{\partial^2 u_3}{\partial x_1^2} + \frac{\partial^2 u_3}{\partial x_2^2} = \frac{1}{c_T^2} \frac{\partial^2 u_3}{\partial t^2} . \qquad \textbf{(11.45)}$$

A solution of (11.49), representing a surface wave, is written in the form

$$u_3 = A e^{-b x_2} \exp [ik(x_1 - ct)], \qquad \textbf{(11.46)}$$

where the real part of b must be positive. By substituting (11.46) into (11.45) we find

$$b = k \left[1 - \left(\frac{c}{c_T} \right)^2 \right]^{\frac{1}{2}} . \qquad \textbf{(11.47)}$$

For a free surface, the boundary condition at $x_2 = 0$ is

$$\frac{\partial u_3}{\partial x_2} = 0 . \qquad \textbf{(11.48)}$$

The boundary condition (11.48) can, however, be satisfied only if either A=0 or b=0. Neither case represents a surface wave.

Experimental data, particularly as gathered from seismological observations, have, however, shown that surface waves may occur along free surfaces. An analytical resolution of this question was provided by Love, who showed that such waves are possible in the half-space covered by a layer of a different material (e.g., Ewing *et al.*, 1957).

11.4. Reflection and Refraction of Waves at a Plane Interface

The presence of a discontinuity in the material properties generally produces a significant influence on systems of wave propagating through the medium. Consider, for example, the propagation of plane harmonic waves in an unbounded medium consisting of two joined elastic half-spaces of different material properties. In such a composite medium, systems of plane waves can be superposed to represent an incident wave in conjunction with reflections and refractions at the interface between the two media. The wave which emanates from an infinite depth in one of the media is called the *incident* wave. An incident wave on an interface would result in additional *reflected and refracted* waves in the region. For the special cases of an elastic half-space which adjoins a medium which does not transmit mechanical waves, the system of waves consists of incident and reflected waves only. In general, all media transmit waves, but, for practical purposes, refraction of elastic waves at an interface of a solid elastic body with air may be neglected. In this case, a reflection of plane waves only at the free surface may be considered.

Basically, as mentioned earlier, two types of body elastic waves may be propagated through a solid medium; namely P- ans S-types of waves.. It is found that, when a wave of either type impinges on a boundary between two media, both reflection and refraction take place. In this section, we study the reflection of both the dilatational and distortional waves at free boundary and also reflection and refraction of these two waves at an interface between two media, whereby each case is reviewed separately. For further studies on this context, the reader is referred to Ewing et al. (1957), Kolsky (1963), Kinslow (1970), Achenbach (1973), Tolstoy (1973), Eringen and Suhubi (1975), Graff (1975), Miklowitz (1978), Miklowitz, and Achenbach (1977), Davis (1988) and McCarthy and Hayes (1989).

11.4.1. DILATATIONAL WAVES AT A FREE BOUNDARY

By Free Boundary we mean a surface in vacuum when there can be no refracted waves. Fig. 11.3 shows the reflection of a dilatational wave at a free surface. In this figure, α_1 is the angle of an incident dilatational wave (of an amplitude A_1). Meantime, α_2 is the angle of the reflected dilatational wave (with an amplitude A_2). Let A_3,, β_2 represent the amplitude and the angle of the reflected distortional wave. As shown in Fig. 11.3, the direction of propagation of the incident dilatational wave is in the x_1 x_2 plane making the angle of incidence α_1 with the x_1- axis, whilst, the free boundary is the x_2 x_3 plane. The following relations between the various angles of incidence and associated with wave velocities may be written (see, e.g., Kolsky, 1963) ,

$$\frac{\sin \alpha_1}{c_L} = \frac{\sin \alpha_2}{c_L} = \frac{\sin \beta_2}{c_T} \qquad \textbf{(11.49)}$$

Thus,

$$\alpha_1 = \alpha_2 \quad \text{and} \quad \frac{\sin \alpha_1}{\sin \beta_2} = \frac{c_L}{c_T}.$$

$$2\,(A_1 - A_2)\cos \alpha_1 \sin \beta_2 - A_3 \cos 2\,\beta_2 = 0\,. \tag{11.50}$$

$$(A_1 + A_2)\cos 2\,\beta_2 \sin \alpha_1 - A_3 \sin \beta_2 \sin 2\,\beta_2 = 0\,. \tag{11.51}$$

Thus, when a dilatational wave is incident on a free surface with an angle α, two waves are generated on reflection: one is a dilatational wave reflected at an angle equal to the angle of incidence α, while the other is a distortional wave reflected at a smaller angle β where $\sin \beta / \sin \alpha = c_T/c_L$.

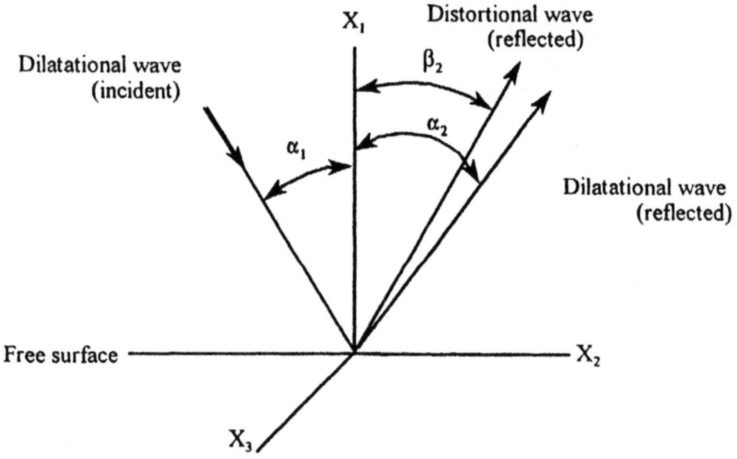

Figure 11.3. Reflection of a dilatational wave at a free surface. The face boundary is the X_2 -X_3 plane.

11.4.2 DISTORTIONAL WAVES AT A FREE BOUNDARY

In a similar analogy to the above presentation, if a distortional wave is incident on a free surface at an angle γ, Figure 11.4, both distortional and dilatational waves are generally reflected. The distortional wave is reflected at the same angle γ while the dilatational wave is reflected at a generally smaller angle δ where $\sin \gamma / \sin \delta = c_T/c_L$.

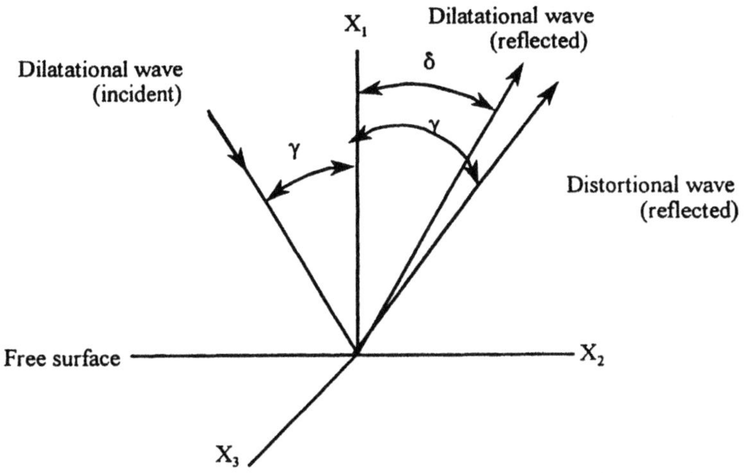

Figure 11.4. Reflection of a distortional wave at a free boundary.

11.5. Wave Propagation in Bounded Elastic Solids

In this section, the propagation of stress waves along a cylindrical bar will be considered first, as this is a problem which has been investigated most fully theoretically and on which there are also some experimental data. Before examining the problem in terms of the exact elastic equations, we shall consider the simple treatment which applies to the propagation of waves the lengths of which are large compared with the diameter of the bar.

There are three different types of vibration which occur in thin rods or bars; these are classified as ***"longitudinal"***, ***"torsional"***, and ***"lateral"***.

In longitudinal vibrations, elements of the rod extend and contract, but there is no lateral displacement of the axis of the rod. In torsional vibrations, each transverse section of the rod remains in its own plane and rotates about its center, with the axis of the rod remaining undisturbed. Meanwhile, lateral vibrations correspond to the flexure of portions of the rod, with elements of the central axis moving laterally during the motion.

In this section the subject of stress wave propagation in bars is first discussed. Then, the approximate theory of stress wave propagation in plates is briefly reviewed. For further information on the subject matter, the reader is referred to Kolsky (1963), Graff (1975), Miklowitz (1978) and McCarthy and Hayes (1989).

11.5.1. STRESS WAVES IN RODS

In all cases of longitudinal, torsional and lateral (flexural) stress waves in bars the approximate description of wave motion has been used by following an approximate solution such as the one developed by Achenbach (1973). In this case, one can find the velocities of different type of waves in a long rod. The exact treatment of harmonic wave motions in an elastic circular cylinder is already rather complicated. For a cylinder with other than circular or an elliptical cross section, however, it becomes rather impossible to carry out an exact solution. Even for a strip of rectangular cross section whose lateral surfaces are free of traction it is not possible to analyze general harmonic wave motions rigorously within the context of the linear theory of elasticity (Davis 1988). For this reason, simplified analytical models have been proposed that provide an approximate description of wave motions in rods of rather arbitrary cross section. In this section, we review some models that are commonly used. These models are based on a priori assumptions with regard to the deformation of the cross-sectional area of the rod, which simplify the description of the kinematics to such an extent that the wave motion can be described by one-dimensional approximate theories. Further, for propagation of time-harmonic waves, it was found that the approximate theories can adequately account for the dispersive behavior of the lowest axisymmetric and flexural models over a limited but significant range of wave-numbers and frequencies.

The governing equations can be obtained either by using variational methods or by straightforward momentum considerations of an element of the rod. The latter approach, however, has the advantage that the physical concepts are conveyed more clearly. For the more complicated theories it is, however, easier to employ the assumed displacement distributions to compute the corresponding kinetic and strain energies for an element of the rod, whereupon Hamilton's principle can be applied to obtain the governing equations. In the following analysis, we present a brief derivation of the equations for the Timoshenko model, and we state only the governing equations for some other models. In all cases, the assumption that the wavelength is long compared to the lateral dimensions of the rod would prevail.

11.5.2. LONGITUDINAL WAVES IN RODS

Longitudinal stress waves are also called extensional waves. In an extensional wave motion, the dominant component of the displacement is in the longitudinal direction. Based on the assumption that the cross-sectional area of the rod remains plane, it can be shown that consideration of the forces acting on an element leads to the equation

$$\frac{\partial^2 u}{\partial x^2} = \frac{1}{c_b} \frac{\partial^2 u}{\partial t^2}, \tag{11.52}$$

where,

$$c_b^2 = \frac{E}{\rho} \tag{11.53}$$

Eq. (11.52) predicts that extensional harmonic waves in the rod are not dispersive.

11.5.3. TORSIONAL STRESS WAVES IN RODS

In the approximate theory, it is assumed that transverse sections remain plane and that the motion consists of a rotation of the sections about the axis. This leads to a wave equation for the angle of rotation with a propagation velocity.

$$\left(\frac{C}{\rho A}\right)^{\frac{1}{2}} \frac{1}{K}. \tag{11.54}$$

In the above equation, *K* is the radius of gyration of a cross section of the rod about its axis, A is the cross-sectional area and C is the torsional rigidity of the rod.

11.5.4. FLEXURAL STRESS WAVES IN RODS

In the approximate theory of flexural motion of rods of an arbitrary but uniform cross section with a plane of symmetry, it is assumed that the dominant displacement component is parallel to the plane of symmetry. It is also assumed that the deflections are small and that cross-sectional areas remain plane and normal to the neutral axis. For a beam, free of lateral loading, the equation of motion is

$$\frac{\partial^2 w}{\partial t^2} + \frac{EI}{\rho A}\frac{\partial^4 w}{\partial x^4} = 0, \tag{11.55}$$

where w is the deflection, I is the moment of inertia of the cross-sectional area A about the neutral axis. Substituting a harmonic wave, the phase velocity is expressed as

$$c = \left(\frac{E}{\rho}\right)^{\frac{1}{2}} \left(\frac{I}{A}\right)^{\frac{1}{2}} k. \tag{11.56}$$

Thus, the phase velocity is proportional to the wave number, which suggests that (11.56) cannot be correct for large wave numbers (short waves). For a circular cylindrical rod, Eqn.

(11.56) reduces to

$$c = \frac{1}{2}\left(\frac{E}{\rho}\right)^{\frac{1}{2}} ka. \tag{11.57}$$

which results in the following frequency

$$\omega = \frac{1}{2}\left(\frac{E}{\rho}\right)^{\frac{1}{2}} k^2 a. \tag{11.58}$$

11.5.5. STRESS WAVES IN A LONG BAR

In this section, elastic stress wave motion in a long bar is considered in view of the work of Zukas et al. (1982).

In the previous sections, we learned that when a material is stressed with a suddenly applied load, the deformations and stresses are not transmitted immediately to all parts of the body. Thus, remote portions may remain undisturbed for some time. Deformations and stresses progress through the material in a form of one or more stress disturbances which travel, in a perfectly elastic material, at a finite velocity form the area of application of the load, this velocity being a characteristic of the material. Such a suddenly applied, or impulsive, load may be produced by a sharp mechanical blow, a detonating explosive, or by impact of a high velocity projectile. Regardless of the method of application, the consequent stress disturbances have identical properties.

In the elementary case, we consider two types of stress pulses generated by an impulsive load. The first, the longitudinal wave, is also called a *dilatational, irrational,* or *primary (P)* wave, the terms being synonymous. In a longitudinal pulse, the particle motion is parallel to the distortional, rotational, secondary (S), or shear wave, the particle motion is normal to the direction of propagation of the pulse and the strain is a shearing strain.
direction of propagation of the pulse and the strain is pure dilatation. In a transverse wave, otherwise called a
Two velocities must be considered: the velocity of propagation c of the disturbance and the particle velocity v.

Particle velocity is defined as the velocity with which a point in the material moves as the disturbance displaces across it. Both the velocity of propagation *c* of the disturbance and the particle velocity *v* enter into the governing equations in distinctly different ways as dealt with in the following section.

11.5.6. GOVERNING WAVE EQUATIONS

The relationship between the longitudinal stress at a point in a body and the longitudinal particle velocity v_L at the point is expressed, in view Newton's second law, as

$$F_L\, d t = (m \upsilon_L) \tag{11.59}$$

Here, F_L is the longitudinal force acting on a given cross section, dt is the time the force acts, m is the mass it acts against, and v_L is the velocity imparted to m by F_L. Since

$$\sigma = \frac{F_L}{A} \qquad m = \rho\, A\, dl \tag{11.60}$$

where dl is the distance the pulse has moved in time dt, equation (11.78) can be written as

$$\sigma\, a\, dt = \rho\, A\, d\, l\, \dot{d}\, \upsilon_L \tag{11.61}$$

or
$$\sigma = \rho \frac{dl}{dt} d\, \upsilon_L$$

but dl / dt is just the speed of the pulse c_L, so that

$$\sigma = \rho\, c_L\, (\Delta\, \upsilon_L) \tag{11.62}$$

In a similar manner it can be shown for the transverse pulse that

$$\tau = \rho\, c_T\, (\Delta\, \upsilon_T) \tag{11.63}$$

where τ is the shear stress, c_T is the velocity of propagation of the transverse disturbance, and $\Delta\, \upsilon_T$ is the change in particle velocity due to shear.

11.5.7. REFLECTION OF WAVES

Any elastic wave will be reflected when it reaches a free surface of the material in which it is traveling. The simplest case occurs when the wave strikes the surface normally. In the case of a longitudinal wave, since the stress normal to the surface, at the surface, must be zero, the reflected pulse must be opposite in sense to the incident pulse (a compression wave would be reflected as tension and vice versa). To illustrate the situation, one considers the displacement due to the incident pulse to be $u_I = f(x_1 - ct)$ moving in the positive x_1 direction. After impingement on a free surface, a reflected wave moves in the negative x_1 direction. Let the displacement for the reflected wave be of the form $u_R = g(x_1 + ct)$. At the free boundary, the net stress must be zero.

Since the stress is given by

$$\sigma_{NET} = \sigma_I + \sigma_R = 0 \quad \text{at } x_1 = l \qquad \textbf{(11.64)}$$

or

$$\begin{aligned} \sigma &= E\varepsilon = E(\partial u_1 / \partial x_1) \\ \sigma_{NET} &= E[\,f'(l - ct) + g'(l + ct)] = 0 \end{aligned} \qquad \textbf{(11.65)}$$

Hence, the shape of the reflected pulse is the same as the shape of the incident pulse, but it is opposite in sign.

$$f'(l - ct) = -g'(l + ct)$$

The net particle velocity may be, also, found by superposition. Thus,

$$\upsilon_{NET} = \upsilon_I + \upsilon_R = \frac{\partial u_{1I}}{\partial t} + \frac{\partial u_{1R}}{\partial t} \qquad \textbf{(11.66)}$$

$$\begin{aligned} &= c(-f' + g') \quad \text{at} \quad x_1 = l \\ &= 2cg' \end{aligned} \qquad \textbf{(11.67)}$$

Hence, the particle velocity and also the displacement in a region where the incident and reflected pulses overlap are twice that for either pulse. At a fixed boundary, we require the displacement and particle velocity to vanish. Thus,

$$\upsilon_{NET} = -cf'(l - ct) + cg'(l + ct)$$

or

$$f'(l - ct) = g'(l + ct) \tag{11.68}$$

The net stress is doubled at a fixed boundary, whilst the net displacement and particle velocity are zero.

$$\sigma_{NET} = E\left(\frac{\partial u_I}{\partial x} + \frac{\partial u_R}{\partial x}\right) = E[f'(l - ct) + g'(l + ct)]$$
$$= 2Ef'(l - ct) \tag{11.69}$$

11.5.8. STRESS WAVES IN BARS OF DISCONTINUOUS CROSS SECTIONS

Following Zurkas et al. (1982), we consider a bar with a change in cross section, as illustrated in Figure 11.5 Assume that a disturbance at the left end of the bar has caused an elastic compressive pulse, with an intensity a, to propagate to the right. At the interface with the second portion of the bar, with different section, the wave will be partly transmitted and partly reflected. Let the transmitted wave amplitude be σ_T, and the reflected wave amplitude be σ_R. Two conditions must be satisfied at the interface:

1. The forces at the interface, in both portions of the bar.
2. Particle velocities at the interface must be continuous.

Taking σ_R and σ_T to be compressive, condition 1 above gives

$$A_1(\sigma_1 + \sigma_R) = A_2\sigma_T \tag{11.70}$$

where A_1, A_2, are the respective cross-sectional areas. Condition 2 above gives

$$\upsilon_1 - \upsilon_R = \upsilon_T \tag{11.71}$$

or, using $\sigma = \rho c v$

$$\frac{\sigma_1}{\rho_1 c_1} - \frac{\sigma_R}{\rho_1 c_1} = \frac{\sigma_T}{\rho_2 c_2} \tag{11.72}$$

Solving for σ_R and σ_T in terms of σ_I, gives

$$\sigma_T = \frac{2 A_1 \rho_2 c_2}{A_1 \rho_1 c_1 + A_2 \rho_2 c_2} \sigma_I \tag{11.73}$$

$$\sigma_R = \frac{A_2 \rho_2 c_2 - A_1 \rho_1 c_1}{A_1 \rho_1 c_1 + A_2 \rho_2 c_2} \sigma_I \tag{11.74}$$

$$\sigma_T = \frac{2 A_1 \sigma_I}{(A_1 + A_2)} \tag{11.75}$$

Consider several implications of the above expressions:

1. If the materials in both bars are identical then $\rho_1 = \rho_2$ and $c_1 = c_2$. Then, If $A_2 > A_1$, then σ_T and σ_R will be of the same type. If $A_2 < A_1$, then, σ_T and σ_R will be of opposite sign.

2. If $A_2 / A_1 \rightarrow 0$, the rod is effectively free and $\sigma_R \rightarrow -\sigma_I$. If $A_2 / A_1 \rightarrow \infty$, the rod is fixed and $b_R \rightarrow b_I$, $b_T \rightarrow 0$.

3. For no wave reflection to occur from the discontinuity in the bar, $\sigma_R = 0 \therefore A_2 \rho_2 c_2 = A_1 \rho_1 c_1$ and $\sigma_T = \sigma_I \sqrt{E_2 \rho_2 / E_1 \rho_1}$ **(11.76)**

4. In (11.76), the coefficient of σ_I is positive. This means that tension will be transmitted as tension and compression as compression. For a situation wherein $\rho_2 c_2 \gg \rho_1 c_1$, or medium 2 is much more rigid than medium 1, Figure 11.5, the stress of the transmitted pulse is approximately twice the stress of the incident wave.

5. In (11.76), the coefficient of σ_I, can be positive or negative depending on if $\rho_1 c_1 < \rho_2 c_2$. If the coefficient is negative, an incident compression-stress is reflected as a tensile stress and vice versa. If the coefficient is positive, the incident compressive stress is reflected as a compressive stress. These results are in complete agreement with the laws of conservation of momentum and kinetic energy.

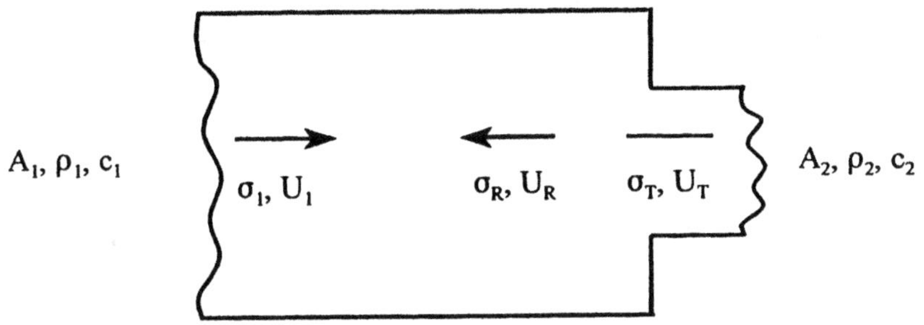

Figure 11.5. Wave reflection and transmission at changes in cross section.

11.5.9. STRESS WAVES IN PLATES

In this section, we briefly review the elastic stress wave propagation phenomena in plates. The reader is referred, in this context, to the work of Rayleigh (1887), Lamb (1917), Graff (1975), Miklowitz and Achenbach (1977), and McCarthy and Hayes (1989).

Similar to the case of elastic wave propagation in rods, we have three different types of waves propagating in a plate, i.e., "***longitudinal***", "***torsional***" and "***flexural***" waves. When we deal with a semi-infinite plate, the wavelength is long compared with the thickness of the plates, and the longitudinal wave velocity C_L is expressed by

$$C_L = \left[\frac{4\mu(\lambda+\mu)}{(\lambda+2\mu)\rho}\right]^{\frac{1}{2}} = \left[\frac{E}{\rho(1-v^2)}\right]^{\frac{1}{2}} \qquad \textbf{(11.77)}$$

The increasing attention to the dynamic behavior of materials and evermore increasing of application of ultrasonics are two very important reasons behind the significance of the wave propagation phenomenon, but, there are, also, a number of other reasons:

First, experimental methods for the generation and detection of high frequency mechanical waves have become available only with the advent of electronic techniques and of high speed photographic recording apparatus. Secondly, the appearance of new materials, such as plastics and polymeric material systems in general, the mechanical properties of which depend very markedly on the time-rate of loading, has led to studies of the mechanical response of such materials to high frequency mechanical waves, with a view to correlating their microscopic structure with their mechanical behavior. Second, engineers have become more and more concerned with the response of conventional engineering materials, such as metals, to large impulsive forces applied for very short periods of time. This interest arises both in military developments and in problems of impact and of shock absorption in engineering structures. A proper understanding of all these problems requires a knowledge of the nature of stress wave propagation in engineering materials.

A number of distinct types of wave propagation in elastic solids have been investigated and, although the phenomena observed in practical situations do not always conform to the idealized mathematical models, the theoretical work has received experimental confirmation in a number of the problems, and the experiments have, in turn, shown effects which have led to further theoretical advances Kolsky (1963). That gives us enough justification for the idealization and assumptions made in the analysis pertaining of wave propagation phenomena. The assumptions of being continuous, isotropic, homogenous and perfectly linear elastic material are never true for media, but in order to prevent very complicated problems that do not have an easy mathematical solution we make those idealizations.

The physical explanation of the propagation of a wave lies in the interaction of the discrete atoms of a material. But two properties of a medium, i. e., *deforrnability* and *inertia,* are essential for the transmission of a mechanical disturbance. All real materials are deformable and possess mass and consequently all real material transmit mechanical waves. *The inertia of a medium first offers resistance to motion, but once the medium is in motion inertia, in conjunction with the resilience of the medium tends, to sustain the motion.* If, after a certain interval the externally applied excitation becomes stationary, the motion of the medium will eventually subside due to frictional losses and a state of deformation will be reached. The importance of dynamic effects depends on the relative magnitudes of two characteristic times: the time characterizing the external application of the disturbance and the characteristic time of transmission of the disturbance across the body. In other words, for low intensity excitations, both the geometry of the entire structure as well as the nature of the material from which it is made play a major role in resisting external forces. As loading intensity increases, the response tends to become highly localized and is more affected by the constitution of the material in the vicinity of load application than the geometry of the total structure. A description of the phenomena in terms of elastic, inelastic, and shock wave propagation becomes appropriate.

Thus, we have to know, indeed, when we can analyze the behavior of a material simply by using strength of material and dynamic theories and under which circumstances we have to use wave propagation phenomena to analyze the behavior of materials. We know that in fact every process of loading is a dynamic case involving wave propagation phenomena. After every loading, disturbances are produced at the place where loading is applied and then propagate toward other areas in the medium. Then, propagation and reflection of waves in the medium continue until the medium reaches the state of static equilibrium. If the rate of applying the load is small compared with the velocity of wave propagation, static equilibrium prevails. On the other hand, if loading is applied at a rate that is fast enough if compared with the wave propagation velocity, then, we have to consider using wave propagation analysis in determining the response of the medium. At this stage we could clarify the point with an example; suppose we have a medium subjected to an external load F(t) applied at point P. We wish to determine the deformation and the distribution of stresses throughout the medium.

We know that we have different types of waves propagating inside the media such as dilatational and distortional waves but we also know that the highest velocity is that of the dilatational wave C_L. Thus, if the external disturbance is applied at time t=0, the disturbed regions at times t_1 and t_2 are surrounded by spheres centered at point P with radii $c_L t_1$ and $c_L t_{2,,}$ respectively. Therefore, the entire body is disturbed at time, $r = r / c_L$, where r is the largest distance within the body, measured from point P. Let us assume that over a time t_a, the loading F(t) has drastically changed. In this case, the dynamic effect are important if t_a and the r/c_L, are of the same order of magnitude. If $t_a \gg r / c_L$ the problem is quasi-static rather than dynamic in nature and inertia effects can be neglected. Thus, for bodies of small dimensions, a wave propagation analysis is called for if t_a is small. If the excitation source is removed, the body returns to rest after a certain time. For excitation sources that are applied and removed, the effects of wave motion are important if the time interval of application is of the same order of magnitude as the characteristic time of transmission of a disturbance across the body. For bodies of finite dimensions, this is the case for loads of explosive origins or for impact loads. For sustained external disturbances, the effects of wave motions need be considered if the externally applied forces are rapidly changing with time.

A very important parameter in wave propagation, is the relative velocity of different wave types. *The velocity of a dilatational wave has the largest magnitude, meantime, the magnitude of a Rayleigh wave is less than that of a distortional wave.* The significant point is that all of these wave velocities are functions solely of the elastic constant of the medium, and, thus, *they are characteristics of the mechanical behavior of the medium.*

11.6. Study Problems

1. What is meant by an "inelastic wave" ? Describe briefly why it is different from an "elastic wave".

2. Describe briefly the following terms:
   Waveguide, node, frequency spectrum, dispersion, and *group velocity*

3. Derive, in a vectorial form, the expression for the "*equation of motion*", in terms of the displacement, for an isotropic elastic material.

4. Based on Problem 3 above, derive the "*wave equation*" for an unbounded isotropic, elastic medium.

5. Derive the governing equation of motion for a homogeneous rod with an elastic modulus E and a constant density ρ.

6. Solve problem #5 above for an inhomogeneous rod with the elastic modulus varies as $E = E_o (1+x)$ and a constant density ρ.

7. Based on Problem 4 above, derive the expressions, in terms of the material elastic parameters, for both the "*dilatational*" and "*rotational*" wave-velocities. Explain briefly the difference in physical significance of the two velocities.

8. Based on Problem 7 above, search the values of the material parameters and determine the magnitudes, of both wave velocities for the following materials: *Aluminum, copper, lead, magnesium, nickel, silver, tin, tungsten and zinc*. Use SI units, and present your results in a an appropriate table format.

9. What is meant by "*rotational*" and "*irrotational*" fields ? Use appropriate mathematical derivations to illustrate your explanation.

10. Explain briefly the difference in physical significance of a "*dispersive wave*" vs. "*nondispersive wave*".

11. (a) What is meant by a "*surface wave*" ?.
(b) Comment briefly on the validity of the following expression: "*Surface waves at a free surface of an elastic half-space are nondispersive*".

12. Explain briefly the difference in physical significance between "*Love*" and "*Rayleigh*" surface waves, then, Comment on the validity of the expression: "*Love waves are dispersive, as opposed to Rayleigh waves which are not dispersive*".

11.7. Problems

The following problems may require some literature search by the student for information not directly available within the material presented in the context of this Chapter.

13. Comment, with an analytical proof, on the following statement: "*The Love waves are dispersive, as opposed to Rayleigh waves which are not dispersive*".

14. Comment, with an analytical proof, on the following statement: "*An elastic wave reflected from a fixed-end bar is entirely unchanged in shape or intensity*".

15. Determine the resulting wave propagation in rod of length l, which is fixed at one end. The rod is subjected to a compressive load P; which is then suddenly removed. Plot

the displacement versus time for the end of the rod. Assume the material of the rod to be linear elastic.

16. Derive the governing equations and boundary conditions for a plate, as based on energy considerations.

17. Derive the frequency equation for the natural frequencies of a clamped, circular plate.

18. Determine the expression for the dilatation in the case of a plane harmonic dilatational wave propagating in an infinite medium.

19. Derive the frequency equation for pure torsional wave in a composite rod. The latter is composed of an inner cylinder of radius, which is attached to an outer shell of an inner radius a and outer radius b. Assume that the shear wave velocity in the inner cylinder to be greater than that in the shell.

20. A semi-infinite plate has traction-free lateral surfaces on $x=\pm a$ and a stress-free edge at y=b. Investigate the reflection of an incident longitudinal plane wave from the boundary. Also, determine the ratios of the pertaining reflection coefficients for the various wave components.

11.8. References

Achenbach, J. D. (1973) *Wave Propagation in Elastic Solids*, Elsevier, New York.

Chou, P. C. (1968) Introduction to wave propagation in composite materials, in *Composite Materials Workshop*, edited by S. W. Tsai, J. C. Halpin and N. J. Pagano, Technomic, Stamford, Ct, pp. 193-216.

Dally, J. W., Durelli, A. J. and Riley, W. F. (1960) Photoelastic study of stress wave propagation in large plates, *Proc. Soc. Exp. Stress Analysis* **17**, 33-50.

Dally, J. W. and Riley, W.F. (1965) *Experimental Stress Analysis*, McGraw-Hill, NewYork.

Dally, J. W. and Riley, W. F. (1967) Initial studies in three-dimensional dynamic photoelasticity, *J. Appl. Mech.* **34**, 405-10.

Dally, J. W. (1968) A dynamic photoelastic study of a doubly loaded half-plane, *Develop. Mech.* **4**, 649-64.

Davis, J. L. (1988) *Wave Propagation in Solids and Fluids*, Springer-Verlag, New York.

Donnell, L. H. (1930) Trans. ASME **52(1)**, 153-67.

Dove, R.C. and Adams, P. H. (1964) *Experimental Stress Analysis and Motion Measurement*, Charles Merril Books, Columbus, Ohio.

Eringen, A. C. and Suhubi, E. S. (1975) *Elastodynamics*, Academic Press, New York.

Ewing, M., Jardestsky, W. S., and Press, F. (1957) *Elastic Waves in Layered Media*, McGraw-Hill, New York

Graff, K.F. (1975) *Wave Motion in Elastic Solids*, Dover Publications, New York.

Hetenyi, M., Ed. (1950) *Handbook of Experimental Stress Analysis*, John Wiley and Sons, New York.

Hillier, K. W. (1960) A review of the progress in the measurement of dynamic elastic properties, *Int. Symp. on Stress Wave Propagation in Materials*, Ed. N. Davids, Inter- science Publishers, London, pp. 183-98.

Keast, D. N. (1967) *Measurements in Mechanical Dynamics*, McGraw-Hill, New York.
Kinslow, R. (1970) *High-Velocity Impact Phenomena*, Academic Press, New York.
Kolsky, H. (1963) *Stress Waves in Solids*, Dover Publications, New York.
Lamb, H. (1904) On the propagation of tremors over the surface of an elastic solid, *Phil. Trans. R. Soc.* **A203**, 1-42.
Lamb, H. (1917) On waves in an elastic plate, *Proc. Roy. Soc.* **A93**, 114-28.
Magrab, E. B. and Blomquist, D.S. (1971) *The Measurement of Time-Varying Phenomena*, Wiley-Interscience, New York.
McCarthy, M. F. and Hayes, M.A. (1989) *Elastic Wave Propagation*, Elsevier, London.
Miklowitz, J. (1978) *Elastic Waves and Waveguides*, North Holland, New York.
Miklowitz, J. and Achenbach, J. D. (1977) *Modern Problems in Elastic Wave Propagation*, Elsevier, New York.
Morse, P. and Feshbach, H. (1953) *Methods of Theoretical Physics*, Vols. I and II, McGraw-Hill, New York.
Pindera, J. T. (1986) New research perspectives opened by isodyne and strain gradient photoelasticity, in *Proceedings of the International Symposium on Photoelasticity, Tokyo*, pp. 193-202.
Rayleigh, J. W. S. (1887) On waves propagated along the plane surface of an elastic solid, *Proc. Lond. Math. Soc.* **17**, 4-11.
Skalak, R. (1957) Longitudinal impact of a semi-infinite circular elastic bar, *J. Appl. Mech.* **34**, 59-64.
Tolosty, I. (1973) *Wave propagation*, McGraw-Hill, New York.
Viktorov, I.A. (1967) *Rayleigh and Lamb Waves: Physical Theory and Applications*, Plenum Press, New York.
Worely, W. J. (Ed.) (1962) *Experimental Techniques in Shock and Vibration*, ASME, New York.
Zukas, J. A., Nicolas, T., Swift, M. F., Greszczuk, L. B. and Curran, D. R. (1982) *Impact Dynamics*, Elsevier, New York.

11.9. Further Reading

Abbott, B. W. and Cornish, R. H. (1965) A Stress wave technique for determining the tensile strength of brittle materials, *Exp. Mech.* **22**, 148-53.
Bailey, P. and Chen, P. J. (1971) On the local and global behaviour of acceleration waves, *Arch. Ration. Mech. Analysis* **41**, 121. Addendum: Asymptotic Behaviour, *ibid* **44**, 212 (1972).
Baker, W. E. and Dove, R.C. (1962) Measurements of internal strains in a bar subjected to longitudinal impact, *Exp. Mech.* **19**, 307-11.
Barker, L. M. and Hollenbach, R. E. (1964) System for measuring the dynamic properties of materials, *Rev. Sci. Inst. 35(6)*, 742-46.
Barker, L. M. and Hollenbach, R. E. (1965) Interfermometer technique for measuring the dynamic mechanical properties of materials, *Rev. Sci. Inst.* **36**, 1617-20.
Barker, L.M. (1968) The fine structure of compressive and release wave shapes in aluminium measured by the velocity interferometer technique, in *Behaviour of Dense Media Under High Dynamic Pressures*, Gordon and Breach, pp. 483-505.
Barton, C.S., Volterra, E.G. and Citron, S.J. (1958) On elastic impacts of spheres on long rods, Proc. *3rd U.S. Natn. Cong. Appl. Mech.*, pp. 89-94.
Becker, E.C.H. and Carl, H. (1962) Transient-loading technique for mechanical impedence measurement, in *Experimental Techniques in Shock and Vibration*, Ed. W.J. Worley, ASME, New York, pp.1-10.
Chree, C. (1889) The equations of an isotropic elastic solid in polar and cylindrical coordinates, Their solutions and applications, *Trans. Camb. Phil. Soc. Math. Phys. Sci.* **6**, 115-17.
Chu, B. T. (1965) Response of various material media to high velocity loadings. I. Linear elastic and viscoelastic materials, *J. Mech. Phys. Solids* **13**, 165-87.
Dally, J. W. and Lewis, D. (1968) A Photoelastic analysis of propagation of Rayleigh waves past a step

change in elevation, *Bull. Seism. Soc. Am.* **58**, 539-63.

Dally, J. W. and Thau, S. A. (1967) Observations of stress wave propagation in a half-plane with boundary loading, *Int. J. Solids Struct.* **3**, 293-307.

Davies, R. M. (1948) A Critical study of the Hopkinson Pressure Bar, *Phil. Trans. R. Soc.* **A240**, 375-457.

Dohrenwend, C. O., Drucker, D. C. and Moore, P. (1944) Transverse impact transients, *Exp. Stress Analysis* **1**, 1-10.

Doyle, J. F. (1989) *Wave Propagation in Structures*, Springer-Verlag, New York.

Dunwoody, J. (1966) Longitudinal wave propagation in a rate dependent material, *Int. J. of Engineering Sci.* **4**, 277-87.

Evans, J. F., Hadley, C. F., Eisler, J. D. and Silverman, D. (1954) A three-dimensional seismic wave model with both electrical and visual observation of waves, *Geophysics* **19**, 120-36.

Karnes, C.H. (1968) The plate impact configuration for determining mechanical properties of materialsat high strain rates, in *Mechanical Behavior of Materials under Dynamic Loads*, U.S. Lindholm, (Ed.), Springer-Verlag, New York, pp. 270-93.

Kolsky, H. and Prager, W. (1964) *Stress Waves in Anelastic Solids*, Springer-Verlag, New York.

Leipholz, H. (1974) *Theory of Elasticity*, Noordhoff, The Netherlands.

Lindholm, U. S. (Ed.) (1968) *Mechanical Behavior of Materials under Dynamic Loading*, Springer-Verlag, New York.

Malvern, L. E. (1969) *Introduction to the Mechanics of a Continuous Medium*, Prentice-Hall, New Jersey.

Medick, M. A. (1961) On classical plate theory and wave propagation, *J. Appl. Mech.* **28**, 223-8.

Fisher, H. C. (1954) Stress pulse in bar with neck or swell, *Appl. Scient. Res.* **A4**, 317-28.

Frederick, J. R. (1965) *Ultrasonic Engineering*, John Wiley and Sons, New York.

Goldsmith, W., Polivka, M. and Yang, T. (1966) Dynamic behaviour of concrete, *Exp. Mech.* **23**, 65-79.

Goodier, J. N., Jahsman, W. E. and Ripperger, E. A. (1959) An experimental surface-wave method for recording force-time curves in elastic impacts, *J. Appl. Mech.* **26**, 3-7.

Gopasamy, K. and Aggarwala, B. D. (1972) Propagation of disturbances from randomly moving sources, *ZAMM* **52**, 31-35.

Gorsky, W. S. (1936) On the transitions in the Cu Au Alloy III. On the influence of strain on the equilibrium in the ordered lattice of Cu Al, *Phys. Zeit Sowjet* **6**, 77-81.

Green, W. A. (1960) Dispersion relations for elastic waves in bars, in *Progress in Solid Mechanics*, Vol. I, edited by I. N. Sneddon and R. Hill, Chapter 5, North Holland Publishing Co., Amsterdam.

Green, W. A. (1964) The growth of plane discontinuities propagating into a homogeneously deformed elastic material, *Arch. Ration. Mech. Anal.* **16**, 79-89.

Harris, C. M. and Crede, E. (1961) *Shock and Vibration Handbook*, Vols. I, II and III, McGraw-Hill, New York.

Hill, R. (1962) Acceleration waves in solids, *J. Mech. Phys. Solids* **10**, 1-16.

Hsieh, D.Y. and Kolsky, H. (1958) An experimental study of pulse propagation in elastic cylinders, *Proc. Phys. Soc.* **71**, 608-12.

Hudson, G. E. (1943) Dispersion of elastic waves in solid circular cylinders, *Phys. Rev.* **63**, 46-51.

Janke, E. and Emde, F. (1945) *Tables of Functions*, Dover Publications, New York.

Jeffrey, A. (1978) Nonlinear wave propagation, *ZAMM* **58**, T38-T56.

Jeffrey, A. and Taniuti, T. (1964) *Nonlinear Wave Propagation*, Academic Press, New York.

Kolsky, H. (1949) An investigation of the mechanical properties of materials at very high rates of loading, *Proc. Phys. Soc.* **B62**, 676-700.

Kolsky, H. (1954) The Propagation of longitudinal elastic waves along cylindrical bars, *Phil. Mag.* **45**, 712-26.

Kolsky, H. (1960) Experimental wave-propagation in solids, in *Structural Mechanics*, edited by J.N. Goodier and N. Hoff, Pergamon Press, Oxford, pp. 233-62.

Kolsky, H. (1965) Experimental studies in stress wave propagation, Proc. Vth U.S. Natn. Congr. Appl. Mech., pp. 21-36.

Kolsky, H. and Prager, W., Eds. (1964) *Stress waves in anelastic solids*, IUTAM Symposium, Brown University, Providence, R.I., April 3-5, 1963, Springer-Verlag, Berlin.

Langhaar, H.L. (1962) *Energy Methods in Applied Mechanics*, John Wiley and Sons, New York.

Lifshitz, J. M. and Kolsky, H. (1965) The propagation of spherical divergent stress pulses in linear viscoelastic solids, *J. Mech. Phys. Solids* **13**, 361-76.

Lindholm, U. S. (1964) Some experiments with the Split Hopkinson Pressure Bar, *J. Mech. Phys. Solids* **12**, 317-35.

Lindsay, R. B. (1960) *Mechanical Radiation*, McGraw-Hill, New York.

Love, A. E. H. (1944) *A Treatise on the Mathematical Theory of Elasticity*, Dover Publications, New York.

Malvern, L. E. (1951) Plastic wave propagation in a bar of material exhibiting a strain rate effect, *Quart. Appl. Math.* **8**, 405-11.

Mason, W. P. and McSkimin, H. J. (1947) Attenuation and scattering of high frequency sound waves in metals and glasses, *J. Acoust. Soc. Amer.* **19**, 464-73.

Medick, M. A. (1961) On classical plate theory and wave propagation, *J. Appl. Mech.* **28**, 223-28.

Meyer, M. L. (1964) On spherical near fields and far fields in elastic and viscoelastic solids, *J. Mech. Phys. Solids* **12**, 77-111.

Orowan, E. (1934) Zür Kristall Plastizität.III. Über den mechanismus des gleitvorganges, *Zeits f. Phys.* **89**, 634-59.

Polanyl, M. (1934) Über eine Art gitterstörung, die einen kristall plastisch machen Könnte, *Zeits f. Phys.* **89**, 660-64.

Prescott, J. (1942) Elastic waves and vibrations of thin rods, *Phil. Mag.* **33**, 703-54.

Press, F. and Oliver, J. (1955) Model study of air-coupled surface waves, *J. Acoust. Soc. Am.* **27**, 45-46.

Rayleigh, J. W. S. (1894) *Theory of Sound*, Dover Publications Inc., New York.

Reinhardt, H. W. and Dally, J. W. (1970) Some characteristics of Rayleigh wave interaction with surface flaws, *Mater. Eval.* **28**, 213-20.

Riley, W. F. and Dally, J. W. (1966) A photoelastic analysis of stress wave propagation in a layered model, *Geophysics* **31**, 881-9.

Ripperger, E. A. (1953) The propagation of pulses in cylindrical bars. An experimental study, *Proc. 1st Midwest Conf. Solid Mech.*, pp. 29-39.

Rubin, J. R. (1954) Propagation of longitudinal deformation waves in a prestressed rod of material exhibiting a strain-rate effect, *J. Appl. Phys.* **25**, 528-36.

Snoek, J. E. (1941) Effect of small quantities of carbon and nitrogen on the elastic and plastic properties of iron, *Physica* **8**, 711-33.

Sokolnikoff, I. S. (1956) *Mathematical Theory of Elasticity*, 2nd edition, McGraw-Hill, New York.

Stoneley, R. (1924) Elastic waves at the surface of separation of two Solids, *Proc. R. Soc.* **A106**, 416-28.

Tatel, H. E. (1954) Note on the nature of a seismogram II., *J. Geophys. Res.* **59**, 289-94.

Thau, S. A. and Dally, J. W. (1969) Subsurface characteristics of the Rayleigh wave, *Int. J. Eng. Sci.* **7**, 37-52.

Thomas, T. Y. (1957) The growth and decay of sonic discontinuities in ideal gases, *J. Math. Mech.* **6**, 455-69.

Thomas, T. Y. (1961) *Plastic Flow and Fracture in Solids*, Academic Press, New York.

Timoshenko, S. P. (1921) On the correction for shear of the differential equation for transverse vibrations of prismatic bars, *Phil. Mag.* Ser. **6(41)**, 744-46.

Timoshenko, S. P. (1928) *Vibration Problems in Engineering*, Van Nostrand, New Jersey.

Truesdell, C. and Toupin, R. A. (1960) The classical field theories, *Handbuch der Physik* **III/1**, Ed. S.Flügge, Springer, Berlin.

Varley, E. and Cumberbatch, E. (1965) Nonlinear theory of wavefront propagation, *J. Ins. Math. and Appl.* **1**, June 1965, 101-12.

Volterra, E. (1955) A one-dimensional theory of wave propagation in elastic rods based on the 'method of internal constraints', *Ing. Arch.* **23**, 410.

Whitham, G. B. (1974) *Linear and Nonlinear Waves*, J. Wiley & Sons, New York.

Wood, D., (1963) in *Response of Metals to High Velocity Deformation*, P.G. Shewmon and O.F. Zackay, (Eds.), Elsevier, New York.

Zemanek, J. (Jr.) and Rudnick, I. (1961) Attenuation and dispersion of elastic waves in a cylindrical bar, *J. Acoust. Soc. Am.* **33**, 1283-8.

Zener, C. (1948) *Elasticity and Anelasticity of Metals*, Univ. Press, Chicago.

Zukas, J.A. (1982) Stress waves in solids, in *Impact Dynamics*, Eds. J.A. Zukas et al., John Wiley & Sons, New York, Chapter 1, pp. 1-27.

TABLE 11.1. Wave propagation in an isotropic, elastic (unbounded) solid. Pertaining relations in terms of displacement.

Displacement field $\mathbf{u}$,

General governing equation:

$$\text{Navier's}\quad \frac{\rho\, \partial^2 \mathbf{u}}{\partial t^2} = \mu \nabla^2 \mathbf{u} + (\lambda + \mu) \nabla \nabla \cdot \mathbf{u}$$

Two propagating waves:

Dilatational (Irrotational); $(\mathbf{u}_{IR}, c_1)$

Rotational (Distortional); $(\mathbf{u}_R, c_2)$

Necessary and sufficient relations for the satisfaction of *Navier's governing equation* (above):

$$\mathbf{u} = \mathbf{u}_{IR} + \mathbf{u}_R$$

Necessary but not sufficient relations for the satisfaction of *Navier's governing equation*:

$$\left[\nabla^2 - \frac{1}{c_1^2} \frac{1}{\partial t^2}\right] \mathbf{u}_{IR} = 0$$

$$\left[\nabla^2 - \frac{1}{c_2^2 \frac{1}{\partial t^2}}\right] \mathbf{u}_R = 0$$

Dilatation Δ:

$$\Delta = \nabla \cdot \mathbf{u};\left[\nabla^2 - \frac{1}{c_1^2} \frac{\partial^2}{\partial t^2}\right] \Delta = 0$$

Rotation ω:

$$\omega = \nabla x u \quad ; \qquad \left[\nabla^2 - \frac{1}{c_2^{\,2}}\frac{\partial^2}{\partial t^2}\right]\omega = 0$$

Sufficient but not necessary relations for the satisfaction of *Navier's governing equation*:

Dilatational (*Irrotational)*

$$\nabla x u = 0; \qquad \left[\nabla^2 - \frac{1}{c_1^{\,2}}\frac{\partial^2}{\partial t^2}\right]\mathbf{u} = 0$$

Rotational (*Distortional*)

$$\nabla \cdot u = 0; \left[\nabla^2 - \frac{1}{c_2^{\,2}}\frac{\partial^2}{\partial t^2}\right]u = 0$$

where $$c_1^{\,2} = \frac{\lambda + 2\mu}{\rho} = \frac{E(1-\nu)}{\rho(1+\nu)(1-2\nu)} = \frac{K + \frac{4}{3}\mu}{\rho}$$

$$c_2^{\,2} = \mu/\rho$$

TABLE 11.2. Wave propagation in an isotropic, elastic (unbounded) solid. Relationships between Navier's equation and other related governing equations in terms of potential.

General governing equation: *Navier's*

$$\rho \frac{\partial^2 u}{\partial t^2} = \mu \nabla^2 \mathbf{u} + (\lambda + \mu)\nabla\nabla \cdot \mathbf{u}$$

Displacement field $\mathbf{u}(\phi, \psi)$

Necessary and sufficient relations for the satisfaction of *Navier's governing equation* (above):

$$\mathbf{u} = \nabla\phi + \nabla \times \psi;\ \nabla \cdot \psi = 0$$

Necessary but not sufficient relations for the satisfaction of *Navier's governing equation*:

$$\nabla\left[\nabla^2 - \frac{1}{c_1^2}\frac{\partial^2}{\partial t^2}\right]\phi + \nabla \times \left[\nabla^2 - \frac{1}{c_2^2}\frac{\partial^2}{\partial t^2}\right]\psi = 0$$

$$\left[\nabla^2 - \frac{1}{c_1^2}\frac{\partial^2}{\partial t^2}\right]\phi = 0;\ \left[\nabla^2 - \frac{1}{c_2^2}\frac{\partial^2}{\partial t^2}\right]\psi = 0$$

$$\left[\nabla^2 - \frac{1}{c_1^2}\frac{\partial^2}{\partial t^2}\right]\nabla^2\phi = 0$$

$$\left[\nabla^2 - \frac{1}{c_2^2}\frac{\partial^2}{\partial t^2}\right]\nabla^2\psi = 0$$

$$\text{where}\quad c_1^2 = \frac{\lambda + 2\mu}{\rho} = \frac{E(1-\nu)}{\rho(1+\nu)(1-2\nu)} = \frac{K + \frac{4}{3}\mu}{\rho}$$

$$c_2^2 = \mu/\rho$$

CHAPTER 12

DYNAMIC PLASTIC BEHAVIOUR

12.1. Introduction

In dealing with static plastic problems, we emphasize that the duration of the experiment is long enough so that the occurring deformation in the material can be considered to be "*time-independent*".

In a dynamic experiment, however, the time duration is very short, to the extent that a relaxation phenomenon might occur. In this context, the length of the period of the dynamic plastic experiment may be compared with the relaxation time of the body considered (Cristescu, 1967). Due to such viscous flow effect, it is often appropriate to include the so-called "rate-effects" when dealing with dynamic plastic problems. *The magnitude of the strain rate at which a given material commences to be rate sensitive varies from a material to another. For a large group of metals, this limiting rate of strain seems to be* $10^2\ sec^{-1}$.

In the present chapter, we deal with the plastic response of engineering materials under dynamic loading, whereby a rate-effect phenomenon might be occur in the material and, hence, the inertia forces would be included in the equation of motion.

The first research contributions in the field of dynamic plasticity include the work of both J. Hopkinson (1872) and B. Hopkinson (1905). Other important contributions in the field were followed during the period 1930 to 1950 (e.g., Donnell, 1930).

The study of dynamic plasticity is of particular interest in a large number of technical fields, e.g., high velocity forming of metals, ballistics in general, response of soils under dynamic loads, etc. All such applications have significantly contributed to the development of the pertaining theory. In this context, the reader is referred to the books, for instance by, Goldsmith (1960) and Cristescu (1967). Reference is, also, made to Davies (1953, 1956), Kolsky (1953), Cristescu (1960a&b, 1970, and 1972), Hopkins (1960, 1961), Craggs (1961), Olszak *et al.* (1963), and Cristescu and Bell (1970), among others.

12.2. The Dynamic Plasticity Problem

The one-dimensional problem in dynamic plasticity is defined as the one in which, in a strict sense, one component of the stress and of particle velocity, as well as a single spatial coordinate are involved. In such case, a one-dimensional stress-strain relation is used. Thus, as dealt with below, one arrives at a single partial differential equation of the first or second order.

The problem of propagation of elastic-plastic waves in thin rods was the first one-dimensional problem to have received most attention in the realm of dynamic plasticity. In this context, special consideration has been given to the "***unloading***" aspect of the problem. The latter is the most difficult point to deal with as only numerical methods have been traditionally successful in locating the "***loading/unloading boundary***". In this, the reader is referred to Ericksen (1955), Hill (1961), Thomas (1961), and Mandel (1962, 1964), among others. For the study of other categories of one-dimensional problems, e. g., problems that involve spherical symmetry, but, one component of particle velocity, reference is made to Hunter (1957), Cristescu (1960 a&b), Goldsmith (1960, 1963), Hopkins (1960), Chadwick (1962), Olszak and Perzyna (1962), Perzyna (1962), Szczepinski (1964) and Wierzbicki (1963), among others.

In order to make the transition to more general multi-dimensional problems, one may consider:

- First, those problems in which several components of the stress and of the velocity are involved, but in which there is a degree of symmetry so that a single spatial coordinate is sufficient to describe the motion. Here, one uses a time-independent constitutive equation that is expressed in a finite form. For such class of problems, the number of equations involved is, in general, manageable, and the problem can be solved without too much difficulty.

- On the other hand, if the viscosity of the material cannot be neglected, the constitutive equations become time-dependent and are often expressed in a differential form. Thus, the pertaining dynamic plastic problem becomes more involved.

In the two cases mentioned above, several plastic waves, travelling with variable velocities, may be involved. These waves could be "***coupled***" or "***partially coupled***". Such coupling effect between several waves, propagating with variable velocities, is the fundamental property that distinguishes plastic waves from ordinary elastic waves. *In general, there are as many plastic waves as components of the particle velocity in the problem considered.*

From the point of view of wave coupling, the constitutive equations may be classified as "*partially coupled*" and "*coupled*" constitutive response equations, *with time-dependency or independency*. In a time-independent problem, both the constitutive equation and the loading/unloading condition are considered to be time-invariant. Difficult mathematical problems are often involved particularly in the case of time-dependent coupled equations, in connection with the numerical methods of integration and with establishing of wave propagation characteristics (e.g., Cristescu, 1967).

In order to simplify the difficulties involved in treating generalized problems in dynamic plasticity, one approach is to consider first those problems which would require a one-dimensional constitutive relation. An example of such class of problems is that concerned with the propagation of waves in extensible strings. In this case, although one constitutive equation is required, there are two kinds of waves, i.e., "*longitudinal*" and "*transverse*", which influence one another during propagation.

12.2.1. THE ONE-DIMENSIONAL, TIME-INDEPENDENT PROBLEM

As introduced earlier, the one-dimensional problem is characterized by a single spatial coordinate. Hence, only single components of stress and strain are considered. Thus, the propagation of longitudinal stress waves in thin rods or wires is the only possible one-dimensional situation. In this case, the influence of the shape of the transverse section of the rod on the propagation of the wave is disregarded, although, the area of this transverse section is taken into account. The rod, in the one-dimensional problem, is considered to be slender, so that the lateral inertia would be neglected. This translates into the assumption that the particles can move freely in the directions transverse to the generatrices of the rod. The coordinate axis will be chosen with the origin O being located at the end of the rod, and the positive direction of the OX-axis is considered to be directed along the rod, Fig. 12.1.

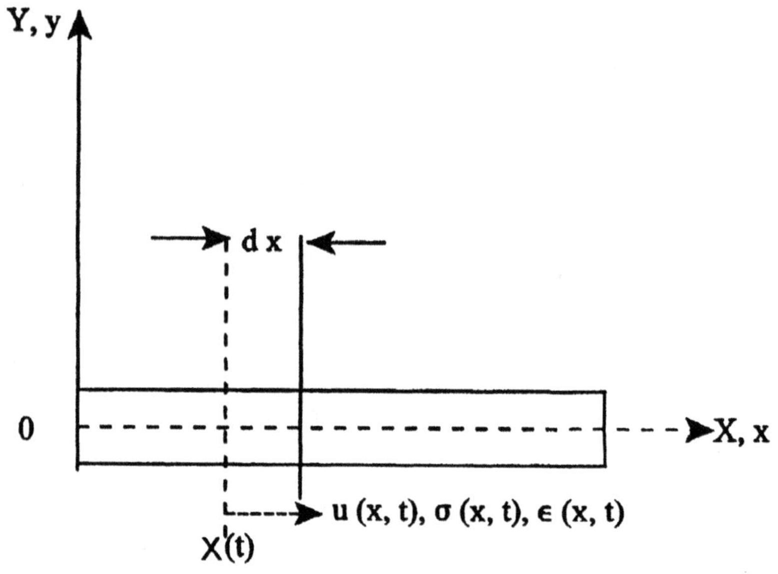

Figure 12.1. The one-dimensional problem: A single spatial coordinate is involved, the influence of the shape of the transverse section of the rod on the propagation of the wave is disregarded (but, the area of this transverse section enters into the calculations), and lateral inertia is neglected.

The loading Problem

Loading. It is defined by the condition that the stress at the end of the rod either increases continuously, or after increasing up to a certain maximum value, remains constant thereafter.

Following Cristescu (1967), we assume that for $t < 0$ the rod is at rest, while for $t=0$ the end of the rod is impacted by a rigid body so that for $t > 0$ the particles of the rod are no longer at rest. We shall assume that the impact occurs in a very short interval of time, so that buckling of the rod would not occur, or at least be negligible. For simplicity, we assume that the cross section of the rod, which is plane before the impact, remains plane also after the impact. This translates into the requirement that all the particles in a given cross section of the rod will displace parallel to the axis of the rod with equal amounts.

With reference to Fig.12.1, we consider an elemental segment of the rod bound, for $t=0$, by the cross-sectional planes at X and $X + dX$. At time t, these planes will have the coordinates $x(t)$ and $x(t)+dx(t)$, *or simply* x *and* $x+dx$, *respectively,* where x is the Lagrangian coordinate (see *Chapter 3*). Considering at $t=0$, the cross sectional area of the bar is A_o and the density is ρ_o, whilst the corresponding quantities at time t are A and ρ, respectively. Thus:

- the conservation of mass equation is

$$\rho A = \rho_o A_o (1 + \epsilon) \tag{12.1}$$

where ϵ denotes the involved measure of strain.

- the equation of motion, for the element dx, is

$$\rho A \frac{\partial^2 u}{\partial t^2} = \frac{\partial F}{\partial x} \tag{12.2a}$$

where $F(x, t)$ is the force acting on the cross-section of the rod of a coordinate x at time t. Equation (12.2a) can be written, alternatively, as

$$\rho A \frac{\partial^2 u}{\partial t^2} = \frac{\partial (A \sigma)}{\partial x} \tag{12.2b}$$

where u is the displacement and σ is the stress on the initial cross-sectional area of the rod. We shall adopt the usual convention for the sign of the stress, i.e., positive in tension and negative in compression. For simplicity, only positive stresses will be considered. Under the additional assumption that A is constant along the rod, i.e., $\partial A/\partial x=0$, the equation of motion

(12.2) can be written as

$$\frac{\partial^2 u}{\partial t^2} = \frac{1}{\rho}\frac{\partial \sigma}{\partial x} \tag{12.3}$$

- the constitutive equation:

In order to proceed with the solution of the dynamic plasticity problem, one must include the constitutive equation for the material of the rod, at the considered experimental conditions, and combine it with the equation of motion (12.3).

Here, we consider that the constitutive equation of the material, during a dynamic experiment, can be written in the following finite form

$$\sigma = f(\epsilon) \tag{12.4}$$

where the function $f(\epsilon)$ is usually a monotonously increasing function of the strain ϵ. It is further assumed that $d\sigma/d\epsilon$ is a monotonously decreasing function of ϵ, e. g., a work-hardening material, Fig. 12.2. During the entire loading process, it is assumed that the same constitutive equation (12.4) applies to every cross-section of the rod.

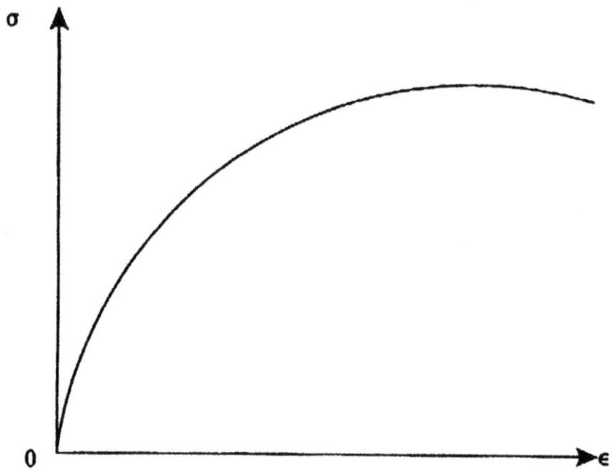

Figure 12.2. One-dimensional stress-strain curve for a work-hardening material (a typical response of majority of metals): $\sigma = f(\epsilon)$ with $f(\epsilon)$ is a monotonously increasing function of ϵ, $d\sigma/d\epsilon$ is a monotonously decreasing function of ϵ and $\sigma\, d^2\sigma/d\epsilon^2 < 0$ for any ϵ.

Combining the constitutive equation (12.4) with the equation of motion (12.3), it follows that

$$\frac{\partial^2 u}{\partial t^2} = \frac{1}{\rho}\frac{d\sigma}{d\epsilon}\frac{\partial^2 u}{\partial x^2} \tag{12.5}$$

Equation (12.5) is the equation of motion of the rod whose response behaviour is described by (12.4). The equation of motion (12.5) is a "*quasi-linear equation of the second order*". This equation may be, also, expressed in the following format of a "*wave equation of first order*".

$$\begin{aligned} \frac{\partial v}{\partial t} &= c^2(\epsilon)\frac{\partial \epsilon}{\partial x} \\ \frac{\partial v}{\partial x} &= \frac{\partial \epsilon}{\partial t} \end{aligned} \tag{12.6}$$

where $v = \dfrac{\partial u}{\partial t}$ is the "***particle velocity***" and

$$\boxed{c(\epsilon) = \sqrt{\frac{1}{\rho}\frac{d\sigma}{d\epsilon}}} \tag{12.7}$$

is the "***velocity of wave propagation***", which is **strain-dependent** for the case of plastic wave propagation, and is governed by the slope of the stress-strain curve of the material. For all kinds of constitutive equations of finite form (12.4) used in practice, $c(\epsilon) \geq 0$, Cristescu (1967). Both the equation of motion (12.5) and the wave equation (12.6) are "*quasi-linear*" equations, i.e., they are linear with respect to the derivatives of the highest order, but their coefficients depend on the involved functions and their first derivatives.

Characteristics of the Equation of Motion. In order to establish the wave equation (12.6), it is necessary to determine the so called "***characteristics of the system***". *The latter are represented by curves in the xOt plane, at the intersection of which* v *and* ϵ *are continuous functions of the arguments* x *and* t, *but, posses discontinuous derivatives.*

In addition to the system of equations (12.6), we consider the following relations (Cristescu, 1967)

$$\frac{\partial v}{\partial s} = \frac{\partial v}{\partial x}\frac{dx}{ds} + \frac{\partial v}{\partial t}\frac{dt}{ds}$$
$$\frac{\partial \epsilon}{\partial s} = \frac{\partial \epsilon}{\partial x}\frac{dx}{ds} + \frac{\partial \epsilon}{\partial t}\frac{dt}{ds} \tag{12.8}$$

where the derivatives dx/ds and dt/ds are computed along one of the "***characteristic directions***", so that $\partial v/\partial s$ and $\partial\epsilon/\partial s$ are in effect "***directional derivatives***" in a characteristic direction. For the purpose of brevity, we denote in (12.8) $\partial v / \partial s$ by dv, etc. Then, this system of equations can be written as

$$d v = \frac{\partial v}{\partial x} dx + \frac{\partial v}{\partial t} dt$$
$$\partial \epsilon = \frac{\partial \epsilon}{\partial x} dx + \frac{\partial \epsilon}{\partial t} dt \tag{12.9}$$

Combining (12.6) and (12.9), one obtains

$$\frac{\partial v}{\partial t} = \frac{c^2(d v d t - d\epsilon d x}{-dx^2 + c^2 dt^2}, \qquad \frac{\partial v}{\partial x} = -\frac{c^2 d\epsilon dt - dv dx}{-dx^2 + c^2 dt^2}$$
$$\frac{\partial \epsilon}{\partial x} = \frac{d\epsilon dx - dv dt}{-dx^2 + c^2 dt^2}$$

which gives the definition of the "***characteristics of the equation of motion***" as

$$\boxed{\frac{dx}{dt} = \pm c(\epsilon)} \tag{12.10}$$

The differential relations satisfied along these lines are referred to as the "***consistency conditions***"; *see* Cristescu (1967). They are written as

$$d v = \pm c(\epsilon)\, d\epsilon \tag{12.11a}$$

or

$$\frac{\partial v}{\partial s_1} = + c(\epsilon)\frac{\partial \epsilon}{\partial s_1}\ , \qquad \frac{\partial v}{\partial s_2} = - c(\epsilon)\frac{\partial \epsilon}{\partial s_2} \tag{12.11b}$$

along the characteristics s_2 and s_1, respectively.

The following points may be concluded concerning the analysis presented above, pertaining to longitudinal wave propagation in a thin rod of a finite constitutive relation of the type (12.4), and with common boundary and initial conditions:

- the wave equation (12.5), or the system (12.6), possesses two distinct, real families of characteristic lines defined by (12.10).
- since (12.5), or (12.6), is quasi-linear, the slopes of the characteristics (12.10) are variable and depend on the strain function. Accordingly, the characteristics are usually two families of curves, which can be determined by the solution of the one-dimensional problem.
- the integration of the wave equation (12.5), or the system (12.6), is equivalent to the integration of the differential equations (12.11) along the characteristic lines (12.10).

Following the presentation above, one can introduce the following definitions:

A wave. A wave may be defined as the solution of the quasi-linear equation of motion (12.5), or, alternatively, the wave equation (12.6), determined within a certain range of variation of the variables x and t, and possessing continuous first and second order derivatives within this range.

A wave front. It is the geometrical locus of the points which separate two waves and moves along the rod with time. Across a wave front, the velocity v and the strain ϵ are continuous, but their derivatives are discontinuous. *Thus, wave fronts coincide with the characteristics of the equation of the motion, although, sometimes, some of the characteristics may not have such mechanical interpretation.*

An "acceleration wave". If the first derivatives of ϵ and v are discontinuous across the wave front, the corresponding wave is referred to as an ***"acceleration wave"***, ***"continuous wave"***, ***"smooth wave"*** or ***"weak wave"***. In this book, we shall adopt the term ***"acceleration wave"***. The fronts of these waves are travelling discontinuity surfaces for first order derivatives of stress and for second order derivatives of displacement.

12.3. Dependence of the Wave Equation and its Characteristics on the Response Behaviour of the Material

With reference to Eqn. (12.7), the variation of the velocity of wave propagation c, as a function of the strain ϵ is governed by the slope of the stress-strain curve, i.e., $d\sigma/d\epsilon$. In other words, the variation of c as a function of ϵ is dependent on the constitutive response of the material under consideration.

(A) Linear Elastic Material

If the impact at the end of the rod is not sufficiently strong, the stress may not yet reach the yield point. In this case, the waves generated at the end of the rod are pure elastic ones. Thus, the stress-strain relation of the rod, in this case, would be still governed by Hooke's law (Fig. 12.3a)

$$\sigma = E\epsilon \tag{12.12}$$

where E is Young's modulus. Recalling (12.7), the velocity of wave propagation is now constant, i.e.,

$$\boxed{c_0 = \sqrt{E/\rho}} \tag{12.13}$$

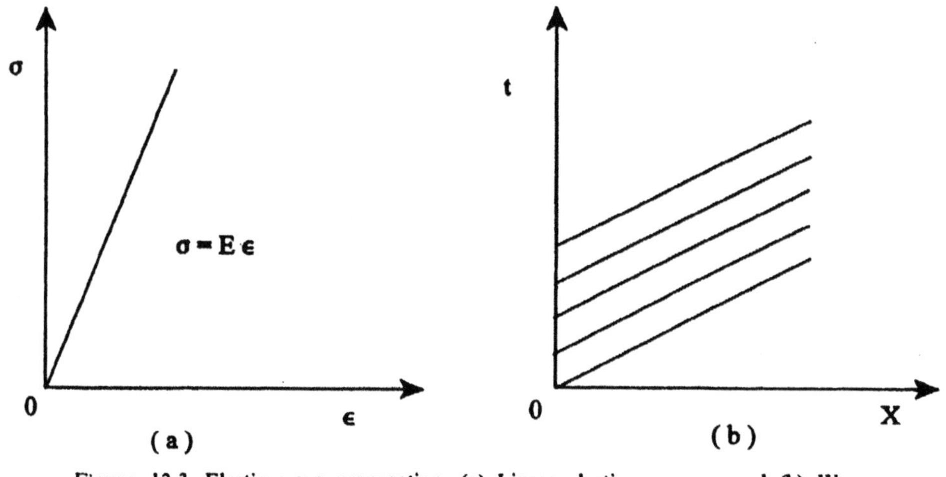

Figure 12.3. Elastic wave propagation. (a) Linear elastic response, and (b) Wave propagation characteristics for a linear elastic material: The characteristic field is composed of parallel straight lines, whereby the velocity of elastic wave propagation is given by $c_o = \sqrt{E/\rho}$. This translates into the fact that the distance between wave fronts, in a linear elastic material, is constant. (Adapted after Cristescu, 1967).

Accordingly, the wave propagation characteristics, in a linear elastic material, take, in view of (12.10), the form of parallel straight lines, Fig. 12.3b. This translates into the fact that the distance between wave fronts, in this material, is constant.

(B) Perfectly Plastic Material

In this case, the stress-strain curve is characterized by $d\sigma/d\epsilon = 0$, Fig.12.4. Thus, in view of (12.7), $c(\epsilon) = 0$, and "*there is no wave propagation in the perfectly plastic rod*". This also happens if certain portions of the stress-strain diagram are parallel to the strain axis. Thus, in these intervals, the wave considered can no longer propagate.

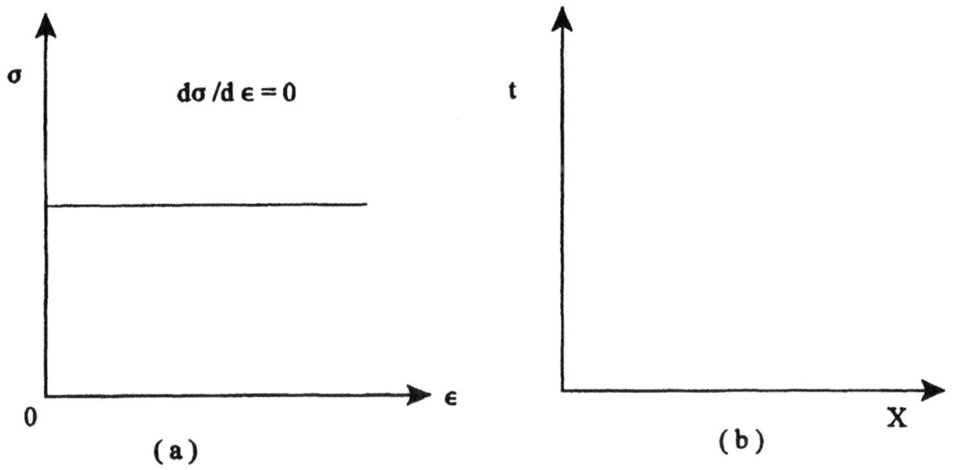

Figure 12.4. Perfectly plastic material: No wave propagation and, thus, no field of characteristics exist in such material; i.e., $c(\epsilon) = 0$.

(C) Work-hardening Material

For most metals, for instance, the stress-strain curve takes the form shown in Fig. 12.2, where $\sigma\, d^2\sigma/d\epsilon^2 < 0$ for any ϵ. Substituting this condition of behaviour into expression (12.7), it can be easily shown that the wave propagation velocity $c(\epsilon)$ decreases when the stress increases, i. e., $dc/d\epsilon < 0$ for any $d\epsilon > 0$. Thus, if one assumes that due to the impact, the stress at the end of the rod, of a work-hardening material, increases continuously, then, the waves generated successively at the end of the rod will propagate with continuously deceasing velocities.

Furthermore, for such materials, the corresponding wave fronts will be represented in a characteristic plane *xOt* by a divergent family of curves, whose slopes dt/dx will increase with the decrease of the wave propagation velocity, in view of (12.10), i. e., with the increase of the stress.. This illustrated in Fig. 12.5.

Thus, the distance between the wave fronts, for a work-hardening material, will increase during their propagation, i. e., the waves will spread.

(D) Materials Governed by Other Stress-Strain Relations

In this category, we consider the following classes of materials

(i) Materials whose stress-strain curve takes a form similar to that illustrated in Fig. 12.6 a, i. e., for which the slope increases continuously, $\sigma\, d^2\sigma/d\epsilon^2 > 0$ for any ϵ.

Such condition may be representative of the response behaviour of particular classes of rubbers, soils, and metals. For such classes of materials, since the slope of the stress-strain curve increases continuously, the velocity of propagation will increase when the stress increases, i.e., $dc/d\epsilon > 0$ for any $d\epsilon > 0$, Eqn. (12.7). Meantime, near the end of the rod, the distance between the wave fronts decreases during propagation, as the slope dt/dx of the representative characteristics decreases, Eqn. (12.10). This is illustrated in Fig. 12.6b. In this case, there is a tendency for wave fronts to be formed.

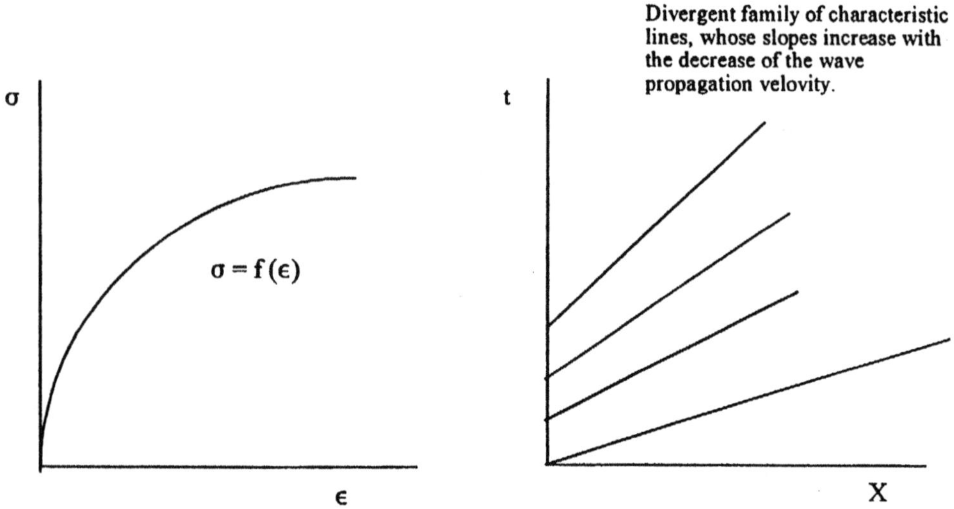

Figure 12.5. Wave propagation in a work-hardening material (typical response of majority of metals). (a) One-dimensional stress-strain curve; $\sigma=f(\epsilon)$ with $f(\epsilon)$ is a monotonously increasing function of ϵ, $d\sigma/d\epsilon$ is a monotonously decreasing function of ϵ and $\sigma\, d^2\sigma/d\epsilon^2 < 0$ for any ϵ. (b) Field of characteristics: Divergent family of characteristic lines whose slopes dt/dx increase with the decrease of the wave propagation velocity.
(Adapted after Cristescu, 1967).

(ii) Materials whose stress-strain diagrams take a form similar to that shown in Fig. 12.7a, whereby the curvature of the stress-strain curve changes at a certain point.

In this case, the wave fronts will first diverge ($dc/d\epsilon < 0$), and then converge ($dc/d\epsilon > 0$). This is illustrated on the plane of characteristics shown in Fig. 12.7b.

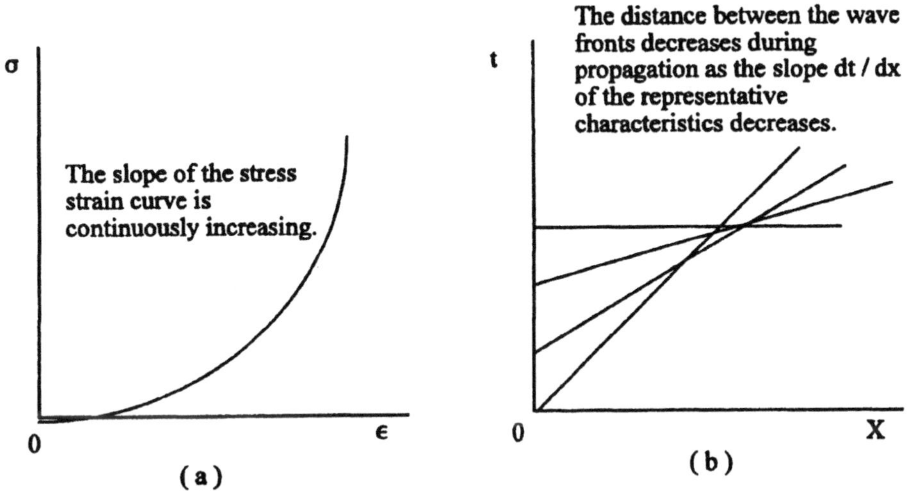

Figure 12.6. (a) A representative of stress-strain curve of soft material (e.g., various classes of rubbers, soils and metals): $\sigma\ d^2\sigma/d\epsilon^2 > 0$ for any ϵ. (b) Field of characteristics corresponding to (a): Divergent family of characeristic lines; $dc/d\epsilon > 0$ for any $d\epsilon > 0$. (Adapted after Cristescu, 1967).

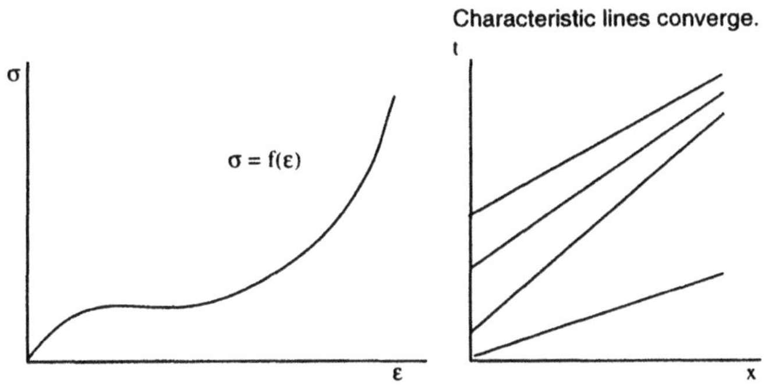

Figure 12.7. (a) A stress-strain curve of variable concavity. (b) Characteristic field corresponding to (a): A convergent family of characteristic lines, with the possibility of forming shock waves. (Adapted after Cristescu, 1967).

Summary

(i) Assuming that the constitutive equation of the material can be written in a finite form (12.4), then, the motion is governed by the second-order equation of motion (12.5), or by the first-order wave equation (12.6). As mentioned earlier, both are "*quasi-linear*" equations, i.e., they are linear with respect to the derivatives of the highest order, but their coefficients depend on the involved functions and the first derivatives of these functions.

(ii) The equations, referred-to under (i) above, are of the "*hyperbolic type*", i. e., either equation will result, for each value of the wave propagation velocity $c(\epsilon)$ in two distinct characteristic lines in the characteristic plane *xOt*.

(iii) Since the equations are quasi-linear, the characteristics will generally be curves with variable slopes. These slopes can be determined by the solution of either of the two equations (12.10). Thus, these curves cannot be drawn *a priori*, but only when the solution of the problem is known. Such solution will depend on the initial and boundary conditions, together with the explicit expression of the constitutive expression (12.4); *see* Cristescu (1967).

EXAMPLE 12.1

We consider in this example an initially undeformed, semi-infinite bar at rest. This case is presented by Cristescu (1967) after Rakhmatulin (1945a&b).

Initial conditions:

$$t = 0 \text{ and } x > 0: \qquad (x, 0) = v(x, 0) = \quad (x, 0) = 0 \qquad \textbf{(12.14)}$$

Boundary conditions:

$$x = 0 \text{ and } t \geq 0: \quad \epsilon(0, t) \text{ or } \sigma(0, t) \text{ or } v(0, t) \text{ are prescribed} \qquad \textbf{(12.15)}$$

The initial conditions (12.14) are satisfied in the region D of Fig. 12.7, while the characteristic line OA represents the first wave front propagated along the rod. A solution can be obtained by integrating the relations (12.11) along the corresponding characteristic lines. These relations then become

$$\begin{aligned} v &= \int c(\epsilon)\, d\epsilon + k_1 = \psi(\epsilon) + k_1(s_2) \\ v &= -\int c(\epsilon)\, d\epsilon + k_2 = -\psi(\epsilon) + k_2(s_1) \end{aligned} \qquad \textbf{(12.16)}$$

where the parameters $k_1(s_2)$ and $k_2(s_1)$ have different constant values on different characteristic lines and are called *"Reiman invariants"*.

With reference to (12.16), all the characteristics of negative slope intersect the characteristic line OA and, hence, commence from the undisturbed region D_1. It follows then that all the constants $k_2 = 0$, and throughout the region D_2, the relation between the velocity and the strain is

$$v = -\Psi(\epsilon) \quad \textbf{(12.17)}$$

Substituting (12.17) into the first relation of (12.16), one concludes that both v and ϵ are constant along the characteristics of positive slope. Accordingly one arrives at the following important conclusion:

When some constant initial state is prescribed, the characteristics of positive slope are a family of straight lines. The equations of these straight lines can be expressed in the form

$$x = c(\epsilon(t^*))(t - t^*) \quad \textbf{(12.18)}$$

where t^* is the time at which the straight line intersects the time-axis, as shown in Fig. 12.7. The slope of this straight line is computed for these values of t^*, i.e., using the boundary conditions. The propagating waves, corresponding to this situation, are called ***"simple waves"***.

It should be emphasized, however, that the above conclusion is valid only if a constant initial state is prescribed, otherwise, the characteristics of positive slope will be, in general, curved lines in the characteristic plane xOt. *Thus, simple waves may appear only in a region adjacent to a constant state region.*

One may further describe the strain at the end of the rod in terms of the time parameter t^*, e. g.,

$$\epsilon = \epsilon(0, t^*) \quad \textbf{(12.19)}$$

then, by eliminating the time parameter t^* between (12.18) and (12.19), a functional relation which defines ϵ in terms of x and t is obtained (Cristescu, 1967).

12.4. The Problem of Instantaneous Impact

Karman and Duwez (1950) considered the special case where the strain was assumed to be a function of the ratio x/t, but not of x and t independently. Thus, from equation (12.17), it follows that the velocity is also a function only of the same ratio. Accordingly, the equation of the characteristic lines (12.18) now becomes

$$x = c(\epsilon)\, t \tag{12.20}$$

The characteristics of positive slope, that correspond to (12.20) pass through the origin. This is the case of ***"instantaneous loading"***, and the corresponding simple waves are referred to as ***"centred simple waves"***.

Karman and Duwez (1950) considered the following boundary conditions:

$$u = v_1 t \quad \text{for} \quad x = 0 \quad \text{and} \quad t \geq 0 \tag{12.21}$$

where v_1 is the constant velocity of impact. These authors sought then different particular solutions which would depend only on the ratio *x/t* and satisfy the equation of motion (12.5), the initial conditions (12.14), and the boundary conditions (12.15). They arrived at the following particular solution

$$u = v_1 \left(t - \frac{x}{c_1} \right) \tag{12.22}$$

where c_1 is an undetermined constant with the dimension of velocity. It is easy to recognize that the strain corresponding to (12.22) is constant, i.e.,

$$\epsilon_1 = - v_1 / c_1 \tag{12.23}$$

In order to find a particular solution to the above problem, in which the strain is a function of the ratio x/t only, i.e.,

$$\epsilon = f(x/t) = f(\xi) \tag{12.24}$$

the function $f(\xi)$ is determined by the condition that the equation of motion (12.5) must be satisfied. Following Cristescu (1967), the displacement is obtained from

$$u = \int_{\infty}^{\xi} \frac{du}{d\xi}\, d\xi$$

Further, using the relation

$$\frac{\partial u}{\partial x} = \frac{du}{d\xi}\frac{\partial \xi}{\partial x} = \frac{1}{t}\frac{du}{d\xi}$$

one obtains

$$u = t\int_{\infty}^{\xi} f(\chi)\, d\chi \tag{12.25}$$

Using the relations

$$\frac{\partial^2 u}{\partial x^2} = f'(\xi)\,\frac{1}{t}\,, \qquad \frac{\partial^2 u}{\partial t^2} = f'(\xi)\,\frac{\xi^2}{t}$$

together with the equation of motion (12.5), one arrives at

$$f'(\xi)\,[\,c^2(\epsilon) - \xi^2\,] = 0 \tag{12.26}$$

Accordingly, under the assumption (12.24), one arrives at the following two particular solutions:

(i) the particular solution (12.22) which results from $f'(\xi) = 0$, and corresponds to a constant strain ϵ_1 and to a constant velocity of impact v_1.

(ii) the particular solution that is obtained from

$$c^2(\epsilon) = x^2/t^2 \tag{12.27}$$

Based on the constitutive equation (12.4), whereby (12.24) is also applicable, the full solution of the problem of instantaneous impact was obtained by Karman and Duwez (1950), *see* Cristescu (1967), as a combination of the two particular solution mentioned above:

(a) for $x > c_0 t : \epsilon = 0$

(b) for $c_1 t < x \le c_0 t$, the relation (12.26) is satisfied, where c_1 is the velocity of propagation of the plastic wave which carries the maximum strain ϵ_1

(c) for $0 < x \le c_1 t$, the strain is constant and is equal to ϵ_1.

Thus, in the problem of instantaneous impact on a rod, of material specified by the constitutive condition (12.4), Figure 12.2, the first elastic wave front propagates with a constant velocity c_0. *The last plastic wave front propagates with a constant velocity* $c_l = c(\epsilon_l)$. *Between these two waves, there is a set of other plastic waves, whose wave fronts are represented as an array of straight lines emitted from the origin of the plane of characteristics. Each of these waves propagates with a certain velocity* $c(\epsilon)$, *where* $c_l < c(\epsilon) < c_0$. *Accordingly, the sudden impact, i.e., the sudden increase of strain at the end of the rod, is transmitted along the rod in the form of "centred simple waves", which bring about a smooth variation of the strain at any other cross-section of the rod. This conclusion is valid for thin rods made of materials whose stress-strain curve is of a continuously decreasing slope.*

In order to determine the velocity c_l and the strain ϵ_l as functions of the velocity of impact v_l, one combines equations (12.22) and (12.25) for the end of the rod to arrive at

$$v_l = u(o,t)/t = -\int_{\infty}^{0} f(\xi)\, d\xi \tag{12.28}$$

The integral of (12.28) represents an area which may be also calculated using the expression

$$v_l = -\int_0^{\epsilon_l} \xi\, d\epsilon = -\int_0^{\epsilon_l} c(\epsilon)\, d\epsilon \tag{12.29}$$

Expression (12.29) establishes a correspondence between the maximum strain ϵ_l and the velocity of impact v_l. Such relation is of significance in mechanical design applications that involve impact loading.

If the stress-strain curve possesses a linear elastic range, then in this domain, all corresponding wave fronts generated by a sudden impact, will run together. In this situation, the first wave front will produce a sudden jump of the strain from $\epsilon=0$ to $\epsilon=\epsilon_y$.

If the impact at the end of the rod is of a small intensity and the stress is within the elastic range, i.e., $\sigma < \sigma_y$, then the velocity of the wave propagation is constant, c_o, which is given by (12.13). Meantime, one can write that

$$v_o = c_o\, \epsilon \tag{12.30}$$

and, by recalling Hooke's law, it follows that

$$\sigma = E\epsilon = \rho\, v\, c_o \tag{12.31}$$

Thus, if the loading of the rod is instantaneous, there will be two regions:

(i) for $x > c_0 t$, $\sigma = \epsilon = v = 0$,
(ii) for $0 \le x \le c_0$ the stress ($\sigma = \rho\, v_1\, c_0$) and the strain ($\epsilon_1 = v_1 / c_0$) are constant.

Thus, in this case, there is a sudden increase of stress, strain and velocity when the wave, travelling with velocity c_0, reaches the corresponding section of the bar.

12.4.1. THE UNLOADING PROBLEM

When the stress at the end of the bar begins to decrease after having increased, unloading starts. During this phase, the formulation of the theory of wave propagation is different from that during loading for the material specimen under consideration. In many situations, after having increased, the stress at the end of the rod decreases to zero. Further, a succession of loading-unloading processes might occur quite often.

The Unloading Constitutive Equation
For a large group of elastic-plastic materials, especially metals, the unloading process is a perfectly elastic one. Thus, it is appropriate to use the following constitutive relation during unloading

$$\sigma = \sigma_m (x) + E\left[\epsilon - \epsilon_m (x)\right] \qquad \textbf{(12.32)}$$

where, for each cross-section x of the rod, σ_1 (x) and ϵ_1 (x) are the maximum stress and strain, respectively. It should be noted that for finite constitutive equations, such as (12.4), both the maximum stress and the maximum strain are reached at the same moment in a given cross-section of the rod.

For an elastic-plastic material, however, the unloading problem is more difficult to deal with than the loading problem. This may be reasoned as follows:

- In each cross-section of the rod, the unloading process commences at a different maximum stress σ_m (x) and a different maximum strain ϵ_m (x). In other words, for each section of the rod, a different constitutive equation (12.4) must be employed.

- Further, at the transition between loading and unloading, one must replace the loading constitutive equation (12.4) by the unloading constitutive equation (12.32).

The Loading/Unloading Boundary. The loading/unloading boundary is defined as the geometrical locus of points in the characteristic plane *xOt*, in which the maximum strain has been reached in each cross-section of the rod. Thus, the loading process occurs below, and the unloading process occurs above this curve (locus). The shape of the loading/unloading boundary depends on the constitutive response of the material, the boundary and the initial conditions. Consider the case when the loading / unloading boundary is represented by a general line below which, is the domain where plastic strain increases, whilst, in the domain above the loading/unloading boundary, the plastic part of strain is constant. This gives for loading/unloading conditions, respectively, the following definitions:

$$\frac{\partial \sigma}{\partial t} > 0 \qquad \text{for loading, i.e.,} \qquad \frac{\partial \epsilon^p}{\partial t} > 0, \tag{12.33a}$$

$$\frac{\partial \sigma}{\partial t} \leq 0 \qquad \text{for unloading, i.e.,} \qquad \frac{\partial \epsilon^p}{\partial t} = 0 \tag{12.33b}$$

Depending on the boundary conditions, the loading and unloading boundary may be sometimes formed as an area. In other words, this boundary is not always a curve in the characteristic plane "*xOt*".

Assuming that the equation for the loading/unloading boundary, in the characteristic plane "*xOt*", can be written as

$$t = f(x) \tag{12.34}$$

(*which is not known a priori*), one may visualize the solution of the problem in a certain cross-section x of the rod as follows: When a time $t = x/c_o$ has elapsed since the beginning of the impact at the end of the rod, the first elastic wave reaches the mentioned section. From this moment onwards, the strain increases continuously until $t = f(x)$, when the first unloading wave reaches this section. Then, the elastic strain decreases to zero. Thus, for a given material, the strain increases and then decreases depending on the boundary conditions, and also on the maximum strain previously reached.

The Equation of Motion in the Unloading Domain

Combining equations (12.3) and (12.32), the equation of motion during unloading can be written as

$$\frac{\partial^2 u}{\partial t^2} = c_o^2 \frac{\partial^2 u}{\partial x^2} + \frac{1}{\rho} \frac{d\sigma_m(x)}{dx} - c_o^2 \frac{d\epsilon_m(x)}{dx} \tag{12.35}$$

where c_o is the constant velocity of propagation, Eqn. (12.13). In (12.35), σ_m and ϵ_m depend only on the coordinate x. Thus, it is necessary to determine the solution in the unloading domain simultaneously with the solution in the loading domain, in order to find $\sigma_m(x)$ and $\epsilon_m(x)$. Meantime, the shape of the loading/unloading boundary can be found only by finding a solution to the problem which satisfies the initial and boundary conditions in the loading and unloading domains simultaneously.

The general solution of (12.35) is

$$u = F_1(c_o t + x) + F_2(c_o t - x) - \frac{1}{E}\int_0^x (\sigma_m - E\epsilon_m)\,dx \tag{12.36}$$

where F_1 and F_2 are arbitrary functions to be determined by the boundary conditions at the end of the rod and the loading/unloading boundary.

Meantime, the characteristics of (12.35) are

$$dx/dt = \pm c_o \tag{12.37}$$

with the following differential equations being satisfied along these characteristics

$$dv = \pm \frac{1}{\rho c_o}\,d\sigma \tag{12.38}$$

This is with the understanding that the upper and lower signs in (12.37) and (12.38) correspond to each other.

The shape of the loading/unloading boundary can be determined by graphical-analytical methods or, alternatively, by numerical methods. However, in the particular case of a linear work-hardening material under sudden loading, the shape of the loading/unloading boundary may be known *a priori (Cristescu, 1967)*.

The loading/Unloading Boundary. Linear Hardening Material under Sudden Loading. In this particular case, there are only two wave fronts, namely, $x = c_o t$ and $x_1 = c_1 t$. The first is the elastic wave front across which the strain jumps from zero to ϵ_y. The second front is the plastic wave front. The latter must pass through each point of the loading/unloading boundary. This boundary must coincide with the straight line $x_1 = c_1 t$. The referred-to line is the plastic wave front which is characterized by a constant strain. On the other hand, it is a loading/unloading boundary along which the strain decreases. Accordingly, this line has contradictory properties. In fact, this line is an idealization of a bundle of parallel straight lines representing plastic wave fronts, the distance between them being very small (Cristescu,

1967).

12.5. Determination of the Loading/Unloading Boundary

As we mentioned earlier, there is no exact analytical method for determining the loading/unloading boundary. To overcome this problem, two alternative methods have been introduced in the literature. These are the graphical-analytical methods and the numerical methods.

12.5.1. THE GRAPHICAL-ANALYTICAL METHOD.

This method is presented in detail by Cristescu (1967). It can be applied to sufficiently long rods (no reflections), initially at rest and undeformed. The rod must also have homogeneous mechanical properties. In addition, the method is only applicable to the case of single loading.

Within the context of the graphical-analytical method, in order to determine the distribution of the plastic strain after impact, one must first establishes the variation of the stress $\sigma_m(x)$ along the loading/unloading boundary. Meantime, the plastic strain may be obtained using the formula

$$\epsilon^p(x) = \left(\sigma_m(x) - \sigma_y\right)\left(\frac{1}{E_1} - \frac{1}{E}\right) \tag{12.39}$$

In this case, the maximum strain is obtained at the impacted end of the rod.

12.5.2. THE NUMERICAL METHOD

This method is based on numerical integration along a network of characteristic lines. The following relations are often employed.

In the loading domain, the characteristics and the differential relations satisfied along them are

$$dx = \pm c(\sigma)\, dt, \qquad dv = \pm \frac{d\sigma}{\rho\, c(\sigma)} \tag{12.40}$$

The corresponding expressions in the unloading domain are

$$dx = \pm c_o\, dt, \qquad dv = \pm \frac{d\sigma}{\rho c_o} \tag{12.41\&42}$$

The loading/unloading conditions, for $\sigma \geq \sigma_y$, are:

$$\begin{aligned}
&\frac{\partial \sigma}{\partial t} > 0 \quad \text{during loading} \quad \left(\text{when } \frac{\partial \epsilon^p}{\partial t} > 0\right) \\
&\frac{\partial \sigma}{\partial t} = 0 \quad \text{on the loading/unloading boundary} \quad \left(\text{when } \frac{\partial \epsilon^p}{\partial t} = 0\right) \\
&\frac{\partial \sigma}{\partial t} < 0 \quad \text{during unloading} \quad \left(\text{when } \frac{\partial \epsilon^p}{\partial t} < 0\right)
\end{aligned} \tag{12.43}$$

EXAMPLE 12.2

This example is taken from Cristescu (1967).
In this example, the following non-dimensional quantities are used:

$$\bar{x} = \frac{x}{c_1 T}, \quad \bar{t} = \frac{t}{T}, \quad \bar{v} = \frac{v}{v_y}, \quad \bar{\sigma} = \frac{\sigma}{\sigma_y}, \quad \bar{\epsilon} = \frac{\epsilon}{\epsilon_y} \tag{12.44}$$

Boundary conditions:
The boundary conditions are chosen in the form

$$\bar{\sigma} = -4\bar{t}\,(1 - \bar{t}\,) - 1 \tag{12.45}$$

Initial conditions:
The initial conditions are selected to be

$$t = 0, \quad x > 0: \quad \bar{\sigma} = -1, \quad \bar{\epsilon} = -1, \quad v = 0 \tag{12.46}$$

Constitutive equations (linear work-hardening Material):
In a non-dimensional form, the constitutive equations are:

During loading; $\quad \bar{\sigma} = 1 + \left(c_1/c_o\right)^2 \left(\bar{\epsilon} - 1\right)$ **(12.47a)**

During unloading; $\bar{\sigma} = \bar{\sigma}_m(x) + \bar{\epsilon} - \bar{\epsilon}_m(x)$ **(12.47b)**

In the non-dimensional $\bar{x}O\bar{t}$ plane, the characteristic network in the loading domain is constructed with $\Delta\bar{x} = \Delta\bar{t} = 0.02$, while in the unloading domain one still has $\Delta\bar{x} = 0.02$ but $\Delta\bar{t} = \left(c_1/c_o\right)\Delta\bar{x}$.

In the solution of the above problem, a difficulty might arise during the transition between the loading domain and the unloading one, as this would involve changing the network of characteristic lines along which the integration is performed during computation. The loops of the two networks are regular, except for those which lie on the loading/unloading boundary. If the solution is computed in the unloading along a characteristic line of positive slope, then at a certain moment, when passing from one loop to the following one, one must estimate the moment at which the loading/unloading boundary crosses the side of the loop (Cristescu, 1967).

Figure 12.8 (Cristescu, 1967) shows the loading/unloading boundary for the present example, as determined using equations (12.45), (12.46) and (12.47) above. As illustrated in the figure, the loading/unloading boundary propagates up to $\bar{x} = 1.68$. In other words, the portion of the rod for which $\bar{x} > 1.68$ remains in an elastic state. The variation of the stress at various sections of the rod, both during loading and unloading are presented in Fig. 12.9.

The variations of the maximum stress and of the plastic strain along the loading / unloading boundary are presented in Figures 12.10 and 12.11, respectively.

The numerical method can be applied for any initial and boundary conditions as well as for any mechanical properties of the material that can be represented by a finite constitutive equation. In addition, the method can be applied if the rod is either semi-finite or finite. In other words, provision can be made in this method for taking into account the effect of the reflected waves on the direct waves and on the loading/unloading boundary, *see* Cristescu, 1967.

12.5.3. SPECIAL CASE: THE PLASTIC / RIGID SOLUTION

In various problems of plasticity theory, it is possible to neglect the elastic strain in comparison with the plastic one. In this case, the material, introduced earlier in Chapter 7, is referred to as "*plastic/rigid*". The propagation of longitudinal waves in thin rods made of such material was studied by Taylor (1948), Lee and Tupper (1954), among others. They considered the situation when a short steel rod makes an impact with a rigid surface, e.g., with a thick armour plate.

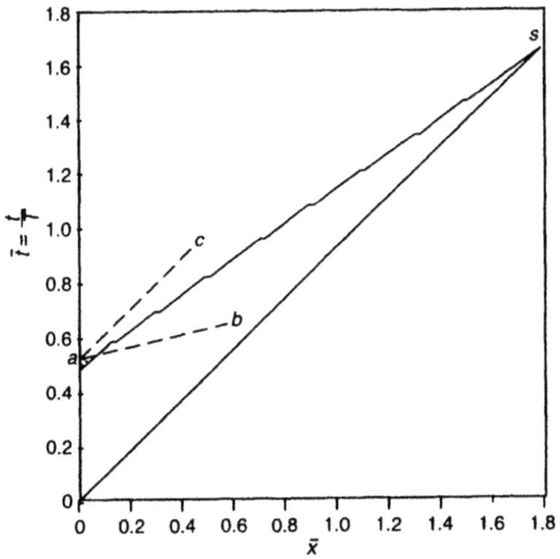

Figure 12.8. The loading/unloading boundary using the numerical method. "Reprinted from Cristescu, N. / Dynamic Plasticity, 1967, pp. 48, with kind permission from Elsevier Science - NL, Sara Burgerhartstraat, 1055 KV Amsterdam. The Netherlands".

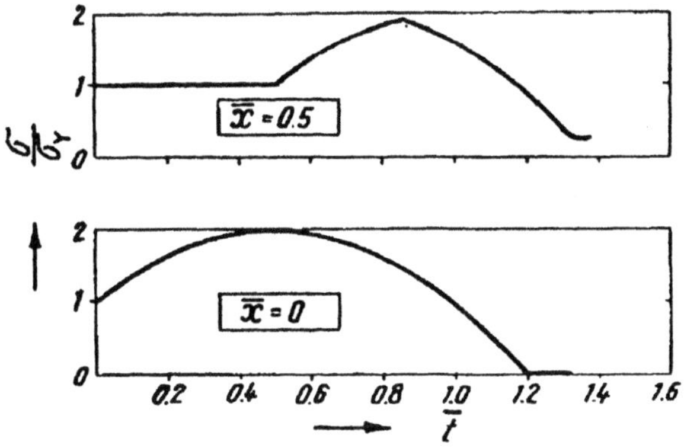

Figure 12.9. Stress profiles as obtained using the numerical method. "Reprinted from Cristescu, N. / Dynamic Plasticity, 1967, pp. 49, with kind permission from Elsevier Science - NL, Sara Burgerhartstraat, 1055 KV Amsterdam. The Netherlands".

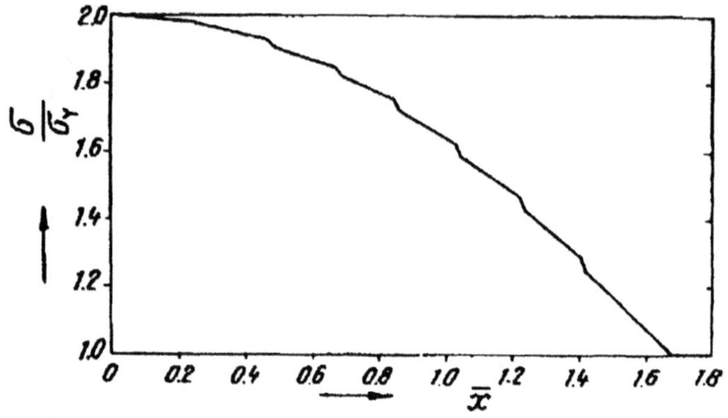

Figure 12.10. Maximum stress profile. "Reprinted from Cristescu, N. / Dynamic Plasticity, 1967, pp. 49, with kind permission from Elsevier Science- NL, Sara Burgerhartstraat, 1055 KV Amsterdam. The Netherlands".

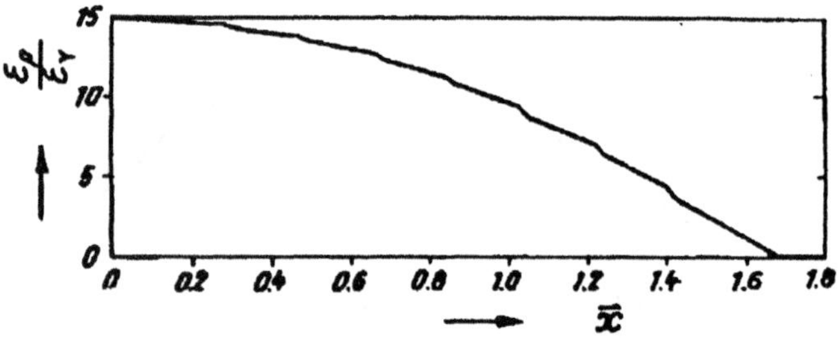

Figure 12.11. Plastic strain profile. "Reprinted from Cristescu, N. / Dynamic Plasticity, 1967, pp. 50, with kind permission from Elsevier Science - NL, Sara Burgerhartstraat, 1055 KV Amsterdam. The Netherlands".

The assumption that the material is plastic/rigid is equivalent to the fact that there are two regions in the rod during wave propagation. The first region, where the plastic wave has not yet passed through, will move as a rigid body towards the fixed target with a velocity *u*. Meantime, the plastic wave front propagates away from the impacted surface with a velocity v. Applying the conservation of mass principle, one obtains

$$(u + v)\, A_o = v\, A \tag{12.48}$$

where A_o is the initial cross sectional area of the rod. Thus, the strain after the plastic wave front has passed may be expressed as

$$\epsilon = (A - A_o)/A = u/(u + v) \tag{12.49}$$

Meantime, if *x* denotes the length of that part of the rod which has not yet been disturbed at the time *t*, one has

$$-\,dx/dt = u + v \tag{12.50}$$

Further, applying the law of conservation of momentum across the wave front, one has

$$\rho(u + v) A_o\, u = (\sigma - \sigma_y)\, A_o \tag{12.51}$$

where σ_y indicates the stress before the shock wave has reached the point under consideration. Thus σ_y represents the yield stress of the plastic/rigid material; as it is understood that the elastic stresses propagate instantaneously in this material.

The equation of motion of the moving rigid part of the rod is written as

$$\rho\, x\, du/dt = -\sigma_y \tag{12.52}$$

Combining (12.49), (12.50) and (12.52), it follows that

$$d(\rho u^2) = -\,\sigma_y\, \epsilon\, dx/x \tag{12.53}$$

Further by combining (12.49), (12.51) and (12.53), one obtains

$$\ell n\, x^2 = \int_{\sigma_1}^{\sigma} \frac{d[(\sigma - \sigma_y)\epsilon]}{\epsilon\, \sigma_y} \tag{12.54}$$

where σ_1 is the maximum stress, corresponding to the moment of impact. Denoting the velocity of the rod at the moment of the impact by U. The latter can be expressed, in relation to the stress σ_1, by combining (12.49) and (12.51), as

$$U = \sqrt{\frac{\epsilon_1 (\sigma_1 - \sigma_y)}{\rho}} \tag{12.55}$$

in which ϵ_1 is the maximum strain corresponding to the stress σ_1.

Following Cristescu (1967), one may use the notation

$$f(\sigma) = \frac{1}{\sigma_y}\left[\sigma(1 + \ell n\, \epsilon) - \int_{\sigma_y}^{\sigma} \ell n\, \epsilon\, d\sigma\right] - \ell n\, \epsilon \tag{12.56}$$

thus, the relation (12.54) becomes

$$\ell n\, x^2 = f(\sigma) - f(\sigma_1) \tag{12.57}$$

Accordingly, since the stress-strain relation is assumed to be known, the expression represented by the function f(σ) may be easily determined from (12.57). Thus, the distribution of the plastic strain can also be obtained.

12.6. Plastic Shock Wave

It is a ***"strong discontinuity wave"***, the fronts of which are surfaces of discontinuity even for stress and first order derivatives of displacement. In this case, the propagating wave is termed "*shock wave*", "*weak shock wave*" or "*wave of stress discontinuity*" or "*stress discontinuity wave*". In our presentation, however, we shall use the term ***"shock wave"***.

In this section, we consider only velocities of impact much smaller than the velocity of propagation of the waves, i. e., the particle velocity is assumed to be much smaller than the wave propagation velocity. Again we consider only one component of the stress and velocity, and we assume that the density of the material to be constant. All dissipative factors, such as lateral inertia. rate effects, thermal conduction, etc. will be again disregarded in the present section. Such dissipative factors oppose the formation of shock waves, or decrease the abruptness of the jump (Cristescu, 1967).

Two possibilities may be envisaged for the development of plastic shock waves in a solid rod:

I) Plastic shock waves may be produced by a sudden impact at the end of the rod, e.g., an explosion or impact with a rigid body. In this case, a bundle of smooth waves with non-diverging fronts may be present. The latter can be however approximated, at the limit, by a propagating shock wave. This is the case most often considered in the literature when shock waves are considered.

II) Plastic shock waves may also be developed due to some particular property of the material of the rod, and/or its particular constitutive equation. In such case, the distance between the wave fronts of smooth waves propagating in the rod decreases during propagation. There is then a tendency for the wave carrying the largest strain to overtake the others. Such situations may occur, for instance, if the stress-strain curve is concave towards the direction of the increasing stress, or alternatively, if certain parameters such as temperature, hydrostatic pressure, etc., might produce a similar effect.

Equations of Motion

As mentioned above, a group of smooth plastic waves may have the tendency, during propagation, to form a shock wave, even if the loading at the end of the rod is not sudden. In this case, the shock wave front does not generally coincide with the envelope of fronts pertaining to the propagating smooth waves. This is due to the fact that across the wave front certain jump conditions must be satisfied. Here, the shock wave front may be obtained from the condition that the jump relations, together with the conditions in front and behind the shock wave front are satisfied.

For the purpose of establishing the equations of motion in Lagrangian coordinates, one considers, following Cristescu (1967), a thin rod moving with a velocity v and initially possessing a uniform strain ϵ. We assume that the rod is in tension, hence, the stress and strain are taken as positive.

At time t, a given section x of the rod is reached by the shock wave front. We shall denote the velocity of propagation of this front by c, with a sense towards the positive direction of the positive Ox- axis. When this front has passed, the particle velocity of the material will be $v + [v]$ and its strain will be $\epsilon + [\epsilon]$.

A superscript "minus" sign will be used to denote the value of a certain function on the shock wave front on the side not yet perturbed, and a superscript "plus" sign to designate the value of the same function, at the same point on the shock wave front, but on the other side, which has already been perturbed. During an interval t, the displacement *u* will vary across the shock wave front either by $(\Delta u)^- = v^- \Delta t + \epsilon^- \Delta x$ or by $(\Delta u)^+ = v^+ \Delta t + \epsilon^+ \Delta x$ depending on which side of the front is considered.

Thus, the "***condition of discontinuity of displacement***" yields the kinematic condition:

$$[v] = -c[\epsilon] \tag{12.58}$$

since

$$(\Delta u)^- = (\Delta u)^+$$

Meantime, the "***momentum equation***" gives the following dynamic condition

$$c\rho [v] = -[\sigma] \tag{12.59}$$

Equations (12.58) and (12.59), together with the constitutive equation of the material of the rod are sufficient for the study of the propagation of shock waves in a thin rod. Equations (12.58) and (12.59) are known as the "***Hugoniot relationhips***", their forms are very similar to the differential equations satisfied along the characteristic lines.

The theory of shock waves, which uses only equations (12.58) and (12.59), considers the process to be adiabatic, and disregards the variation of the internal energy as a result of the impact. This theory is often called "***the elementary theory of shock waves***". The dealt-with shock waves in this case are referred to as "***weak shock waves***".

On the other hand, when the impact velocity or the jump in the applied pressure is very high, the variation of the internal energy or temperature cannot be neglected. In this case, the considered shock waves are referred to as "***strong shock waves***".

Combining equations (12.58) and (12.59), one obtains the velocity of propagation of the shock wave as

$$\boxed{c = \sqrt{\frac{1}{\rho}\frac{[\sigma]}{[\epsilon]}}} \tag{12.60}$$

This velocity of propagation does not generally coincide with the velocity of propagation of smooth plastic waves. This happens only for those portions of the stress-strain relation which are linear. In particular in the elastic range, the velocity (12.60) coincides with the constant elastic velocity of propagation c_o *where equations (12.58) and (12.59) for* $c = c_o$ *are well known in elasticity theory.*

In equations (12.58) - (12.60), a plus or minus sign must be associated with the velocity of propagation c, depending on the direction in which the wave propagates. This system of equations have been applied extensively in the literature. This is essentially due to the fact that computations using this system of equations are much simpler than the corresponding equations associated with a smooth wave propagation problem. That is, one has to solve an algebraic system of equations, instead of a system of partial differential equations, see, e. g., White and Van Griffis (1947, 1948) and Cristescu (1967).

Rods with Variable Yield Stress
The propagation of longitudinal waves in bars with variable yield stress was considered by Rakhmatulin (1946, 1950); *see* Cristescu (1967). In this work, it is assumed that initially the rod possesses the same yield stress value at any of its cross-sections, but after a longitudinal impact at one of its ends, its yield stress varies along the rod, i.e., as a function of the longitudinal coordinate of the cross-section x.

Thus, each impact which produces a plastic strain would modify the yield stress of the part of the rod which is plastically deformed by that impact. In this context, the following two situations may arise:

a) *if the previous impact was applied at the same end of the rod, the yield stress decreases along the rod.*

b) *if the previous impact was applied at the opposite end, the yield stress increases along the rod.*

Constitutive Equations. Denoting the variable yield stress, along the axis of the rod, by σ_y (x), the following stress-strain relations can be written

$$\begin{aligned} \sigma &= E\,\epsilon \quad &\text{for} \quad \epsilon < \sigma_y(x) \\ \sigma - \sigma_y &= g(.) \quad &\text{for} \quad \epsilon > \epsilon_y(x) \end{aligned} \qquad \textbf{(12.61)}$$

where g is a function of the argument $(\epsilon - \epsilon_y)$.

Equations of Motion. In this case, the velocity of propagation c depends on the argument $(\epsilon - \epsilon_y)$. It is expressed as

$$\boxed{c = \sqrt{\frac{1}{\rho}\,\frac{\partial\,\sigma(\epsilon - \epsilon_y)}{\partial\,[\epsilon - \epsilon_y]}}} \qquad \textbf{(12.62)}$$

Meantime, the resulting equations of motion are

$$\frac{\partial^2 u}{\partial t^2} = \pm c_o^2 \frac{\partial^2 u}{\partial x^2} \qquad \text{for} \qquad \epsilon < \epsilon_y(x)$$
$$\frac{\partial^2 u}{\partial t^2} = c^2(\epsilon - \epsilon_y)\left[\frac{\partial^2 u}{\partial x^2} - \frac{d\epsilon_y}{dx}\right] + c_o^2 \frac{d\epsilon_y}{dx} \qquad \text{for} \qquad \epsilon > \epsilon_y(x) \tag{12.63}$$

The characteristics of equation (12.63) are given, and the differential relations satisfied along them are expressed, respectively, as

$$dx = \pm c\,dt, \quad du_t = \pm c\,du_x + [c^2 - c_o^2]\frac{d\epsilon_y}{dx}dt \tag{12.64}$$

Rakhmatulin (1950), *see* Cristescu (1967), presented the following description of wave propagation for the case when the yield stress of the rod increases.

The first elastic wave front will propagate with a constant velocity c_o, and the last one will propagate with a variable velocity. In the characteristic plane, the domain of elastic deformation is separated from the domain of plastic deformation by the boundary $x = f(t)$. Across this boundary, the second derivatives of the displacement suffer discontinuities. On the boundary itself, the strain is $\epsilon_y(x)$.

Using the characteristics (12.64), Rakhmatulin (1950), *see* Cristescu (1967), gave a method of determining approximately the boundary $x = f(t)$. Meantime, the solution of the problem in the unloading domain is similar to that for homogeneous rods.

Rods with Non-Homogeneous Mechanical Properties
A more general problem is the propagation of longitudinal waves in rods possessing non-homogeneity related to various mechanical properties, e. g., variations, as functions of coordinates, in, e.g., the elastic modulus, yield point, work-hardening modulus and density. In general, the rod might have a stress-strain response behaviour that varies from a point to another within the rod.

The propagation, in elastic rods, of waves which gradually change the elastic properties was considered by Juhasz (1949). Meantime, the propagation of longitudinal waves in rods which are both elastically and plastically non-homogeneous was studied by Perzyna (1962). Here, the stress-strain relation is of the form $\sigma = f(\epsilon, x)$. For example, in the case of linear work-hardening material, this relation becomes:

For loading:

$$\sigma(x,t) = a(x)\,\epsilon(x,t) \qquad \text{for} \qquad \epsilon(x,t) \le \epsilon_y(x)$$
$$\sigma(x,t) = b(x)\,\epsilon(x,t) + [\,a(x) - b(x)\,]\,\epsilon_y(x) \qquad \text{for} \qquad \epsilon(x,t) \ge \epsilon_y(x) \tag{12.65}$$

For unloading:

$$\sigma(x,t) = \sigma_y(x) + [\,\epsilon_m(x) - \epsilon_y(x)]\,b(x) - \left[\epsilon_m(x) - \frac{\partial u}{\partial x}\right] a(x) \tag{12.66}$$

where, in the above two expressions, a(x) and b(x) are material parameters which are x-dependent,

Meantime, the *equation of motion* can be written in the form

$$\rho(x)\,\partial^2 u(x,t)/\partial t^2 = \partial\sigma(x,t)/\partial x \tag{12.67}$$

The reader is referred, in this context, to Cristescu, (1967) and Olszak (1959).

Bars with Variable Cross-Section

Here, the problem of the propagation of elastic-plastic waves in semi-infinite rods with variable cross-section is considered as one-dimensional with the same assumptions as given earlier. Denoting the variable cross-sectional area of the rod by *A(x)*, the following equation of motion, for the case of a linear work-hardening material, can be written

$$\frac{\partial \sigma}{\partial x} + \frac{A'(x)}{A(x)}\sigma = \rho\,\frac{\partial^2 u}{\partial t^2} \tag{12.68}$$

Cristescu (1967) presents the boundary conditions pertaining to this case as follows:

Initially the end of the rod is subjected to a sudden pressure, which subsequently continues to increase. It can accordingly be assumed that the first elastic and plastic wave fronts are strong discontinuity fronts which propagate with velocities c_o and c_1, respectively. Combining, along these fronts, the differential relations satisfied along the characteristics and the condition of continuity of the displacement, one obtains the law of variation of the strain along these lines. Beyond this point, in the loading domain, the characteristic method is suggested.

12.7. Dynamic Plasticity under a State of Combined Stress

We consider in this section the problem of dynamic plasticity under the effect of combined stress as was treated by Cristescu (1965).

12.7.1. THE PROBLEM

Cristescu (1965) considered a thin walled tubular specimen of initial length ℓ_0. The end $x = \ell_0$ of the tube is assumed to be fixed, while the other end $x = 0$ is put dynamically into a combined motion: a tension and a torsion. This motion of the end section $x = 0$ of the tube is transmitted along the tube by intermedia of waves. This mechanism of propagation was analysed by Cristescu (1965) for several kinds of constitutive equations, in order to give the possibility to choose the appropriate constitutive equation which may be used for a certain material specimen under specific loading conditions.

In the analysis, presented below, cylindrical coordinates of reference x, r, θ are used, whereby the Ox- axis being directed along the symmetry axis of the tube. The components of the displacements in the axial and circumferential directions are denoted, respectively, by u and v. Due to the small thickness of the wall of the tube, the components σ_r, σ_{rx}, $\sigma_{r\theta}$ are assumed to be small and negligible by comparison with σ_{xx} and $\sigma_{\theta x}$. Meanwhile, the entire problem is considered to be axisymmetrical, so that all derivatives with respect to θ would be considered to be zero $(\partial / \partial\theta = 0)$. Due to this last assumption and as the radial motion is disregarded, the single coordinate which is involved in computation is the axial coordinate x. The analysis aims at the determination of the rotation of various transverse sections of the tube.

Equations of Motion

Using the assumptions mentioned above and taking into account that only two stress components are assumed to be different from zero (these will be denoted by $\sigma_{xx} = \sigma$ and $\sigma_{\theta x} = \tau$), the equations of motion are

$$\begin{aligned} \frac{\partial \sigma}{\partial x} + \chi_x &= \rho \frac{\partial u_t}{\partial t} \\ \frac{\partial \tau}{\partial x} + \chi_\theta &= \rho \frac{\partial \upsilon_t}{\partial t} \end{aligned} \tag{12.69}$$

where χ_x and χ_θ are body forces components, ρ is the density and u_t and υ_t, whereby the subscript t refers to the derivative with respect to time, are the components of the velocities (deformation rates) in the axial and circumferential directions.

Strain-Displacement Relations

Denoting by u, υ and w the displacement components in the x, θ and r directions, respectively, and taking into account the assumed axial symmetry, the strain components may be expressed as

$$\epsilon_{rr} = \frac{\partial w}{\partial r}\,, \quad \epsilon_{\theta\theta} = \frac{w}{r}\,, \quad \epsilon_{xx} = \frac{\partial u}{\partial x}\,(=\epsilon)$$

$$\epsilon_{\theta x} = \frac{1}{2}\frac{\partial \upsilon}{\partial x}\,(=\gamma)\,, \quad \epsilon_{xr} = \frac{1}{2}\left(\frac{\partial u}{\partial r} + \frac{\partial w}{\partial x}\right), \quad \epsilon_{r\theta} = \frac{1}{2}\left(\frac{\partial \upsilon}{\partial r} - \frac{\upsilon}{r}\right) \tag{12.70}$$

Out of these strain components, two components are dominant, namely ϵ_{xx} and $\epsilon_{\theta x}$. These components are denoted in the analysis by ϵ and γ, respectively.

The first invariant of the strain tensor is

$$\epsilon_m = \frac{\epsilon_{rr} + \epsilon_{\theta\theta} + \epsilon_{xx}}{3} \tag{12.71}$$

while the second invariant of the stress deviator is, for the case under consideration,

$$I_2' = \frac{1}{2}\sigma_{ij}'\sigma_{ij}' = \frac{\sigma^2}{3} + \tau^2. \tag{12.72}$$

where σ'_{ij} are the components of the stress deviator.

Two particular motions corresponding to two specific sets of boundary conditions are considered.

- The first case corresponds to a uniaxial longitudinal compressive motion, by the assumption that everywhere $\upsilon = 0$. In this case, the system of equations (12.69) reduces to a single equation of motion

$$\frac{\partial \sigma}{\partial x} + \chi_x = \rho\frac{\partial u_t}{\partial t} \tag{12.73}$$

- The other particular case corresponds to $u = 0$; then, instead of (12.69), one has

$$\frac{\partial \tau}{\partial x} + \chi_\theta = \rho\frac{\partial \upsilon_t}{\partial t} \tag{12.74}$$

which describes a uniaxial shearing motion.

The Constitutive Equations
In order to express the constitutive equations, one assumes that the strain-rate components can be decomposed in an elastic component and a plastic one, i.e.,

$$\dot{\epsilon}'_{ij} = \dot{\epsilon}'^{E}_{ij} + \dot{\epsilon}'^{P}_{ij} \tag{12.75}$$

where ϵ'_{ij} are the rate of strain deviator tensor. The elastic part of the strain rate is assumed to satisfy the Hooke's law, i.e.,

$$2\,G\dot{\epsilon}'^{E}_{ij} = \dot{\sigma}'_{ij} \tag{12.76}$$

and

$$\dot{\sigma}_m = 3K\dot{\epsilon}_m \tag{12.77}$$

where σ'_{ij} are the components of the stress deviator, σ_m is the mean stress and G and K are the elastic constants; namely the shear and bulk moduli, respectively.

Concerning the plastic component of the strain, Cristescu (1965) assumed quite general constitutive relations which are able to emphasize not only plastic inviscid properties, but also viscoplastic effects. In a general form, such constitutive equations are written as (see, also, Cristescu, 1967)

$$\dot{\epsilon}'^{P}_{ij} = A_{ijkl}\,\dot{\sigma}'_{kl} + B_{ij} \tag{12.78}$$

where by giving various expressions for the coefficients A_{ijkl} and B_{ij}, one may obtain several constitutive equations used in dynamic plasticity.

For the particular stress state under consideration, equation (12.78) can be written as

$$\begin{aligned} \dot{\epsilon}'^{P}_{xx} &= \varphi_{11}\,\dot{\sigma}'_{xx} + \varphi_{12}\,\dot{\sigma}'_{\theta x} + \psi_1 \\ \dot{\epsilon}'^{P}_{\theta x} &= \varphi_{21}\,\dot{\sigma}'_{xx} + \varphi_{22}\,\dot{\sigma}'_{\theta x} + \psi_2 \end{aligned} \tag{12.79}$$

From equations (12.70) and (12.75) to (12.77), together with the simplified notation mentioned before, equations (12.79) can be written, respectively, as

$$\frac{\partial u_t}{\partial x} = \left(\frac{2}{3} \varphi_{11} + \frac{1}{E} \right) \dot{\sigma} + \varphi_{12} \dot{t} + \psi_1$$
$$\frac{\partial \upsilon_t}{\partial x} = \frac{4}{3} \varphi_{21} \dot{\sigma} + \left(2\varphi_{22} + \frac{1}{G} \right) \dot{t} + 2\psi_2 \qquad \textbf{(12.80)}$$

taking into account that $\sigma_m = \frac{\sigma}{3}$, $E = \frac{9KG}{(3K+G)}$.

For simplicity, Cristescu (1965) adopted in his analysis the notations

$$\sigma_{11} = \frac{1}{3}\left(2\,\varphi_{11} + \frac{1}{G} + \frac{1}{3K} \right), \quad \alpha_{12} = \varphi_{12}, \quad \beta_1 = \psi_1$$
$$\sigma_{21} = \frac{4}{3}\varphi_{21}, \quad \alpha_{22} = 2\,\varphi_{22} + \frac{1}{G}, \quad \beta_2 = 2\,\psi_2 \qquad \textbf{(12.81)}$$

so that the constitutive equations (12.80) can be written, respectively, in the following simple form

$$\frac{\partial u_t}{\partial x} = \alpha_{11}\,\dot{\sigma} + \alpha_{12}\,\dot{t} + \beta_1$$
$$\frac{\partial \upsilon_t}{\partial x} = \alpha_{21}\,\dot{\sigma} + \alpha_{22}\,\dot{t} + \beta_2 \qquad \textbf{(12.82)}$$

Velocities of Propagation

Following Cristescu (1965), one computes, then, the characteristic lines of the systems (12.69&12.82) to obtain four families of characteristic lines satisfying the differential equation

$$\rho^2 (\alpha_{11}\,\alpha_{22} - \alpha_{12}\,\alpha_{21})\,(dx)^4 - \rho\,(\alpha_{11} + \alpha_{22})(dx)^2\,(dt)^2 + (dt)^4 = 0 \qquad \textbf{(12.83)}$$

Thus, one obtains two velocities of propagation which are furnished (Cristescu, 1965) by

$$\left.\begin{matrix} C_{TL}^2 \\ C_{LT}^2 \end{matrix}\right\} = \frac{\alpha_{11} + \alpha_{22} \pm \sqrt{(\alpha_1 - \alpha_{22})^2 + 4\,\alpha_{12}\,\alpha_{21}}}{2\rho(\alpha_{11}\,\alpha_{22} - \alpha_{12}\,\alpha_{21})} \qquad \textbf{(12.84)}$$

where the subscripts "TL" and "LT" refer, respectively, to a circumferential and a longitudinal motion.

Differential Relations Satisfied along the Characteristic Lines
In order to perform an integration scheme and to study the properties of the waves involved, it is necessary to write the differential relations satisfied along the characteristic lines.

For the system of equations (12.69&12.82) the differential relations satisfied along the characteristic lines (12.83) can be determined as

$$\begin{aligned} &\mp \rho c(1-\rho\alpha_{22}c^2)\,du_t \mp \rho^2\alpha_{12}c^3\,d\upsilon_t + (1-\rho\alpha_{22}c^2)d\sigma \\ &+\rho\alpha_{12}c^2\,d\tau + [\rho\beta_1 c^2(1-\rho\alpha_{22}c^2) + \beta_2\alpha_{12}\rho^2 c^4 \\ &\qquad \mp c(1-\rho\alpha_{22}c^2)\chi_x \mp \rho c^3\alpha_{12}\chi_\theta]\,dt = 0 \end{aligned} \tag{12.85}$$

where for c, appearing in (12.85), one of the two expressions (12.84) may be replaced, and "*d*" stands for the "*interior derivative*" along a characteristic line. It is sometimes useful to write the equations (12.85) in another form, i.e.,

$$\begin{aligned} &\rho\alpha_{21}c^2\,du_t + (1-\rho\alpha_{11}c^2)\,d\upsilon_t \mp \alpha_{21}c\,d\sigma \\ &\qquad + \frac{1-\rho\alpha_{11}c^2}{\mp\rho c} + [\mp\beta_1\rho\alpha_{21}c^3\beta_2 c(1-\rho\alpha_{11}c^2) \\ &\qquad + \alpha_{21}c^2\chi_x + \frac{1-\rho\alpha_{11}c^2}{\rho}\chi_\theta]\,dt = 0 \end{aligned} \tag{12.86}$$

All the unknown functions involved in the problem are present in all the differential relations (12.85), and thus all the four waves are at the same time shearing waves and longitudinal waves. This conclusion holds, also, for the general constitutive equation (12.82).

Coupling of the two Types of Waves
From (12.69) and (12.82), the dynamic jump conditions across a wave front can be obtained as

$$\begin{aligned} \left[\frac{\partial u_t}{\partial x}\right] &= \alpha_{11}\left[\frac{\partial\sigma}{\partial t}\right] + \alpha_{12}\left[\frac{\partial\tau}{\partial t}\right] \\ \left[\frac{\partial \upsilon_t}{\partial x}\right] &= \alpha_{21}\left[\frac{\partial\sigma}{\partial t}\right] + \alpha_{22}\left[\frac{\partial\tau}{\partial t}\right] \end{aligned} \tag{12.87}$$

and

$$\rho\left[\frac{\partial u_t}{\partial t}\right]=\left[\frac{\partial \sigma}{\partial x}\right],\qquad \rho\left[\frac{\partial \upsilon_t}{\partial t}\right]\left[\frac{\partial \tau}{\partial x}\right] \tag{12.88}$$

where [.] designates the jump of the function inside the same brackets. Meanwhile, the kinematic jump conditions are expressed as

$$\left[\frac{\partial \psi}{\partial t}\right]=-c\left[\frac{\partial \psi}{\partial x}\right] \quad \text{for} \quad \psi=u_t,\upsilon_t,\sigma,\tau \tag{12.89}$$

Combining (12.88), (12.89), and the result in (12.87), Cristescu (1965) obtained the following two (equivalent) conditions

$$\begin{aligned}\left[\frac{\partial u_t}{\partial x}\right](1-\rho\alpha_{11}c^2)&=\rho\alpha_{12}c^2\left[\frac{\partial \upsilon_t}{\partial x}\right]\\ \left[\frac{\partial \upsilon_t}{\partial x}\right](1-\rho\alpha_{22}c^2)&=\rho\alpha_{21}c^2\left[\frac{\partial u_t}{\partial x}\right]\end{aligned} \tag{12.90}$$

From (12.90), it is evident that generally for (12.82), because $\alpha_{12}\neq 0$ and $\alpha_{21}\neq 0$, both kinds of waves are coupled: both produce a longitudinal and a circumferential motion. This is the reason why the velocities of propagation (12.84) were denoted by LT and TL.

In view of the relations (12.90), one can conclude that, if one of the following three possibilities arises

$$c_{LT}{}^2 \begin{smallmatrix}<\\=\\>\end{smallmatrix} \frac{1}{\rho(\alpha_{11}+\alpha_{12})} \tag{12.91}$$

one has

$$\left[\frac{\partial u_t}{\partial x}\right] \begin{smallmatrix}<\\=\\>\end{smallmatrix} \left[\frac{\partial \upsilon_t}{\partial x}\right] \tag{12.92}$$

respectively. In the first case, the shearing motion is dominant with respect to the longitudinal motion, while in the third case a reverse situation arises. The equality in (12.91) and (12.92) occurs in isolated points or for a particular type of (theoretical) loading when everywhere $\dot{\sigma} = \dot{\tau}$, which is referred-to by Cristecu (1965) as "***diagonal loading***". Similarly, one obtains

$$C_{TL}^2 \gtreqless \frac{1}{\rho\,(\alpha_{21} + \alpha_{22})} \tag{12.93}$$

Initial Conditions.
One may assume that at $t < 0$ the tube is at rest and that some constant uniformly distributed stress state is present, i.e.,
For instance: $\sigma_0 = \tau_0 = 0$.

$$\left.\begin{array}{l} 0 \le x \le \ell_0 \\ \\ t<0 \end{array}\right\} \begin{array}{l} u_t = \upsilon_t = 0 \\ \\ \sigma = \sigma_0, \quad \tau = \tau_0 \end{array} \tag{12.94}$$

Boundary Conditions
Following Cristescu (1965), the boundary conditions may be prescribed as follows:
The end $x = \ell_0$ of the tube is fixed;

$$\left.\begin{array}{l} x = \ell_0 \\ \\ t \ge 0 \end{array}\right\} u_t = \upsilon_t = 0 \tag{12.95}$$

while the end $x = 0$ is put into a combined motion

$$\left.\begin{array}{l} x = 0 \\ t \ge 0 \end{array}\right\} u_t = U(0,t), \quad \upsilon_t = V(0,t). \tag{12.96}$$

Applying the conditions (12.94 to 12.96), one integrate the system of equations (12.69&12.82) in order to obtain

$$u_t(x,t), \quad \upsilon_t(x,t), \quad \sigma(x,t), \quad \tau(x,t)$$

Then, the longitudinal and circumferential motions of the points lying on a certain circle $x = x^*$ can be determined from

$$u(x^*,t) = \int_0^t u_t(x^*,\bar{t})\,d\bar{t}, \quad \theta(x^*,t) = \frac{1}{r}\int_0^t \upsilon_t(x^*,\bar{t})\,d\bar{t},$$

where r is the radius of the tube. Summing up the displacement u (x, t) along the tube (fixed t) we obtain the variation of the length of the tube, at the considered time t.

The approach presented above can be used to study the dynamic plastic behaviour of various materials under combined loading (bi-axial loading of tubes). The entire picture of wave propagation and reflection can be described by integrating the previously mentioned equations. Cristescu (1965) demonstrated the application of his approach, summarized above, to the following particular representations of constitutive relations.

(A) The perfectly-elastic case.

It is the simplest possible case.

$$E\frac{\partial u_t}{\partial x} = \dot{\sigma}$$
$$G\frac{\partial \upsilon_t}{\partial x} = \dot{\tau} \tag{12.97}$$

which is obtained from equations (12.82) for

$$\varphi_{11} = \varphi_{12} = \varphi_{21} = \varphi_{22} = \psi_1 = \psi_2 = 0$$

(B) A more complicated constitutive equation corresponds to

$$\varphi_{11} = \varphi_{12} = \varphi_{21} = \varphi_{22} = \psi_1 = \psi_2 = 0 \tag{12.98}$$

For instance the Hohenemser-Prager (1932) constitutive equation is of this form

$$\dot{\varepsilon} = \frac{1}{E}\dot{\sigma} + \frac{1}{3\eta}\left(1 - \frac{k}{\sqrt{I_s^{(2)}}}\right)\sigma$$
$$\dot{\gamma} = \frac{1}{2G}\dot{\tau} + \frac{1}{2\eta}\left(1 - \frac{k}{\sqrt{I_s^{(2)}}}\right)\tau \tag{12.99\&12.100}$$

(C) The last particular case (12.98) may be considered as a special case of the more general case.

The constitutive equations (12.82) can now be written as

$$\frac{\partial u_t}{\partial x} = \left(\frac{2}{3} \varphi_{11} + \frac{1}{E} \right) \dot{\sigma} + \psi_1$$

$$\frac{\partial \upsilon_t}{\partial x} = \left(2 \varphi_{22} + \frac{1}{G} \right) \dot{t} + \psi_2 \qquad (12.101)$$

Such constitutive equations are referred-to as "*quasi-linear uncoupled*" constitutive equations.

(D) If all the coefficients φ_{ij} are different from zero but

$$\psi_1 = \psi_{21} = 0 \qquad (12.102)$$

the constitutive equations will be called quasi-linear coupled constitutive equations. An example of such constitutive equations is the Prandtl-Reuss constitutive equation

$$2G\,\dot{\epsilon}'_{ij} = \dot{\sigma}'_{ij} + \frac{2G}{H'} \frac{\sigma'_{kl}\,\dot{\sigma}'_{kl}}{\sigma'_{mn}\,\sigma'_{mn}}\,\sigma'_{ij} \qquad (12.103)$$

where the work-hardening law takes the form

$$\sigma'_{ij}\,\sigma'_{ij} = H\left(\int \sigma'_{kl}\, d\,\epsilon'^{P}_{kl} \right) \qquad (12.104)$$

and H' is the derivative of H with respect to its argument.

For the case under consideration, Eqn. (12.103) is expressed as

$$\dot{\varepsilon} = \left[\frac{1}{E} + \frac{4}{27}\frac{1}{H' I_2'}\sigma^2\right]\dot{\sigma} + \frac{2}{3}\frac{1}{H' I_2'}\sigma\tau\dot{\tau}$$

$$\dot{\gamma} = \frac{2}{9}\frac{1}{H' I_2'}\sigma\tau\dot{\sigma} + \left[\frac{1}{2G} + \frac{1}{H' I_2'}\tau^2\right]\dot{\tau} \tag{12.105}$$

(E) Finally the last special case, which will be considered is an example of the general constitutive equation of the type (12.76&12.78), is of the form

$$\dot{\varepsilon}'_{ij} = \frac{\dot{\sigma}'_{ij}}{2G} + \frac{1}{2\eta}\left(1 - \frac{k}{\sqrt{I_2'}}\right)\sigma'_{ij} + \frac{\mathscr{F}(I_2')}{2 I_2'}\sigma'_{kl}\dot{\sigma}'_{kl}\sigma'_{ij} \tag{12.106}$$

This can be obtained (Cristescu, 1967) by assuming that the rate of strain component can be decomposed as

$$\dot{\varepsilon}_{ij} = \dot{\varepsilon}_{ij}^{E} + \dot{\varepsilon}_{ij}^{VP} + \dot{\varepsilon}_{ij}^{P} \tag{12.107}$$

where $\dot{\varepsilon}_{ij}^{VP}$ is the viscoplastic rate of strain component and $\dot{\varepsilon}_{ij}^{P}$ is the plastic inviscid rate of strain component. In (12.106) k is a plastic constant, η is the viscosity coefficient while the function $\mathscr{F}$ describes the work-hardening properties of the material.

In particular, for the problem under consideration, Eqn. (12.106) can be written as

$$\dot{\varepsilon} = \left[\frac{1}{E} + \frac{4}{27}\frac{\mathscr{F}(I_2')}{I_2'}\sigma^2\right]\dot{\sigma} + \frac{2}{3}\frac{\mathscr{F}(I_2')}{I_2'}\dot{\sigma}\tau\dot{\tau} + \frac{1}{3\eta}\left(1 - \frac{k}{\sqrt{I_2'}}\right)\sigma$$

$$\dot{\gamma} = \frac{2}{9}\frac{\mathscr{F}(I_2')}{I_2'}\sigma\tau\dot{\sigma} + \left[\frac{1}{2G} + \frac{\mathscr{F}(I_2')}{I_2'}\tau^2\right]\dot{\tau} + \frac{1}{2\eta}\left(1 - \frac{k}{\sqrt{I_2'}}\right)\tau \tag{12.108}$$

The above constitutive equations describe the main properties emphasized by the general form of the constitutive equation (12.78). If necessary, other effective examples of the constitutive equations (12.78) can be considered, which would emphasize other possible mechanical properties.

12.8. Transition to Dynamic Thermoelasticity

The presence of a thermal field, due to heating of the body, is often considered as non-homogeneity. The situation is more complicated when the thermal field is variable, and, also, when the temperature variation is sudden at the boundary of the body. A more difficult problem, which is often encountered in practice is that in which variations in both temperature and stress are prescribed at the boundary of the body.

When a sudden variation of the temperature is prescribed at the boundary of the body, then a stress field will be produced as a result of thermal dilatation. The abruptness of the temperature variation will involve inertia forces. This, by consequence, results in stress waves which will propagate through the body. Meantime, if the temperature at the boundary surpasses a certain limit, the body will pass from the elastic to the plastic state. Thus, the resulting stress waves can be either elastic or elastic-plastic. In addition, it is likely that various plastic loading domains will appear due to the existence of the thermal field in the body, just as supplementary plastic regions can appear due to the non-homogeneity of the body (e. g., Raniecki, 1964 and Cristescu, 1967).

12.9. References

Chadwick, P. (1962) Propagation of spherical plastic-elastic disturbances from an expanded cavity, *Quart. J. Mech. Appl. Math.* **15**, 349-76.

Craggs, J. W. (1961) Plastic waves, *in Progress in Solid Mechanics*, I.N. Sneddon and R. Hill (eds.), Vol. II (1), North-Holland, Amsterdam, pp. 141-97.

Cristescu, N. (1960a) European contributions to dynamic loading and plastic waves, *in Plasticity*, Second Symp. on Naval Structural Mechanics, Lee, E. H. and Symonds, P. S.(Eds.), Providence, 1960, Pergamon Press, New York, pp 385-442.

Cristescu, N. (1960b) Some observations on the propagation of plastic waves in plates, in *Plasticity*, Second Symp. on Naval Structural Mechanics, Lee, E. H. and Symonds, P. S. (eds.), Providence, 1960, Pergamon Press, New York, pp. 501-10.

Cristescu, N. (1965) Dynamic plasticity under combined stress, in *Behaviour of Materials under Dynamic Loading*, ed. U. S. Lindholm, 4 *ASME* **42**, pp. 329-42.

Cristescu, N. (1967) *Dynamic Plasticity*, North-Holland, Amsterdam

Cristescu, N. (1970) The unloading in symmetric longitudinal impact of two elastic-plastic bars, *Int. J. Mech. Sci.* **12**, 723-38.

Cristescu, N. (1972) A procedure for determining the constitutive equations for materials exhibiting both time-dependent and time-independent plasticity, *Int. J. Solids Structures* **8**, 511-31.

Cristescu, N. and Bell, J. F. (1970) On unloading in the symmetrical impact of two aluminum bars, Battelle Colloquium on Inelastic Behavior of Solids, Columbus, Ohio, Sept. 1969, Mcgraw Hill Book Co., pp. 397-419.

Davies, R.M. (1953) Stress waves in solids, *Appl. Mech. Rev.* **6**, 1-3.

Davies, R.M. (1956) Stress waves in solids, *British J. Appl. Phys.* **7**, 203-9.

Donnell, L. H. (1930) Trans. ASME **52** (1), 153-67.

Ericksen, J. L. (1955) *J. Math. Phys.* **34**, 74-9.

Goldsmith, W. (1960) *Impact*, Arnold, London.

Goldsmith, W. (1963) Impact: The collision of solids, *Appl. Mech. Rev.* **16**, 855-66.

Hill, R. (1961) Discontinuity relations in mechanics of solids, *in Progress in Solid Mechanics*, I.N. Sneddon and R. Hill (eds.), North Holland, Amsterdam, Vol. II, pp. 245-76.

Hohenemser, K., Von and Prager, W. (1932) Über die ansätze der mechanik isotroper kontinua, *Z. Angew. Math. Mech.* **12(4)**, 216-226.

Hopkins, H. G. (1960) Dynamic expansion of spherical cavities in metals, in *Progress in Solid Mechanics*, Sneddon, I. N. And Hill, R. (eds.), North Holland, Amsterdam, Vol. 1, pp. 83-164.

Hopkins, H.G. (1961) Dynamic anelastic deformation of metals, *Appl. Mech. Rev.* **14**, pp. 417-31.

Hopkinson, B. (1905) *Proc. Roy Soc. A*, 74498--506.

Hopkinson, J. (1872) *Collected scientific papers*, Vol. 11, pp.316.

Hunter, S. C. (1957) The propagation of spherical symmetric disturbances in ideally plastic materials, *in* Proceedings Conf. on Properties of Materials at High Rates of Strain, *Inst. Mech. Engrs.*, London, pp. 147-55.

Juhasz, K. J. De (1949) Graphical analysis of impact of bars stressed above the elastic range, *J. Franklin Inst. Part I*, **248**, 15-48.; Part II, **248**, pp. 113-42.

Karman, T. and Duwez, P. (1950) The propagation of plastic deformation in solids, *J. Appl. Phys.* **21**, 987-994.

Kolsky, H. (1953) *Stress Waves in Solids*, Clarendon Press, Oxford.

Lee, E.H. and Tupper, S. J. (1954) Analysis of plastic deformation in a steel cylinder striking a rigid target, *J. Appl. Mech.* **21**, 63-70.

Mandel, J. (1962) Ondes plastiques dans un milieu indefini a trois dimensions, in "*Extrait du Seminaire de Plasticite", Publ. Sci. Techn. Min. Air. NT* 116 Paris, pp 151-78.

Mandel, J. (1964) Propagation des surfaces de discontinuite dans un milieu elastoplastique, IUTAM Symp. on Stress Waves in Anelastic Solids, Kolsky, H. And Prager, W. (Eds.), Providence, 1963, pp. 331-40.

Olszak, W., ed. (1959) *Non-homogeneity in Elasticity and Plasticity*, Warsaw Symp. 1959, Pergamon Press, London.

Olszak, W., Mroz, Z. and Perzyna, P. (1963) *Recent Trends in the Development of the Theory of Plasticity*, Pergamon Press, Oxford.

Olszak, W. and Perzyna, P. (1962) Propagation of spherical waves in a non-homogeneous elastic-viscoplastic medium, in "*La propagation des ebralements dans les milieux hetrogenes*", *Colloque CNRS*, Marseille, 1961, CNRS, Paris, pp. 67-78.

Perzyna, P. (1962) Propagation of shock wave in elastic-visco-plastic medium of definite non-homogeneity type, *Arch. Mech. Stos.* **14**, 93-111.

Rakhmatulin, H.A. (1945a) *Prikl. Mat. Meh.* **9**, 91-100.

Rakhmatulin, H.A. (1945b) *Prikl. Mat. Meh.* **9**, 449-62

Rakhmatulin, H.A. (1946) *Prikl. Mat. Meh.* **10**, 333-46.

Rakhmatulin, H.A. (1950) *Prikl. Mat. Meh.* **14**, 65-74.

Raniecki, B. (1964) *Proc. Vibr. Probl.* **5**, 319-47.

Szczepinski, W. (1964) *Arch. Mech. Stos.* **16**, 1207-14.

Taylor, G. I. (1948) *Proc. Roy. Soc. Ser. A* **194**, 289.

Thomas, T. Y. (1961) *Plastic Flow and Fracture in Solids*, Academic Press, New York.

White, M.P. and Van Griffis, L. (1947) Permanent strain in uniform bar due to longitudinal impact. *J. Appl. Mech.*, **14**, 337-43.

White, M.P. and Van Griffis, L. (1948) The propagation of plasticity in uniaxial compression, *J. Appl. Mech.* **15 (3)**, 256-60.

Wierzbicki, T. (1963) *Arch. Mech. Stos.* **15**, 297-308.

12.10. Further Reading

Abramson, H.N., Plass, H.J. and Ripperger, E.A. (1958) Stress wave propagation in rods and beams, in

Advances in Applied Mechanics, Dryden, H.L. and Von Karman, Th. (eds.), Academic Press, New York, Vol. V, pp. 111-94.

Aggarwal, H.R., Soldate, A.M., Hook, J.F. and Miklowitz, J. (1964) Bilinear theories in plasticity and an application to two-dimensional wave propagation, *J. Appl. Mech.* **31**, Ser. E, 181-8.

Allen, W., Mapes, J. and Mayfield, E. (1955) Shock waves in air produced by waves in plate, *J. Appl. Phys.* **26**, 125-6.

Allen, W.A., Mayfield, E.B. and Marrison, H.L. (1957) Dynamics of projectile penetrating sand, *J. Appl. Phys.* **28**, 370-6.

Alter, B.E.K. and Curtis, C.W. (1956) Effect of strain rate on propagation of plastic strain pulse along lead bar, *J. Appl. Phys.* **27**, 1079-85.

Alverson, R.C. (1956) Impact with finite acceleration time of elastic and elastic-plastic beams, *J. Appl. Mech.* **23**, Ser. E., 411-15.

Back, P.A.A and Campbell, J.D. (1957) The behaviour of a reinforced plastic material under dynamic compression, in Proceedings, *Conference on Properties of Materials at High Rates of Strain*, London, 1957, Inst. Mech. Engrs., London, pp. 221-8.

Backman, M.E. (1964) Form of the relation between stress and finite elastic and plastic strains under impulsive load, *J. Appl. Phys.* **35**, 2524-33.

Baker, W.E. (1960a) Elastic-plastic response of thin spherical shells to internal blast loading, *J. Appl. Mech.* **27**, Ser. E, 139-44.

Baker, W.E., (1960b). Modelling of large transient elastic and plastic deformations of structures subjected to blast loading. *J. Appl. Mech.* **27**, Ser. E, 521-7.

Bancroft, D., Peterson, E.L. and Minshall, F.S. (1956) Polymorphism of iron at high pressure, *J. Appl. Phys.* **27**, 291-8.

Baker, L.M., Lundergan, C.D. and Herman, W. (1964) Dynamic response of aluminium, *J. Appl. Phys.* **35**, 1203-12.

Baron, M.L., Bleich, H.H. and Weidlinger, P. (1961) Dynamic elastic-plastic analysis of structures, *J. Eng. Mech. Div., Am. Soc. Civil Engrs.* **87**, 23-42.

Baron, H.G. and. Henn, R.H. (1964) Spring-back and metal flow in forming shallow dishes by explosives. *Int. J. Mech. Sci.* **6**, 435-44.

Bell, J.F. (1958) Normal incidence in the determination of large strain through the use of diffraction greetings, in Proceedings, 3rd U.S. Nat. Congr. Appl. Mech., pp. 489-93.

Bell, J.F. (1960a) Propagation of large amplitude waves in annealed aluminium, *J. Appl. Phys.* **31**, 277-88.

Bell, J.F. (1960b) Study of initial conditions in constant velocity impact, *J. Appl. Phys.* **31**, 2188-95.

Bell, J.F. (1961a) Experimental study of unloading phenomena in constant velocity impact, *J. Mech. Phys. Solids* **9**, 1-15.

Bell, J.F. (1961b) Further experimental study of unloading phenomena in constant velocity impact, *J. Mech. Phys. Solids* **9**, 261-78.

Bell, J.F. (1961c) Experimental study of interrelation between theory of dislocations in polycrystalline media and finite amplitude wave propagation in solids, *J. Appl. Phys.* **32**, 1982-93.

Bell, J.F. (1962a) Experimental study of dynamic plasticity at elevated temperatures, *Experimental Mechanics* **2**(6), 181-6.

Bell, J.F. (1962b) Experiments on large amplitude waves in finite elastic strain, in Proceedings, Int. Symposium on Second-Order Effects in Elasticity, Plasticity and Fluid Dynamics, Haifa, Pergamon Press, Oxford, 173-86.

Bell, J.F. (1963) Single, temperature-dependent stress-strain law for dynamic plastic deformation of annealed face centered cubic metals, *J. Appl. Phys.* **34**, 134-41.

Bell, J.F. (1964a) The initiation of finite amplitude waves in annealed metals, IUTAM Symposium, *Stress Waves in Anelastic Solids*, Providence, 1963, Springer-Verlag, Berlin, pp. 166-82.

Bell, J.F. (1964b) Generalized large deformation behaviour for face-centred cubic solids. High purity copper, *Philos. Magazine* **10(103)**, pp. 107-126.

Bell, J.F. and Suckling, J.H. (1962) The dynamic overstress and the hydrodynamic transition velocity in the symmetrical free flight plastic impact of annealed aluminium, in Proceedings, 4th U.S. Nat. Congr. Appl. Mech., pp. 877-83.

Bell, J.F. and Werner, M.W. (1962) Applicability of Taylor theory of polycrystalline aggregate to finite amplitude wave propagation in annealed copper, *J. Appl. Phys.* **33**, 2416-25.

Berger, J., Camion, T., and Bourreau, H. (1962) Études des conditions d'interférence de deux chocs de même intensité dans un solide, in "Les ondes de détonation", Colloque CNRS, Gif-sur-Yvette, 1961, CNRS, Paris, 481-4.

Bernhard, R.K. (1954) Bibliography and publications on soil dynamics, in *Symposium on Dynamic Testing of Soils*, 65th Annual Meeting, Atlantic City, N.J., 1953, ASTM Special Technical Publ., No. 156.

Bianchi, G. (1964) Some experimental and theoretical studies on the propagation of longitudinal plastic waves in a strain-rate-dependent material, Kolsky, H. and Prager, W. (eds), IUTAM Symposium, *Stress Waves in Anelastic Solids*, Providence, 1963, Springer-Verlag, Berlin, pp. 101-17.

Biot, M.A. (1956a) Theory of propagation of elastic waves in fluid-structured porous solid, *J. Coust. Soc. Am.* **28**, 168-78.

Biot, M.A. (1956b) Thermoelasticity and irreversible thermodynamics, *J. Appl. Phys.* **27**, 240-53.

Bleich, H.H. and Nelson, I. (1966) Plane wave in elastic-plastic half-space due to combined surface pressure and shear, *J. Appl. Mech.* **33**, Ser. E, 149-58.

Bodner, S.R. and Kolsky, H. (1958) Stress wave propagation in lead, in Proceedings, 3rd U.S. Nat. Congr. Appl. Mech., pp. 495-501.

Bodner, S.R. and Speirs, W.G. (1963) Dynamic plasticity experiments on aluminium cantilever beams at elevated temperature, *J. Mech. Phys. Solids* **11**, 65-78.

Bodner, S.R. and Symonds, P.S. (1962) Experimental and theoretical investigation of the plastic deformation of cantilever beams subjected to impulsive loading, *J. Appl. Mech.* **29**, 719-28.

Bowden, F.P. and Field, J.E. (1964) Brittle fracture of solids, by liquid impact, by solid impact and by shock. *Proc. Roy. Soc. London* A **282**, 331-52.

Brinkworth, B.J. (1957) The propagation of strain in textile cables under longitudinal impact, in Proceedings, Conf. on Properties of Materials at High Rates of Strain, London, 1957, The Inst. of Mech. Engrs., London, pp. 184-9.

Broberg, K.B. (1955) Studies on scabbing of solids under explosive attack, *J. Appl. Mech.* **22**, 317-23.

Brown, A. (1964) Quasi-dynamic theory of containment, *Int. J. Mech. Sci.* **6**, 257-62.

Brown, A.F.C. and Vincent, N.D.G. (1941) Relationship between stress and strain in the tensile impact test, *Proc. Inst. Mech. Engrs.* **145**, pp. 126-34.

Bryan, G.M. and Pugh, E.M. (1962) Cratering of lead by oblique impacts of hyper velocity steel pellets, *J. Appl. Phys.* **33**, 734-8.

Butcher, B.M. and Karnes, C.H. (1966) Strain-rate effects in metals, *J. Appl. Phys.* **37**, 402-11.

Campbell, J.D. (1953a) Investigation of plastic behaviour of metal rods subjected to longitudinal impact, *J. Mech. Phys. Solids* **1**, 113-23.

Campbell, J.D. (1953b) Dynamic yielding of mild steel, *Acta Metallurgica* **1**(b), 206-10.

Campbell, J.D. (1954) Yield of mild steel under impact loading, *J. Mech. Phys. Solids* **3**, 54-62.

Campbell, J.D. (1957) Discussion, in Proceedings, *Conf. on Properties of Materials at High Rates of Strain*, London, 1957, The Inst. of Mech. Engrs., London, pp. 353.

Campbell, J.D. and Duby, J. (1957) Delayed yield and other dynamic loading phenomena in a medium-carbon steel, in Proceedings, *Conf. on Properties of Material at High Rates of Strain*, London, 1957, The Inst. of Mech. Engrs., London, pp. 214-20.

Campbell, J. D., and March, K.J. (1962) Effect of grain size on delayed yielding of mild steel, *Phil. Magazine* **7**, 933-52.

Campbell, J. D, Simmons, J.A. and Dorn, J.E. (1961) On the dynamic behaviour of a Frank-Read source, *J. Appl. Mech.* **28**, Ser. E, 447-53.

Campbell, W.R. (1952) Determination of dynamic stress-strain curves from strain waves in long bars, *Proc.*

Soc. Stress Anal. **10**, 113-24.
Carrington, W.E. and Gayler, M.L.V. (1948) The use of flat-ended projectiles for determining dynamic yield stress. III. Changes in microstructure caused by deformation under impact at high striking velocities, *Proc. Roy. Soc. A* **194**, 323-31.
Chadwick, P. (1962) Propagation of spherical plastic-elastic disturbances from an expanded cavity, *Quart. J. Mech. Appl. Math.* **15**, 349-76.
Chadwick, P., Cox, A.D. and Hopkins, H.G. (1964) Mechanics of deep underground explosions, *Phil. Trans. Roy. Soc. London A* **256**, 235-300.
Chiddister, J.L. and Malveen, L.E. (1963) Compression-impact testing of aluminium at elevated temperatures, *Proc. Soc. Exptl. Stress Anal.* **20**, 81-90.
Chou, P.C. (1961) Perforation of Plates by high-speed projectiles, in *Developments in Mechanics Lay*, J.E. and Malvrn, L.E. (eds.), North Holland, Amsterdam, Vol. **1**, pp. 286-95.
Chou, P.C. (1962) Visco-plastic flow theory in hyprvelocity perforation of plates, in *Proc. Fifth Symp. Hypervelocity Impact*, Denver, Colorado, 1962, Vol. 1, Part 1, pp. 307-28.
Chou, P.C. and Allison, F.E. (1966) Strong plane shock produced by hyper velocity impact and late-stage equivalence, *J. Appl. Phys.* **37**, 853-60.
Clark, A.B.J. (1956) Static and dynamic calibration of photoelastic model material, *Proc. Soc Exptl. Stress Anal.* **14**, 195-204.
Clifton, R.J. and Bodner, S.R. (1966) Analysis of longitudinal elastic-plastic pulse propagation, *J. Appl. Mech.* **33**, Ser. E, 248-55.
Coleman, B.D. (1964a) Thermodynamics of materials with memory, *Arch. Ration. Mech.Anal.* **17**, 1-46.
Coleman, B.D. (1964b) On the thermodynamics, strain impulses, and viscoelasticity, *Arch. Rational Mech. Anal.* **17**, 230-54.
Coleman, B.D. and Mizel, V.I. (1963) Thermodynamics and departure from Fourier's law of heat conduction, *Arch. Rational Mech. Anal.* **13**, 245-61.
Coleman, B.D. and Mizel, V.I. (1964) Existence of caloric equations of state in Thermodynamics, *J. Chem. Phys.* **40**, 1116-25.
Coleman, B.D. and Noll, W. (1960) An approximation theorem for functionals with applications in Continuum mechanics, *Arch. Ration Mech. Anal.* **6**, 355-70.
Coleman, B.D. and Noll, W. (1963) The thermodynamics of elastic materials with heat conduction and viscosity, *Arch. Rational Mech. Anal.* **13**, 167-78.
Coleman, B.D. and Noll, W. (1964) Material symmetry and thermostatic inequalities in finite elastic deformation, *Arch. Rational Mech. Anal.* **15**, 87-111.
Cook, M.A. (1959) Mechanism of cratering in ultra-high velocity impact, *J. Appl. Phys.* **30**, 725-35.
Cotter, B.A. and Symonds, P.S. (1955) Impulsive loading of elastic-plastic beams, *Proc. ASME* **81**, pp. 21.
Cottrell, A.H. and Bilby, B.A. (1949) Dislocation theory of yielding and strain aging of iron, *Proc. Phys. Soc. London A* **62**, 49-62 .
Courant, R. and Friedrichs, K.O. (1948) *Supersonic Flow and Shock Waves*, Interscience, New York.
Cox, A.D., Eason, G. and Hopkins, H.G. (1961) Axially symmetric plastic deformation in soils, *Phil. Trans. Roy. Soc. London A* **254**, 1-45.
Cox, A.D. and Morland, L.W. (1959) Dynamic plastic deformation of simply-supported square plates, *J. Mech. Phys. Solids* **7**, 229-41.
Craggs, J.W. (1965) Rate-dependent theory of plasticity, *Int. J. Eng. Sci.* **3**, 21-26.
Cristescu, N. (1957a) Sur la propagation des ondes plastiques dans les barres d'acier, in Proc. 9th Int. Congr. Appl. Mech., Bruxelles, 1956, Vol.8, pp. 191-201.
Cristescu, N. (1964a) Some problems in the mechanics of extensible strings, IUTAM Symp., *Stress Waves in Anelastic Solids*, Kolsky, H. and Prager, W. (eds.), Providence, 1963, Springer-Verlag, Berlin, pp. 118-32.
Cristescu, N. (1965) The braking of high speed moving bodies by extensible strings, *Proc. 4th Int. Cong. on Rheology*, Providence, 1963, John Wiley, New York, Vol. 3, pp. 59-78.

Cristescu, N. (1966a) About the propagation of elastic/plastic waves in thin reds, in *Dynamics of Machines*, Proc. Conf. Liblice-Praga, 1963 (Bratislava), pp. 87-96.
Cristescu, N. (1966b) Some dynamic problems in one-dimensional elastic/visco/plastic bodies, Proc. XIth Int. Congr. Appl. Mech., Munchen 1964, Springer-Verlag, Berlin, pp. 390-5.
Cristescu, N. and Predeleanu, M. (1965) Dynamic stress waves in viscoelastic plastic bodies, *Proc. 4th Int. Cong. on Rheology*, Providence, 1963, John Wiley, Vol. 3, pp. 79-93.
Curran, D.R. (1965) Residual strains in shock-loaded aluminium, *J. Appl. Phys.* **36**, 2591-2.
David, E., Schall, R. and Schardin, H. (1964) Visualization of wave propagation in impulse-loaded bars, in Proceedings, IUTAM Symp., *Stress Waves in Anelastic Solids*, Providence, 1963, Springer-Verlag, Berlin, pp. 183-92.
Davids, N. (1960b) Some problems of transient analysis of waves in plates, in Proceedings, Int.Symp., *Stress Wave Propagation in Materials*, Pennsylvania, 1959, Interscience Publ., New York, pp. 271-7.
Davies, R.M. (1953) Stress waves in solids, *Appl. Mech. Rev.* **6**, 1-3.
Davies, R.M. (1956) Stress waves in solids, *British J. Appl. Phys.*, **7**, 203-9.
Davis, R.S. and Jackson, K.A. (1960) On the deformation associated with compression shock in crystalline solids, in Proceedings, *Plasticity*, Second Symp. on Naval Structural Mechanics, Providence, 1960, Pergman Press, New York, pp. 100-6.
De Costello, E. (1957) Yield strength of steel at an extremely high rate of strain, in Conf. Proc., *Properties of Materials at High Rates of Strain*, London, 1957, The Institution of Mech. Engrs., London, pp. 13-21.
De Groot, S.R. and Mazur, P. (1962) *Non-equilibrium Thermodynamics*, North Holland, Amsterdam.
Dietz, A.G.H. and Eirich, F.K., eds. (1960) *High Speed Testing*, Proc. Symp., Boston, Massachusetts, 1958, Interscience, New York, Vol. 1.
Dietz, A.G.H. and Eirich, F.K., eds. (1961) *High Speed Testing*, Proc. Symp., Boston, Massachusetts, 1960, Interscience, New York, Vol. II.
Dietz, A.G.H. and Eirich, F.K., eds. (1962) *High Speed Testing*, Proc. Symp., Boston, Massachusetts, 1961, Interscience, New York, Vol. III.
Dillon Jr., O.W. (1965) Experiemental data on aluminimum as an unstable solid, in *Proceedings, Fourth Int. Congr. on Rheology*, Providence 1963, John Wiley, New York, Part 2, pp. 377-94.
Dillon Jr., O.W. (1966) Waves in bars of mechanically unstable materials, *J. Appl. Mech.* **33**, Ser. E, pp. 267-74.
Duvall, G.E. (1961) Some properties and applications of shock waves, in Conf. Proc., *Response of Metals to High Velocity Deformation*, Shewmon, P.G. and Zackay, V.F. (eds.), Colorado 1960, Interscience, London, pp. 165-201.
Duvall, G.E. (1962) Shock wave stability in solids, in "Les ondes de détonation", Colloque CNRS, 1961, CNRS, Paris, pp. 337-53.
Duvall, G.E. (1964) Propagation of plane shock waves in a stress-relaxing medium, in Proceedings, IUTAM Symposium, *Stress Waves in Anelastic Solids*, Providence, 1963, Springer-Verlag, Berlin, pp. 20-32.
Eichelberger, R.J. (1960) Effects of very intense stress waves in solids, in Proc. Int. Symp., *Stress Wave Propagation in Materials*, Pennsylvania, 1959, Interscience, New York, pp. 133-66.
Eichelberger, R.J. (1965) Hypervelocity impact, in Colloquium on *Behaviour of Materials under Dynamic Loading*, Winter Ann. Meeting of ASME, Chicago, Illinois, ASME, New York, pp. 155-87.
Ely, R.E. (1960) Review of a high-speed tensile testing program for thermoplastics, in Symp. Proc., High Speed Testing, Dietz, A.G.H. and Eirich, F.K. (eds.), Boston, Massachusetts, Interscience, New York, pp. 3-24.
Freudenthal, A.M. (1958) The mathematical theories of the inelastic continuum, in *Handbuch der Physik*, Band VI, Springer-Verlag, Berlin.
Gilman, J.J. (1960) Physical nature of plastic flow and fracture, in *Plasticity*, Proc. Second Symp. on Naval Structural Mechanics, Providence 1960, Pergamon Press, New York, pp. 44-99.
Gilman, J.J. (1966) Progress in the micro mechanical theory of plasticity, in Proc. 5th U.S. National Congr.

Applied Mechanics, University of Minneapolis, 1966, ASME, New York.

Glass, C.M., Moss, G.L. and Golaski, S.K. (1961) Effects of explosive loading on single crystals and polycrystalline aggregate, in Conf. Proc., *Response of Metals to High Velocity Deformation*, Shewmon, P.G. and Zackay, V.F. (eds.), Colorado, 1960, Interscience, New York, pp. 115-41.

Goldsmith, W. and Austin, C.F. (1964) Some dynamic characteristics of rocks, IUTAM Symp., *Stress Waves in Anelastic Solids*, Providence, 1963, Kolsky, H. and Prager, N. (eds.), Springer-Verlag, Berlin, pp. 277-303.

Green, W.A. (1960) Dispersion relations for elastic waves in bars, *Progress in Solid Mechanics*, Sneddon, I.N. and Hill, R. (eds.), North Holland, Amsterdam, Vol. I, pp. 225-61.

Grigorian, S.S. (1964) On some simplifications in the description of the motion of a soft soil, IUTAM Symp., *Stress Waves in Anelastic Solids*, Kolsky H. and Prager, W. (eds.), 1963, Springer-Verlag, Berlin, pp. 304-13.

Grine, M. (1962) Discussion, in "Les ondes de détonation", Colloque CNRS, 1961, CNRS, Paris, pp. 351-2.

Hadamard, J. (1903) *Leçons sur la Propagation des Ondes*, Hermann, Paris.

Hauser, F.E., Simmons, J.A. and Dorn, J.E. (1961) Strain rate effects in plastic wave propagation, in Conf. Proc., *Response of Metals to High Velocity Deformation*, Shewmon, P.G. and Zackay, V.F. (eds.), Colorado, 1960, Interscience, New York, pp. 93-110.

Hill, R. (1961) Discontinuity relations in mechanics of solids, in *Progress in Solid Mechanics*, North Holland, Amsterdam, Vol. II, pp. 245-76.

Holtzman, A.H. and Cowan, G.R. (1961) The strengthening of austenitic manganese steel by plane shock waves, in Conf. Proc., *Response of Metals to High Velocity Deformation*, Shewmon, P.G. and Zackay, V.F. (eds.), Colorado, 1960, Interscience, New York, pp. 447-81.

Hopkins, H.G. (1961) Dynamic anelastic deformation of metals, Appl. Mech. Rev. **14**, pp. 417-31.

Hopkins, H.G. (1964) Mechanical waves and strain-rate effects in metals, IUTAM Symposium, *Stress Waves in Anelastic Solids*, Kolsky, H. and Prager, W. (eds.), Springer-Verlag, Berlin, pp. 133-48.

Hunter, S.C. (1957) The propagation of spherical symmetric disturbances in ideally plastic materials, in Conf. Proc., *Properties of Materials at High Rates of Strain*, London, 1957, Inst. Mech. Engrs., London, pp. 147-55.

Hunter, S.C. and Johnson, I.A. (1964) The propagation of small amplitude elastic-plastic waves in pre-stressed cylindrical bars, IUTAM Symp., *Stress Waves in Anelastic Solids*, Providence, 1963, Springer-Verlag, Berlin, pp. 149-65.

Jones, J.W. (1961) Tensile testing of elastomers at ultra-high strain rates, in Conf. Proc., *High Speed Testing*, Dietz, A.G.H. and Eirich, F.R. (eds.), Boston, Mass., 1960, Interscience, London, Vol. II, pp. 29-35.

Kaliski, S. and Osiecki, J. (1959) The unloading wave in a body with variable unloading modulus, in Proceedings, *Symp. on Non-Homogeneity in Elasticity and Plasticity*, Warsaw, 1959, Pergamon Press, New York, pp. 403-13.

Keil, A.H. (1960) Problems of plasticity in naval structures: Explosive and impact loading, Proc. 2nd Symp., *Plasticity*, Lee, E.H. and Samoyeds, P.S. (eds.), Naval Structural Mechanics, Providence, 1960, Pergamon Press, New York, pp. 23-39.

Kolsky, H. and Prager, W., eds. (1964) *Stress Waves in Anelastic Solids*, IUTAM Symp., Providence, Springer-Verlag, Berlin.

Krafft, J.M. (1961) Instrumentation for high-speed strain measurement, in Conf. Proc., *Response of Metals to High Velocity Deformation*, Shewmon, P.G. and Zackay, V.F. (eds.), Colorado, 1960, Interscience, London, pp. 9-48.

Lee, E.H. and Liu, D.T. (1964) An example of the influence of yield on high pressure wave propagation, Kolsky, H. and Prager, W. (eds.), IUTAM Symposium, *Stress Waves in Anelastic Solids*, Providence, 1963. Springer-Verlag, 1963, pp. 239-54.

Lenskii, V.S. (1957) The construction of a dynamical tension-compression diagram based on the wave propagation theory, in Conf. Proc., *Properties of Materials at High Rates of Strain*, London, 1957, Inst. Mech. Engrs., London, pp. 173-6.

Leni-Civita, T. (1932) Charactéristiques de systèmes différentiels et propagation des ondes, Félix Alcan, Paris.

Lewis, G.M (1957) A method of investigation of the stress-strain behaviour of fibres at very high rates of extension, in *Conf. Proc. Properties of Materials at High Rates of Strain*, London, 1957, Inst. Mech. Engrs., London, pp. 190-4.

MacDonald, R.J., Carlson, R.L. and Lankford, W.T. (1956) Apparatus for determination of stress-strain properties at high rates of strain, Proc. Soc. Exp. Stress Anal. **14**, pp.163-70.

Maiden, C.J. and Campbell, J.D. (1958) Static and dynamic strength of carbon steel at low temperatures, *Phi. Mag.* **3**, 872-85.

Mallory, H.D. (1955) Propagation of shock waves in aluminium, *J. Appl. Phys.* **26**, 555-9.

Malvern, L.E. (1951a) Plastic wave propagation in a bar of material exhibiting a strain rate effect, *Quart. Appl. Math.* **8**, 405-11.

Malvern, L.E. (1951b) Propagation of longitudinal waves of plastic deformation in bar of material exhibiting strain rate effect, *J. Appl. Mech.* **18**, 203-8.

Mandel, J. (1962) Ondes plastiques dans un milieu indéfini à trois dimensions, in *Extrait du séminaire de plasticité*, Publ. Sci. Techn. Min. Air. NT 116 (Paris), 151-78.

Mandel, J. (1964) Propagation des surfaces de discontinuité dans un milieu élastoplastique, IUTAM Symp., *Stress Waves in Anelastic Solids*, H. Kolsky and W. Prager (eds)., Providence, 1963, Springer-Verlag, 331-40.

Marsh, K.J. and Campbell, J.D. (1963) Effect of strain rate on post-yield flow of mild steel, J. *Mech. Phys. Solids* **11**, 49-63.

Mason, P. (1960) Longitudinal wave propagation in stretched polymers, *J. Appl. Phys.* **31**, 1706-8.

Maurer, W.C. and Rinehart, J.S. (1960) Impact crater formation in rock, in *Hyper-velocity Impact* **3**, 1-20.

McQueen, R.G. and Marsh, S.P. (1960) Equation of state for nineteen metallic elements from shock-wave measurements to two Megabars, *J. Appl. Phys.* **31**, 1253-69.

Mellor, P.B. (1956) Stretch forming under fluid pressure, *J. Mech. Phys. Solids* **5**, 41-56.

Mellor, P.B. (1960) Ultimate strength of thin-walled shells and circular diaphragms subjected to hydrostatic pressure, *Intern. J. Mech. Sci.* **1**, 216-28.

Mentel, T.J. (1955) Plastic deformations due to dynamic loading of beam with attached mass, *Can. J. Technol.* **33**, 237-55.

Miklowitz, J. (1947) The initiation and propagation of the plastic zone along a tension specimen of nylon, *J. Colloid Sci.* **2**, 193-215.

Minshall, F.S. (1961) The dynamic response of iron and iron alloy to shock waves, in *Response of Metals to High Velocity Deformation*, Proc. Conf., P.G. Shewmon and V.F. Zackay (eds.), Colorado, 1960 (Interscience Publ., New York-London), 249-72.

Morland, L.W. (1959) Propagation of plane irrototional waves through elastoplastic medium, *Phil. Trans. Roy. Soc. London A* **251**, 341-83.

Morrison, J.A. (1956) Wave propagation in rods of Voigt material and viscoelastic materials with three-parameter models, *Quart. Appl. Math.* **14**, 153-69.

Munday, G. and Newitt, D.M. (1963) Deformation of transversely loaded disks under dynamic loads, *Phil. Trans. Roy. Soc. London A* **256**, 1-30.

Nadai, A. (1963) *Theory of Flow and Fracture of Solids*, McGraw-Hill, New York, Vol. 2, Part 4.

Naghdi, P.M. and Murch, S.A. (1963) On the mechanical behaviour of viscoelastic/plastic solids, *J. Appl. Mech.* **30**, Ser. E, 321-28.

Noll, W. (1958) A mathematical theory of the mechanical behaviour of continuous media, *Arch. Rational Mech. Anal.* **2**, 197-226.

O'Brien, J.L. and Davis, R.S. (1961) On the fracture of solids under impulsive loading conditions, in Response of Metals to High Velocity Deformation, Proc. Conf. Colorado, 1960, P.G. Shewmon and V.F. Zackay (eds.), Interscience Publ., New York, 371-88.

Olszak, W., ed. (1959) *Non-homogeneity in Elasticity and Plasticity*, Warsaw Symp. 1959, Pergamon Press,

London.

Olszak, W., Mroz, Z. and Perzyna, P. (1963) *Recent Trends in the Development of the Theory of Plasticity*, Pergamon Press, Oxford.

Olszak, W. and Perzyna, P. (1962) Propagation of spherical waves in a non-homogeneous elastic-viscoplastic medium, in *La propagation des ébranlements dans les milieux hétérogènes*, Colloque CNRS, Marseille, 1961 (Ed. CNRS, Paris), 67-78.

Owens, R.H. and Symonds, P.S. (1955) Plastic deformation of free ring under concentrated dynamic loading, *J. Appl. Mech.* **22**, 523-9.

Parkes, E. W. (1955) Permanent deformation of cantilever struck transversely at its tip, *Proc. Roy. Soc. A* **228**, 462-76.

Partridge, W.S., Vanfleet, H.B. and Whited, C.R. (1958) Crater formation in metallic targets, *J. Appl. Phys.* **29**, 1032; 1332-6.

Paul, W. and Warschauer, D.M., eds. (1963) *Solids under Pressure*, McGraw-Hill, New York.

Perzyna, P. (1958) Dynamic load bearing strength of circular plate, *Arch. Mech. Stos.* **10**, 635-47.

Perzyna, P. (1959) The problem of propagation of elastic-plastic waves in a non-homogeneous bar, in W. Olszak (ed)., Non-homogeneity in Elasticity and Plasticity, Symp., Warsaw, 1959, Pergamon Press, London, 431-8.

Perzyna, P. (1962) Propagation of shock wave in elastic-visco-plastic medium of definite non-homogeneity type, *Arch. Mech. Stos.* **14**, 93-111.

Perzyna, P. (1963) Constitutive equation for rate sensitive plastic materials, *Quart. Appl. Math.* **20**, 321-32.

Perzyna, P. and Wierzbicki, T. (!964) Temperature-dependent and strain-rate sensitive plastic materials, *Arch. Mech. Stos.* **16**, 135-43.

Petterson, D.R., Stewart, G.M., Odell, F.A. and Maheux, R.C. (1960) Dynamic distribution of strain in textile materials under high speed impact, *Textile Res. J.* **30**, 411-21.

Phillips, A. and Sun, C.T. (1965) Finite expansion of a spherical cavity in a locking medium, in *Proc. 4th Intern. Congr. on Rheology*, E.H. Lee (ed.), Providence, 1963, John Wiley, New York, Part 2, 61-76.

Pian, T.H.H. (1960) Dynamic response of thin shell structures, in *Plasticity*, Proc. 2nd Symp. on Naval structural mechanics, E.H. Lee and P.S. Symonds (eds.), Providence, 1960, Pergamon Press, New York, 443-52.

Pilsworth Jr., M. N. and Hoge, H.J. (1965) Rate-dependent response of polymers to tensile impact, *Textile Res. J.* **35**, 129-39.

Pironneau, Y. (1953) The deformation by repeated blows of single crystals of Al-Cu Alloy, *C.R. Acad. Sci., Paris* **236**, 46-8.

Plass Jr., H.J. (1960). A theory of longitudinal plastic waves in rods of strain-rate dependent material, including effects of lateral inertia and shear, in: *Plasticity*, Proc. 2nd Symp., *Naval structural mechanics*, E.H. Lee and P.S. Symonds (eds.), Providence, 1960, Pergamon Press, New York, 453-74.

Prager, W. (1949) Recent developments in Mathematical Theory of Plasticity, *J. Appl. Phys.* **20**, 235-41.

Prager, W. (1957) Ideal locking materials, *Trans. Soc. Rheol.* **1**, 169-75.

Prandtl, L. (1928) Hypothetical model for the kinetic theory of solid bodies, *Z. Angew. Math. Mech.* **8**, 85-106.

Pugh, E.M., Heine-Geldern, R. v., Fouer, S. and Mutschler, E.C. (1952) Glass cracking caused by high explosive, *J. Appl. Phys.* **23**, 48-53.

Ramberg, W. and Irwin, L.K. (1956) Longitudinal impact tests of long bars with a slingshot machine, in: *Proc. Symp. on Impact Testing*, 58 Ann. Meeting ASTM, Atlantic City, N.J., 1955, ASTM Special Techn. Publ., no. 176.

Rawlings, B. (1963a) The present state of knowledge of the behaviour of steel structures under the action of impulsive loads. *Civil Eng. Trans. of Inst. of Engrs., Australia*, Sept. 1963, 89-103.

Rawlings, B. (1963b) Dynamic behaviour of mild steel in pure flexure, *Proc. Roy. Soc. A* **275**, 528-43.

Rawlings, B. (1964a) Dynamic plastic analysis of steel frames, *J. Struct. Div., Am. Soc. Civil Engrs.* **90**, no.

ST3, 265-82.

Rawlings, B. (1964c) A computer analysis of structures under impulsive loading, in Preliminary Publ., 7th Congr. for Bridge and Struct. Engr., Rio de Janeiro, 45-54.

Rawlings, B. (1964d) Impact tests on model steel frames, Proc. Inst. Civ. Engrs., 29, 389-414.

Rawlings, B. (1964e) Mode changes in frames deforming under impulsive loads, *J. Mech. Engr. Sci.* **6**, 327-36.

Rawlings, B. (1965) Dynamic changes of mode in rigid plastic structures, *J. Eng. Mech. Div., Am. Soc. Civil. Engrs.* **91**, No. EM2, 1-20.

Recht, R.F. and Ipson, T.W. (1963) Ballistic deformation dynamics, *J. Appl. Mech.* **30**, Ser. E., 384-90.

Rice, M.H., McQueen, R. G. and Walsh, J.M. (1958) Compression of solids by strong shock waves, in: *Solid State Physics*, Seitz, F. and Turnbull, D. (eds.), Academic Press, New York, Vol. 6, 1-63.

Rinehart, J. S. (1960) The role of stress waves in the comminution of brittle, rocklike materials, in: *Stress wave propagation in materials*, N. Davids ed., Intern. Symp., Pennsylvania, 1959, Interscience Publ., New York, 247-68.

Rinehart, J. S. (1964) Transient stress wave boundary interactions, in H. Kolsky and W. Prager (eds.), *IUTAM Symp, Stress Waves in Anelastic Solids*, Providence, 1963, Springer-Verlag, Berlin, 193-206.

Rinehart, J. S. and Pearson, J. (1954) *Behaviour of Metals under Impulsive Loads*, Publ. Amer. Soc. for Metals, Cleveland, Ohio.

Rinehart, J. S. and Pearson, J. (1963) *Explosive Working of Metals*, Pergamon Press, London.

Ringleb, F.O. (1957) Motion and stress of elastic cable due to impact, *J. Appl. Mech.* **24**, 417-25.

Riparbelli, C. (1955) Discussion: Analysis of plastic deformation in a steel cylinder striking a rigid target, *J. Appl. Mech.* **22**, 131-2.

Ripperger, E. A. (1960) Experimental studies of plastic wave propagation in bars, in *Plasticity*, Proc. 2nd Symp. on Naval structural mechanics, E. H. Lee and P.S. Symonds (eds.), Providence, 1960, Pergamon Press, New York, 475-85.

Rolsten, R.F. (1965) Hyper-velocity crater depth and target strength, *AIAA J.* **3**, 2149-51.

Rubin, R. J. (1954) Propagation of longitudinal deformation waves in prestressed rod of material exhibiting a strain-rate effect, *J. Appl. Phys.* **25**, 528-31.

Salvadori, M. G. and Di Maggio, F. (1953) On the development of plastic hinges in rigid plastic beams, *Quart. Appl. Math.* **11**, 223-320.

Salvadori, M .G., Skalak, R. and Weidlinger, P. (1960) Waves and shocks in locking and dissipative media, *J. Eng. Mech. Div., Am. Soc. Civil Engrs.* **86**, no. EM2, Part 1, 77-105.

Salvadori, M.G., Skalak, R. and Weidlinger, R. (1961) Spherical waves in plastic locking medium, *J. Eng. Mech. Div., Am. Soc. Civil Engrs.* **87**, no. EM1, 1-11.

Savin, G. N. (1964) Dynamics of an anelastic string of variable length, in IUTAM Symp., *Stress Waves in Anelastic Solids*, H. Kolsky and W. Prager (eds.), Providence, 1963, Springer-Verlag, Berlin, 78-86.

Schardin, H. (1960) Cinematography of stress waves, in Intern. Symp., *Stress Wave Propagation in Materials*, N. Davids (ed.), Pennsylvania, 1959, Interscience Publ., New York, 289-302.

Schiffman, R. L. (1965) The viscoelastic compression of soil-water systems, in *Proc. 4th Intern. Congr. Rheology*, E. H. Lee (ed.), Providence, 1963, John Wiley, New York, Part 2, 397-422.

Seiler, J. A., Cotter, B.A. and Symonds, P.S. (1956) Impulsive loading of elastic-plastic beams, *J. Appl. Mech.* **23**, 515-21.

Seiler, J.A. and Symonds, P. S. (1954) Plastic deformation in beams under distributed dynamic loads, *J. Appl. Phys.* **25**, 556-63.

Shewmon, P. G. and Zackay, V. F., eds. (1961) Response of metals to high velocity deformation, Proc. Conf. Colorado, 1960, Interscience Publ., New York.

Shield, R. T. (1955) On Coulomb's law of failure in soils, *J. Mech. Phys. Solids* **4**, 10-16.

Skalak, R. and Weidlinger, P. (1961) Attenuation of stress waves in bi-linear materials, *J. Eng. Mech. Div., Am. Soc. Civil Engrs.* **87**, no. EM3, Part 1, 1-12.

Smith, J. C. (1966) Wave propagation in three-element linear spring and dashpot model filament, *J. Appl.*

Phys. **37**, 1697-1704.

Smith, J. C., Blandford, J. M. and Schiefer, H. F. (1960) Stress-strain relationships in yarns subjected to rapid impact loading, *Textile Res. J.* **30**, 752-60.

Smith, J. C., Fenstermaker, C. A. and Shouse, P. J. (1962) Behaviour of filamentous materials subjected to high-speed tensile impact, in: *Proc. Symp. on Dynamic Behaviour of Materials*, Special Techn. Publ., no. 336, 47-69.

Smith, J. C., Fenstermaker, C. A. and Shouse, P. J. (1963) Stress-strain relationships in yarns subjected to rapid impact loading, *Textile Res. J.* **33**, 919-34.

Smith, J. C., Fenstermaker, C. A. and Shouse, P. J. (1965) Stress-strain relationships in yarns subjected to rapid impact loading, *Textile Res. J.* **35**, 743-57.

Smith, C. S. and Fowler, C. M. (1961) Further metallographic studies on metals after explosive shock, in: *Response of metals to high velocity deformation*, P.G. Shewmon and V. F. Zackay, eds., Proc. Conf. Colorado, 1960, Interscience Publ., New York, 309-35

Smith, J. C., Mc Crackin, F.L. and Schiefer, H.F. (1955) Stress-strain relationships in yarns subjected to rapid impact loading, *J. Res. Nat. Bur. Std.* **55**, 19-28.

Smith, J. C., Mc Crackin, F. L. and Schiefer, H. F. (1958) Stress-strain relationships in yarns subjected to rapid impact loading. Wave propagation in long textile yarns impacted transversely, *J. Res. Nat. Bur. Std.* **60**, 517-34.

Smith, G. F. and Rivlin, R. S. (1957) The anisotropic tensors, *Quart. Appl. Math.* **15**, 308-14.

Smith, R.C. (1961) Studies of effects of dynamic preloads on mechanical properties of steel, *Proc. Soc. Exptl. Stress Anal.* **18**, 153-59.

Smith. R.C., Pardue, T. E. and Vigness, I. (1956) Mechanical properties of certain steels indicated by axial dynamic load tests, *Proc. Soc. Exptl. Stress Anal.* **13**, 183-97.

Spencer, A. J. M. (1960) Dynamic plane deformation of ideal plastic-rigid solid, *J. Mech. Phys. Solids* **8**, 262-79.

Sperranzza, J. (1962) Propagation of large amplitude waves in pure lead, in Proc. 4th U.S. Nat. Congr. Appl. Mech., Vol. 2, 1123-29.

Sternglass, E. J. and Stuart, D.A. (1953) Experimental study of propagation of transient longitudinal deformations in elastoplastic media, *J. App. Mech.* **20**, 427-34.

Stone, W. K., Schiefer, H. F. and Fox, G. (1955) Stress-strain relationships in yarns subjected to rapid impact loading, *Textile Res. J.* **25**, 520-34.

Strella, S. (1960) High-rate tension testing of plastics, in Proc. Symp. Boston, Mass., Dietz, A.G.H., and Eirich, F.R. (eds.), *High speed testing*, 1958, Interscience Publ., New York - London, Vol. 1, pp. 27-40.

Stresan, R.H. and Napadensky, H.S. (1962) Observation and measurement of behaviour of porous materials when rapidly compressed, *Proc. Soc. Exptl. Stress Anal.* **19**, 15-22.

Sutcliffe, S. (1963) Strong shock formation in bilinear hardening media, *J. Eng. Mech. Div., Am. Soc. Civil Engrs.* **89**, no. EM2, Part 1, 1-12.

Symonds, P.S. and Leth, C. (1954) Impact of finite beams of ductile metal, *J. Mech. Phys. Solids* **2**, 92-102.

Symonds, P.S. and Mentel, T.J. (1958) Impulsive loading of plastic beams with axial constraints. *J. Mech. Phys. Solids* **6**, 186-202.

Symonds, P.S. and Ting, T.C.T. (1964) Longitudinal impact on viscoelastic rods. Approximate methods and comparisons, *J. Appl. Mech.* **31**, 611-20.

Tapley, B.D. (1962) The propagation of plastic waves in finite specimens of a strain-rate dependent material, in Proc. 4th U.S. Nat. Congr. Appl. Mech., Vol. 2, 1137-46.

Tapley, B.D. and Plass, Jr., H.J. (1961) The propagation of plastic waves in a semi-infinite cylinder of a strain-rate-dependent material, in *Developments in mechanics*, Lay, J.E. and Malvern, L.E. (eds.), North-Holland Publ. Comp., Amsterdam, Vol. 1, 257-67.

Taylor, D.B.C. (1954) Dynamic straining of metals having definite yield points, *J. Mech. Phys. Solids* **3**, 38-46.

Taylor, D.B.C. (1957) Non-uniform yield in a mild steel under dynamic straining, in *Proc. Conf. Properties of Materials at High Rates of Strain*, Inst. Mech. Engrs., London, 229-38.

Taylor, D.B.C. and Malvern, L.E. (1961) Dynamic stress and deformation in a mild steel at normal and low temperatures, in Proc. Conf. Colorado, 1960, Shewmon, P.G. and Zackay, V.F. (eds.), *Response of Metals to High Velocity Deformation*, Interscience Publ., New York, 77-91.

Taylor, G.I. (1946) Testing of materials at high rates of loading, *J. Inst. Civil Engrs.* **2**, 486-519.

Taylor, G.I. (1958a) Propagation of earth waves from an explosion, in G. I. Taylor scientific papers, Vol. 1, Mechanics of solids, G.K. Batchelor (ed.), Cambridge Univ. Press, 456-63.

Taylor, G.I. (1958b) The plastic wave in wire extended by an impact load, in Taylor scientific papers, Vol. 1, Mechanics of solids, G.K. Batchelo (ed.)., Cambridge Univ. Press, 467-79.

Taylor, J.W. (1965) Dislocation dynamics and dynamic yielding, *J. Appl. Phys.* **36**, 3146-50.

Taylor, J.W. and Rice, M.H. (1963) Elastic-plastic properties of iron, *J. Appl. Phys.* **34**, 364-71.

Terzaghi, K. (1943) *Theoretical Soil Mechanics*, John Wiley, New York.

Thomas, T.Y. (1961) *Plastic Flow and Fracture in Solids*, Academic Press, New York.

Thomson, R.G. (1965) Analysis of hyper velocity perforation of a visco-plastic solid including the effects of target-material yield strength, *NASA* Techn. Rep., R-221.

Thomson, W.T. (1954) Impulsive response of beams in elastic and plastic regions, *J. Appl. Mech.* **21**, 271-8.

Thomson, W.T. (1955) Approximate theory of armor penetration, *J. Appl. Phys.* **26**, 80-82.

Ting, T.C.T. (1964) The plastic deformation of a cantilever beam with strain rate sensitivity under impulsive loading, *J. Appl. Mech.* **31**, Ser. E, 38-42.

Ting, T.C.T. (1965) Large deformation of rigid, ideally plastic cantilever beam, *J. Appl. Mech.* **32**, Ser. E, 295-302.

Ting, T.C.T. and Symonds, P.S. (1962) Impact of a cantilever beam with strain rate sensitivity, in Proc. 4th U.S. Nat. Congr. Appl. Mech., 1153-65.

Ting, T.C.T. and Symonds, P.S. (1964) Longitudinal impact on viscoelastic rods. Approximate methods and comparisons, *J. Appl. Mech.* **31**, Ser. E., 199-207.

Travis, F.W. and Johnson, W. (1963) The explosive forming of cones, in Proc. 3rd Intern. Conf. Machine tool design and research, Birmingham, 1962, Pergamon Press, Oxford, 341-64.

Truesdell, C. and Noll, W. (1965) The non-linear field theories of mechanics, in Flügge, S. ed., *Encyclopedia of physics*, Vol. III/3, Springer-Verlag, Berlin.

Truesdell, C. and Toupin, R. A. (1960) The classical field theories, in *Handbuch der Physik*, herausgegeben von S. Flügge, Springer-Verlag, Berlin, Band III/1, 226-744.

Turnbow, J.W. and Ripperger, E.A. (1959) Strain-rate effects on stress-strain characteristics of aluminium and copper, in Proc. 4th Midwestern Conf. Solid Mech., Univ. Texas.

Uzhik, G.V. and Voloshenko-Klimovitskii, Yu. Ya. (1957) On the elastic-plastic strain of steels under longitudinal impact, in *Proc. Conf. Properties of materials at high rates of strain*, London, Inst. Mech. Engrs., London, 239-43.

Vafiadakis, A.P., Johnson, W. and Donaldson, I.S. (1964) High-rate forming. *Proc. Inst. Mech. Engrs.* **179**, Part I, 222-34.

Van Valkenburg, M.E., Glay, W.G. and Huth, J.H. (1956) Impact phenomena at high speeds, *J. Appl. Phys.* **27**, 1123-9.

Van Wely, F.E. (1966) Explosive forming, Intern. Symp. on *High Energy Rate Forming*, Prague, Sept. 1966.

Vigness, I., Krafft, J.M. and Smith, R.C. (1957) Effect of loading history upon the yield strength of a plain carbon steel, in: *Proc. Conf. Properties of Materials at High Rates of Strain*, Inst. Mech. Engrs., London, 138-46.

Vincent, B.J., Gee, R. and Hunter, S.C. (1957) A study of rapid micro-yielding in mild steel by scleroscopic methods, in: *Proc. Conf. Properties of Materials at High Rates of Strain*, Inst. Mech. Engrs., London, 51-65.

Vitman, F.F. and Zlatin, N.A. (1949) Resistance against deformation of metals at velocities from 10^{-6} to 10^{2} m/sec. II, *J. Techn. Phys.* **19**, 315-26.

Vitman, F.F., Zlatin, N.A. and Ioffe, B.S. (1949) Resistance against deformation of metals at velocities from 10^{-6} to 10^{2} m/sec. I, *J. Techn. Phys.* **19**, 300-14.

Wackerle, J. (1962) Shock-wave comparison of Quartz, *J. Appl. Phys.* **33**, 922-37.

Walsh, J.M. and Christian, R.H. (1955) Equations of state of metals from shock wave measurements, *Phys. Rev.* **97**, 1544-56.

Walsh, J.M., Rice, M.H., Mc Queen, R.G. and Yarger, F.L. (1957) Shock-wave comparisons of twenty-seven metals. Equations of state of metals. *Phys. Rev.* **108**, 196-216.

Wang, A.J. (1955) The permanent deflections of a plastic plate under blast loading, *J. Appl. Mech.* **22**, 375-6.

Wang, A.J. and Hopkins, H.G. (1954) On plastic deformation of built-in circular plates under impulsive loading, *J. Mech. Phys. Solids* **3**, 22-37.

Watstein, D. (1956) Properties of concrete at high rates of loading, in: *Proc. Symp. on Impact Testing*, Atlantic City, N.J., 1955, ASTM, New York, 156-69.

Wen, R.K. and Torides, T. (1964) Discrete dynamic models for elasto-inelastic beams, *J. Eng. Mech. Div., Am. Soc. Civil Engrs.* **90**, No. EM5, 71-102.

Whiffin, A.C. (1948) The use of flat-ended projectiles for determining dynamic yield stress. II. Test on various metallic materials, *Proc. Roy Soc. A* **194**, 300-22.

White, M.P. (1949) On impact behaviour of material with yield point, *J. Appl. Mech.* **16**, 39-52.

White, M.P. (1955) Discussion. Analysis of plastic deformation in a steel cylinder striking a rigid target, *J. Appl. Mech.* **22**, 132-33.

White, M.P. and Van Griffis, L. (1947) Permanent strain in uniform bar due to longitudinal impact, *J. Appl. Mech.* **14**, 337-43.

Wick, C.H. (1960) *Chipless Machining*, The Industrial Press, New York, Ch. 16.

Wilson, F.W., ed. (1964) *High-Velocity Forming of Metals*, Prentice-Hall, Inc., Englewood Cliffs, N.J.

Woo, D.M. (1964) Analysis of axisymmetric forming of sheet metal and hydrostatic bulging process, *Intern. J. Mech. Sci.* **6**, 303-17.

Wood, D.S. (1952) On longitudinal plane waves of elastic-plastic stress in solids, *J. Appl. Mech.* **19**, 521-5.

Wright, R.N. and Hall, W.J. (1964) Loading effects in structural steel design, *J. Struct. Div., Am. Soc. Civil Engrs.* **90**, No. ST5, 11-37.

Yamaguchi, Y. (1965) Stress-strain-time relations in filamentous high polymeric materials, in Proc. 4th Intern. Congr. on Rheology, Lee, E.H. (ed.), Providence, 1963, John Wiley, New York, Part. 3, 283-99.

Yang, C.Y., Hansen, R.J. and Reinschmidt, K.F. (1963) Dynamic response of elastic-viscous-plastic columns, *J. Eng. Mech. Div., Am. Soc. Civil Engrs.* **89**, No. EM3, 43-57.

Yonezawa, H. (1959) Plastic deformations of beams under impulsive load, in Proc. 8th Japan Nat. Congr. Appl. Mech., Tokyo, 1958, 141-06.

Zaid, M. and Paul, B. (1957) Mechanics of high speed projectile perforation, *J. Franklin Inst.* **264**, 117-26.

Zaid, M. and Paul, B. (1958) Normal perforation of thin plate by truncated projectiles, *J. Franklin Inst.* **265**, 317-35.

Zielgler, H. (1963) Some extremum principles in irreversible thermodynamics with application to continuum mechanics, in: *Progress in solid mechanics*, Sneddon, I.N. and Hill, R. (eds.), North-Holland Publ. Comp., Amsterdam, Vol. IV, 93-193.

Zukas, E.G. and Fowler, C.M. (1961) The behaviour of iron and steel under impulsive loading, in Proc. Conf. Colorado, 1960, *Response of Metals to High Velocity Deformation*, Shewmon, P.G. and Zackay, V.F. (eds.), Interscience Publ., New York-London, 343-68.

Zvolinskii, N.V. and Rykov, G.V. (1964) Reflection of a plastic wave at an obstacle, IUTAM Symp., *Stress waves in Anelastic Solids,* Kolsky, H. and Prager, W.(eds.), Providence, 1963, Springer-Verlag, Berlin, 207-21.

CHAPTER 13

CHARACTERIZATION OF LINEAR VISCOELASTIC RESPONSE USING A DYNAMIC SYSTEM APPROACH

13.1. Introduction

The characterization of the relaxation and creep response functions of viscoelastic materials has always been a main research topic in *Viscoelasticity*. The construction of a precise model for describing the rheological response of the material and the establishment of an efficient and accurate method for determining the pertaining material response functions from experimental data have been basic tasks in this research field.

To achieve this objective, both analytical and experimental methods have been developed, and can be classified into the following three categories.

a) Quasi-static Methods

These methods are directly based on the definition of the creep, or the relaxation, function. By these methods, one usually conducts a series of quasi-static creep or relaxation experiments, in which, respectively, a constant stress or constant strain is loaded onto the specimen, and the time-dependent response of the material is measured. The pertaining creep or relaxation function is then determined from the experiment data (*Chapter 8*).

Quasi-static methods, as described above, are simple, but they have a vital shortcoming that a very long period of time is usually required so that the creep, or relaxation, properties of the material are to be fully demonstrated. This requirement of a very long time scale often constitutes a distinct obstacle for using quasi-static methods to determine the creep and relaxation functions of a viscoelastic material.

b) Time-temperature Superposition Methods

To overcome the inconvenience of long time testing periods in the quasi-static methods, the so-called time-temperature superposition (TTS) method has been developed. The basis of this method is that the time and temperature effects on a linear viscoelastic material are directly interrelated. At a low testing temperature, the creep, or the relaxation, experiment requires a long period of time, while at a higher temperature, it takes a relatively shorter period of time. By using the TTS method, the creep or relaxation experiments are conducted at elevated temperatures for relatively short periods of time, then, the measurements are transformed to obtain the corresponding viscoelastic properties at the required low temperature (usually the room temperature).

The TTS method would demand, however, a set of complex temperature control facilities. Another difficulty related to this method stems from the fact that nonlinearity in the viscoelastic response of the material could be introduced at high temperature.

c) Dynamic Methods

Dynamic methods are based on the results of dynamic experiments performed on viscoelastic materials. In these methods, the experiments are often conducted by applying a sinusoidal loading on the viscoelastic material and the resulting experimental response data are gathered in a frequency domain to obtain pertaining response spectra. The creep or relaxation functions are, then, obtained from the analysis of these spectra (*Chapter 8*).

The most distinct advantage of the dynamic methods over the quasi-static and TTS methods is that the dynamic experiments are relatively easy to conduct within a short period of time and without the need of complex experimental facilities. Therefore, the use of dynamic methods are attracting recently more and more attention from the researchers in the field.

Gibson, *et al.* (1990), for instance, presented a method by which experimental dynamic data are used to determine both dynamic and quasi-static viscoelastic response behaviour of the material. In their method, the complex moduli are obtained first from vibration measurements by employing the Fast Fourier Transform technique. Then, the quasi-static creep or relaxation properties are calculated from the already determined complex moduli by a numerical integration algorithm.

13.2. Dynamic System Identification Methods

In this chapter, a linear viscoelastic material is considered as a dynamic system. From this point of view, a dynamic system identification method is presented for the determination of the relaxation or creep function of the material from dynamic experimental measurements. *First*, the relation between the relaxation or creep function and the frequency response function of the system is established by assuming a model of rational function of polynomials for the frequency response function. *Second*, a discrete-time system analysis method is introduced to identify the order and parameters of the model. Within the context of this approach, the presentation of this chapter deals distinctly with the following two topics:

(i) Characterization of the response behaviour of linear viscoelastic materials by incorporating the measurements of the time-rate of the input, together with those of the output signals. This is carried out in Section 13.3.

(ii) Extending the identification model introduced under (i) above to include the instantaneous response behaviour at time t=0. This is presented in Section 13.4.

The presented approach is illustrated, at its various stages, by numerical examples.

13.3. Discrete-time System Analysis as Based on the Time-rate of the Input Signal

For a linear viscoelastic behaviour, the general expressions of the constitutive relationship can be written for both relaxation and creep, respectively, as follows

$$\sigma(t) = \int_{-\infty}^{\infty} \frac{d\varepsilon(\tau)}{d\tau} R(t - \tau) d\tau \qquad (13.1)$$

$$\varepsilon(t) = \int_{-\infty}^{\infty} \frac{d\sigma(\tau)}{d\tau} C(t - \tau) d\tau \qquad (13.2)$$

where $\sigma(t)$ is the stress, $\varepsilon(t)$ is the strain, $R(t-\tau)$ is the relaxation function and $C(t-\tau)$ is the creep function of the material.

Equations (13.1) and (13.2) above are one-dimensional representations of the "*Boltzmann Superposition Principle*" or "*Hereditary Law*" introduced earlier in Chapter 8. The presented equations (13.1) and (13.2) are governing, respectively, the stress-relaxation and creep response of the linear viscoelastic material. Both the relaxation function R(t) and creep function C(t) are usually defined for $t \geq 0$, whereby for $t<0$, one has

$$\begin{aligned} R(t) &= 0 \\ C(t) &= 0 \end{aligned} \qquad t<0 \qquad (13.3)$$

Mathematically, equations (13.1) and (13.2) have the same structure and both can be written in the following general form:

$$y(t) = \int_{-\infty}^{\infty} x(\tau)g(t-\tau)d\tau \tag{13.4}$$

where, in the case of stress-relaxation,

$$x(t) = \frac{d\varepsilon(t)}{dt}$$
$$g(t) = R(t) \tag{13.5}$$
$$y(t) = \sigma(t)$$

and, in the creep phase,

$$x(t) = \frac{d\sigma(t)}{dt}$$
$$g(t) = C(t) \tag{13.6}$$
$$y(t) = \varepsilon(t)$$

For simplicity, we shall refer to the situation identified by (13.5) as "***the relaxation experiment***", and to the situation designated by (13.6) as "***the creep experiment***". The function g(t) appearing in both (13.5) and (13.6) is referred to as the "***characteristic function***" of the material. *Thus, from a system theory point of view, Eq.(13.4) represents a relationship between an input x(t) and a corresponding output y(t) of the system with g(t) being the system characteristic function.*

Therefore, if one considers the viscoelastic material specimen as a dynamic system, then, the characterization of its rheological response would involve a process of identification of its characteristic function g(t) *from dynamical measurements.*

Taking Fourier transform of y(t) and x(t) and denoting

$$Y(i\omega) = \frac{1}{2\pi}\int_{-\infty}^{\infty} y(t)e^{-i\omega t}dt \tag{13.7}$$

and

$$X(i\omega) = \frac{1}{2\pi}\int_{-\infty}^{\infty} x(t)e^{-i\omega t}dt \tag{13.8}$$

then, by substituting Eqn. (13.4) into Eqn. (13.7), one can write that

$$Y(i\omega) = \frac{1}{2\pi}\int_{-\infty}^{\infty} e^{-i\omega\zeta}g(\zeta)d\zeta\int_{-\infty}^{\infty} x(\tau)e^{-i\omega\tau}d\tau \tag{13.9}$$

where ζ is the time parameter $(t-\tau)$. Thus, by combining (13.8) and (13.9), one has the following relation in frequency domain

$$Y(i\omega) = 2\pi\ H(i\omega)\ X(i\omega) \tag{13.10}$$

where

$$H(i\omega) = \frac{1}{2\pi}\int_{-\infty}^{\infty} e^{-i\omega t}g(t)dt \tag{13.11}$$

With reference to Eqn.(13.11), $H(i\omega)$ is the Fourier transform of the system characteristic function $g(t)$. Meantime, the ***"frequency response function"*** of the system is identified, with reference to (13.10), as

$$F(i\omega) = 2\pi\ H(i\omega) \tag{13.12}$$

In terms of the frequency response function (13.12), the response equation of the system in frequency domain, Eqn. (13.10), becomes

$$Y(i\omega) = F(i\omega)X(i\omega) \tag{13.13}$$

Denoting the inverse Fourier transform of the frequency response function $F(i\omega)$ by $f(t)$, then, in view of expressions (13.11) and (13.12), it follows that

$$f(t) = \int_{-\infty}^{\infty} F(i\omega)e^{i\omega t}d\omega = \int_{-\infty}^{\infty} 2\pi\ H(i\omega)e^{i\omega t}d\omega = 2\pi g(t) \tag{13.14}$$

Therefore, from (13.12), or.(13.14) implies that the frequency response function of the dynamic system is the Fourier transform of the characteristic function g(t) of the system multiplied by 2π.

To model the response behaviour of a linear viscoelastic material, we assume that the "***frequency response function***" of the corresponding dynamic system has the following form

$$F(i\omega) = \frac{a}{(i\omega)^p + b_1(i\omega)^{p-1} + \dots + b_{p-1}(i\omega) + b_p} \tag{13.15}$$

where a and $b_1, b_2, \dots, b_p$ are constant parameters.

13.3.1. SYSTEM CHARACTERISTIC FUNCTION

Corresponding to Eqn.(13.15), the system characteristic function g(t) is derived as follows:

Assume the following *p-th. order algebraic equation*:

$$s^p + b_1 s^{p-1} + \dots + b_{p-1}s + b_p = 0 \tag{13.16}$$

has roots $\xi_1, \xi_2, \dots, \xi_p$, then, the frequency response function, Eqn. (13.15), can be expressed as

$$F(i\omega) = \frac{a}{(i\omega-\xi_1)(i\omega-\xi_2)\cdots(i\omega-\xi_p)} \tag{13.17}$$

Further, the above equation can be expressed in a partial fraction form as

$$F(i\omega) = \sum_{m=1}^{p} \frac{A_m}{i\omega - \xi_m} \quad ; \qquad (m=1,2, \cdots , p) \tag{13.18}$$

where A_m (m=1,2,...,p), corresponding to roots ξ_m (m=1,2,...,p), are calculated by

$$A_m = \frac{a}{\prod\limits_{\substack{j=1,\ m=1 \\ j\neq m}}^{p} (\xi_m - \xi_j)} \quad ; \qquad (j,\ m = 1,\ 2, \cdots , p \tag{13.19}$$

Taking the inverse Fourier transform of Eqn. (13.18), one obtains

$$\begin{aligned} f(t) &= \int_{-\infty}^{\infty} F(i\omega)e^{i\omega t}d\omega \\ &= \int_{-\infty}^{\infty} \sum_{m=1}^{p} \frac{A_m}{i\omega-\xi_m} e^{i\omega t}d\omega \\ &= \sum_{m=1}^{p} \int_{-\infty}^{\infty} \frac{A_m}{i\omega-\xi_m} e^{i\omega t}d\omega \\ &= \sum_{m=1}^{p} A_m\ e^{\xi_m t}u(t) \end{aligned} \tag{13.20}$$

$$(m=1,2, \cdots , p)$$

where u(t) is the Heaviside step function (*Appendix B*) defined by

$$u(t) = \begin{cases} 0 & t<0 \\ 1 & t\geq 0 \end{cases} \tag{13.21}$$

Thus, Eqn. (13.20) can be expressed as

$$f(t) = \begin{cases} \sum\limits_{m=1}^{p} A_m\ e^{\xi_m t} & t\geq 0 \\ 0 & t<0 \end{cases} \tag{13.22}$$

$$(m=1,2, \cdots , p)$$

From equations (13.5), (13.14) and (13.20), the relaxation function R(t), in a dynamic relaxation experiment, is expressed as

$$R(t) = \frac{1}{2\pi} f(t)$$
$$= \frac{1}{2\pi}\sum_{m=1}^{p} A_m e^{\xi_m t} u(t)$$
$$= \begin{cases} \frac{1}{2\pi}\sum_{m=1}^{p} A_m e^{\xi_m t} & t \geq 0 \\ 0 & t<0 \end{cases} \tag{13.23}$$

$$(m=1,2, \cdots, p)$$

On the other hand, if the experiment is a dynamic creep experiment, then, the system characteristic function g(t) represents the creep function C(t), Eqn.(13.6). Thus, the expression for the creep function C(t), corresponding to (13.23), can be written as

$$C(t) = \frac{1}{2\pi}\sum_{m=1}^{p} A_m e^{\xi_m t} u(t)$$
$$= \begin{cases} \frac{1}{2\pi}\sum_{m=1}^{p} A_m e^{\xi_m t} & t \geq 0 \\ 0 & t<0 \end{cases} \tag{13.24}$$

$$(m=1,2, \cdots, p)$$

Referring back to Eqn. (13.15), if the frequency response function of a dynamic system in frequency domain is expressed by this equation, then, the dynamic behaviour of the system may be assumed to be governed by the following differential equation in the time domain

$$\frac{d^p}{dt^p} y(t) + b_1 \frac{d^{p-1}}{dt^{p-1}} y(t) + \cdots + b_p y(t) = a x(t) \tag{13.25}$$

where a and $b_1, b_2, \ldots, b_p$ are the same constant parameters appearing in Eqn. (13.15). Taking Fourier transform of Eqn.(13.25), one has, with reference to (13.7),

$$(i\omega)^p Y(i\omega) + b_1 (i\omega)^{p-1} Y(i\omega) + \cdots + b_p Y(i\omega) = a X(i\omega) \tag{13.26}$$

Thus, by recalling the definition of frequency response function, that is

$$F(i\omega) = \frac{Y(i\omega)}{X(i\omega)} \tag{13.27}$$

it is clear that one arrives back to Eqn. (13.15).

Eqn. (13.25) can be further written in the following operational form

$$D(t)\ y(t) = a\ x(t) \tag{13.28}$$

where

$$D(t) = \frac{d^p}{dt^p} + b_1 \frac{d^{p-1}}{dt^{p-1}} + \cdots + b_p \tag{13.29}$$

However, in order to determine the frequency response function of the system, it is necessary to establish the values of the parameters a; b_1, b_2,..., b_p and the order p, Eqn. (13.26), or from (13.28) the measurements of the input $x(t)$ and the output $y(t)$.

13.3.2. DETERMINATION OF THE PARAMETERS a AND b_1, b_2,..., b_p OF THE MODEL EQUATION (13.26)

In practice, x(t) and y(t) are usually given in the form of discrete-time signals. That is, the continuous-time signals x(t) and y(t) are sampled into discrete series. Let us assume that the sampling interval is ΔT. Thus, from experimental measurements, one obtain the following two discrete time series; representing, respectively, the output and input of the system:

$$\left.\begin{aligned} y_i &= y(\Delta T \cdot i) \\ x_i &= x(\Delta T \cdot i) \end{aligned}\right\} \quad (i = 0,\ 1,\ 2,\cdots) \tag{13.30}$$

With reference to equations (13.5) and (13.6), a signal of the discrete-time input series $\{x_i\}$, i=0,1,2,..., represents, in the formulation below, the time-rate of strain in a dynamic relaxation experiment, or, alternatively, the time-rate of stress in a dynamic creep experiment. In an actual experiment, the loading signal is always known, thus, the corresponding time-rate of the signal can be easily obtained. In this, one assumes (e.g., Cadzow, 1973) that the relation between the two discrete-time series $\{x_i\}$ and $\{y_i\}$, i=0,1,2,..., is governed by a discrete-time system of the p-th. order. That is

$$y_i + \beta_1 y_{i-1} + \cdots + \beta_{p-1} y_{i-p-1} + \beta_p y_{i-p} = \alpha x_i \qquad (i=0,1,2,\cdots) \tag{13.31}$$

where α and $\beta_1, \beta_2,..., \beta_p$ are constant parameters.

Let

$$By_i = y_{i-1} \quad ; \; (i=0,1,2,...)$$

and

$$\Phi(B) = 1+\beta_1 B+...+ \beta_p B^p$$

where B is a "*single-step delay operator*". Eqn. (13.31) can, thus, be written as

$$\Phi(B)y_i = \alpha x_i \qquad ; \qquad (i=0,1,2,\cdots) \tag{13.32}$$

Taking the z-transform of Eqn. (13.32), one has (e.g., Cadzow, 1973),

$$Y(z) = H_d(z)\, X(z)$$

where Y(z), X(z) are, respectively, the z-transforms of the discrete-time series $\{y_i\}$ and $\{x_i\}$. In this equation, $H_d(z)$ is the "*transfer function of the discrete-time system*" expressed by

$$H_d(z) = \frac{\alpha}{1 + \beta_1 z^{-1} + \beta_2 z^{-2} + \cdots + \beta_p z^{-p}} \tag{13.33}$$

In analogy to (13.16), we consider the p-th order algebraic equation:

$$1 + \beta_1\, s^{-1} + \cdots + \beta_p\, s^{-p} = 0 \tag{13.34}$$

Denoting the roots of (13.34) by $\lambda_1,\lambda_2,...,\lambda_p$,, then, the transfer function of the discrete-time system, Eqn. (13.33), can be written as

$$H_d(z) = \frac{\alpha}{(1-\lambda_1 z^{-1})\cdots(1-\lambda_p z^{-1})} = \sum_{m=1}^{p} \frac{B_m}{1-\lambda_m z^{-1}} \tag{13.35}$$

$$(m=1,2,\cdots, p)$$

where B_m (m=1,2,...,p), corresponding to roots λ_m (m=1,2,...,m), are calculated by

$$\mathbf{B}_m = \frac{\alpha}{\prod_{\substack{j=1,\, m=1 \\ j \neq m}}^{p} (1-\lambda_j \lambda_m^{-1})} \quad ; \qquad (j,\ m = 1,\ 2,\ \cdots,\ p \tag{13.36}$$

Taking the inverse z-transform of Eqn.(13.35), one obtains the following characteristic series of the discrete-time system

$$h_d(k) = \sum_{m=1}^{p} B_m \lambda_m^k\, u[k] \ ; \qquad (k = 0,\ 1,\ 2,\cdots \ ; \ m = 1,2,\cdots,\ p) \tag{13.37}$$

where u[k] is a discrete-time unit step function defined by

$$u[k] = \begin{cases} 0 & k<0 \\ 1 & k = 0,\ 1,\ 2,\cdots \end{cases} \tag{13.38}$$

By using the function $h_d(k)$, Eqn. (13.37), the relation between the input and output of the discrete-time system can be expressed as

$$y_i = \sum_{k=0}^{\infty} h_d(k)\, x_{i-k} \quad ; \qquad (i,\ k = 0,1,2,\cdots) \tag{13.39}$$

In searching for the relation between the characteristic function g(t) of a continuous-time system and the characteristic series $h_d(k)$ of the corresponding discrete-time system, one may approximate Eqn. (13.4) by

$$\begin{aligned} y(\Delta T \cdot i) &\approx \sum_{k=0}^{\infty} g(k \cdot \Delta T)\, x[\Delta T \cdot (i-k)] \cdot \Delta T \\ &= \frac{1}{2\pi} \sum_{k=0}^{\infty} f(k \cdot \Delta T)\, x[\Delta T \cdot (i-k)] \cdot \Delta T \end{aligned} \tag{13.40}$$

$$(i,\ k = 0,\ 1,\ 2,\cdots)$$

Combining the above equation with Eqn. (13.30), one has

$$y_i \approx \frac{1}{2\pi}\sum_{k=0}^{\infty} f(k \cdot \Delta T)\, x_{i-k} \cdot \Delta T \quad ; \qquad (i,\ k = 0,1,2,\cdots)$$

(13.41)

Comparing equations (13.39) and (13.41), the following equation may approximately hold

$$\frac{1}{2\pi} f(k \cdot \Delta T)\Delta T \approx h_d(k) \quad ; \qquad (k=0,1,2,\cdots) \tag{13.42}$$

Thus, from equations (13.22), (13.37) and (13.42), it follows that

$$\begin{aligned} \frac{1}{2\pi}\sum_{m=1}^{p} A_m\, e^{\xi_m k\Delta T}\, \Delta T &= \sum_{m=1}^{p} B_m \lambda_m^k \\ \frac{1}{2\pi}\Delta T\, A_m &= B_m \\ e^{\xi_m \Delta T} &= \lambda_m \end{aligned} \tag{13.43}$$

$$(m=1,2,\cdots,p \quad ; \quad k=0,1,2,\cdots)$$

Eqn. (13.43), above, is the relation which determines the parameters of the model equation (13.15). If the parameters α and β_m, $(m=1,2,...,p)$ of the discrete-time system are determined from the discrete-time series of input signals $\{x_i\}$ and of the corresponding output signals $\{y_i\}$, then, the parameters a and b_m, $(m=1,2,...,p)$ for the continuous model equation (13.15), or alternatively (3.26), can be calculated, as illustrated below, by using Eqn. (13.43).

13.3.3. DETERMINATION OF THE PARAMETERS α, β_m, $m=1,2,...,p$ OF THE DISCRETE-TIME SYSTEM EQUATION (13.31)

In the following, we discuss the method to determine the order p and parameters α and β_m $(m=1,2,...,p)$ of the discrete-time system.

Choose arbitrarily an order p and parameters $\hat{\alpha}$, $\hat{\beta}_m$ $(m=1,2,...,p)$ of a convenient discrete -time system. Then, with reference to Eqn. (13.31), it follows that

$$y_i + \hat{\beta}_1 y_{i-1} + \ldots + \hat{\beta}_p y_{i-p} = \hat{\alpha} x_i + e_i \qquad (i = 0,1,2,\cdots)$$

(13.44)

where e_i (i=0,1,2,...) is the combined error in choosing the values of $\hat{\alpha}$, β_m (m=1,2,...,p) and the order p. The error e_i can be expressed as

$$e_i = y_i - (-y_{i-1}, \cdots, -y_{i-p}, x_i)\begin{Bmatrix} \hat{\beta}_1 \\ \vdots \\ \vdots \\ \hat{\beta}_p \\ \hat{\alpha} \end{Bmatrix}$$

$$= y_i - \{W_i\}^T \{\hat{\beta}_m\} \tag{13.45}$$

$$(i = 0, 1, 2, \cdots \ ; \ m=1,2,\cdots,p)$$

where

$$\{W_i\}^T = (- y_{i-1} , \cdots , y_{i-p} , x_i)$$

$$\{\hat{\beta}_m\}^T = (\hat{\beta}_1 , \ \cdots , \hat{\beta}_p , \hat{\alpha}) \tag{13.46}$$

$$(i=0,1,2,\cdots \ ; \ m=1,2,\cdots,p)$$

By minimizing, i.e.,

$$e^2 = \frac{1}{N}\sum_{i=p+1}^{N} e_i^2$$

$$= \frac{1}{N}\sum_{i=p+1}^{N} (y_i - \{W_i\}^T \{\hat{\beta}_m\}) (y_i - \{\hat{\beta}_m\}^T \{W_i\}) \tag{13.47}$$

$$(i=0,1,2,\cdots \ ; \ m=1,2,\cdots,p)$$

one has

$$(\frac{1}{N}\sum_{i=p+1}^{N} \{W_i\}\{W_i\}^T) \{\hat{\beta}_m\} = \frac{1}{N}\sum_{i=p+1}^{N} y_i \{W_i\} \tag{13.48}$$

$$(i=0,1,2,\cdots \ ; \ m=1,2,\cdots,p)$$

Thus, for every choice of an order p, a corresponding $\{\beta_m\}$ can be determined by Eqn. (13.48). Then, from Eq.(13.47), the error corresponding to the choice of the order p can be calculated. Since e^2 is a function of the order p, the choice of the order of the discrete-time system can be made by the requirement that it would result in a minimum e^2.

13.3.4. NUMERICAL EXAMPLES

To test the analytical model developed in the previous section, a number of numerical illustrations are carried out below. The formalism of these illustrations is outlined as follows:

1. For a given system, one calculates the response under certain dynamic loading by a numerical method. Here, the Runge-Kuta method (e.g., Morris, 1983) is employed. Consequently, two discrete-time series (One is the input to the system and the other is its response) are obtained.

2. Assuming that no other knowledge about the system is given except the two discrete-time series mentioned by Eqn. (13.30), one applies the already introduced *Dynamic System Identification Method* to the two discrete-time series. In this, one determines first, the parameters characteristic of the transfer function of the discrete-time system, then, establishes the corresponding continuous system.

Example 13.1

Consider the first order system

$$\dot{y} + 5y = x(t) \tag{13.49}$$

under an input represented by: $x(t) = 100 \sin (t^{1.5})$.

By comparing (13.49) to (13.25), the parameters of the system (13.49) are given by:

$$p=1;\ b_1=5;\ a=1.0$$

With an input $x(t) = 100 \sin (t^{1.5})$, which may be the rate of strain or stress, one can obtain two discrete-time series of input and output as plotted, with $\Delta T = 0.01$, in Figures 13.1 and 13.2, respectively, One uses, then, discrete-time systems (DTS's) of different orders, Eqn. (13.32), to model the system. The errors pertaining to three different discrete-time systems were calculated and are listed in Table 13.1.

TABLE 13.1. Errors in determining three different discrete-time systems

Order	First	Second	Third
Error	0.852409E-02	0.253536E+00	0.634591E-1

From Table 13.1, the DTS of first order is the system with minimum error, therefore, one chooses the first order DTS to model the continuous system governed by a first order differential equation of (13.49). The parameters of this first-order DTS are listed in Table 13.2.

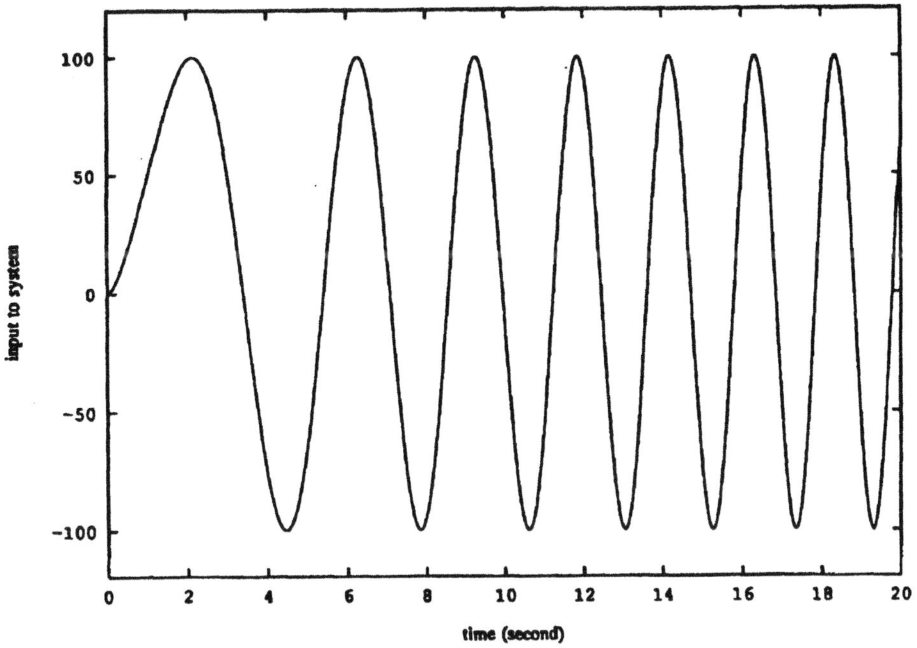

Figure 13.1 Input: $x(t) = 100 \sin(t^{1.5})$ with $\Delta T = 0.01$. "Reprinted from *Int. J. Pres. Ves. & piping* **61**, Yu P. and Haddad, Y.M., A dynamic system identification method for the characterization of the rheological response of a class of viscoelastic materials, 87-97, 1995, with kind permission from Elsevier Science Ltd, The Boulevard, Langford Lane, Kidlington 0X5 1GB, UK".

TABLE 13.2. Parameters characteristic of first-order DTS

Parameter	β	α	λ
Value	-0.952681E+00	0.951930E-02	0.95268

where β *and* α are parameters of the discrete-time system, (13.31), and λ is the root of the characteristic equation (13.34) of the considered first-order DTS. The corresponding transfer function $H_d(z)$ and the parameter B can be obtained by using equations (13.35) and (13.36). Then, according to Eqn.(13.43), the parameters, A_m , ξ_m (m=1) of the corresponding continuous-time system can be calculated.

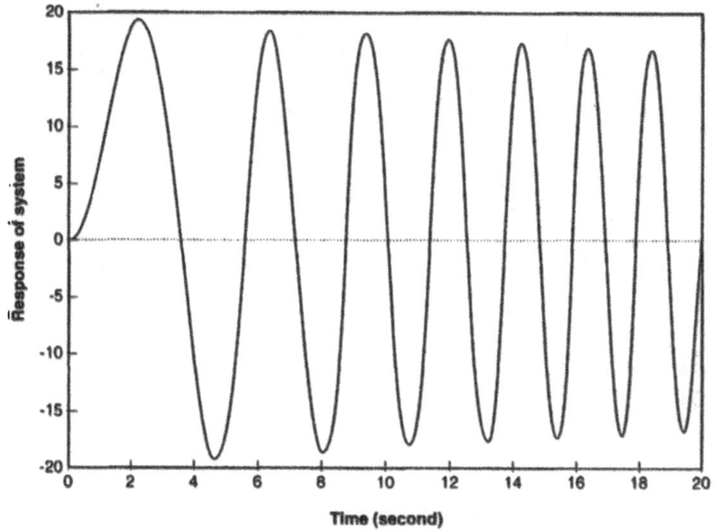

Figure 13.2 Output $y(t)$ corresponding to the input $x(t)$ in Fig. 13.1. First-order system $\dot{y} + 5y = x(t)$ with parameters $a = 1.0$, $b_1 = 5$ and an order $p = 1$. "Reprinted from *Int. J. Pres. Ves. & piping* 61, Yu P. and Haddad, Y.M., A dynamic system identification method for the characterization of the rheological response of a class of viscoelastic materials, 87-97, 1995, with kind permission from Elsevier Science Ltd, The Boulevard, Langford Lane, Kidlington 0X5 1GB, UK".

Figure 13.3 shows the exact and the estimated response given by the 1st order DTS. Figure 13.4 shows the exact and the estimated system characteristic function g(t) from the 1st order DTS.

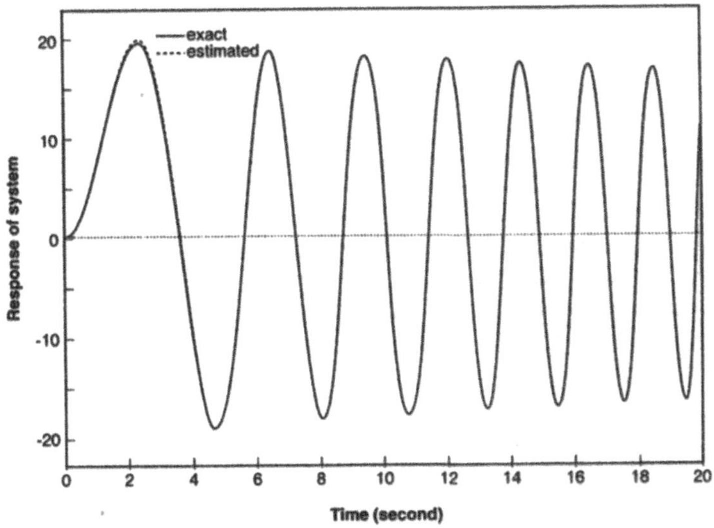

Figure 13.3 The exact and the estimated responses from the first-oder DTS. First-order system: $\dot{y} + 5y = x(t)$ with parameters $a = 1{\cdot}0$, $b_1 = 5$, $p = 1$ and input $x(t)$ of Fig. 13.1. "Reprinted from *Int. J. Pres. Ves. & piping* **61**, Yu P. and Haddad, Y. M., A dynamic system identification method for the characterization of the rheological response of a class of viscoelastic materials, 87-97, 1995, with kind permission from Elsevier Science Ltd, The Boulevard, Langford Lane, Kidlington 0X5 1GB, UK".

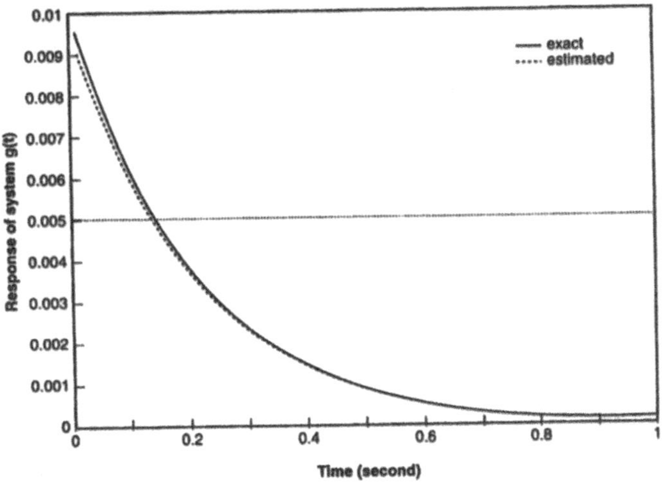

Figure 13.4 The exact and the estimated system characteristic functions $g(t)$ from the first-order DTS. First-order continuous time system: $\dot{y} + 5y = x(t)$ with parameters $a = 1{\cdot}0$, $b_1 = 5$, $p = 1$ and input $x(t)$ of Fig. 13.1. "Reprinted from *Int. J. Pres. Ves. & piping* **61**, Yu P. and Haddad, Y. M., A dynamic system identification method for the characterization of the rheological response of a class of viscoelastic materials, 87-97, 1995, with kind permission from Elsevier Science Ltd, The Boulevard, Langford Lane, Kidlington 0X5 1GB, UK".

Example 13.2

Consider the second order system

$$\ddot{y} + 25\dot{y} + 100y = x(t) \quad \textbf{(13.50)}$$

With an input $x(t) = 100 \sin(t^{1.5})$, Fig. 13.1, which may be the rate of strain or stress. The corresponding output discrete-time series is plotted in Fig. 13.5, with $\Delta T = 0.01$. The errors for discrete-time systems of different orders are listed in Table 13.3.

TABLE 13.3

Order	First	Second	Third	Fourth
Error	0.254506E-02	0.971892E-05	0.234843E-04	0.877769E-03

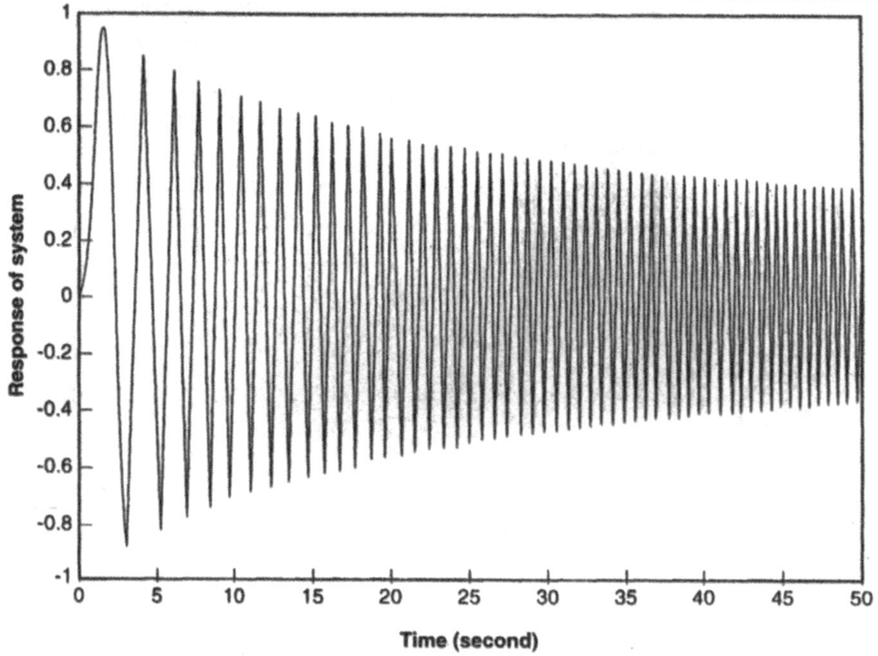

Figure 13.5 Output $y(t)$ corresponding to the input $x(t)$ in Fig. 13.1. Second-order system: $\ddot{y} + 25\dot{y} + 100y = x(t)$ with parameters $a = 1{\cdot}0$, $b_1 = 25$, $b_2 = 100$ and $p = 2$. "Reprinted from *Int. J. Pres. Ves. & piping* **61**, Yu P. and Haddad, Y. M., A dynamic system identification method for the characterization of the rheological response of a class of viscoelastic materials, 87-97, 1995, with kind permission from Elsevier Science Ltd, The Boulevard, Langford Lane, Kidlington 0X5 1GB, UK".

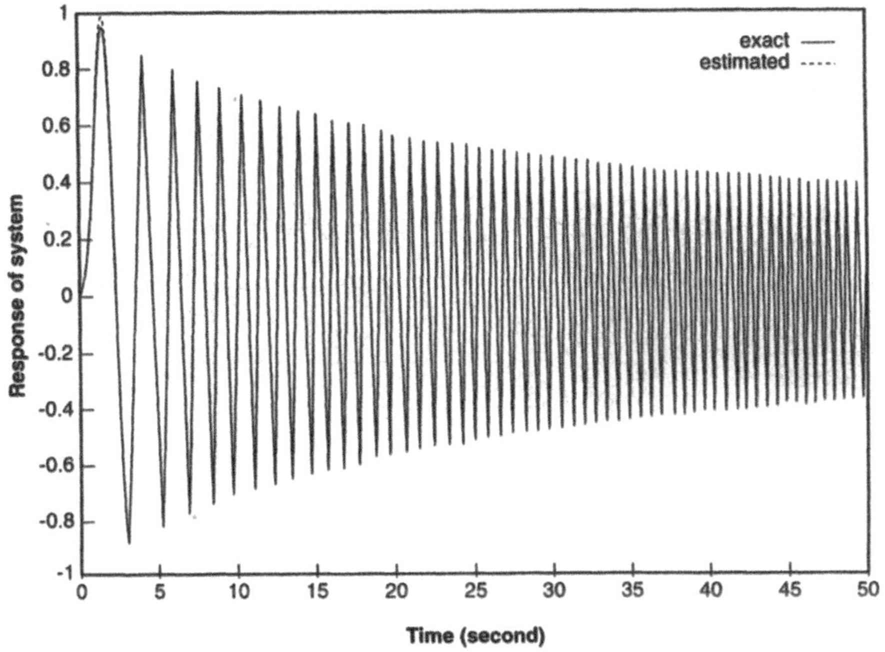

Figure 13.6 The exact response and the estimated responses from the second-order DTS. Second-order system: $\ddot{y} + 25\dot{y} + 100\,y = x(t)$ with parameters $a = 1{\cdot}0$, $b_1 = 25$, $b_2 = 100$, $p = 2$ and input $x(t)$ of Fig. 13.1. "Reprinted from *Int. J. Pres. Ves. & piping* **61**, Yu P. and Haddad, Y. M., A dynamic system identification method for the characterization of the rheological response of a class of viscoelastic materials, 87-97, 1995, with kind permission from Elsevier Science Ltd, The Boulevard, Langford Lane, Kidlington 0X5 1GB, UK".

From Table 13.3, the DTS of second order is the system with minimum error, the parameters of which are given in Tables 13.4 and 13.5.

TABLE 13.4

Parameter	β_1	β_2	α
Value	-0.175611E+01	0.764908E+00	0.906642E-04

TABLE 13.5

Parameter	λ_1	λ_2
Value	0.95601	0.80010

where λ_1 and λ_2 are two roots of the characteristic equation of the corresponding DTS, Eqn.(13.34)

Figure 13.6 shows the exact and the estimated response given by the 2nd order DTS. Figure 13.7 shows the exact and the estimated system characteristic function g(t) for the 2nd order DTS.

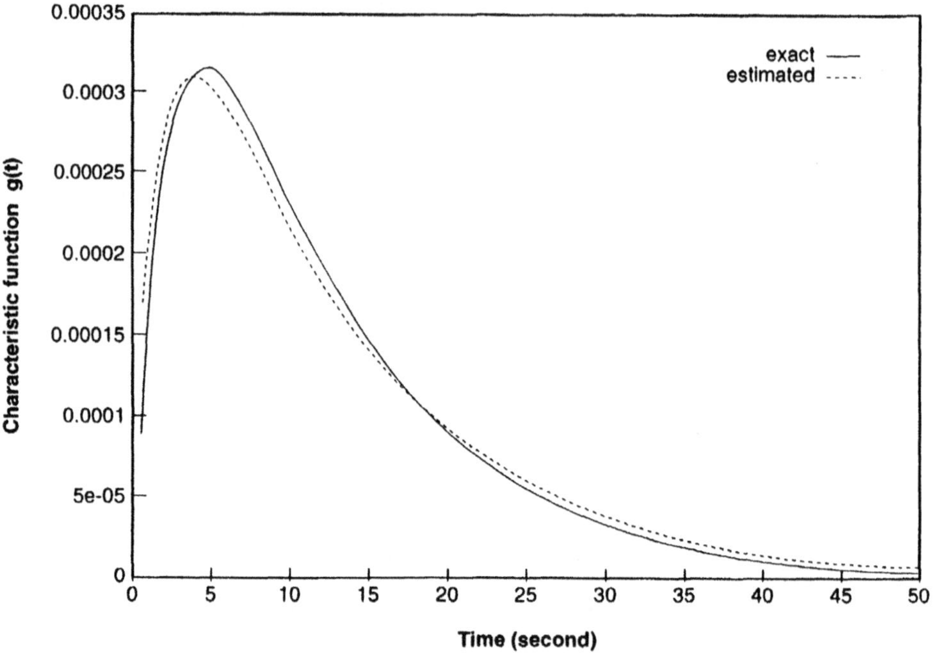

Figure 13.7 The exact and the estimated system characteristic functions $g(t)$ from the second-order DTS. Second-order system: $\ddot{y} + 25\dot{y} + 100\,y = x(t)$ with parameters a = 1. 0, $b_1 = 25$, $b_2 = 100$, $p = 2$ and input $x(t)$ of Fig. 13.1. "Reprinted from *Int. J. Pres. Ves. & piping* 61, Yu Ping and Haddad, Y. M., A dynamic system identification method for the characterization of the rheological response of a class of viscoelastic materials, 87-97, 1995, with kind permission from Elsevier Science Ltd, The Boulevard, Langford Lane, Kidlington 0X5 1GB, UK".

13.4. Extension of the Model to Include the Instantaneous Response Behaviour

In Section 13.3, the linear viscoelastic material was considered as a dynamic system whereby an analytical model was presented for the determination of the creep and relaxation functions of the material from dynamic experimental measurements. First, the relation between the viscoelastic material function and the frequency response function of the system was established by assuming a model of rational function of polynomials for the frequency response function. Then, a discrete-time system analysis method was introduced to identify the order and parameters of the model. The introduced method requires two discrete-time series, i.e., the time-rate of the input signal and the corresponding output signal. The instantaneous response of the system was not, however, taken into consideration. In the present section, the approach of Section 13.3 is followed with the consideration that the viscoelastic material function is discontinuous at the time t=0. Here, the relation between the relaxation or creep function and the transfer function of the system is established by using the Laplace transform method. A discrete-time system analysis method is introduced to identify the order and parameters of the model. The method requires, similar to the earlier treatment of Section 13.3, two discrete-time series, i.e., the time-rate of input signal and the corresponding output signal. Numerical examples are given to illustrate the application of the proposed analytical model.

13.4.1. THE MODEL

In the present section, the same idea of Section 13.3 is adopted, but with the pertaining viscoelastic experiment is considered to begin at time $t = 0$. Therefore, Laplace transform is seen to be more suitable for the analysis of the viscoelastic problem by using the initial value theorem. In this context, the Laplace transform pair (e.g., Fodor, 1965)

$$\mathcal{L}(s) = \int_0^{\infty} \mathcal{F}(t)\, e^{-st} dt$$

$$\mathcal{F}(t) = \frac{1}{2\pi i} \int_{r-i\infty}^{r+i\infty} \mathcal{L}(s)\, e^{st} ds \qquad \textbf{(13.51)}$$

is employed in the course of the presented analysis of this section. Here, in equation (13.51), $\mathcal{F}(t)$ is a time function, $\mathcal{L}(s)$ is the Laplace transform of $\mathcal{F}(t)$ and r is a real constant.

Denoting the Laplace transforms of $y(t)$, $x(t)$ and $h(t)$ by $Y(s)$, $X(s)$ and $H(s)$, respectively, and taking Laplace transform of Eqn.(13.8), one obtains the following relationship between the input and output

$$Y(s) = H(s)\ X(s) \qquad \textbf{(13.52)}$$

where the theorem of Laplace transform of the convolution of two signals has been used (e.g., Fodor, 1965).

In view of the previous analysis, in the dynamical relaxation experiment case, $H(s)$ is interpreted as:

$$H(s) = \int_0^\infty R(t)\, e^{-st} dt \tag{13.53}$$

and, in the dynamic creep experiment case, *H(s)* is identified as:

$$H(s) = \int_0^\infty C(t)\, e^{-st} dt \tag{13.54}$$

Before a model is assumed for the function *H(s)*, we have to analyze the special behaviour of the relaxation and creep functions at $t=0$. We know that both *R(t)* and *C(t)* are not continuous functions at $t=0$. Each is equal to zero at $t=0^-$, and is equal to a finite value at $t=0^+$.

From the initial value theorem of Laplace transform (e.g., Fodor, 1965), one has

$$\begin{aligned} &\lim_{s\to\infty} s\, H(s) = R(0^+) \quad ; \quad \text{for the relaxation case} \\ &\lim_{s\to\infty} s\, H(s) = C(0^+) \quad ; \quad \text{for the creep case} \end{aligned} \tag{13.55}$$

Because each of $R(0^+)$ and $C(0^+)$ is not equal to zero, we can assume the following rational function for *H(s)*

$$\begin{aligned} H(s) &= \frac{b_1 s^{p-1} + b_2 s^{p-2} + \dots + b_p}{s^p + a_1 s^{p-1} + \dots + a_p} \\ &= \frac{Q(s)}{P(s)} \end{aligned} \tag{13.56}$$

where,

$$Q(s) = b_1 s^{p-1} + b_2 s^{p-2} + \ldots + b_{p-1} s + b_p$$
$$P(s) = s^p + a_1 s^{p-1} + \ldots + a_{p-1} s + a_p \tag{13.57}$$

in which, b_1, b_2..., b_p, and a_1, a_2, ..., a_p are constant parameters, and p is the order of the polynomial. To satisfy the condition (13.55), the order of polynomial $Q(s)$ has to be one less than the order of $P(s)$.

On the basis of the assumption of Eqn.(13.56), the response relation (13.52) is expressed as

$$P(s)\, Y(s) = Q(s)\, X(s) \tag{13.58}$$

Taking the inverse Laplace transform of (13.58), one obtains the following model in time-domain

$$\frac{d^p}{dt^p} y(t) + a_1 \frac{d^{p-1}}{dt^{p-1}} y(t) + \cdots + a_p y(t)$$
$$= b_1 \frac{d^{p-1}}{dt^{p-1}} x(t) + b_2 \frac{d^{p-2}}{dt^{p-2}} x(t) + \cdots + b_p x(t) \tag{13.59}$$

With reference to (13.56), one assumes the characteristic equation

$$\xi^p + a_1 \xi^{p-1} + \cdots + a_{p-1} \xi + a_p = 0 \tag{13.60}$$

with roots ξ_1 , ξ_2, ..., ξ_p , then, the transfer function $H(s)$, (13.56), may be written in the following partial fraction form (e.g., Fodor, 1965)

$$H(s) = \frac{b_1 s^{p-1} + b_2 s^{p-2} + \cdots + b_p}{(s - \xi_1)(s - \xi_2) \cdots (s - \xi_p)}$$
$$= \frac{A_1}{s - \xi_1} + \frac{A_2}{s - \xi_2} + \cdots + \frac{A_p}{s - \xi_p} \tag{13.61}$$
$$= \sum_{m=1}^{p} \frac{A_m}{s - \xi_m}$$

where $A_m (m = 1, 2, \dots p)$ can be calculated by

$$A_m = \lim_{s \to \xi_m} H(s)\,(s - \xi_m)$$

$$= \frac{b_1 \xi_m^{p-1} + b_2 \xi_m^{p-2} + \dots + b_{p-1} \xi_m + b_p}{\prod_{\substack{k=1,\ m=1 \\ k \neq m}}^{p} (\xi_m - \xi_k)} \tag{13.62}$$

By taking the inverse Laplace transform of Eqn.(13.61) and noting that the inverse Laplace transform of $1/(s-\xi)$ is $e^{\xi t}$, one can obtain the time-domain model for $h(t)$ as

$$h(t) = \sum_{m=1}^{p} A_m\, e^{\xi_m t} \tag{13.63}$$

13.4.2. DETERMINATION OF THE CHARACTERISTIC PARAMETERS OF THE PROPOSED MODEL.

In the previous section, we established the model for the characterization of the response behaviour of a linear viscoelastic material with the inclusion of the instantaneous response at time $t = 0$. In this section, we discuss the determination of the parameters characteristic of the model by using the dynamic experimental measurements. Thus, in the pertaining dynamical relaxation experiment, we assume to obtain two discrete-time series of the time-rate of loading strain, and the stress response, respectively, as

$$\begin{aligned} &\dot{\varepsilon}(t_0),\ \dot{\varepsilon}(t_1),\ \cdots,\ \dot{\varepsilon}(t_{N-1}),\ \dot{\varepsilon}(t_N) \\ &\sigma(t_0),\ \sigma(t_1),\ \cdots,\ \sigma(t_{N-1}),\ \sigma(t_N) \end{aligned} \tag{13.64}$$

Similarly, in the pertaining dynamical creep experiment, one assumes to obtain two discrete-time series of the time-rate of stress loading and the strain response, respectively, as

$$\begin{aligned} &\dot{\sigma}(t_0),\ \dot{\sigma}(t_1),\ \cdots,\ \dot{\sigma}(t_{N-1}),\ \dot{\sigma}(t_N) \\ &\varepsilon(t_0),\ \varepsilon(t_1),\ \cdots,\ \varepsilon(t_{N-1}),\ \varepsilon(t_N) \end{aligned} \tag{13.65}$$

Further, with reference to Eqn.(13.5), or (13.6), one may express the input and output of the dealt-with experiment, in the form of generalized discrete-time series, as:

$$x(t_0),\ x(t_1),\ \cdots,x(t_{N-1}),\ x(t_N)$$
$$y(t_0),\ y(t_1),\ \cdots,y(t_{N-1}),\ y(t_N) \tag{13.66}$$

Now, we have to analyze the two discrete-time series of Eqn.(13.66) in order to determine the parameters A_m, ξ_m $(m = 1, 2, \ldots p)$ in Eqn.(13.63). In doing so, and in correspondence to the continuous-time differential function Eqn.(13.59), we introduce the following discrete-time system (e.g., Cadzow, 1970&1973).

$$\begin{aligned} &y_k + \beta_1 y_{k-1} + \cdots + \beta_{p-1} y_{k-p+1} + \beta_p y_{k-p} \\ &= \alpha_0 x_k + \alpha_1 x_{k-1} + \cdots + \alpha_{p-1} x_{k-p+1} \end{aligned} \tag{13.67}$$

$$(k = 0, 1, 2, \ldots)$$

where α_0, α_1, ..., α_{p-1} and β_1, β_2 ..., β_p and p are constant parameters. Denoting

$$Dy_k = y_{k-1} \tag{13.68}$$

and

$$\begin{aligned} \varphi(D) &= 1 + \beta_1 D + \cdots + \beta_p D^p \\ \theta(D) &= \alpha_0 + \alpha_1 D + \cdots + \alpha_{p-1} D^{p-1} \end{aligned} \tag{13.69}$$

Eqn.(13.67) can be written as

$$\varphi(D)\ y_k = \theta(D)\ x_k \tag{13.70}$$

Representing the z-transform (*see* Appendix D) of $\{y_k\}$ and $\{x_k\}$ by $Y(z)$ and $X(z)$, respectively, and taking the z-transform of Eqn.(13.70), one has

$$\begin{aligned} \varphi(z^{-1})\ Y(z) &= \theta(z^{-1})\ X(z) \\ Y(z) &= \frac{\theta(z^{-1})}{\varphi(z^{-1})}\ X(z) \\ &= H_d(z)\ X(z) \end{aligned} \tag{13.71}$$

where $H_d(z)$ is called the "*transfer function*" of the "*discrete-time system*", expressed by

$$H_d(z) = \frac{\theta(z^{-1})}{\varphi(z^{-1})} = \frac{\alpha_0 + \alpha_1 z^{-1} + \cdots + \alpha_{p-1} z^{-(p-1)}}{1 + \beta_1 z^{-1} + \cdots + \beta_{p-1} z^{-(p-1)} + \beta_p z^{-p}} \qquad \textbf{(13.72)}$$

Assume that the characteristic equation

$$\varphi(\lambda^{-1}) = 1 + \beta_1 \lambda^{-1} + \cdots + \beta_p \lambda^{-p} = 0 \qquad \textbf{(13.73)}$$

has roots $\lambda_1, \lambda_2, \ldots \lambda_p$. Then, Eqn.(13.72) can be written as

$$H_d(z) = \frac{\alpha_0 + \alpha_1 z^{-1} + \cdots + \alpha_{p-1} z^{-p+1}}{(1 - \lambda_1 z^{-1})(1 - \lambda_2 z^{-1}) \cdots (1 - \lambda_p z^{-1})} = \frac{B_1}{1 - \lambda_1 z^{-1}} + \frac{B_2}{1 - \lambda_2 z^{-1}} + \cdots + \frac{B_p}{1 - \lambda_p z^{-1}} = \sum_{m=1}^{p} \frac{B_m}{1 - \lambda_m z^{-1}} \qquad \textbf{(13.74)}$$

where B_m $(m = 1, 2, \ldots, p)$ are calculated by

$$B_m = \lim_{z \to \lambda_m} H_d(z)\,(1 - \lambda_m z^{-1}) = \frac{\alpha_0 + \alpha_1 \lambda_m^{-1} + \cdots + \alpha_{p-1} \lambda_m^{-p+1}}{\prod\limits_{\substack{k=1,\, m=1 \\ k \neq m}}^{p} (1 - \lambda_k \lambda_m^{-1})} \qquad \textbf{(13.75)}$$

Taking the inverse z-transform of Eqn.(13.74), one obtains the system characteristic series or weighting sequence of the discrete-time system as

$$h_d(i) = \sum_{m=1}^{p} B_m \lambda_m^i u(i) \tag{13.76}$$

$$(i = 0, 1, 2, \dots)$$

By using the function $h_d(i)$, Eqn.(13.76), the relation between the input and output of the discrete-time system can be expressed as

$$y_i = \sum_{k=0}^{\infty} h_d(k) x_{i-k} \; ; \quad (i, k = 0,1,2,\cdots) \tag{13.77}$$

In searching for the relation between the characteristic function h(t), Eqn.(13.63), of a continuous-time system and the characteristic series $h_d(i)$, Eqn.(13.76), of the corresponding discrete-time system, we approximate Eqn.(13.4) by

$$\begin{aligned} y(\Delta T \cdot i) &\approx \sum_{k=0}^{\infty} h(k \cdot \Delta T) \, x[\Delta T \cdot (i - k)] \cdot \Delta T \\ &= \sum_{k=0}^{\infty} h(k \cdot \Delta T) \, x[\Delta T \cdot (i-k)] \cdot \Delta T \end{aligned} \tag{13.78}$$

$$(i, k = 0, 1, 2, \cdots)$$

From Eqn. (13.78), one has

$$y_i \approx \sum_{k=0}^{\infty} h(k \cdot \Delta T) \, x_{i-k} \cdot \Delta T \quad ; \quad (i, k = 0, 1, 2, \cdots) \tag{13.79}$$

Substituting equations (13.63) and (13.76) into equations (13.77) and (13.79), respectively, then, by comparing equations (13.77) and (13.79), the following equation will approximately hold

$$\begin{aligned} A_m &= \frac{1}{\Delta T} B_m \\ \xi_m &= \frac{1}{\Delta T} \ln \lambda_m \end{aligned} \qquad (m = 1, 2, \cdots, p) \tag{13.80}$$

To determine the parameters B_m, λ_m ($m = 1, 2, \dots p$) appearing in (13.76) for the discrete-time system, the parameters $\alpha_0, \alpha_1, ..., \alpha_{p-1}; \beta_1, \beta_2, ..., \beta_p$ in the model Eqn.(13.67) have to be determined first. Choosing arbitrarily an order p and parameters $\alpha_0, \alpha_1, ..., \alpha_{p-1}; \beta_1, \beta_2, ..., \beta_p$ of a discrete-time system and substituting them into Eqn.(13.67), one has

$$\begin{aligned} & y_i + \hat{\beta}_1 y_{i-1} + \cdots + \hat{\beta}_p y_{i-p} \\ & = \hat{\alpha}_0 x_i + \hat{\alpha}_1 x_{i-1} + \cdots + \hat{\alpha}_{p-1} x_{i-p+1} + e_i \end{aligned} \tag{13.81}$$

$$(i = 1, 2, ..., N)$$

where $e_i (i = 1, 2, ..., N)$ are the values of error in the chosen values of the parameters $\alpha_0, \alpha_1, ..., \alpha_{p-1}; \beta_1, \beta_2, ..., \beta_p$ and the order p. The errors $e_i (i = 1, 2, ..., N)$ can, then, be expressed as

$$\begin{aligned} e_i &= y_i - (-y_{i-1}, \cdots, -y_{i-p}, x_i, \cdots, x_{i-p+1}) \begin{Bmatrix} \hat{\beta}_1 \\ \vdots \\ \hat{\beta}_p \\ \hat{\alpha}_0 \\ \hat{\alpha}_1 \\ \vdots \\ \hat{\alpha}_{p-1} \end{Bmatrix} \\ &= y_i - \{w_i\}^T \{\hat{\beta}\} \end{aligned} \tag{13.82}$$

$$(i = 1, 2, \cdots, N)$$

where,

$$\begin{aligned} \{w_i\}^T &= (-y_{i-1}, \cdots, -y_{i-p}, x_i, x_{i-1}, \cdots, x_{i-p+1}) \\ \{\hat{\beta}\}^T &= (\hat{\beta}_1, \cdots, \hat{\beta}_p, \hat{\alpha}_0, \hat{\alpha}_1, \cdots, \hat{\alpha}_{p-1}) \end{aligned} \tag{13.83}$$

and "T" represents the transpose of a matrix.

By minimizing the sum of the square of errors, i.e.,

$$e^2 = \frac{1}{N} \sum_{i=p+1}^{N} e_i^2$$
$$= \frac{1}{N} \sum_{i=p+1}^{N} (y_i - \{w_i\}^T \{\hat{\beta}\}) (y_i - \{w_i\}^T \{\hat{\beta}\}) \tag{13.84}$$

one has

$$\left(\frac{1}{N} \sum_{i=p+1}^{N} \{w_i\}^T \{w_i\} \right) \{\hat{\beta}\} = \frac{1}{N} \sum_{i=p+1}^{N} y_i \{w_i\} \tag{13.85}$$

Thus, for every choice of an order p, a corresponding $\{\beta\}$ can be determined by Eqn.(13.85). Accordingly, the pertaining sum of the square of errors can be calculated form Eqn.(13.84). The choice of the order *p* of the discrete-time system is determined by the requirement that it would result in a minimum sum of the square of errors.

To test the analytical model developed above, a number of numerical examples are to follow.

Example 13.3

Consider the first order system

$$\dot{y} + 0.2y = x(t) \tag{13.86}$$

Let the input *x(t)*, onto the system, be

$$x(t) = 100 \sin(t^{1.5})$$

With reference to equation (13.59), the parameters of this system are listed in Table 13.6.

TABLE 13.6

p	a_1	b_1
1	0.2	1.0

Then, the corresponding parameters A_m and ξ_m (m = 1) which are defined in

Eq.(13.61) can be calculated, respectively, from equations (13.60) and (13.62), and are listed in Table 13.7.

TABLE 13.7

A_1	ξ_1
1.0	-0.2

By employing the Runge-Kutta method (e.g., Morris, 1983), with $\Delta T=0.01$, one obtains two discrete-time series of input and output, which are shown, respectively, in Figures 13.1 and 13.8. Then, one uses discrete-time systems (DTS) of different orders, Eqn.(13.67), to model the system. The error e^2, Eqn.(13.84), for three discrete-time systems (DTS's) of different order are given in Table 13.8.

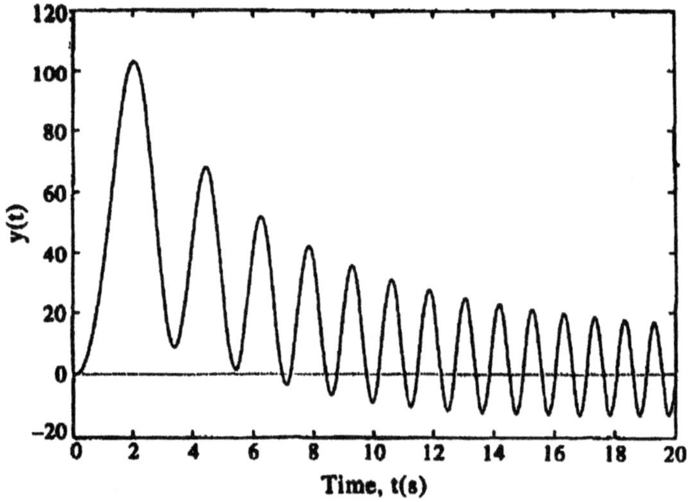

Figure 13.8 Output $y(t)$ corresponding to the input $x(t)$ of Fig. 13.1. First-order system $\dot{y}+0\cdot 2y=x(t)$ with parameters $a_1=1{\cdot}0$, $b_1=0{\cdot}2$, and order $p=1$. "Reprinted from *Int. J. Pres. Ves. & piping* 67, Yu P. and Haddad, Y. M., On the dynamic system identification of the response behaviour of linear viscoelastic materials, 45-54, 1996, with kind permission from Elsevier Science Ltd, The Boulevard, Langford Lane, Kidlington 0X5 1GB, UK".

Table 13.8

Order	First	Second	Third
Error	0.526948E+00	0.205991E+2	0.206344E+01

From Table 13.8, the discrete-time system of first order is the system with minimum error, therefore, one chooses this DTS to model the system . The parameters of this DTS as determined from equations (13.73) and (13.85) are listed in Table 13.9.

TABLE 13.9

Parameter	β_1	α_0	λ_1
Value	-0.998250E+00	0.997577E-02	0.99825

where β_1, α_0 are parameters of the discrete-time system defined in Eqn.(13.67) and λ_1 is the root of the characteristic equation (13.73) of the DTS. Because this DTS is of first order, its transfer function, Eqn.(13.72), is written as

$$H_d(z) = \frac{B_1}{1 - \lambda_1 z^{-1}} \tag{13.87}$$

where B_1 is calculated, according to Eqn.(13.75), as

$$B_1 = \alpha_0 = 0.997577E\text{-}02$$

Then, according to Eqn.(13.80), the parameters of the corresponding continuous-time system can be determined as

$$\begin{aligned} A_1 &= \left(\frac{1}{\Delta T}\right) B_1 = 0.997577 \\ \xi_1 &= \left(\frac{1}{\Delta T}\right) \ln \lambda_1 = -0.1852 \end{aligned} \tag{13.88}$$

For comparison, we list the exact and estimated values of these parameters in Table 13.10.

TABLE 13.10

Parameter	A_1	ξ_1
Exact	1.0	-0.2
Estimated	0.997577	-0.1852

Thus, the estimated system characteristic function, (13.63) is written as

$$h(t) = A_1\ e^{\xi_1 t}$$
$$= 0.997577\ e^{-0.1852t} \tag{13.89}$$

Figure 13.9 shows the exact and the estimated responses given by the first-order discrete-time system.
Figure 13.10 shows the exact and the estimated system characteristic function *h(t)* obtained from the first-order discrete-time system.

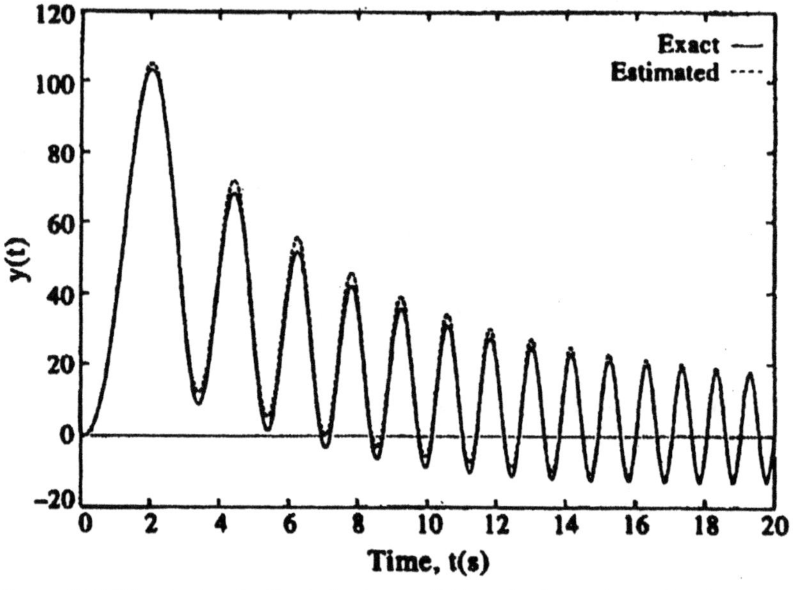

Figure 13.9 The exact and estimated output *y (t)* from the first-order DTS. First-order system $\dot{y} + 0{\cdot}2y = x(t)$ with parameters $a_1 = 1{\cdot}0$, $b_1 = 0{\cdot}2$, order $p = 1$, and input *x(t)* of Fig. 13.1. "Reprinted from *Int. J. Pres. Ves. & piping* 67, Yu P. and Haddad, Y. M., On the dynamic system identification of the response behaviour of linear viscoelastic materials, 45-54, 1996, with kind permission from Elsevier Science Ltd, The Boulevard, Langford Lane, Kidlington 0X5 1GB, UK".

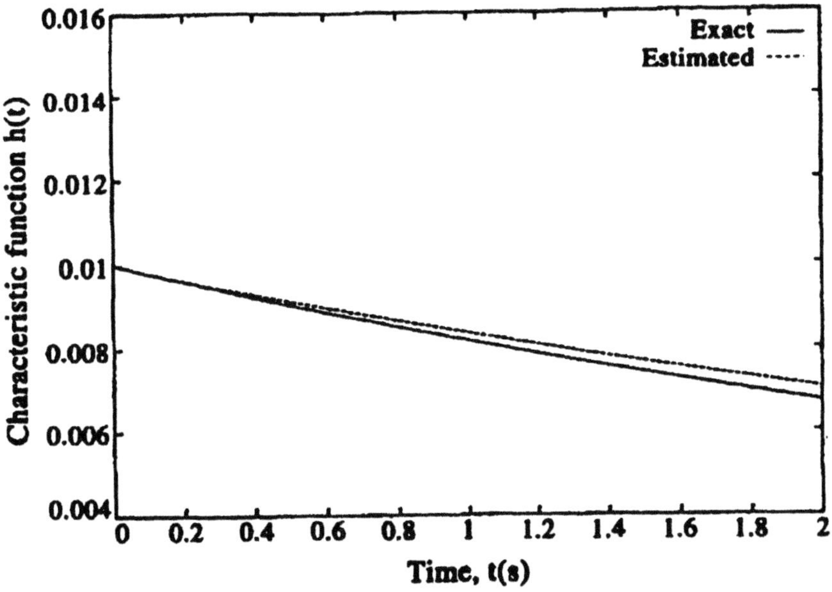

Figure 13.10 The exact and estimated characteristic function *h (t)* from the first-order DTS. First-order sytem $\dot{y} + 0 \cdot 2y = x(t)$ with parameters $a_1 = 1{\cdot}0$, $b_1 = 0{\cdot}2$, order $p = 1$, and input *x (t)* of Fig. 13.1. "Reprinted from *Int. J. Pres. Ves. & piping* 67, Yu P. and Haddad, Y. M., on the dynamic system identification of the response behaviour of linear viscoelastic materials, 45-54, 1996, with kind permission from Elsevier Science Ltd, The Boulevard, Langford Lane, Kidlington 0X5 1GB, UK".

Example 13.4

Consider the second order system

$$\ddot{y} + a_1\dot{y} + a_2 y = b_1\dot{x} + b_2 x \tag{13.90}$$

The input x(t), onto the system above, is assumed to be given by

$$x(t) = \sin[(\beta t^{\gamma-1} - \omega_0)t] \tag{13.91}$$

where β, ω_0 and γ are constant parameters.

Meantime, the frequency of the input signal is assumed to be

$$\omega(t) = \beta\, t^{\gamma-1} + \omega_0 \tag{13.92}$$

The parameters in equations (13.90) and (13.92) are assumed as shown, respectively, in Tables 13.11 and 13.12.

TABLE 13.11

a_1	a_2	b_1	b_2
51.000	50.000	11	60

TABLE 13.12

β	γ	ω_0
0.5	2.0	30.0

Solving equation (13.90) by using the Runge-Kutta numerical method (e.g., Morris, 1983), we obtain two discrete-time series $\{x_k;\ k = 0,1,2,...,N\}$ and $\{y_k;\ k = 0,1,2,...,N\}$ which are shown, respectively, in Figures 13.11 and 13.12 with $\Delta T = 0.01$. Here, the parameter N represents the number of discrete points.

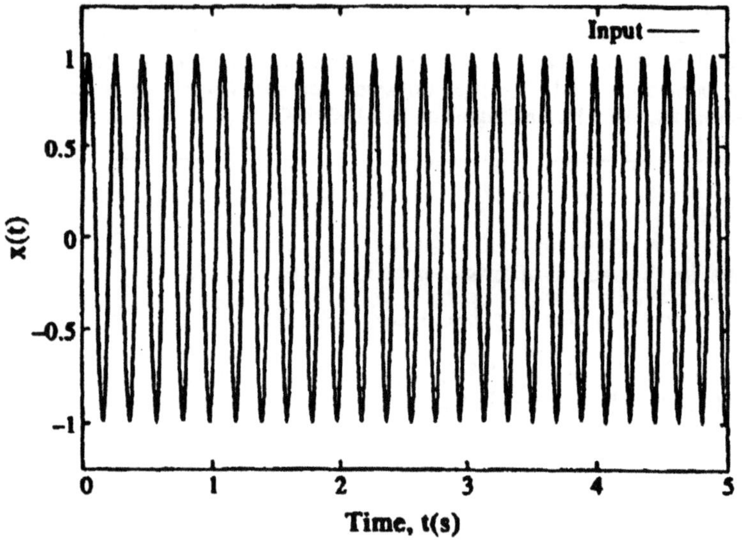

Figure 13.11 Input $x(t) = \sin[(\beta t^{\gamma-1} - \omega_0)t]$ with $\beta = 0{\cdot}5$, $\gamma = 2{\cdot}0$ and $\omega_0 = 30{\cdot}0$ to system $\ddot{y} + a_1\dot{y} + a_2 y = b_1\dot{x} + b_2 x$ with $a_1 = 51{\cdot}0$, $a_2 = 50{\cdot}0$, "Reprinted from *Int. J. Pres. Ves. & piping* 67, Yu P. and Haddad, Y. M., On the dynamic system identification of the response behaviour of linear viscoelastic materials, 45-54, 1996, with kind permission from Elsevier Science Ltd, The Boulevard, Langford Lane, Kidlington 0X5 1GB, UK".

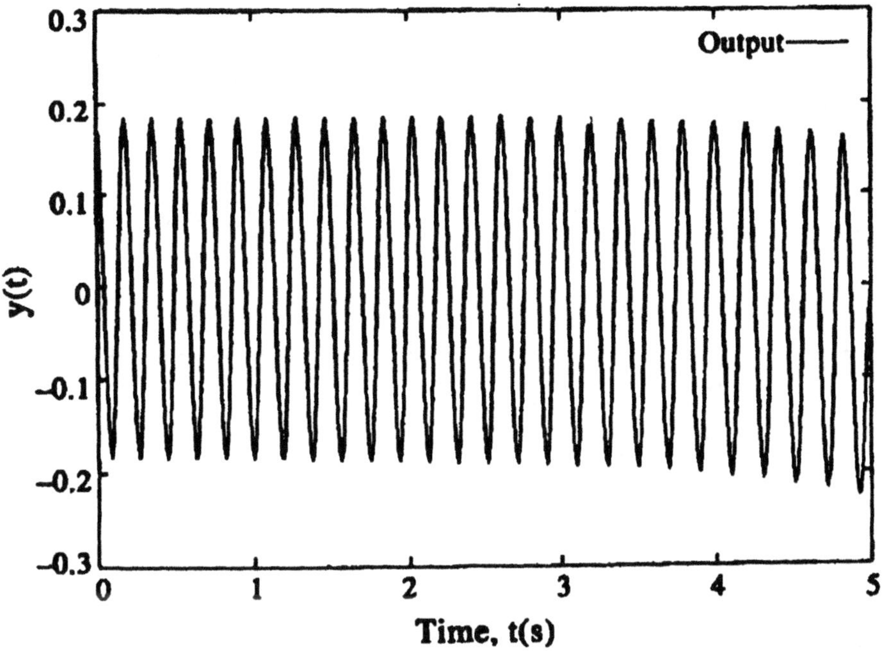

Figure 13.12 Output *y (t)* corresponding to the input shown in Fig. 13.11. "Reprinted from *Int. J. Pres. Ves. & piping* 67, Yu P. and Haddad, Y. M., On the dynamic system identification of the response behaviour of linear viscoelastic materials, 45-54, 1996, with kind permission from Elsevier Science Ltd, The Boulevard, Langford Lane, Kidlington 0X5 1GB, UK".

Using discrete-time systems (DTS) of different orders, Eqn.(13.67), to model the system, the error e^2, Eqn.(13.84), for discrete-time systems of different orders are shown in Table 13.13.

TABLE 13.13

Order	First	Second	Third	Fourth
Error	0.287038E-2	0.236837E-05	0.211482E-04	0.284848E-04

Based on the results shown in Table 13.13, we choose the second order DTS, of minimum error, to model the system. The parameters characterizing Eqn.(13.67) are determined, by (13.85), as presented in Table 13.14.

TABLE 13.14

parameters	β_1	β_2	α_1	α_2
Value	-1.601352	0.605265	0.109850	-0.104986

The roots of the characteristic equation Eqn.(13.73) of the discrete-time system of the second order are then determined and are given in Table 13.15.

TABLE 13.15

Roots	λ_1	λ_2
Value	0.989929	0.611423

Accordingly, the pertaining parameters B_m (m = 1, 2) corresponding to Eqn.(13.75) can be determined and are shown in Table 13.16.

TABLE 13.16

B_1	B_2
0.0099283	0.09992170

The parameters of Eqn.(13.62) can be finally identified form Eqn.(13.80) and are given in Table 13.17.

TABLE 13.17

Parameter	A_1	A_2	ξ_1	ξ_2
Estimated	0.992830	9.992171	-1.012206	-49.196643
Exact	1.00000	10.00000	-1.00000	-50.00000

Figure 13.13 shows the exact and estimated output y(t).
Figure 13.14 shows the exact and estimated values of characteristic function h(t).

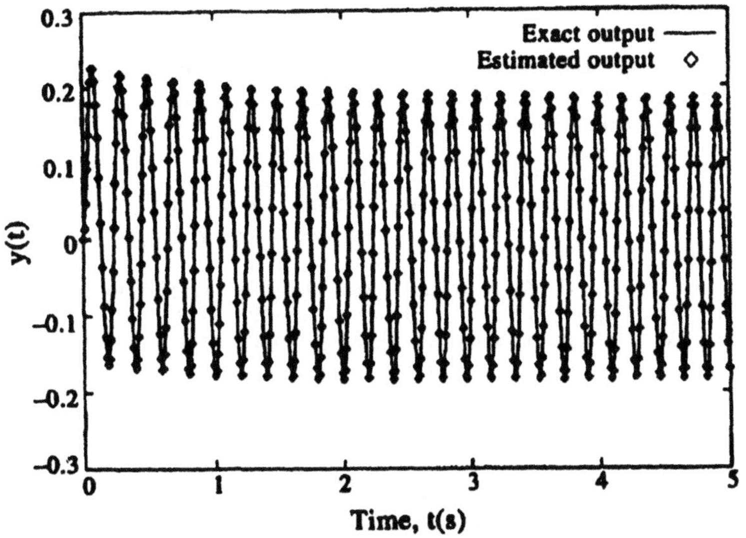

Figure 13.13 The exact and estimated output *y (t)* corresponding to the input shown in Fig. 13.11. "Reprinted from *Int. J. Pres. Ves. & piping* 67, Yu P. and Haddad, Y. M., On the dynamic system identification of the response behaviour of linear viscoelastic materials, 45-54, 1996, with kind permission from Elsevier Science Ltd, The Boulevard, Langford Lane, Kidlington 0X5 1GB, UK".

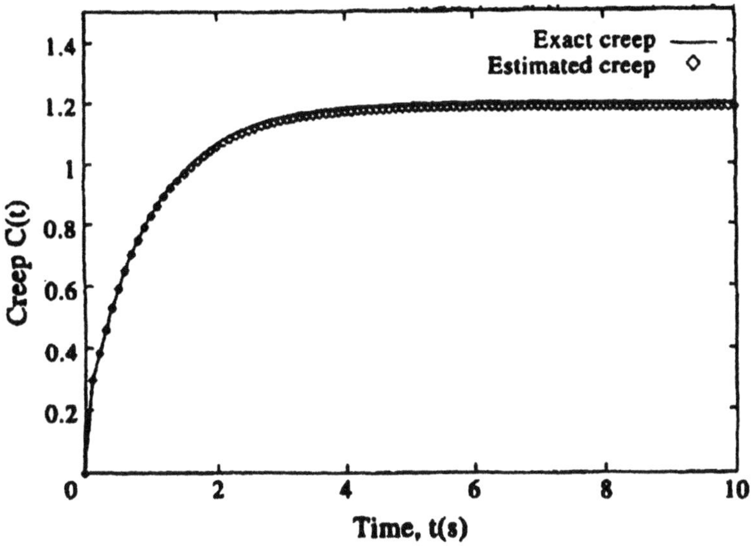

Figure 13.14 The exact and estimated creep curves corresponding to the input shown in Fig. 13.11. "Reprinted from *Int. J. Pres. Ves. & piping* 67, Yu P. and Haddad, Y. M., On the dynamic system identification of the response behaviour of linear viscoelastic materials, 45-54, 1996, with kind permission from Elsevier Science Ltd, The Boulevard, Langford Lane, Kidlington 0X5 1GB.

13. 5. References

Cadzow, J.A. (1970) *Discrete-Time and Computer Control System*, Prentice-Hall, Inc., New York.

Cadzow, J.A. (1973) *Discrete-Time Systems* , Prentice Hall, New Jersey.

Fodor, G. (1965) *Laplace Transforms in Engineering*, Publishing House of the Hungarian Academy of Sciences, Budapest

Gibson, R. F., Hwang, S. J. & Sheppard, C. H. (1990) Characterization of creep in polymer composites by use of frequency-time transformations, *J. Comp. Mat.* **24**, 441-53.

Morris, L.J. (1983) *Computation Method in Elementary Numerical Analysis*, John Wiley & Sons, New York.

Yu P. and Haddad, Y.M. (1995) A Dynamic system identification method for the characterization of the rheological response of a class of viscoelastic materials, *Int. J. Pres. Ves. & Piping* **61**, 87-97.

Yu P. and Haddad, Y. M. (1996) On the dynamic system identification of the response behaviour of linear viscoelastic materials, *Int. J. Pres. Ves. & Piping* **67**, 45-54.

13. 6. Further Reading

Christensen, R.M. (1971) *Theory of Viscoelasticity: An Introduction*, Academic Press, New York.

Freeman, H. (1965) *Discrete-Time Systems*, John Wiley & Sons Inc., New York. Gross, B. (1953) *Mathematical Structure of the Theory of Viscoelasticity*, Hermann, Paris.

Haddad, Y.M. (1988) On the theory of the viscoelastic solid, *Res Mechanics* **25**, 225-59.

Haddad, Y.M. (1995) *Viscoelasticity of Engineering Materials*, Kluwer, Dordrecht.

Kleinbaum, D.G. and Kupper, L.L. (1978) *Applied Regression Analysis and Other Multivariable Methods*, Wadsworth, Mass.

Ljung, L.(1987) *System Identification: Theory for the User*, Prentice Hall, New York.

Orbey, N. and Bed, J. M. (1991) Determination of the relaxation system from oscillatory shear data, *J. Rheol.* **35**, 1035-49.

Pandit, S.M. and Wu, S.M. (1983) *Time Series and System Analysis with Applications*, John Wiley & Sons Inc., New York.

Pandit, S.M. (1991) *Model and Spectrum Analysis: Data Dependent Systems in State Space*, John Wiley & Sons, Inc., New York.

Smith, J.D. (1989) *Vibration Measurement and Analysis*, Butterworth, New York.

Yu P. and Haddad, Y. M. (1995) A discrete-time, dynamic system method for the identification of the viscoelastic properties of a class of materials, *Int. J. Pres. Ves. & Piping* **62**, 291-301.

CHAPTER 14

VISCOELASTIC WAVES AND BOUNDARY VALUE PROBLEM

14.1. Introduction

When a localized disturbance is applied suddenly in a medium, it will soon propagate to other parts of this medium. This simple fact constitutes a general basis for the interesting subject of "wave propagation". Well-cited examples of wave propagation in different media include, for instance, the transmission of sound in air, the propagation of seismic disturbances in the earth, the transmission of radio waves, among others. In the particular case when the suddenly applied disturbance is mechanical, e.g., a suddenly applied force, the resulting waves in the medium are due to stress effects and, thus, these waves are referred to as "*stress waves*". Our attention in this chapter is focussed on the propagation of stress waves in viscoelastic solid media. In our representation, we consider the solid medium to be a continuum. Hence, the mechanics of wave motion in the medium will be dealt with from a continuum mechanics point of view. The basic concepts of continuum mechanics have been presented in Chapter 2. In such continuum, the solid medium, the disturbance is generally considered to spread outward in a three-dimensional sense (Graff, 1975). A wavefront is considered to be associated with the outward propagating disturbance. Consequently, particles of the medium that are located ahead of the wavefront are assumed to have experienced no motion, meantime, particles that are located behind the front are visualized to have experienced motion and may continue to vibrate for some time.

14.2. Internal Friction and Dissipation

Real materials are never perfectly elastic. Thus, when a material specimen is subjected to dynamic loading, part of its mechanical energy is converted into heat. The various micro-structural mechanisms by which the mechanical energy is converted into heat is conventionally referred to as "*internal friction*" Kolsky, 1963). Due to the complexity of the microstructures, several microscopic and macroscopic dissipative mechanisms exist in the material. The extent of energy loss would generally depend on the input load characteristics, the environmental conditions, as well as the inherent and macroscopic properties of the material specimen.

An internal dissipative mechanism in case of polycrystalline solids, for instance, is due to the variation in crystallographic orientation of neighbouring grains. This results in nonuniformity of the distribution of local strains when the material specimen is loaded. This is in addition to the nonuniformity of local strains that may be caused by imperfections in the material (e.g., micro- cracks, fissures, flaws, foreign inclusions and grain boundaries). Consequently, a nonuniform temperature field may exist and thermal currents of varying magnitudes would flow within the crystal lattice. Other microscopic mechanisms could be also responsible for the transfer of energy into heat. One mechanism is due to dislocations, that is the movement of regions of disarray in the crystals (see, e.g.,Orowan, 1934; Polanyi, 1934 and Bradfield, 1951). An additional microscopic mechanism is due to the motion of solute atoms in the crystal lattice on the application of external loading (e.g., Gorsky, 1936 and Snoek, 1941). A possible microscopic mechanism which attenuates stress waves in polycrystalline solids is "scattering" (e.g., Kolsky, 1963). This mechanism may occur in a polycrystalline solid when the incident wavelength becomes comparable with the grain size. In this, Mason and McSkimin (1947), for instance, found that when the wave length is long compared to the grain size, the attenuation is inversely proportional to fourth power of the wavelength (see Rayleigh, 1894).

On the macroscopic level, the following effects of internal friction are particularly important.

(i) Static Hysteresis

This is primarily due to the inelastic characteristics of the material. In this case, when a material specimen is taken through a stress cycle, it would show a "hysteresis loop", that is the stress-strain curve for an increasing stress input does not retrace its earlier downward path, if the material specimen were reloaded in an exact manner reflecting the unloading. The area enclosed by this loop represents mechanical energy which has been dissipated into heat. Although this effect may seem to be insignificant for some materials under static loading, it could be a pronounced factor in the attenuation of stress waves travelling in such materials. In the latter case, each layer of the material is taken through a loading cycle. For sinusoidal oscillations, for example, the number of hysteresis cycles is dependent on the frequency and the latter may be of the order of millions per second.

(ii) Viscous Loss

Such a loss is particularly noticeable in case of polymers with organic long chain molecules. The internal forces here are of a viscous nature and imply that the mechanical behaviour of such materials is a function of the rate of strain (see, e.g., Tobolosky, Powell and Eyring, 1943; Alfrey, 1948 and Kolsky, 1963). In case of viscoelastic materials, it is recognized that stress waves whose periods are close to the relaxation times of the material are severely attenuated when passing through it (Kolsky, 1963). In metallic materials, however, the dissipative mechanism tends to be more related to their macroscopic thermal properties (Zener, 1948).

(iii) Stress Wave Motion Effect

In this, the compression and dilatation due to the stress wave motion in the material produce temperature gradients. Thus, the finite thermal conductivity of the solid would be an influential mechanism by which the mechanical energy of waves may dissipate as thermal energy.

Internal friction in solids is often defined by the so-called "specific loss" or, alternatively, "specific damping" of the specimen. It is denoted by the symbol D and is conventionally expressed by

$$D = \frac{\Delta W}{W} \tag{14.1}$$

In the above relation, ΔW is the energy dissipated upon subjecting the specimen to a stress cycle and W is the elastic energy stored in the specimen during this cycle. The magnitude of D depends on the amplitude and the speed of the cycle, other boundary conditions, as well as the past history of the specimen. The reader is referred to Kolsky (1963) for other definitions of internal friction and its measurement.

Mechanical dissipation is particularly pronounced in case of viscoelastic materials, particularly those of high polymeric origin. In most of these materials, the presence of mechanical dissipation can effectively change the nature of wave motion in them. In addition to the significant mechanical dissipation that can occur in viscoelastic materials, it is well-recognized that these materials are "***dispersive***". In view of the latter property, phase velocity of a wave propagating in a viscoelastic material will depend on wave frequency. More specifically, waves of high frequency will propagate in viscoelastic materials with a greater phase velocity than if these waves have a low frequency. Consequently, a mechanical disturbance would continually change in shape during its motion in a viscoelastic medium. Further, the attenuation of high frequency waves in viscoelastic materials is greater than that of waves of low frequency. In case of sinusoidal waves, for instance, the above two characteristics of wave motion in a viscoelastic medium would translate into a differential absorption as well as a differential dispersion of the Fourier components of the pulse (Kolsky, 1963).

14.3. Viscoelastic Wave Motion

As realized in Section 2 of this chapter, the constitutive equation for a particular material must be combined with the equations of motion in order to solve a specific problem concerning the wave propagation in such material. In contrast to the situation in linear elasticity, the viscoelastic constitutive equation, even in the linear case, is complex by virtue of the existence of integro-differential terms in this equation and the time-dependency of the viscoelastic material functions involved. This added complexity has limited quite significantly the progress in dynamic viscoelasticity in general. Consequently, the majority of problems that have been successfully treated concerning viscoelastic wave phenomena have been limited to simple material representation. A large number of viscoelastic wave propagation problems, within the linear response behaviour of the material, have been attempted by different researchers using a correspondence with an available or deductible solution of an analogous linear elastic problem.

Kolsky (1956,1960) presented a comprehensive review of the subject of viscoelastic waves in solids from both theoretical and experimental points of view. In his treatment of the subject matter, Kolsky employed the superposition property of solutions in linear viscoelasticity through the application of Fourier analysis. Kolsky (1960) considered, for instance, the motion of a longitudinal disturbance along a thin filament. In this context, the equation of motion along the filament is expressed by

$$\frac{\partial \sigma}{dx} = \rho\left(\frac{\partial^2 u}{\partial t^2}\right) \tag{14.2}$$

where σ is the longitudinal stress, x is the distance along the filament, u is the longitudinal displacement and ρ is the density. For a sinusoidal wave propagating in a linear visco-elastic solid, the stress is related to the strain through a complex modulus representation (see Chapter 3),

$$\sigma = \left(E_1 + i E_2\right)\epsilon = \left(E_1 + i E2\right)\frac{\partial u}{\partial x} \tag{14.3}$$

Combining (14.2) and (14.3), then

$$\left(E_1 + i E_2\right)\frac{\partial^2 u}{\partial x^2} = \rho\left(\partial^2 u/\partial t^2\right) \tag{14.4}$$

The solution of (14.4) for a propagating sinusoidal wave of frequency $\omega/2\pi$, whose displacement at the origin is $u_0 \cos \omega t$, is expressed as

$$u(x) = u_0 \exp(-\alpha t) \cos\left[\omega(t - x/c)\right] \tag{14.5}$$

where,

$$c = \left(E^*/\rho\right)^{1/2} \sec \delta/2 \tag{14.6}$$

$$\alpha = \omega/c \tan \delta/2 \tag{14.7}$$

$$E^* = E_1^2 + E_2^2 \tag{14.8}$$

$$\tan\delta = E_2/E_1 \tag{14.9}$$

On the assumption that, for most polymers, $\tan\delta \ll 1$, then, $\sec\delta/2 \approx 1$ and $\tan\delta/2 \approx \frac{1}{2}\tan\delta$. Thus, (14.6) and (14.7) are, respectively, reduced to

$$c = \sqrt{E^*/\rho} \quad \text{and} \quad \alpha = \left(\frac{\omega}{2c}\right)\tan\delta \tag{14.10}$$

where c and α are referred to as "propagation constants". Accordingly, if the values of the moduli E_1 and E_2 (or E^* and $\tan\delta$) are known from experiment over a sufficient frequency range, the displacement of the disturbance along the filament may be calculated by (14.5) with the use of (14.10).

From an experimental point of view, two types of disturbance inputs are often considered for the study of wave propagation in materials, i.e., sinusoidal waves and pulse inputs (e.g., Hillier, 1949, 1960, Hillier and Kolsky, 1949 and Kolsky, 1960).

14.3.1. SINUSOIDAL INPUTS

For this type of disturbance input, continuous trains, of small amplitude of vibration, are propagated along filaments of the material. As introduced in the foregoing, if the displacement input on one end of the specimen is $u_0 \cos\omega t$, then, the displacement at a distance x along the filament is given by (14.5). Hence, by measuring the amplitude and phase of the vibration at different points along the filament, the propagation constants c and α can be determined from (14.5). Consequently E^* and $\tan\delta$ (or E_1 and E_2) as functions of frequency $\omega/2\pi$ are found from (14.6) or (14.7). Hillier and Kolsky (1949) and Ballon and Smith (1949), for instance, have used this method for the determination of the dynamic properties of viscoelastic materials such as rubber and plastics in the range of 10^2 - 10^3 cycles per second (e.g., Kolsky, 1960).

For a linear viscoelastic solid, provided that E_1 is not changing too rapidly with frequency, one may write (Ferry and Williams, 1952) that

$$\frac{dE_1}{d\omega} \approx \frac{2E_2}{\pi\omega} \tag{14.11}$$

which can be written in view of (14.9) as

$$\frac{d(\log E_1)}{d(\log \omega)} \approx \left(\frac{2}{\pi}\right) \tan \delta \tag{14.12}$$

For most polymers, at temperatures near their transition from the rubber-like to the glassy-like temperature, $\tan \delta$ varies comparatively little with frequency (see, e.g., Nashif et al., 1965). For this case, one may assume, that $\tan \delta$ is constant (i.e. independent of frequency). Under the latter assumption, Eqn. (14.12) may be integrated to give

$$E_1 \approx E_1(\omega_0)\exp\left[\left(\frac{2}{\pi}\right) \tan \delta \log\left(\frac{\omega}{\omega_0}\right)\right] \tag{14.13}$$

where $E_1(\omega_0)$ is the value of E_1 at a fixed reference frequency $\omega_0/2\pi$. Further, if one assumes, as mentioned before, that $\tan \delta << 1$, one can express the propagation constants α and c, with reference to (14.6) to (14.10), as

$$c \approx \sqrt{E_1/\rho} \quad \text{and} \quad \alpha = \frac{\omega}{2c} \tan \delta \tag{14.14}$$

Meantime, equation (14.13) is approximated further as

$$E_1 \approx E_1(\omega_0)\left[1 + \left(\frac{2}{\pi}\right) \tan \delta \log\left(\frac{\omega}{\omega_0}\right)\right] \tag{14.15}$$

whereby the exponential term in (14.13) has been expanded asymptotically and the first two terms in the expansion are retained. Accordingly, one writes with reference to (14.14) that

$$c \approx c_0\left[1 + \{2(\tan \delta)/\pi\} \log\left(\frac{\omega}{\omega_0}\right)\right]^{1/2} \tag{14.16a}$$

where,

$$c_0 = \sqrt{E_1(\omega_0)/\rho} \tag{14.16b}$$

Fig. 14.1 (Kolsky, 1960, Experimental results after Hillier, 1949) supports a linear relation between c and $\log \omega$ for polyethylene in the frequency range shown.

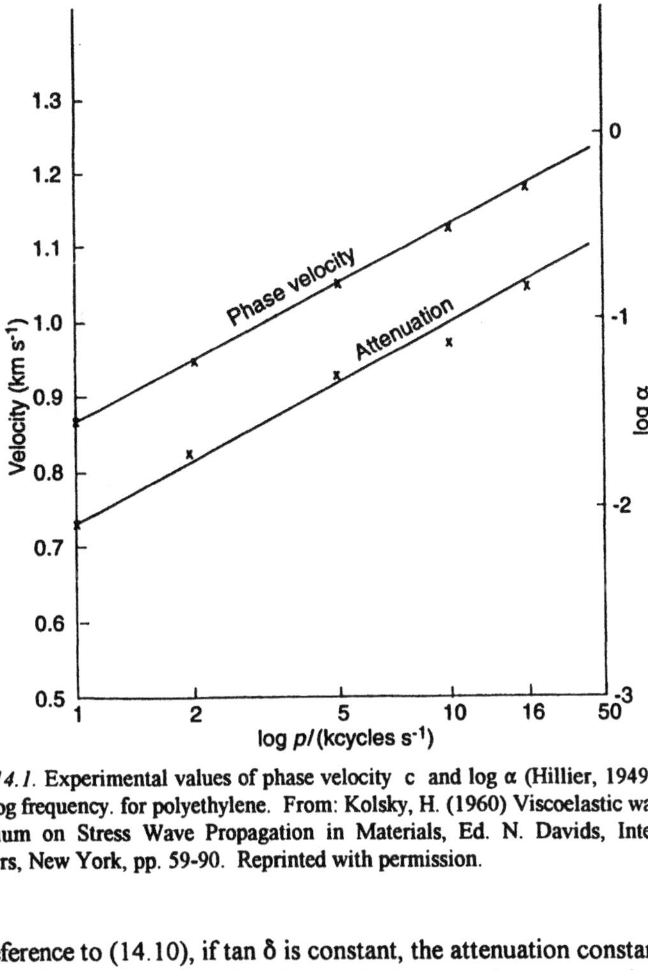

Figure 14.1. Experimental values of phase velocity c and log α (Hillier, 1949) plotted against log frequency. for polyethylene. From: Kolsky, H. (1960) Viscoelastic waves, Int. Symposium on Stress Wave Propagation in Materials, Ed. N. Davids, Interscience Publishers, New York, pp. 59-90. Reprinted with permission.

With reference to (14.10), if $\tan\delta$ is constant, the attenuation constant α would be proportional to ω/c. Further, since the phase velocity c varies comparatively slowly with frequency, over a limited frequency range, Eqn. (14.16), one may expect that the attenuation constant α, Eqn. (14.14), to be proportional to the frequency. Accordingly, $\log\alpha$ should vary linearly with $\log\ \omega$ as shown by the graph by Hillier (1949) in Fig. 14.1.

For the study of viscoelastic wave propagation at higher frequencies, pulsed ultrasonic methods are used (see, for example, Ivey et al, 1949 and Cunningham and Ivey, 1956). The experimental practice of the technique could vary quite significantly, however, in principle, a finite number of sinusoidal cycles of frequency $\omega/2\pi$ are introduced in one end of the specimen and the resultant wave motion is recorded at a number of points along the length of the specimen. From a measurement of transient time and amplitude ratio, estimates of the phase velocity $c(\omega)$ and the attenuation constant $\alpha(\omega)$ can be made. The ultrasonic technique has the advantage of being relatively simple. It is particularly powerful for investigating the wave propagation properties in elastic materials. In case of viscoelastic materials, however, the technique unfortunately suffers from certain theoretical difficulties of interpretation as pointed out by Kolsky (1960): The time of transit of the pulse depends on the group velocity of the wave packet, and for a dispersive medium, this is, in general, different from the phase velocity c. In the absence of attenuation, these two velocities can be related (Kolsky, 1963), however, in a medium which is dissipative as well as dispersive, the relation between group velocity and phase velocity is not clear yet.

14.3.2. PULSE INPUTS

Few experimental research efforts have been focussed on the study of pulse propagation in viscoelastic materials. In the early work of Hillier (1949) and Hillier and Kolsky (1949,1956) steady-state longitudinal vibrations were induced in prest retched filaments (0.06 cm. in diameter) of polythene, neoprene and nylon by means of a transducer element attached to one end of the material specimen. The experimental studies were carried out within low frequency range < 16 kc/s. The response of the filament was determined at various points along its length by means of a crystal pick-up. In this, measurements were taken of the variations in the vibration amplitude and the phase. After allowing for the effect of pick-up (see Hunter, 1960), the experimental results included both phase velocity and attenuation at a number of frequencies. Kolsky (1956), presented experimental results after Hillier (1949) which show the phase velocity and attenuation in polythene (I.C.I. Alkathene grad 20) against frequency for experiments carried out at 10°C.

Kolsky (1954 a and b, 1956) has carried out a number of experiments on the change of the shape of longitudinal stress pulses as they travel along rods of various plastics. These pulses were produced by the detonation of small explosive charges with initial durations of about two or three microseconds. Figure 14.2 shows oscillograph records which were obtained by Kolsky (1960) with rods of polymethyl methacrylate and polyethylene. As noted by Kolsky (1960), with the polyethylene specimen, after two or three reflections, the length of the pulse had become more than twice the length of the specimen, with the result that the movement of the ends of the specimen become continuous. Figure 14.3 (due to Kolsky, 1960) shows the curves of particle velocity with the passage of time for pulses which had propagated in polyethylene rods 30, 60 and 90 cm in length. It can be seen in the figure that the pulses become progressively flatter, but retain an asymmetrical shape.

14.4. Wave Propagation in Semi-Infinite Media

In this section, we deal with the problem of determining the stress distribution in a semi-infinite viscoelastic rod subject to dynamic loading. The problem was examined by Lee and Morrison (1956). In Lee and Morrison's work, the stress and velocity distributions associated with the propagation of an impulsively applied velocity and stress along viscoelastic rods, as presented by different mechanical models, were determined. Morrison (1956) also considered analytically the wave propagation in a viscoelastic rod of the Voigt model type and also studied viscoelastic materials with three-parameter models. In an earlier work, Hillier (1949), see also Hillier and Kolsky (1949), studied the motion of longitudinal sinusoidal waves along a viscoelastic filament assuming a Maxwell solid, a Voigt solid and a three-element model representations. Lee and Kanter (1953), considered the stress distribution in a rod of Maxwell material subjected to a mechanical impact. Glauz and Lee (1954), on the other hand, used the method of characteristics to determine the stress in a viscoelastic material made of a four-parameter model.

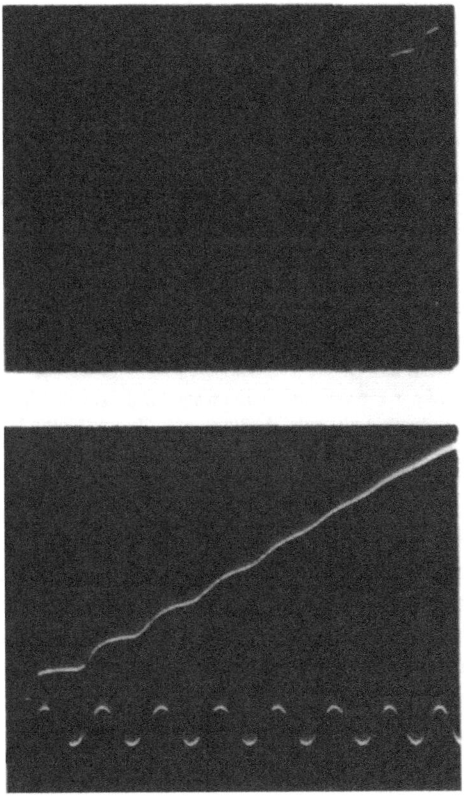

Figure 14.2. (a) Oscillograph record of displacement at end of polymethyl methacrylate rod 46 cm long and 1.25 cm diam. when 5 mg charge of lead has been detonated at opposite end. Period of timing wave is 500 microseconds. (b) Oscillograph record, similar to (a), for polyethylene rod 20 cm long and 1.25 cm diam. From: Kolsky, H. (1960) Viscoelastic Waves, Int. Symposium on Stress Wave Propagation in Materials, Ed. N. Davids, Interscience Publishers, London, pp. 59-90. Reprinted with permission.

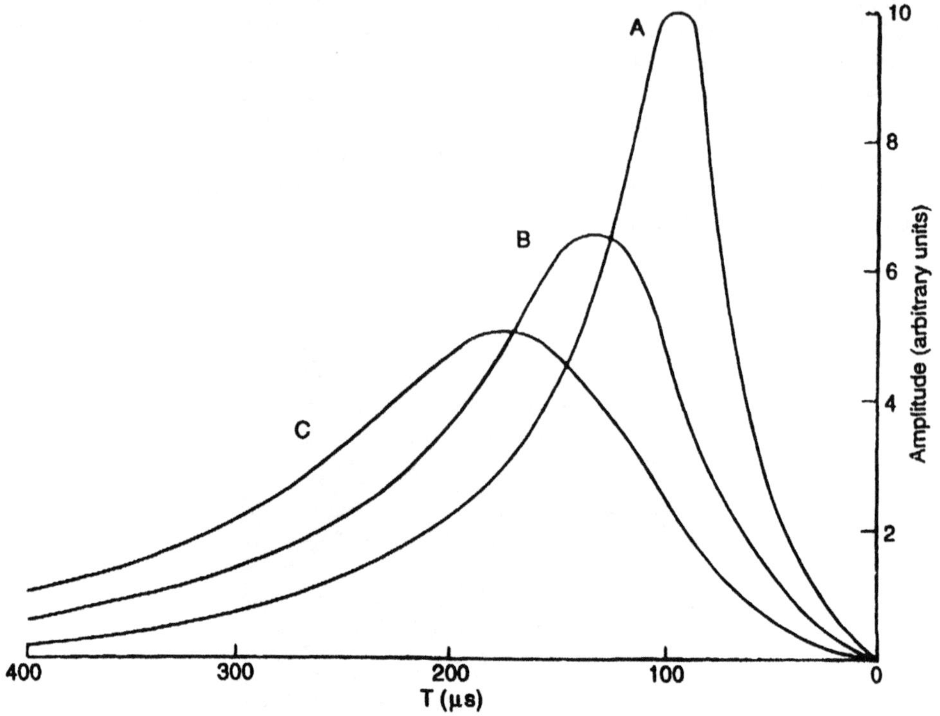

Figure 14.3. Curves of particle velocity distributions for pulses that have travelled through different lengths of polyethylene rods. A, 30 cm; B, 60 cm; C, 90 cm. From: Kolsky, H. (1960) Viscoelastic Waves, Int. Symposium on Stress Wave Propagation in Materials, Ed. N. Davids, Interscience Publishers, London, pp. 59-89. Reprinted with permission.

Consider a semi-infinite rod as shown in Fig. 14.4 where $x \geq 0$, with the x-coordinate measured along the length of the rod. In this figure, $x(t)$ denotes the position of a section of the rod at time t and $u(x,t)$ is the displacement of this section in the direction of increasing x. Let, $\sigma(x, t)$ denote the nominal compressive stress transmitted across the section x of the rod at time t. $\epsilon(x, t)$ designate the nominal compressive strain corresponding to $\sigma(x,t)$ above. ρ is the mass density of the material the governing equation of motion, in the absence of body forces, in the x-direction is

$$\rho \frac{\partial^2 u}{\partial t^2} = -\frac{\partial \sigma}{\partial x}$$

or, in a more compact form,

$$\rho\, u_{tt} = -\sigma_x \tag{14.17}$$

where a subscript denotes partial differentiation with respect to the corresponding variable.

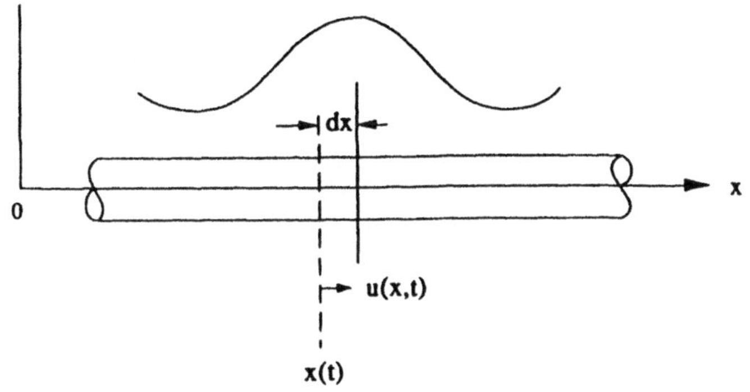

Figure 14.4. Stress wave propagation in a semi-infinite rod.

The nominal compressive strain $\epsilon(x,t)$ is written in terms of the displacement u as

$$\varepsilon = -u_x \tag{14.18}$$

The particle velocity v (x, t) is expressed in terms of the displacement u as

$$v = u_t \tag{14.19}$$

The semi-infinite rod is considered to be initially unstrained and at rest when, at $t=0$, the end $x=0$ is subjected to a mechanical impact (disturbance) with either a constant stress or a constant velocity. In either case, the stress or the velocity at the end $x=0$ is specified. The governing boundary conditions are

$$\sigma = \sigma_0 H(t) \quad \text{or} \quad v = v_0 H(t), \qquad \text{at} \quad x = 0 \tag{14.20}$$

where σ_0 is the applied constant stress, v_0 is the applied constant velocity and $H(t)$ is the Heaviside step function, that is

$$H(t) = \begin{cases} 1 & \text{for } t>0 \\ 0 & \text{for } t<0 \end{cases}$$

In addition to equations (14.17) to (14.20), the constitutive equation for the particular viscoelastic material must be included in the process of determining the stress distribution in the semi-infinite rod.

14.5. The Wave Equation in Linear Viscoelasticity as Based on Boltzmann's Superposition Principle

Consider a homogeneous, isotropic rod and let x_i $(i=1,2,3)$ denote the Cartesian coordinates of any material particle p in the deformed (current) state. For the longitudinal motion of the rod in the x_1-direction, the displacement is expressed as

$$\mathbf{u}(\mathbf{x},t) = u_1(x_1,t)\,\mathbf{e}_1 \tag{14.21}$$

where $\mathbf{e}_1$ is the unit vector component along the x_1-axis. It is assumed here that the displacement $\mathbf{u}(\mathbf{x},t)$ is a continuous function of $\mathbf{x},t$ for all $\mathbf{x},t$. In this case, $\epsilon_{11}(x_1,t)$ and $\sigma_{11}(x_1,t)$ will be the only corresponding nonvanishing components of the strain and stress, respectively, where

$$\varepsilon_{11}(x_1,t) = \frac{\partial u_1(x_1,t)}{\partial x_1} \tag{14.22}$$

and the stress is connected to the strain via the Boltzmann's hereditary creep and relaxation constitutive equations introduced in Chapter 12 (see, also, Haddad, 1995) Recalling the latter two equations, one may write in the same order that

$$\epsilon(x,t) = E^{-1}\left[\sigma(x,t) + \int_0^t F(t-\tau)\sigma(x,\tau)d\tau\right] \tag{14.23}$$

and

$$\sigma(x,t) = E\left[\varepsilon(x,t) + \int_0^t R(t-\tau)\epsilon(x,\tau)d\tau\right] \tag{14.24}$$

Meantime, the equation of motion can be written, in the absence of body forces as

$$\rho\frac{\partial^2 u(x,t)}{\partial t^2} = \frac{\partial\sigma(x,t)}{\partial x} \tag{14.25}$$

Combining now (14.23) and (14.25), the creep wave equation, in the absence of body forces, can be written as

$$\frac{\partial^2 u(x,t)}{\partial x^2} - \frac{1}{c^2}\frac{\partial^2 u(x,t)}{\partial t^2} = \frac{1}{c^2}\int_0^t F(t-\tau)\frac{\partial^2 u(x,t)}{\partial \tau^2}d\tau \tag{14.26}$$

$$\text{where } c^2 = E/\rho \tag{14.27}$$

Further, with reference to (14.24) and (14.25), the relaxation wave equation, in the absence of body forces, is

$$\frac{1}{c^2}\frac{\partial^2 u(x,t)}{\partial t^2} - \frac{\partial^2 u(x,t)}{\partial x^2} = \int_0^t R(t-\tau)\frac{\partial^2 u(x,t)}{\partial x^2}d\tau \tag{14.28}$$

with $c^2 = E/\rho$

The Laplace transforms of the wave equations (14.26) and (14.28) read, respectively, as follows

$$\frac{d^2}{dx^2}\bar{u}(x,s) = \left(\frac{s}{c}\right)^2\{1+\bar{F}(s)\}\bar{u}(x,s) \tag{14.29}$$

and

$$\frac{d^2}{dx^2}\bar{u}(x,s) = \left(\frac{s}{c}\right)^2\{1+\bar{R}(s)\}^{-1}\bar{u}(x,s) \tag{14.30}$$

Considering, for instance, equation (14.29) corresponding to the creep case, the general transform solution can be written (Graffi, 1982) as

$$\begin{aligned}\bar{u}(x,s) &= A(s)\exp\left\{\frac{xs}{c}\sqrt{1+\bar{F}(s)}\right\}\\ &+ B(s)\exp\left\{\frac{-xs}{c}\sqrt{1+\bar{F}(s)}\right\}\end{aligned} \tag{14.31}$$

where $A(s)$ and $B(s)$ are functions of the Laplace parameter s. Both $A(s)$ and $B(s)$ are to be determined.

At this point, we shall assume that the rod is semi-infinite in extent and initially undisturbed in the sense that

$$u(x,0) = \frac{\partial u(x,0)}{\partial t} = 0 \tag{14.32}$$

The following boundary conditions are further assumed

$$\begin{aligned} &u(0,t) = u_0(t) \quad &;\ t \geq 0\\ &\lim_{x\to\infty} u(x,t) = 0 \quad &;\ t \geq 0\end{aligned} \tag{14.33}$$

In this case, one must impose that $A(s)=0$ in (14.31) in order to avoid the exponential increase with x of the first term in this equation. Thus, $B(s)$ in (14.31) will assume the value of the Laplace transform of the input $u_0(t)$ and (14.31) becomes

$$\bar{u}(x,s) = \bar{u}_0(s)\exp\left\{\frac{-xs}{c}\sqrt{1+\bar{F}(s)}\right\} \tag{14.34}$$

the inversion of the Laplace transforms in (14.34) leads to

$$u(x,t) = H(t-x/c)\, e^{\alpha x/c} \Big\{ u_0(t-x/c) + \int_0^{t-x/c} F(x,t-\tau-x/c) u_0(\tau) d\tau \Big\} \tag{14.35}$$

in which $H(\cdot)$ is the Heaviside step function and where the translation and convolution formulae for the Laplace transform (see Appendix C) were used. Another representation of the solution (14.35) is due to Mainardi and Turchetti (1975,1979). Mainardi and Nervosi (1980) have also considered the inclusion of such presentation in their treatment of transient waves in a viscoelastic rod. Similar treatment may be considered for the relaxation case based on equation (14.30).

14.6. The Wave Propagation Problem as Based on the Correspondence Principle

In this section, a presentation is given, following Chao and Achenbach (1964), on the utilization of the correspondence principle to solve wave propagation problems in linear viscoelasticity when the solutions of the corresponding elastic problems are known.

The constitutive equations for an isotropic, elastic solid are given in Chapter 6 With reference to these equations, it is recognized that for an isotropic, elastic solid, two independent constants completely define the stress-strain relations. If the shear modulus μ and the bulk modulus K, for instance, are chosen, the constitutive equation, for a linear elastic solid, can be written in the following tensorial form

$$\sigma_{ij} = (K - 2/3\,\mu)\epsilon_{kk}\,\delta_{ij} + 2\mu\,\epsilon_{ij} \tag{14.36}$$

On the other hand, the constitutive relations for the creep of an isotropic, viscoelastic solid can be written as

$$2\mu\,\epsilon'_{ij}(t) = \sigma'_{ij}(t) + \int_{-\infty}^{t} F_1(t-\tau)\frac{d\sigma'_{ij}(\tau)}{d\tau}\,d\tau \tag{14.37a}$$

$$3K\,\epsilon_{ii}(t) = \sigma_{ii}(t) + \int_{-\infty}^{t} F_2(t-\tau)\frac{d\sigma_{ii}(\tau)}{d\tau}\,d\tau \tag{14.37b}$$

in which $F_1(\cdot)$ and $F_2(\cdot)$ are the creep functions governing, respectively, the shear and dilatational behaviours of the medium. The treatment of stress-wave propagation in a viscoelastic solid which obeys the constitutive relations leads to complicated mathematical analysis in that the solution of partial integro-differential equations is involved. Volterra (1931) considered the problem by adopting a functional analysis approach, but it seems that the results of the theory have found, so far, little application in the study of the dynamic behaviour of viscoelastic materials (see, e.g., Kolsky, 1963).

Chao and Achenbach (1964) discussed the application of Laplace transform to viscoelastic wave propagation problems using the well-known correspondence principle (Bland, 1960 and Schapery, 1974). It was shown by these authors that under the restricted condition of constant Poisson's ratio, a class of viscoelastic problems may be solved provided that the solution of the corresponding elastic problem is known. Applying Laplace transform to (14.37a) and (14.37b), with some additional manipulation, yields

$$\bar{\sigma}_{ij} = \left[\mathrm{K}(s) - \frac{2}{3}\bar{u}(s)\right]\bar{\epsilon}_{kk}\delta_{ij} + 2\bar{\mu}(s)\varepsilon_{ij} \tag{14.38a}$$

where

$$\bar{\mu}(s) = \frac{\mu}{1 + s\bar{F}_1(s)} = \mu\,\beta(s) \tag{14.38b}$$

and

$$\bar{K}(s) = \frac{K}{1 + s\bar{F}_2(s)} = K\,\gamma(s) \tag{14.38c}$$

where s is the Laplace transform parameter.

Meantime, the Laplace transform of the stress equation of motion , in the absence of body forces, can be written as,

$$\bar{\sigma}_{ij,j} = \rho s^2 \bar{u}_i \tag{14.39}$$

where ρ is the mass density of the material.

Combining (14.38) and (14.39) yields the governing differential equations for the transformed displacements of a viscoelastic medium, that is

$$\left(\bar{K} + \frac{1}{3}\bar{\mu}\right)\bar{u}_{j,ji} + \bar{u}\,u_{i,jj} = \rho s^2 \bar{u}_i \tag{14.40}$$

Decomposing the displacement vector $\mathbf{u}$ into dilatational and rotational parts, i.e.

$$\bar{\mathbf{u}} = \bar{\mathbf{v}} + \bar{\boldsymbol{\omega}} \qquad (14.41)$$

where

$$\bar{v}_{i,i} = 0 \quad \text{and} \quad \bar{\omega}_{i,j} = \omega_{j,i} \qquad (14.42)$$

Accordingly, the transformed equations (14.40) will be satisfied if

$$\bar{v}_{i,jj} = \frac{s^2}{\bar{c}_1^2}\bar{v}_i \qquad (14.43)$$

$$\bar{\omega}_{i,jj} = \frac{s^2}{\bar{c}_2^2}\bar{\omega}_i \qquad (14.44)$$

where $\bar{c}_1$ and $\bar{c}_2$ are the transformed velocities for the dilatational and rotational waves respectively, i.e.

$$\bar{c}_1^2 = \bar{K}(s) + \frac{4}{3}\bar{\mu}(s) \qquad (14.45)$$

$$\bar{c}_2^2 = \frac{\bar{\mu}(s)}{\rho} \qquad (14.46)$$

The same treatment may be applied for the isotropic, elastic medium if the constitutive equation (14.36) is used instead of (14.38a). On the other hand, the analogous equations to (14.43) and (14.44) for the isotropic, elastic body are obtained if $\bar{\mu}(s)$ and $\bar{K}(s)$ in (14.38b) and (14.38c) are replaced by the elastic moduli μ and K respectively. The above treatment was presented by Chao and Achenbach (1964) with the following conclusion: The Laplace transforms of the solutions for a viscoelastic wave propagation problem can be obtained from the Laplace transforms of the solutions for the elastic problem with the same boundary and initial conditions by replacing the shear modulus μ by its Laplace transform $\bar{\mu}(s)$ and the bulk modulus K by its Laplace transform $\bar{K}(s)$.

The above conclusion is, in essence, a form of the well-known correspondence principle; that is the problem of obtaining solutions concerning the response behaviour of a linear viscoelastic solid is reduced to a problem of inverting the Laplace transforms of the corresponding elastic solutions.

Chao and Achenbach (1964), see, also, Achenbach amd Chao, 1962, considered the application of the above approach to the study of the displacement and stress fields inside an infinite, viscoelastic body of a constant Poisson's ratio. In their treatment, the authors assumed the input force to be time-dependent and concentrated at one point. Two illustrative examples were subsequently given. In the first example, the displacement components in the radial and the vertical directions on the surface of a viscoelastic half-space loaded suddenly by a vertical force of constant magnitude were evaluated. In the second example, the radial stress for the problem of the expanding spherical cavity in an infinite viscoelastic medium was dealt with.

14.7. Nonlinear Viscoelastic Wave Propagation

Although considerable research efforts have been made over the last decades towards characterization of the nonlinear viscoelastic nature of materials (e.g., Haddad, 1995), interest in the study of wave propagation in such materials did not develop until recently. Most of the studies on wave propagation in nonlinear viscoelastic materials dealt essentially with the one-dimensional motion within the context of the general constitutive theory of materials with fading memory. These studies have considered the propagation of both acceleration and shock waves in viscoelastic media with the objective of establishing the governing conditions for their growth or decay. Such governing conditions implied the existence of steady waves in the dissipative viscoelastic media. An initial study in the area of nonlinear wave propagation is due to Malvern (1951). Malvern's approach is concerned with the motion of a plastic wave in a ductile material (e.g. a metal with a strain memory effect). As a special case, however, Malvern considered the motion of such type of wave in a model of viscoelastic solid. The modes of propagation of acceleration waves in different media have been studied, among others, by Truesdell and Toupin (1960), Thomas (1961), Hill (1962), Varley and Cumberbatch (1965), Coleman, Gurtin and Herrera (1965), Coleman and Gurtin (1965) and Bailey and Chen (1971). Varley (1965) discussed the mode of propagation of an arbitrary acceleration wave as it advances into a finitely strained viscoelastic material which, until the arrival of the front is undergoing any admissible deformation. The viscoelastic material is seen in Varley's work to be generally inhomogeneous and anisotropic. Coleman, Gurtin and Herrera (1965) and Coleman and Gurtin (1965) dealt comprehensively with the theory of nonlinear viscoelastic wave propagation in a series of research papers. In the first two papers of the series, the authors dealt with the propagation of shock and acceleration fronts in materials with memory resting on the assumption that the stress is a functional of the history of the deformation gradient with the exclusion of any thermal influences. In subsequent two papers (Parts III and IV of the series), Coleman and Gurtin (1965) have allowed the stress to be affected not only by the history of strain, but also by the history of a thermodynamic variable such as the temperature (see, also, Coleman, 1964 and Coleman and Gurtin, 1966). An extension of this work to include mild discontinuities was considered by Coleman,

Greenberg and Gurtin (1966). The problem of propagation of steady shock waves in nonlinear thermoviscoelastic solids has been also considered, for instance, by Ahrens and Duvall (1966), Greenberg (1967), Chen, Gurtin and Walsh (1970), Schuler and Walsh (1971), Dunwoody (1972), Huilgol (1973) and Nunziato and Walsh (1973). In this, Nunziato and Walsh (1973) expressed the governing equations in terms of material response functions which can be determined from shock wave, thermophysical, and bulk response data. The results of the analysis were compared with experimental steady wave studies concerning the solid polymer, polymethyl methacrylate (PMMA). The existence and propagation of steady waves in a class of dissipative materials were considered also, among others, by Greenberg (1968) and Schuler (1970).

On the experimental side, research in the field of shock wave physics has made it possible to produce high amplitude strain waves. Barker and Hollenbach (1970) and Schuler (1970) used a gas gun (Barker and Hollenbach, 1964 & 1965) to produce a planar impact between two plates. This has been parallel with the development of advanced recording and measurement techniques such as laser inter- ferometry (Barker and Hollenbach, 1964, 1965 & 1970 and Barker, 1968). Such experimental efforts were particularly effective in the production of one-dimensional strains of very large amplitude, meantime, they allowed wave motion to be observed with high resolution and accuracy. Chen and Gurtin (1972 a and b) discussed the use of experimental results concerning steady shock waves to predict the acceleration wave response of nonlinear viscoelastic materials. Meantime, Nunziato and Sutherland (1973) used acoustic waves for the determination of stress relaxation functions of a class of polymeric materials.

Schuler, Nunziato and Walsh (1973) presented a comprehensive review of some theoretical and experimental developments in the domain of nonlinear viscoelastic wave propagation. Confining their attention to the case of one-dimensional strain, they reviewed theories of shock and acceleration wave propagation in materials with memory and discussed the theoretical predictions with some experimental results for the polymeric solid PMMA. In this, these authors were particularly influenced by the work of Coleman, Gurtin and Herrera (1965 a and b), Coleman and Gurtin (1965) and Chen and Gurtin (1970). We follow closely the work of these authors in the following presentation.

14.7.1 KINEMATICS AND BALANCE LAWS IN ONE-DIMENSIONAL MOTION

Kinematics

In the case of one-dimensional motion, we identify the spatial position of a material point (particle) at time t by the coordinate $x(X,t)$. The counterpart of this position coordinate in the reference configuration, R, is $X(x,t)$. It is assumed that the coordinate function $x(X,t)$ is continuous for all X and t. The corresponding displacement function $u(X,t)$ is, thus, a continuous function of X,t for all X and t. Assuming suitable smoothness of the motion (Schuler, Nunziato and Walsh, 1973) the particle velocity is expressed by

$$v(t) = \frac{\partial}{\partial t} u(X,t) = \partial_t u(X,t) \tag{14.47}$$

and the compressive strain is expressed by

$$\varepsilon(t) = \frac{-\partial}{\partial X} u(X,t) = -\partial_x u(X,t) \tag{14.48}$$

A wave propagating in such continuous medium may be seen (Coleman and Gurtin, 1965) as a family of points $\hat{X}(t)$, $-\infty < t < \infty$, where $\hat{X}(t)$ is the material point in the reference configuration R at which the wave front is to be found at time t. Thus, the spatial position of the wave may be expressed by

$$\hat{x}(t) = \hat{x}\left(\hat{X}(t), t\right) \tag{14.49}$$

with $\hat{x}(t)$ designating the spatial position of the wave at time *t*.

The "***wave velocity***" *V(t)* at time *t* is defined by

$$V(t) = \frac{d}{dt}\hat{x}(t) = \frac{d}{dt}\hat{x}\left(\hat{X}(t), t\right) \tag{14.50}$$

The wave velocity *V(t)* is identified with respect to an external fixed frame of reference (i.e., as seen by an external observer at rest).

Meantime, the wave "***intrinsic velocity***" *U(t)* is defined as the velocity of propagation of the wave front relative to the material in the reference configuration. It is expressed as

$$U(t) = \frac{d}{dt}\hat{X}(t) \tag{14.51}$$

where $\hat{X}(t)$, as defined earlier, is the coordinate of the material point in the reference configuration *R* at which the wave front is to be found at time *t*.

The "***material trajectory***" of the wave front is given here the notation $\Omega(t)$. It is defined as the set of ordered pairs $\hat{X}(t)$, (t), $-\infty < t < \infty$.

Jump Discontinuity. Coleman, Gurtin and Herrera (1965), following the standard notation used earlier by Truesdell and Toupin (1960), advanced that if a function f(X,t) has a jump discontinuity at $X = \hat{X}(t)$, one may define the jump in f(X,t), labeled below by [f], across the trajectory of the wave $\Omega(t)$ at time t by

$$[f] = \lim_{X \to X(t)^-} f(X,t) - \lim_{X \to X(t)^+} f(X,t) \tag{14.52a}$$

This expression may be also written in the form

$$[f] = f^- - f^+ \tag{14.52b}$$

where, with the wave intrinsic velocity $U(t) > 0$, f^+ and f^- are the limiting values of the function $f(X,t)$ immediately ahead and behind the wave front respectively. The associated "***condition of compatibility***" to is expressed (Truesdell and Toupin, 1960) as

$$[\partial_t f] = -U[\partial_X f] \tag{14.53}$$

In the present section, the function $f : f(X, t)$ is used to designate the kinematical function $x(X, t)$ or one of its derivatives.

Balance Laws

Mass Balance. In one-dimensional motion, the mass balance is expressed with reference to (1.2:3) as

$$\rho(X,t)/\rho_0 = \frac{1}{1 - \varepsilon(X,t)} \tag{14.54}$$

where $\rho(X, t)$ is the current mass density and ρ_0 is the mass density in the reference configuration of the material specimen.

Balances of Linear Momentum and Energy. With the exclusion of external body forces, heat conduction and external heat supply, the balances of linear momentum and energy are expressed respectively as

$$\frac{d}{dt}\int_{x_\alpha}^{x_\beta} \rho_0 \, v(x,t)dx = \sigma(x_\beta,t) - \sigma(x_\alpha,t) \tag{14.55}$$

and

$$\frac{d}{dt}\int_{x_\alpha}^{x_\beta}\left[\frac{1}{2}\rho_0\, v^2(x,t) + e(x,t)\right]dx = \sigma(x_\beta,t)\,v(x_\beta,t) - \sigma(x_\alpha,t)\,v(x_\alpha,t) \tag{14.56}$$

where σ is the one-dimensional stress and e is the internal energy per unit volume.

Clausius-Duhem Inequality (Second Law of Thermodynamics)
It can be expressed as (see, also, Chapter 4)

$$\frac{d}{dt}\int_{x_\alpha}^{x_\beta}\rho\,\varsigma(x,t) \geq 0 \tag{14.57}$$

where ς is the specific entropy per unit mass.

14.7.2. MATERIAL RESPONSE FUNCTIONS

Following Schuler, Nunziato and Walsh (1973), we consider a strain jump ϵ suddenly applied to a material point which has been unstrained for all past times, i.e.,

$$\epsilon(\tau) = \begin{cases} \epsilon & \tau = 0 \\ 0 & \tau<0,\ \tau>0 \end{cases} \tag{14.58}$$

The instantaneous stress, denoted by σ_I, corresponding to the strain jump, is expressed in the following functional format

$$\sigma_I(\epsilon) = F(\epsilon) \tag{14.59}$$

In (14.59), the constitutive functional $F(\epsilon)$ is assumed to be twice continuously differentiable, i.e., $\partial_\epsilon F\{\epsilon(t-\tau)\}$ and $\partial_\epsilon^2 F\{\epsilon(t-\tau)\}$ exist where the partial differentiation is with respect to the present value of strain $\epsilon(t)$; see, e.g., Haddad (1995).

The stress-relaxation function corresponding to the strain history (14.58) is designated by $R(\epsilon;\tau)$. Meantime, the ***"instantaneous tangent modulus"*** is designated by $E_I(\epsilon)$ where

$$E_I(\varepsilon) = \frac{d\,\sigma_I(\varepsilon)}{d\varepsilon} = R(\varepsilon;0) \tag{14.60a}$$

Similarly, the "***instantaneous second-order modulus***", $\overset{\vee}{E_i}$, is defined by

$$\overset{\vee}{E}_I(\epsilon) = \frac{d^2\sigma_I(\epsilon)}{d\epsilon^2} \tag{14.60b}$$

On the other hand, the equilibrium response of the material may be expressed as

$$\sigma_E(\epsilon) = F(\epsilon_E) \tag{14.61}$$

From which, the "***equilibrium tangent modulus***" is given as

$$E_E(\epsilon) = \frac{d\sigma_E(\epsilon)}{d\varepsilon} = R(\epsilon;\tau) \tag{14.62}$$

Thus, the "***equilibrium second-order modulus***", $\overset{\vee}{E_i}(\epsilon)$, is identified by

$$\overset{\vee}{E}_E(\epsilon) = \frac{d^2\sigma_E(\epsilon)}{d\epsilon^2} \tag{14.63}$$

Schuler, Nunziato and Walsh (1973) have imposed certain curvature conditions on the constitutive functional $F(\epsilon)$ of (14.59). They advanced that these conditions would hold valid for most of viscoelastic materials. These conditions may be presented as follows:

For all ϵ on $(0,1)$ and all τ on $(0,\infty)$,

$$\begin{aligned} &\text{(i)}\quad \sigma_I(\epsilon) > \sigma_E(\epsilon) > 0 \\ &\quad\;\; E_I(\epsilon) > E_E(\epsilon) > 0 \\ &\quad\;\; \overset{\vee}{E}_I(\epsilon) > 0,\ \overset{\vee}{E}_E(\epsilon) > 0 \end{aligned} \tag{14.64}$$

The inequalities under (i) above imply that the instantaneous and equilibrium stress-strain curves are strictly convex from below and the instantaneous response curve lies everywhere above the equilibrium curve as shown in Fig. 14.5. It is assumed in the latter figure that $\sigma_I(0) = \sigma_E(0) = 0$. Meantime, the inequalities under (ii) above affirm that for all strain histories on (0,1), the stress relaxation function is positive and a monotonically decreasing function of the elapsed time τ. Gurtin and Herrera (1965) have also discussed these inequalities, (ii) above, within the context of linear viscoelasticity.

$$\text{(ii)} \quad R(\epsilon(t-\tau),\tau)>0, \quad R'(\epsilon(t-\tau),\tau)\leq 0$$

$$\text{where} \quad R'(\epsilon(t-\tau),\tau) = \frac{\partial}{\partial\tau}R((t-\tau),t) \tag{14.65}$$

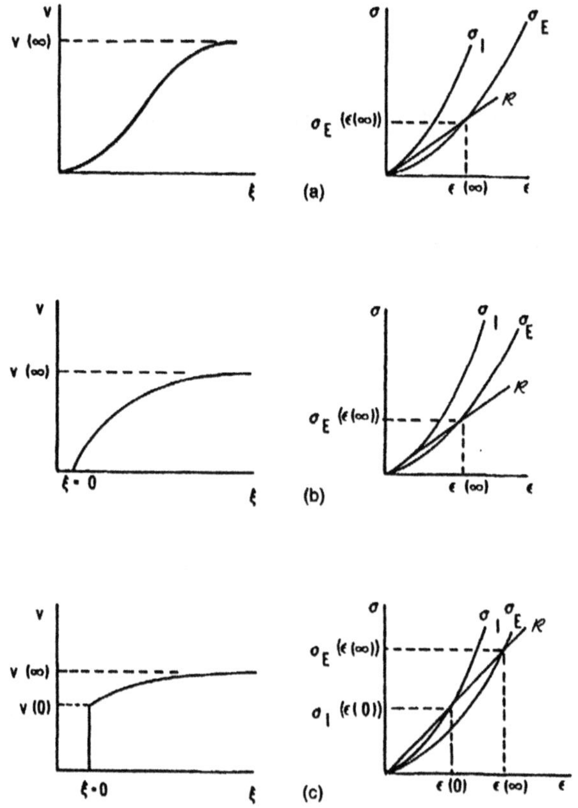

Figure 14.5. Steady wave solutions (Curvature conditions imposed on the instantaneous and equilibrium stress-strain curves). "Reprinted with permission from Int. J. Solids Structures, Vol.9, Schuler, K.W., Nunziato, J.W. and Walsh, E.K., Recent Results in Nonlinear Viscoelastic Wave Propagation, pp. 1237-81, 1973, Pergamon Press Ltd."

14.8. Acceleration Waves

The subject of acceleration wave propagation in nonlinear materials with fading time-memory has been considered by, amongst others, Truesdell and Toupin (1960), Varley (1965), Coleman, Gurtin and Harrera (1965), Coleman and Gurtin (1965), Bailey and Chen (1971) and Schuler, Nunziato and Walsh (1973).

Definition
Coleman and Gurtin (1965) advanced that if the following two conditions are satisfied, then $X(t)$, $-\infty < t < \infty$, is an acceleration wave.

A-1: $x(X,\tau)$, $\dot{x}(X,\tau)$ and the deformation gradient F are continuous functions of X and τ jointly for all X and τ, while $\ddot{x}(X,\tau)$, $\partial F(X,\tau)/\partial X$, $\dot{F}(X,\tau)$ have jump discontinuities across the wave material trajectory, but are continuous in X and τ jointly everywhere else.

A-2: the past history of the deformation gradient, $F_r^t(X_i)$, is a smooth function of X and t with respect to the norm $||\cdot||_r \cdot F_r^t(X,\cdot)$ is the restriction on the history of the deformation gradient $F(t-\tau)$ to its domain of definition $(0,\infty)$.

The condition (**A-2**) limits the wildness of the past history for the material with memory.

Coleman and Gurtin (1965) considered, within the general linear theory of simple materials with fading memory, the case when an acceleration wave which since $t=0$ has been propagating into a region which had been previously at rest in a fixed homogeneous configuration R. For this case, it was remarked by these authors that the hypothesis **A-2** above follows from hypothesis **A-1** for all $X > X(0)$ and $t > 0$. In other words, whenever the acceleration front is entering a homogeneous medium at rest, the hypothesis **A-1** would generally suffice to ensure that $F_r^t(X,\cdot)$ is a smooth function of X and t with respect to the norm $||\,.\,||_r$.

Thus, following the condition **A-1** above, the compatibility condition (11.53) and taking $f(X,t) = \epsilon(X,t)$, one can write at $X = X(t)$ that

$$[\dot{v}] = U[\dot{\epsilon}] = U^2[\partial_x \epsilon] \tag{14.66}$$

Following (8.8:5), the stress and the internal energy must also be discontinuous at $X = \hat{X}(t)$. Thus, with reference to (8.8:8), the balance of linear momentum asserts that

$$[\sigma] = \rho_0 U[v], \quad [\partial_x \sigma] = -\rho_0[\dot{v}] \tag{14.67}$$

Also, the balance of energy implies that

$$-U[e + \frac{1}{2}\rho_0 v^2] = [\sigma v], \quad [\dot{e}] = [\sigma\dot{\epsilon}] \tag{14.68}$$

(e.g., Schuler, Nunziato and Walsh, 1973).

The jump in the particle acceleration is conventionally taken as the amplitude of the wave front. Denoting the latter at any instant of time by a(t), then

$$a(t) = [\,\dot{v}\,](t) \tag{14.69}$$

Based on purely kinematical considerations, Coleman and Gurtin (1965) affirmed that the amplitude of an acceleration wave obeys the following relationship

$$2\frac{da}{dt} - \frac{a}{U}\frac{dU}{dE} = [\ddot{x}] - U^2\frac{\partial\dot{F}}{\partial X} \tag{14.70}$$

where, as introduced earlier $U = U(t)$ is the intrinsic wave velocity and E_I is the instantaneous tangent modulus. As an alternate expression to (14.70), Coleman and Gurtin (1965) advanced the following equation for the amplitude of an acceleration wave using condition **A-1** and the balance of momentum equation (14.67).

$$2\frac{da}{dt} = \frac{a}{U}\frac{dU}{dt} + \frac{1}{\rho_0}\left[\frac{\partial^2\sigma}{\partial t\,\partial X}\right] - U^2\left[\frac{\partial\dot{F}}{\partial X}\right] \tag{14.71}$$

Meantime, Coleman, Gurtin and Herrera (1965) indicated that the intrinsic velocity U of an acceleration wave satisfies the equation

$$U^2 = R(0)/\rho_0 \tag{14.72}$$

In this equation, R(0) is the instantaneous tangent modulus corresponding to the history $F^{(t)}(X(t),.)$ i.e.,

$$R(0) = \frac{d\sigma_I(\varepsilon)}{d\epsilon}\Big|_{\epsilon=0} \tag{14.73}$$

Thus, eqn. (14.72) implies that the intrinsic velocity of an acceleration wave $U=U_0$ is a constant given by

$$\rho_0 U_0^2 = R(0) \tag{14.74}$$

Following the above conditions, Schuler, Nunziato and Walsh (1973), following Coleman and Gurtin (1965), affirmed that the amplitude of an acceleration wave is given by

$$\frac{da}{dt} = -\beta a + \frac{\beta}{\gamma} a^2 \tag{14.75}$$

where β and γ are constants given by

$$\beta = \frac{-R'(0)}{2(E_I)_0} = \text{constant}, \quad \gamma = \frac{R'(0)U_0}{(\check{E}_I)_0} = \text{constant} \tag{14.76}$$

and

$$(\check{E})_0 = \frac{d^2\sigma_I(\epsilon)}{d\epsilon^2}\Big|_{\epsilon=0} \tag{14.77}$$

The solution of (14.75) can be written as

$$a(t) = \frac{\gamma}{\left(\frac{\gamma}{a(0)} - 1\right)\exp(\beta t) + 1} \tag{14.78}$$

For a given material, β and γ are constants.

Assuming that the hypothesis of the above theorem holds and suppose that

$$R(0)>0, \quad R'(0)<0, \quad (E_I)_0 \neq 0 \tag{14.79}$$

One concludes, with reference to (14.78), following Coleman and Gurtin (1965), that:

I. If either

(i) $|a(0)|$ is less than $|\gamma|$, or

(ii) $\operatorname{sgn} a(0) = \operatorname{sgn} (\check{E}_I)_0$

then $a(t) \to 0$ monotonously as $t \to \infty$

II. If $a(0) = \gamma$, then $a(t) = a(0)$

III. If both

(i) $|a(0)|$ is greater than $|\gamma|$, and

(ii) $\operatorname{sgn} a(0) = -\operatorname{sgn}(\overset{\vee}{E}_{I})_0$

then $|a(t)| \to 0$ monotonously and in a finite time t_∞ given by

$$t_\infty = -\frac{1}{\beta}\ln\left(1-\frac{\gamma}{a(0)}\right) \tag{14.80a}$$

It is apparent, from the above discussion, that $|\gamma|$ plays the role of the "critical amplitude" of the input acceleration. The latter may be denoted by a_c which is expressed with reference to Eqn. (14.76) as

$$a_c = |\gamma| = \left|\frac{R'(0)U_0}{\left(\overset{\vee}{E}_{I}\right)_0}\right| = \text{constant} \tag{14.80b}$$

Thus (14.80) implies that, assuming (14.79):

(i) if the amplitude of the input acceleration is sufficiently small (< critical amplitude) or if the amplitude has the same sign as the instantaneous second-order modulus, then an acceleration wave obeying (11.76) is gradually damped out. In this, the internal dissipation of the material is expected to be the governing factor in the mode of wave motion.

(ii) if, however, the amplitude of the input acceleration is greater than the critical amplitude and has its sign opposite to that of the instantaneous second-order modulus, then the wave would achieve an infinite amplitude in a finite time, i.e., a shock wave may be produced. In this, the nonlinearity of the instantaneous stress-strain curve would be the controlling factor (*see* Schuler, Nunziato and Walsh, 1973).

As noted by Coleman and Gurtin (1965), the presence of internal damping, manifested by a strictly negative value of $R'(0)$, does not always imply that a singular surface moving into a homogeneous region must be damped out.

In the linear theory of simple materials with fading memory, the stress-relaxation function $R(\tau)$, with

$$R'(\tau) = \frac{dR(\tau)}{d\tau},$$

is a material function independent of the strain history $\epsilon(t-\tau)$. In the physical application of this theory, it is generally expected that (Gurtin and Herrera, 1964)

$$R(0)>0, \quad R'(0) \leq 0$$

Coleman, Gurtin and Herrera (1965) ruled out, however, the possibility when R(0)=0 or R'(0) > 0 in the applicability of relation (11.76) above. Meantime, Coleman and Gurtin (1965), considered the applicability of (14.76) in the following two cases:

(i) $R'(0) = 0$:

In this case, it is advanced that the time-dependency of the amplitude a(t) of the acceleration wave is expressed as

$$a(t) = \frac{a(0)}{1+\xi\, a(0)t}, \qquad \xi = \frac{\overset{\vee}{E}_{I}}{2UR(0)} \tag{14.81}$$

where $\overset{\vee}{E}_I$ is the instantaneous second-order modulus. Two situations may be considered here:

(i.1) $\overset{\vee}{E}_I \neq 0$, then (11.81) implies, since $\rho_0 > 0$, that if a (0) has the same sign as $\overset{\vee}{E}_I$, then, $|a(t)| \to 0$ monotonously in a finite time.

(i.2) $\overset{\vee}{E}_I \neq 0$, then (11.78) would reduce to

$$a(t) = a(0)\exp\left(\frac{R'(0)t}{2R(0)}\right) \tag{14.82}$$

which may be generally valid for a large class of linear viscoelastic materials. A special class of materials with R'(0)=0 is the class of perfectly elastic materials for which (14.80) is known to be applicable (Thomas, 1957, Green, 1964 and Coleman and Gurtin, 1965).

$$\textbf{(ii)} \qquad R'(0) < 0$$

In this case, it follows from (14.76) that the amplitude of the acceleration wave $a(t) \to 0$ as $t \to \infty$ regardless of the sign of $a(0)$.

14.9. Shock Waves

The subject of shock wave propagation in nonlinear materials with fading time-memory has been considered, by Duvall and Alverson (1963), Coleman and Gurtin (1965), Coleman, Gurtin and Herrera (1965), Chen and Gurtin (1970, 1972a) and Huilgol (1973), amongst others.

Coleman, Gurtin and Herrera (1965) asserted that the following two conditions must be satisfied for the wave $X(t)$, $-\infty < t < \infty$, to be called a shock wave in a material with memory.

S-1: the coordinate function $x(X, t)$ to be continuous in both X and t jointly while the deformation gradient $F(X,t) = \dfrac{\partial}{\partial X} x(X,t)$ and the time-derivative of the coordinate $\dot{x}(X,t) = \dfrac{\partial x(X,t)}{\partial t}$ have jump discontinuities across the wave material trajectory $\Omega(t)$ but are continuous in X and t jointly everywhere else.

S-2: the past history of the material is not too wild. For this purpose, it is assumed that the past history of the deformation gradient $F_r^t(X,\cdot)$ to be a smooth function of X and t with respect to the norm $\|\cdot\|_r$. In this, $F_r^t(X,\cdot)$ is the restriction on the history of the deformation gradient $F(t-\tau)$ to its domain of definition $(0, \infty)$.

Thus, following condition **S-1** above, the compatibility condition (14.53) together with $f(X, t) = u(X, t)$ affirm that at $X = \hat{X}$,

$$[v] = -U[\epsilon] \tag{14.83}$$

Where U is the intrinsic velocity of the shock wave. In view of (14.83), either the jump in the particle velocity $[v]$ or the jump in the strain $[\epsilon]$ may be taken as a measure of the amplitude of the shock. Meantime, the equations (14.67) and (14.68) concerning, respectively, the balance of momentum and the balance of energy are also valid for the case of shock waves.

Coleman, Gurtin and Herrera (1965) showed that the intrinsic velocity U of a shock wave satisfies the relation

$$U^2 = \frac{E_I[F]}{\rho_0} \tag{14.84}$$

where $E_I[F]$ is the instantaneous secant modulus (14.60a) cor-responding to the history just before the arrival of the shock and a jump of amount $[F]$ where F is the deformation gradient. Considering now the case of a compressive wave propagating into a region at rest and unstrained for all past times (Chen and Gurtin, 1970 and Schuler, Nunziato and Walsh, 1973), i.e., for $X > X(t)$, $\epsilon(t - \tau) = 0$, $0 \leq \tau < \infty$ and

$$[\epsilon] = \epsilon^- > 0, \quad [\partial_x \epsilon] = (\partial_x \epsilon)^-$$

Thus, the corresponding stress jump is expressed in view of the definition of the instantaneous stress σ_I as $[\sigma] = \sigma_I(\varepsilon^-)$.

This implies, in view of (14.83) and (14.84) that the intrinsic velocity can be expressed by

$$U^2 = \frac{\sigma_I(\epsilon^-)}{\rho_0 \epsilon^-} = \frac{E_I(\epsilon^-)}{\rho_0} \tag{14.85}$$

which, in view of the second inequality of the convexity conditions (14.64), implies that

$$\frac{\rho_0 U^2}{E_I(\epsilon^-)} < 1 \tag{14.86}$$

The inequality (14.86) above affirms (Schuler,Nunziato and Walsh, 1973) that the shock velocity is *subsonic* with respect to the material behind the wave front. From (14.85), it can be seen that the shock velocity depends on the strain amplitude ϵ^-. Furthermore, one can write with reference to (14.85) that

$$\frac{dU}{dt} = \frac{(1 - \hat{\mu}) E_I(\epsilon^-)}{2\rho_0 U \epsilon^-} \frac{d\epsilon^-}{dt} \tag{14.87}$$

where

$$\hat{\mu} = \frac{\rho_0 U^2}{E_I(\epsilon^-)} < 1 \qquad (14.88)$$

given earlier by (14.86).

In view of (14.84), one concludes that the time rate of change of the shock velocity U is proportional to that of the amplitude of the strain behind the front ϵ^-. Following this and using the assumed characteristics of the deformation gradient F, Chen and Gurtin (1970) derived the following "***shock amplitude equation***"

$$\frac{d\epsilon^-}{dt} = U\frac{(1-\hat{\mu})}{1+3\hat{\mu}}\left\{\hat{\lambda} - (\partial_x \epsilon)^-\right\} \qquad (14.89)$$

where

$$\hat{\lambda} = \frac{R'(\epsilon^-;0)\,\epsilon^-}{U E_I(\epsilon^-)(1-\hat{\mu})} \qquad (14.90)$$

In (11.90), $R'(\epsilon^-;0)$ is the initial slope of the stress-relaxation function corresponding to the jump strain input (14.58). It is evident, in view of (14.64), (14.65) and (14.90), that $\lambda \le 0$.

Thus, with reference to the shock amplitude expression (14.89), one may conclude that the growth or decay behaviour of the shock wave front would depend on the strain gradient immediately behind the front (Schuler, Nunziato and Walsh, 1973). That is

$$\begin{aligned} &\text{(i) if } \hat{\lambda} < (\partial_x \epsilon)^-, \quad \frac{d\epsilon^-}{dt} < 0 \\ &\text{(ii) if } \hat{\lambda} = (\partial_x \epsilon)^-, \quad \frac{d\epsilon^-}{dt} = 0 \\ &\text{(iii) if } \hat{\lambda} > (\partial_x \epsilon)^-, \quad \frac{d\epsilon^-}{dt} > 0 \end{aligned} \qquad (14.91)$$

In view of the above, Schuler et al (1973) referred to λ as the "***critical strain gradient***". These authors expressed, the shock amplitude equation (14.90) in terms of particle velocity as

$$\frac{dv^-}{dt} = \frac{(1-\hat{\mu}^2)}{(1+3\hat{\mu})}\left\{(\dot{v})^- - U^2\left|\hat{\lambda}\right|\right\} \tag{14.92}$$

In this equation, $(\dot{v})^-$ is the particle acceleration immediately behind the shock front and $U^2|\lambda|$ is "the critical acceleration". It is evident from (14.92) that

(i) the wave front grows if $(\dot{v})^- > U^2|\hat{\lambda}|$
(ii) is steady if $(\dot{v})^- = U^2|\lambda|$
(iii) decays if $(\dot{v})^- < U^2|\lambda|$ (14.93)

Further, it can be shown (Chen and Gurtin, 1970) that equation (14.92) reduces for the case of weak shock waves to the following simple expression

$$\frac{dv^-}{dt} = -\beta v^- \tag{14.94}$$

where β is a constant given by

$$\beta = -\frac{R'(0;0)}{2(E_I)_0} \tag{14.95a}$$

in which,

$$(E_I)_0 = \frac{d\sigma_I(\epsilon)}{d\epsilon} \tag{14.95b}$$

The solution of the differential equation (14.94) is

$$v^-(t) = v^-(0)\exp(-\beta t) \tag{14.96}$$

which asserts that the amplitude of a weak shock wave decays exponentially to zero as $t \to \infty$. Such response is identical to that predicted by the linear theory of viscoelasticity (see, e.g., Lee and Kanter, 1953; Chu, 1962; Coleman and Gurtin, 1965 and Valanis, 1965).

14.10. Thermodynamic Influences

14.10.1. ACCELERATION WAVES

It is noted by Schuler et al. (1973) that thermal effects have no influence on the propagation of acceleration waves in nonconducting materials with memory. Accordingly, on the assumption that a particular material to be a thermal nonconductor (which could be reasonable for a large class of polymeric solids), the study of the propagation of acceleration waves in such material would provide no information about the thermodynamic influence on its mechanical response. Thus, for a thermal nonconducting material, if the relaxation function and second-order modulus are taken at a fixed entropy, then the velocity of every acceleration wave would satisfy (14.72). Furthermore, the amplitude of an acceleration wave entering a region at rest, unstrained and at uniform temperature would satisfy (14.78) with the material constants appearing in this equation being given by (14.76) and (14.77).

In case of conducting materials, however, thermodynamic influences on the propagation of acceleration waves in viscoelastic materials are pronounced. In this, the reader is referred, for instance, to Coleman and Gurtin (1966).

14.10.2. SHOCK WAVES

Thermodynamic effects on the propagation of shock waves in nonconducting materials have been considered by Coleman and Gurtin (1966) and Chen and Gurtin (1972b). Meantime, studies on shock wave propagation in materials with memory which conduct heat have been carried out, for instance, by Achenbach, Vogel and Herrmann (1966) and by Dunwoody (1972).

14.10.3. AN ILLUSTRATIVE EXAMPLE

Determination of the Stress Relaxation Function from Shock Wave Data

Nunziato and Sutherland (1973) considered plate impact experiments to study the one-dimensional dynamic response of PMMA. They considered the characteristic time scale for such experiments to be 10^{-2} - 1 μsec. Denoting, over this time scale, the relaxation function by $R(\tau)$ where $R(0)$ is equivalent to the value of the relaxation function at 10^{-2} μsec and $R(\infty)$ is corresponding to the relaxation function at 1 μsec. Nunziato and Sutherland (1973) obtained the stress-relaxation function $R(\tau)$ shown in Fig. 14.6. From Fig. 14.6, one has:

$$\hat{R}(0) = 90.1 \text{ kbars}, \qquad \hat{R}(\infty) = 88.0 \text{ kbars} \qquad \textbf{(14.97)}$$

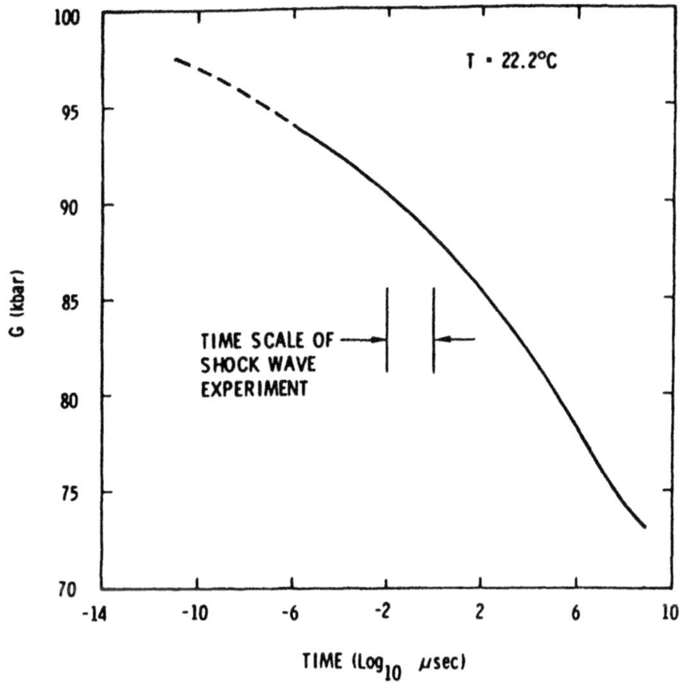

Figure 14.6. Longitudinal stress relaxation function for PMMA. From: Nunziato, J.W. and Sutherland, H.J. (1973) Acoustical Determination of Stress Relaxation Functions in Polymers, J. Appl. Phys., 44(1), 184-87 (American Institute of Physics). Reprinted with permission

Meantime, the characteristic relaxation time λ is evaluated by considering the relation

$$\hat{R}(\tau)\big|_{\tau=\lambda} = \frac{R_0 - R_\infty}{e} + R_\infty \tag{14.98a}$$

in which e is the Naperian base. From (14.97) and (14.98a), one concludes from Figure 14.6 that the characteristic relaxation time is

$$\lambda = 0.22\ \mu\text{sec} \tag{14.98b}$$

Schuler (1970) considered the propagation of steady one-dimensional shock waves in PMMA. The relaxation behaviour of PMMA was characterized by a nonlinear constitutive relation of the form

$$\sigma(x,t) = \sigma_E(\epsilon) + \int_0^\infty R(\epsilon;\tau)\left[\frac{1+\epsilon(t-\tau)}{1+\epsilon(t)} - 1\right]d\tau \tag{14.99}$$

where $\epsilon = \frac{\partial u}{\partial x}$ is the longitudinal strain, $\sigma_E(\epsilon)$ is the equilibrium response function and $R(\epsilon;\tau)$ is the generalized stress relaxation function. The latter is assumed to have the form

$$R(\epsilon;\tau) = R(\epsilon;0)\exp(-\tau/\lambda) \tag{14.100}$$

Evaluating (14.99) for the strain jump

$$\begin{aligned} \epsilon(t-\tau) &= \epsilon, \quad & \tau &= 0 \\ &= 0 & \tau &> 0 \end{aligned}$$

then, it follows from (14.100) that

$$R(\epsilon;0) = -(1+\epsilon)^2\left[\sigma_I(\epsilon) - \sigma_E)\right]/\lambda\epsilon(2+\epsilon) \tag{14.101}$$

From an analysis of steady waves (see, e.g., Greenberg, 1968, Schuler 1970 and Schuler et al., 1973), one has

$$\begin{aligned} \sigma_I(\epsilon) &= \rho_0 U_I^2(\epsilon)\,\epsilon \\ \sigma_E(\epsilon) &= \rho_0 U_E^2(\epsilon)\,\epsilon \end{aligned} \tag{14.102}$$

where $U_I(\epsilon)$ and $U_E(\epsilon)$ are least square polynomial functions of the steady shock velocity as a function of strain ϵ at the shock front and at the tail of the wave (Nunziato and Sutherland, 1973). For small strains, the nonlinear constitutive relation (14.99) would reduce to the corresponding constitutive relation in the linear case with (see Nunziato and Walsh, 1973)

$$\hat{R}(\tau) = \frac{d\sigma_I(0)}{d\epsilon} + 2\int_0^\tau R(0;\tau)\,d\tau \tag{14.103}$$

Combining, now, (14.100) and (14.103) it follows that

$$\frac{d\sigma_I(0)}{d\epsilon} = \rho_0 U_I^2(0) \tag{14.104}$$

consequently,

$$\hat{R}(\tau) = (R_0 - R_\infty)\exp(-\tau/\lambda) + R_\infty \tag{14.105}$$

where

$$R_0 = \rho_0 U_I^2(0) \quad \text{and} \quad R_\infty = \rho_0 U_E^2(\infty)$$

It is noted that $U_I(0)$ and $U_E(0)$ are the zero strain intercepts of the $U_I(\epsilon)$ and $U_E(\epsilon)$ curves. Using data reported by Schuler (1972), *see* Nunziato and Sutherland (1973), and Barker and Hollenbach (1970), Nunziato and Sutherland (1973) concluded that

$$\begin{aligned} R_0 &= 90.2 \text{ kbars}, \\ R_\infty &= 88.2 \text{ kbars} \end{aligned} \tag{14.106a}$$

Schuler (1970), by fitting the observed steady wave profiles, found that

$$\lambda = 0.25 \ \mu\text{sec} \tag{14.106b}$$

which is comparable to the value given earlier by (14.98b). With the above data, the relaxation function (14.105) is given in Fig. 14.7 (after Nunziato and Sutherland, 1973). As demonstrated, there is reasonable agreement with the relaxation function deduced from acoustic dispersion data.

14.11. Study Problems

1. Define briefly the following terms:
   (a) Internal friction.
   (b) Static hysteresis.
   (c) Viscous loss.
2. Derive the wave equation in a linear viscoelastic material as based on Boltzmann's superposition principle.
3. Derive the wave equation in a linear viscoelastic material as based on the

correspondence principle.
4. What is meant by the expression: "*a material with a time fading memory*".
5. Define an "*acceleration wave*" as applied to wave propagation in a viscoelastic material.
6. Define a "shock wave" as applied to wave propagation in a viscoelastic material.
7. Comment on the kinematic conditions that would be required to be satisfied for a shock wave to form in both linear and nonlinear viscoelastic material.
8. Following Problem 7 above, what are the required kinematic conditions that would be required for a shock wave to form in a ductile material (*Consult Chapter 12*).
9. What are the kinematic conditions for an acceleration wave to form in a viscoelastic
10. Following Problem 9 above, what are the required kinematic conditions that would be required for an acceleration wave to form in a ductile material (Consult Chapter 12).

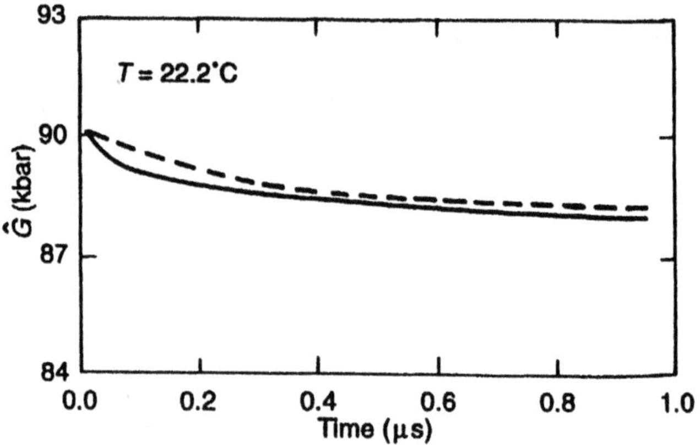

Figure 14.7. Stress-relaxation function for PMMA appropriate for shock wave experiments: - - - , Schuler (1970, unpublished data), _____, this work. From: Nunziato, J.W. and Sutherland, H.J. (1973) Acoustical Determination of Stress Relaxation Functions in Polymers, J. Appl. Phys., 44(1), 184-187 (American Institute of Physics). Reprinted with permission.

14.12. Transition to the Viscoelastic Boundary Value Problem

In view of the time-dependency of the response behaviour in viscoelasticity which is further complicated by the form of the constitutive relations and, hence, the associated boundary conditions, serious attempts to solve viscoelastic boundary value problems have lagged considerably behind those in classical elasticity. It is only in the last four decades that viscoelastic boundary value problems have been actively considered. At the beginning, researchers have given attention to the solution of the simpler viscoelastic problems that have analogues in classical elasticity whereby the viscoelastic solution may be expressed directly in terms of the analogous elastic problem. Research efforts have been advanced since then to tackle more difficult viscoelastic boundary value problems with or without correspondence to the theory of elasticity.

14.12.1. CLASSIFICATION OF VISCOELASTIC BOUNDARY VALUE PROBLEMS

Since the response behaviour of a viscoelastic material is time-dependent, it thus follows that no real static viscoelastic problem exists. However, in a large number of cases, it may be admissible
(*see* Hunter, 1967) to neglect the acceleration terms in the equations of motion. In such case, the viscoelastic boundary value problem is referred to as "***quasi-static***" or "***quasi-stationary***". As Hunter (1967), for instance, pointed out, the only "true static" problems in viscoelasticity are those corresponding to the equilibrium limit of complete stress relaxation. A"quasi-static" viscoelastic boundary value problem is often classified from the point of view of the time-dependency of its boundary regions. In this, the following two categories are often dealt with.

(A) ***Quasi-static problems with fixed (time-independent) boundary conditions.*** For this category, the loading history is assumed to be known for all time over a fixed part of the boundary, while the displacement history is specified for the remaining part. This type of problem is generally solvable using a correspondence with an analogous elastic problem, i.e., by employing the so called "***correspondence principle***" (to be introduced in this Section). This is essentially due to the possibility of obtaining Laplace (or Fourier) transforms of the boundary conditions as illustrated later in this Section.

(B) ***Quasi-static problems with mixed boundary regions which are time-dependent.***
This category of viscoelastic boundary value problems is not generally susceptible to solution by the correspondence principle as it may be impossible to obtain appropriate transforms of the boundary conditions. Examples of such type of problems may include contact problems where the load on the indenter is varying or the indenter is moving into the viscoelastic material specimen with an indentation of varying geometry.

Much less research work has been carried out on inertial and dynamic viscoelastic boundary value problems. In this domain, a large portion of the research has concentrated primarily on viscoelastic wave propagation problems that involve only one space variable. Chao and Achenbach (1964) and Gurtin and Herrera (1964), among others, considered the use of the correspondence principle for the solution of viscoelastic wave propagation problems of this type. In general, however, viscoelastic waves may propagate in three dimensions with different magnitudes of attenuation and dispersion (e.g., Lockett, 1962). In this case, an associated boundary value problem may not be solvable via a dynamic correspondence principle (Hunter, 1967).

Research efforts to solve thermoviscoelastic boundary value problems have been often distracted by the fact that mechanical properties of viscoelastic materials are sensitive to temperature variations. This is complicated further by the heat generated in the viscoelastic material during deformation. The formulation of the governing equations has been proven, thus, to be difficult. Morland and Lee (1960), for instance, have considered the case of a thermo-rheologically simple solid (e.g., Schwarzl and Staverman, 1952 and Hunter, 1961) in the absence of internally generated heat and thermodynamic coupling effects. Morland and Lee (1960) applied, then, the resulting equations to the quasi-static problem of an incompressible long cylinder subject to radial temperature gradient and internal pressure. Muki and Sternberg (1961) have dealt with the thermal stresses in viscoelastic materials with temperature dependent properties and considered transient stress problems in plane slabs and spheres subject to temperature variation. Rogers and Lee (1962) have considered the solution of the quasi-static thermoviscoelastic problem of a sphere with an internally ablating cavity. Sternberg and Gurtin (1963,1964) considered the uniqueness of the theory of thermorheologically simple ablating viscoelastic solids.

A classification of boundary value problems in viscoelasticity is presented in Fig. 14.8. For comprehensive studies of the subject matter, the reader is referred further to Read (1950), Lee *et al.* (1959), Sternberg (1964), Predeleanu (1965), Rogers (1965), Lee (1966) and Golden and Graham (1988); among others.

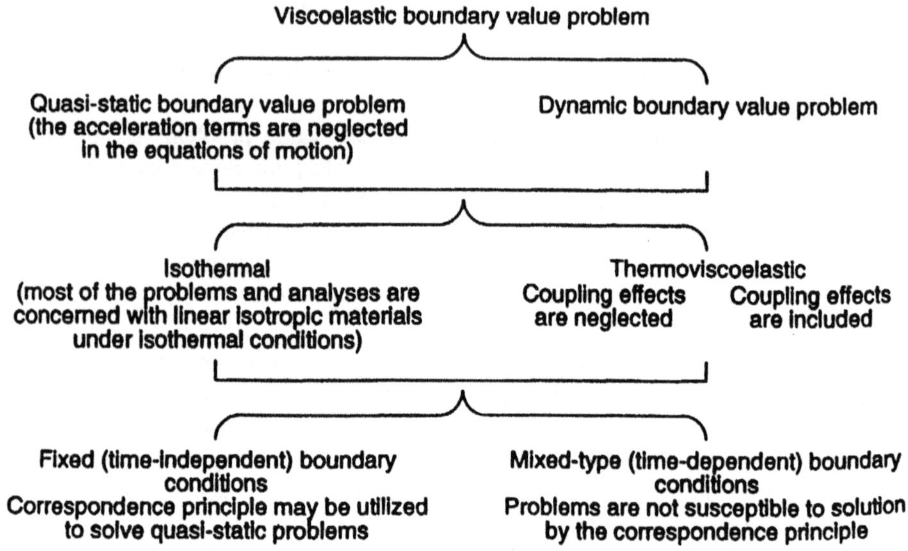

Figure 14.8. Classification of viscoelastic boundary value problems.

14.12.2. FORMULATION OF THE VISCOELASTIC BOUNDARY VALUE PROBLEM

In compliance with the principles of continuum mechanics, the motion of a viscoelastic (continuum) body is generally governed by the laws of conservation of mass and momentum, the stress-strain constitutive relations, the boundary conditions and the initial conditions. As demonstrated in the remainder of this section, the formulation of this set of governing conditions is determined by the type of the boundary value problem considered.

Isothermal, Linear Viscoelastic Boundary Value Problem

In this class of boundary value problem, all the geometrical assumptions of infinitesimal elasticity theory are implied. These would usually include the assumptions of small deformations and small strains, the boundary conditions applied to undisturbed surfaces and the neglect of any convective terms in the acceleration. In this class of viscoelastic boundary value problem, only the viscoelastic stress-strain relations would differ from the linear elastic constitutive equations. All other governing conditions would follow directly from linear elasticity with proper inclusion of the time-dependency of the pertaining variables.The governing set of conditions for an isothermal, linear viscoelastic boundary value problem are as follows:

(i) Initial conditions

We assume that the body is initially undisturbed. In other words, it is initially stress free and in mechanical equilibrium. Thus, the initial conditions are,

$$u_i(t) = 0, \quad \varepsilon_{ij}(t) = 0, \quad \sigma_{ij}(t) = 0; \quad -\infty<t<0 \tag{14.107}$$

where u_i designate the components of the displacement vector in a rectangular Cartesian coordinate system.

(ii) Boundary conditions

The boundary B of the body is considered to be composed of two parts B_σ and B_u. That is

$$B = B_\sigma + B_u$$

where B_σ denotes the part of the boundary of the body over which the components of the stress σ are prescribed; and B_u indicates the remaining part of the boundary over which the components of the displacement $\mathbf{u}$ are specified. The boundary conditions may be assigned in the form of magnitudes of:

- traction vector components T_i over B_σ such that

$$\sigma_{ij}(x,t)\,n_j = T_i(x,t); \quad x \ \ on \ \ B_\sigma \tag{14.108a}$$

where n_j are the components of the outward unit normal to B_σ.

- displacement vector components U_i over B_u as

$$u_i(x,t) = U_i(x,t); \quad x \ \ on \ \ B_u \tag{14.108b}$$

The boundary conditions (14.108) are assumed to be fixed, that is, both the traction vector components T_i and the displacement vector components U_i are considered to be prescribed for all t.

(iii) Balance of linear momentum
One of the following two situations may be considered:

- *A quasi-static problem*. In this case, the equilibrium equation is

$$\sigma_{ij,j} + \chi_i = 0 \tag{14.109}$$

- *A dynamic problem*. In this case, the equation of motion is

$$\sigma_{ij,j} + \chi_i = \rho \frac{\partial^2 u_i}{\partial t^2} \tag{14.110}$$

where, in (14.109) and (14.110), χ_i are the body force components per unit volume.

(iv) Linear strain-displacement relations

$$\varepsilon_{ij}(t) = \frac{1}{2}\left(u_{i,j}(t) + u_{j,i}(t)\right) \tag{14.111}$$

in which a comma indicates partial differentiation with respect to the coordinates x_i of the material particle.

(v) Stress-strain relations.
General linear constitutive equations for a viscoelastic material with an arbitrary degree of anisotropy may be expressed in the form of Boltzmann superposition integral (e.g., Haddad, 1995)

Creep case:

$$\varepsilon_{ij}(t) = \int_0^t C_{ijkl}(t-\tau)\frac{\partial\sigma_{kl}(\tau)}{\partial\tau}d\tau \tag{14.112}$$

where $C_{ijkl}(t-\tau)$ are the components of the creep function.

Relaxation case:

$$\sigma_{ij}(t) = \int_0^t R_{ijkl}(t-\tau)\frac{\partial\varepsilon_{kl}(\tau)}{\partial\tau}d\tau \tag{14.113}$$

where $R_{ijkl}(t-\tau)$ are the components of the relaxation function.

The constitutive relation for an isotropic material can be reduced to two pairs of operators, one for the stress-strain deviatoric constitutive relation which covers shear response and one for average hydrostatic tension and dilatation. In this case, a differential operator law in the following form may be used (Lee, 1960)

$$\begin{aligned} P_1(D)\sigma'_{ij}(t) &= Q_1(D)\epsilon'_{ij}(t) \\ P_2(D)\sigma_{kk}(t) &= Q_2(D)\epsilon_{kk}(t) \end{aligned} \tag{14.114}$$

where σ'_{ij}, ϵ'_{ij} are the stress and strain deviators defined, respectively, by

$$\begin{aligned} \sigma'_{ij} &= \sigma_{ij} - \frac{1}{3}\delta_{ij}\sigma_{kk} ; \qquad \sigma'_{ii} = 0 \\ \epsilon'_{ij} &= \epsilon_{ij} - \frac{1}{3}\delta_{ij}\epsilon_{kk} ; \qquad \epsilon'_{ii} = 0 \end{aligned} \tag{14.115}$$

In Equation (14.114), P_1, P_2, Q_1 and Q_2 are polynomials of the time-derivative operator $D = \partial/\partial t$.

Alternatively, the stress-strain relations (14.114) may be used in either of the following constitutive forms:

Creep case:

$$\epsilon'_{ij} = \sigma'_{ij}(t) * dC_1(t); \qquad \epsilon_{kk}(t) = \sigma_{kk}(t) * dC_2(t) \tag{14.116}$$

where $C_1(t)$ and $C_2(t)$ are the creep functions in pure shear and pure dilatation, respectively (e.g., Haddad, 1995).

Relaxation case:

$$\sigma'_{ij}(t) = \epsilon'_{ij}(t) * dR_1(t); \quad \sigma_{kk}(t) = \epsilon_{kk}(t) * dR_2(t) \tag{14.117}$$

where $R_1(t)$ and $R_2(t)$ are the relaxation functions in pure shear and pure dilatation, respectively.

For an isotropic viscoelastic material, an approximate form of the constitutive relation in the relaxation case may be expressed (Hunter, 1967) as,

$$\sigma_{ij}(t) = \delta_{ij} \int_0^t \lambda(t-\tau) \frac{\partial \epsilon_{kk}(\tau)}{\partial \tau} d\tau + 2 \int_0^t \mu(t-\tau) \frac{\partial \epsilon'_{ij}(\tau)}{\partial \tau} d\tau \tag{14.118}$$

where $\lambda(t)$ and $\mu(t)$ are appropriate relaxation functions.

In terms of deviatoric and dilatational components, the isotropic constitutive equation in the relaxation case can be further written as,

$$\begin{aligned} \sigma'_{ij}(t) &= 2 \int_0^t \mu(t-\tau) \frac{\partial \varepsilon'_{ij}(\tau)}{\partial \tau} d\tau \\ \sigma_{ii}(t) &= 3 \int_0^t k(t-\tau) \frac{\partial \varepsilon_{jj}(\tau)}{\partial \tau} d\tau \end{aligned} \tag{14.119}$$

where $\sigma'_{ij}(t)$ and $\epsilon'_{ij}(t)$ are, respectively, the deviatoric stress and the deviatoric strain components and $\mu(t)$ and $k(t)$ are the relaxation functions in pure shear and pure dilatation, respectively.

In the isothermal linear boundary value problem, the three balance of linear momentum equations (14.109), or (14.110), the six strain-displacement relations (14.111) and the six stress-strain constitutive equations, e.g. (14.119), constitute a set of fifteen field equations for the fifteen dependent variables u_i, ϵ_{ij} and σ_{ij} under the prescribed boundary conditions $T_i(\mathbf{x},t)$ and $U_i(\mathbf{x},t)$, (14.108), and the assumed initial conditions (14.107).

14.12.3. UNIQUENESS OF SOLUTION

An important question concerning the solution of a boundary value problem in continuum mechanics is whether the formulated problem has a solution and whether the solution is unique or not (Fung, 1965). On physical grounds, this question may be dealt with by reference to the thermodynamics of the problem involved. On mathematical grounds, however, this question must be answered by the theory of partial differential equations. A satisfactory solution to the problem in hand must comply with both the laws of physics and principles of mathematics. In solving boundary value problems of static equilibrium within classical elasticity, for example, one may proceed in the following sequence: (i) one solves the equations of equilibrium for the stresses σ_{ij} . (ii) the constitutive response equations are then solved for the strains ϵ_{ij} by using the stress components σ_{ij} obtained from (i). Here, an infinite set of solutions may be found. However, the unique solution would be singled out by employing, for instance, the conditions of compatibility (*Chapter 3*).

The existence and uniqueness of solution theorems in classical elasticity have been extended by Gurtin and Sternberg (1962) to the class of linear boundary value problems in viscoelasticity. This was carried out in light of an earlier work by Volterra (1909).

For the most direct case of isothermal, isotropic, linear viscoelastic boundary value problem under a quasi-static condition, Christensen (1971), following Gurtin and Sternberg (1962), presented a uniqueness condition of solution. This condition may be stated, in view of the set of governing equations presented earlier, as follows:

The isotropic, quasi-static, viscoelastic boundary value problem governed by the initial conditions (14.107), the boundary conditions (14.108), the equations of equilibrium (14.109), the strain-displacement relations (14.111) and the stress-strain equations (14.119) possesses a unique solution provided that the initial values of the relaxation functions appearing in the constitutive equations (14.119) satisfy the conditions

$$\mu(0)>0 \quad \textit{and} \quad k(0)>0 \qquad \textbf{(14.120)}$$

For a proof of the uniqueness theorem stated above, the reader is referred to Christensen (1971). Other versions of uniqueness theorems for the above class of boundary value problems are given by Onat and Breuer (1963), Edelstein and Gurtin (1964), Odeh and Tadjbakhsh (1965), Barberan and Herrera (1966) and Lubliner and Sackman (1967), amongst others.

14.12.4. CORRESPONDENCE PRINCIPLE. THE ELASTIC-VISCOELASTIC ANALOGY

For a large number of technical viscoelastic problems, it is possible to relate mathematically the solution of a linear, viscoelastic boundary value problem to an analogous problem of an elastic body of the same geometry and under the same initial and boundary conditions. This is carried out by transforming the governing equations of the viscoelastic problem to be mathematically equivalent to those governing a corresponding elastic problem. In this, both Laplace and Fourier transforms are often used. Accordingly, one would be able to employ the tools of the theory of elasticity to solve different boundary value problems in linear viscoelasticity.

The above analogy is referred to as the "***correspondence principle***". It implies that elastic analysis procedures may be utilized to derive transformed viscoelastic solutions (see, for instance, Lee, 1955, Morland and Lee, 1960 and Schapery, 1967). Lee (1955) demonstrated the correspondence principle for isotropic media at constant temperature. Meantime, Morland and Lee (1960) considered the application of the correspondence principle for isotropic materials with temperature variations. Biot (1958) argued that the correspondence principle may be also applied to anisotropic materials due to the symmetry of the relaxation modulus tensor, i.e. $R_{ijk\ell}(t) = R_{k\ell ij}(t)$.

Isothermal, Linear Viscoelastic Boundary Value Problem

(i) *Initial conditions.* The body is assumed to be initially undisturbed. Thus, the initial conditions (14.107) will hold.

(ii) *Boundary conditions.* The Laplace-transformed forms of the boundary conditions (14.108a) and (14.108b) are, respectively,

$$\bar{\sigma}_{ij}(x,s)n_j = \bar{T}_i(x,s) \qquad x \text{ on } B_\sigma \tag{14.121a}$$

$$\bar{u}_i(x,s) = \bar{U}_i(x,s) \qquad x \text{ on } B_u \tag{14.121b}$$

where s is the Laplace transform variable and the "*overbar*" designates the Laplace-transform of the variable, i.e.,

$$\bar{T}_i(x,s) = \int_0^\infty T_i(x,t)e^{-st}dt \tag{14.122a}$$

and

$$\bar{U}_i(x,s) = \int_0^\infty U_i(x,t)e^{-st}dt \tag{14.122b}$$

(*see* Appendix C)

(iii) *Balance of linear momentum.* The quasi-static case is dealt with here. Recalling (14.109), multiplying it by e^{-st} and integrating over $-\infty<t<\infty$, then, the Laplace transform of the equilibrium equation is

$$\bar{\sigma}_{ij,j} + \bar{\chi}_i = 0 \tag{14.123}$$

(iv) *Linear strain displacement relations.* The Laplace-transformed strain-displacement relation (14.111) is,

$$\bar{\varepsilon}_{ij} = \frac{1}{2}(\bar{u}_{i,j} + \bar{u}_{j,i}) \tag{14.124}$$

(v) *Stress-strain relations.* The constitutive equations (14.112) and (14.113) can be transformed by the rule of convolution integrals (*see* Schapery, 1967) to yield, respectively, the algebraic relations:

$$\bar{\varepsilon}_{ij} = \tilde{C}_{ijkl}\,\bar{\sigma}_{kl} \tag{14.125}$$

$$\bar{\sigma}_{ij} = \tilde{R}_{ijkl}\bar{\varepsilon}_{kl} \tag{14.126}$$

where C_{ijkl} and R_{ijkl} are the s-multiplied (Laplace) transforms of the creep and relaxation functions, respectively, i.e.,

$$\tilde{C}_{ijkl} = s\,\bar{C}_{ijkl} \tag{14.127a}$$

$$\tilde{R}_{ijkl} = s\,\bar{R}_{ijkl} \tag{14.127b}$$

$$\left[\tilde{C}_{ijkl}\right] = \left[\tilde{R}_{ijkl}\right]^{-1} \tag{14.128}$$

The quantities $\widetilde{C}_{ijkl}$ and $\widetilde{R}_{ijkl}$ are interrelated operational functions, and both are completely symmetric. Thus, in view of the thermodynamic theory, the transformed constitutive equations in terms of these operational functions are identical to those of an elastic body with compliance F_{ijkl} and modulus R_{ijkl} and of the same degree of geometric symmetry (*see* Schapery, 1967).

For the case of an isotropic material, the constitutive equation (14.118) may be used. The Laplace transform of this equation is (*see* Hunter, 1967).

$$\bar{\sigma}_{ij}(s) = \lambda(s)\bar{\varepsilon}_{ii}(s)\delta_{ij} + 2\mu(s)\bar{\varepsilon}'_{ij}(s) \tag{14.129}$$

where σ_{ij} and ϵ'_{ij} are the Laplace transforms of σ_{ij} and ϵ'_{ij}, respectively. In this equation, the transform moduli $\lambda(s)$ and $\mu(s)$ are defined by,

$$\lambda(s) = k(s) - \frac{2}{3}\mu(s) \tag{14.130}$$

where $k(s)$ is the Laplace transform of the relaxation function in pure dilatation, i.e., with reference to Eqn. 14.117,

$$k(s) = s\int_0^\infty R_1(t)e^{-st}dt \tag{14.131a}$$

and $\mu(s)$ is the Laplace transform of the relaxation function in pure shear, i.e.,

$$\mu(s) = s\int_0^\infty R_2(t)e^{-st}dt \tag{14.131b}$$

In the general case of nonhomogeneous material, the field quantities σ_{ij}, ϵ_{ii}, and ϵ'_{ij} of (14.129) are usually functions of both the transform parameter s and the position vector $\mathbf{x}$. However, the transform moduli $\lambda(s)$ and $\mu(s)$ are functions of the transform variable s only (Hunter, 1967).

The corresponding format to (14.129) in linear elasticity is the constitutive equation,

$$\sigma_{ij} = \lambda\varepsilon_{ii}\delta_{ij} + 2\mu\varepsilon'_{ij} \tag{14.132}$$

where λ and μ are the Lamé constants. Such analogy reflects the basis of the correspondence principle.

The set of Laplace-transformed relations, (14.121), (14.123), (14.124), together with the transformed constitutive equations (14.125) and (14.126), or alternatively (14.129), constitutes an "*associated*" elastic problem corresponding to the original (quasi-static) viscoelastic boundary value problem for the same geometry and subject to surface tractions $T_i = T_i(\mathbf{x},s)$, displacements $u_i = u_i(\mathbf{x},s)$ and body forces $\chi_i = \chi_i(\mathbf{x},s)$. The task then would be to solve this analogous elastic problem (Laplace transformed of the original viscoelastic problem) to determine the transformed components of the stress σ_{ij} and/or the transformed components of the displacement u_i throughout the body.

A Laplace inversion procedure would follow afterwards to determine the components of the stress and displacement in the original viscoelastic boundary value problem. The reader is referred, in this context, to Sips (1951), Brull (1953), Lee (1955,1960), among others.

Although the presentation above uses Laplace transform procedure, a similar treatment could be accomplished using Fourier transform (e.g., Read, 1950).

Remarks on the Use of the Correspondence Principle to Solve Linear Viscoelastic Boundary Value Problems

In the course of solving a linear viscoelastic boundary value problem using the correspondence principle, one might consider some simplifications in order to ease the difficulty which might arise in the inversion of the resulting Laplace transforms. For instance (*see* Hunter, 1967),

- In a large number of boundary value problems it may be unnecessary to invert the resulting Laplace transform for all positions on the boundary (i.e., x on B_σ or B_u) if the stress and/or displacement are only required at one particular position.

- In some situations, the integral value of the stress and/or displacement is required to be determined rather than individual values of these variables. In such case, it might be easier if one establishes the relevant integral property before the inversion process.

- The inversion procedure can be simplified significantly if one assumes a constant Poisson's ratio model and particularly if the body forces χ_i are neglected. In this case, if B_σ is considered to be stress-free, then, for the same boundary conditions, the resulting displacement field at any given instant of time would be identical to the displacement field of the corresponding elastic problem. A similar example here is when B_u=0 and B_σ=B, i.e., the traction vector is specified everywhere on the total boundary, then, the resulting viscoelastic stress field would be identical with the stress field of the corresponding elastic problem.

EXAMPLE 14. 1: ***Torsional Quasi-Static Twisting of a Linear, Viscoelastic Cylinder.***

This example (Hunter, 1967) considers the determination of the time-dependent twisting moment and displacement of a solid cylinder of radius a and length ℓ made of linear viscoelastic material.

Let u_r, u_θ and u_z represent the displacement components in cylindrical polar components. Let θ denote the angle of twist at $z=\ell$ and M designate the twisting moment.

From the theory of elasticity, the displacement field of an elastic solid cylinder under the action of a twisting moment and a stress free cylindrical surface condition is expressed (see, e.g., Love, 1944), as,

$$u_\theta = \theta rz/l, \; u_r = u_z = 0 \tag{14.133}$$

For an analogous quasi-static viscoelastic problem with prescribed displacements as

$$u_\theta = \begin{cases} r\theta, & z=l \\ 0, & z=0 \end{cases}$$

and a stress free cylindrical surface, the displacement field is given by (14.133) with θ now is time-dependent variable. In this case, the only non-vanishing strain component is,

$$\varepsilon_{\theta z} = \frac{1}{2}\theta(t)\frac{r}{l} \tag{14.134}$$

The non-vanishing (transformed) stress corresponding to the above strain becomes

$$\bar{\tau}_{\theta z} = \bar{\mu}(s)\,\bar{\theta}(s)\frac{r}{l} \tag{14.135}$$

where $\bar{\mu}(s)$ is the transformed shear modulus.

Thus, the total (transformed) couple required to maintain the (transformed) angle of twist $\bar{\theta}(s)$ can be expressed as,

$$\overline{M} = 2\pi \int_0^a \overline{\tau}_{\theta z} r^2 dr = \frac{\pi a^4}{2l} \overline{\mu}(s)\overline{\theta}(s) \tag{14.136}$$

which may be inverted in either of the relaxation form,

$$M(t) = \frac{\pi a^4}{2l} \int_0^t \dot{\theta}(t') G(t-t') dt' \tag{14.137a}$$

or, in the creep form

$$\theta(t) = \left(\frac{\pi a^4}{2l}\right)^{-1} \int_0^t \dot{M}(t') G^{-1}(t-t') dt' \tag{14.137b}$$

where G(t-t') is the relaxation function of the material in shear.

Equations (14.137) provide a quasi-static linear viscoelastic solution of the presented problem for prescribed $\theta(t)$ or M(t).

EXAMPLE 14.2: ***Impact of a Flat Circular Punch on a Linear, Viscoelastic Half-Space.***

Elastic Solution

According to the theory of elasticity (*see* Hunter, 1967) the solution of the problem of the normal indentation of an elastic half-space by a flat ended rigid-circular punch of radius a gives a pressure distribution described as (Boussinesq, 1885).

$$p(r) = \begin{cases} \dfrac{4\mu a d}{1-\nu}(a^2 - r^2)^{-1/2}, & r<a \\ 0 \quad , & r>a \end{cases} \tag{14.138}$$

where d is the depth of the penetration and ν is Poisson's ratio of the elastic half-space.

Further, the total load is given by,

$$F = 2\pi \int_0^a prdr = \frac{8\pi a}{1-r}\mu d \tag{14.139}$$

Viscoelastic Solution

For the linear viscoelastic case, Eqn. (14.139) becomes,

$$\bar{F} = \frac{8\pi a}{1-\nu}\bar{u}(s)\,\bar{d} \quad : \quad \nu = constant \tag{14.140}$$

where ν is, as presented above, the Poisson's ratio of the viscoelastic half-space. Accordingly, given F(t) or d(t), Eqn. (14.140), when inverted, gives d or F respectively.

Extension of the Problem above to Include Impact

For the impact problem, the load F is expressed by Newton's second law of motion

$$F = -m\,\ddot{d}$$

where m is the mass of the indenter and d is its acceleration. On taking Laplace transform of the above expression, then, the (transformed) force is written as

$$\bar{F} = -m(s^2\ddot{d} - \ \ ) \tag{14.141}$$

where v is the initial impact velocity at the initial conditions: d=0 at t=0.

Solving (14.140) and (14.141) for d gives

$$\ddot{d} = \left[s^2 + \frac{8\pi a}{m(1-\nu)}\mu(s) \right]^{-1} \tag{14.142}$$

With some physical approximation (Hunter, 1967), the inversion of (14.142) gives

$$d = \frac{v}{\omega}\left(1 + \frac{\tan\delta}{\pi}\right)\exp\left[-\frac{1}{2}(\omega\tan\delta)t\right]\sin(\omega t - \tan\delta)$$
$$+ \frac{v\tan\delta}{2\omega}\exp\frac{-2\omega t}{\pi} \tag{14.143a}$$

where ω is the solution of

$$\frac{\omega(1-\nu)}{8\pi a}\omega^2 = \mu_1(\omega) \tag{14.143b}$$

In (14.143b), $\mu_1(\omega)$ is the real part of the complex shear modulus and, in (14.143a), $\tan\delta = \mu_2(\omega)/\mu_1(\omega)$. For the values of ω where the viscoelasticity is significant, $\mu_2(\omega) << \mu_1(\omega)$ and $\tan\delta << 1$.

Further, the impact terminates at a time given by $\ddot{d}=0$ with solution

$$\omega t = d - \frac{1}{2}\left[1 + \left(\frac{2}{\pi\zeta}\right)^2\right]\tan\delta + 0(\tan^2\delta)$$

when the indenter velocity is $-v(1 - \gamma\tan\delta)$ where

$$\gamma = \frac{1}{2}\pi - \left(1 + \zeta^{-2}\right) / \pi = 1.205$$

This results in a "*coefficient of restitution*" ζ given by

$$\zeta = 1 - 1.2\tan\delta \tag{14.144}$$

Thus, the energy absorbed by the solid is

$$E = 1.2\, mv^2 \tan\delta \tag{14.145}$$

(*see* Hunter, 1967).

14.12.5. QUASI-STATIC VISCOELASTIC MIXED BOUNDARY VALUE PROBLEMS

Earlier in Subsection 14.12.3, the set of conditions governing an isothermal, linear viscoelastic boundary value problem has been introduced. In Subsection 14.12.4, the correspondence principle was presented to solve a boundary value problem of this type, subject to the condition that B_σ and B_u are independent of time where those are the parts of the boundary upon which stress vector components and displacement components, respectively, are specified. This is necessitated by the requirement that the assumed boundary conditions at a point to be time invariant so that the integral transform methods would be applicable.

> *Consequently, an elastic-viscoelastic correspondence principle does not exist when the parts of the boundary B_σ and B_u are functions of time, i.e., when the boundary conditions at the particular point in question may involve with the passage of time both stress and displacement vectors.*

A representative boundary value problem of the latter type is the time-dependent indentation of a viscoelastic half-space by a curved rigid indenter. In this case, as the indenter is loaded and the depression into the viscoelastic half-space is progressing, there are some points on the boundary of the indentation region that, at first, may have traction free boundary conditions, but later could have displacement followed by stress boundary conditions. In other words, a portion of the boundary is the boundary B_u part of the time and is the boundary B_σ at other times, so that the half-space would conform to the geometry of the indenter in the contact region. Studies concerning this problem were presented, for instance, by Lee and Radok (1960), Hunter (1960), Graham (1965,1967), Calvit (1967) and Ting (1966,1968). Other examples of mixed boundary value problems are, for instance, those involving rolling of rigid bodies over a viscoelastic half-space (e.g.,Hunter, 1961 and Morland, 1962, 1967) and ablation problems in which phase change could cause the boundaries of a viscoelastic medium to change size and shape. An example of this problem is the case of a spinning rocket's filling burning internally (e.g., Cornelliussen *et al.*, 1961). A similar problem of an internally ablating sphere was considered by Rogers and Lee (1962). Other examples of boundary value problem where integral transform methods are invalid are nonisothermal problems in which the mechanical properties are assumed to be temperature-dependent. A number of boundary value problems of the latter types have been solved, but it appears that no systematic methods of solution are available.

EXAMPLE 14.3: ***Deformation of a Uniform Viscoelastic Beam by a Curved Rigid Indenter.*** (*see* Christensen, 1971)

The schematics of the problem are shown in Figure 14.9. As indicated,

$P(t)$ is the force applied to the indenter,
$d(t)$ is the vertical displacement of the indenter, and

a (t) is is half-length of the contact region (It is considered a basic unknown of the problem).

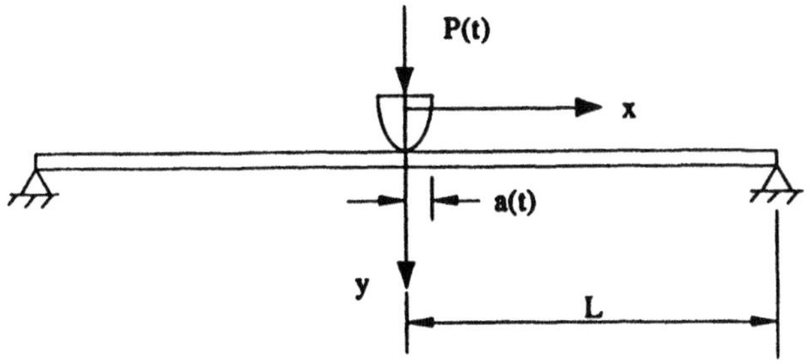

Figure 14.9. Deformation of a uniform beam by a curved rigid indenter.

The indenter is assumed to have a cubic profile expressed by

$$y = d(t) - b|x|^3 \tag{14.146}$$

where b is a given constant.

In this problem, classical beam theory with simply supported end conditions are assumed. Inertia effects are neglected. Contact is considered to begin at t=0.

Based on the above assumptions, elasticity theory gives

$$EI\frac{d^4w}{dx^4} = q(x) \tag{14.147}$$

where I is the moment of inertia of the cross-section of the beam, w is the transverse displacement and $q(x)$ is the lateral load.

Meantime, a viscoelastic beam theory gives

$$I\int_0^t R(t-\tau)\frac{\partial}{\partial\tau}\left[\frac{\partial^4 w(x,\tau)}{\partial x^4}\right]d\tau = q(x,\tau) \tag{14.148}$$

in which $R(t-\tau)$ is the uniaxial relaxation function.

In the contact region, x<a(t), the deflection of the beam must conform to the geometry of the indenter, then, with reference to Eqn. (14.146),

$$w(x,t) = d(t) - bx^3; \quad x<a(t) \ \ and \ \ t \geq 0 \tag{14.149}$$

Outside the contact region, x>a(t), the lateral load vanishes and Eqn. (14.148) is satisfied by

$$(x,t) = b_1(t) + b_2(t)x + b_3(t)x^2 + b_4(t)x^3; \quad x>a(t) \ \ and \ \ t \geq \tag{14.150}$$

where $b_1(t)$, $b_2(t)$, $b_3(t)$ and $b_4(t)$ are functions of time required to be determined.

Implying the condition that the shear resultants on the ends of the beam balance the applied load $P(t)$ gives, in view of (14.150),

$$12I\int_0^t R(t-\tau)\frac{db_4(\tau)}{d\tau}d\tau = P(t) \tag{14.151}$$

The end conditions are:

$$at - x = L, \ w = \frac{\partial^2 w}{\partial x^2} = 0$$

which can be specified by (*see* Figure 14.9)

$$b_1(t) + b_2(t)L + b_3(t)L^2 + b_4(t)L^3 = 0 \tag{14.152}$$

and

$$2b_3(t) + 6b_4(t)L = 0 \tag{14.153}$$

The continuity conditions at the edge of the contact region $x=a$ imply that, w, $\partial w/\partial x$ and $\partial^2 w/\partial x^2$ to be continuous. Accordingly, equations (14.149) and (14.150) give

$$b_1(t) + b_2(t)a(t) + b_3(t)a^2 \\ + b_4(t)a^3(t) = d(t) - ba^3(t) \tag{14.154}$$

$$b_2(t) + 2b_3(t)a(t) + 3b_4(t)a^3(t) = -3ba^2(t) \tag{14.155}$$

and

$$2b_3(t) + 6b_4(t)a(t) = -6ba(t) \tag{14.156}$$

Relations (14.151) to (14.156) give six nonlinear equations:

(i) If the load $P(t)$ is considered to be known, then, the six equations would be solved for the six unknowns $b_1(t)$, $b_2(t)$, $b_3(t)$, $b_4(t)$, $a(t)$ and $d(t)$.

(ii) Alternatively, if the displacement of the indenter $d(t)$ is taken to be specified, then, the above-mentioned equations can be solved for the unknowns $C_1(t)$, $C_2(t)$, $C_3(t)$, $C_4(t)$, $a(t)$ and $P(t)$. These two cases are considered separately by Christensen (1971).

EXAMPLE 14.4: ***A Spherical Indenter on a Viscoelastic Half-Space***

As a second example of a viscoelastic boundary value problem with mixed-typed boundary conditions, the problem of indentation of a viscoelastic half-space by a rigid spherical indenter is considered. Previous studies on this type of problem were carried out, for instance, by Lee and Radok (1960), Hunter (1960) and Graham (1965). The analysis presented below follows that of Hunter (1967) after Graham (1965). Reference is, also, made to Christensen (1971).

The indenter is considered to be applied at the origin of a rectangular Cartesian coordinate system (x,y,z) and its motion is vertical in the z-direction. The shear stresses over the entire boundary of the half-space are assumed to be identically equal to zero. In the contact region, however, the normal component of the displacement of the boundary is considered to conform to the shape of the indenter. Let,

R	is the radius of the indenter
$\delta(t)$	is the depth of penetration
a(t)	is the radius of the contact surface
$z = 0$	is the surface of the (viscoelastic) half-space
$r^2 = x^2+y^2$	

The Problem
It is to determine the stress distribution under the spherical indenter and the relation between the depth of penetration $\delta(t)$ and the radius of the contact surface $a(t)$ subject to the following boundary conditions:

$$At\ z=0, \quad \begin{cases} u_z = \delta(t) - \dfrac{1}{2R}(x^2+y^2)H(t), & r \le a(t) \qquad (a) \\ \sigma_z = 0, & r > a(t) \qquad (b) \\ \tau_{rz} = 0 & \qquad\qquad (c) \end{cases} \tag{14.157}$$

where H(t) is the Heaviside step function (*Appendix B*).

Elastic Solution
The starting point of Graham's solution is taken as (*see* Hunter, 1967) the Boussinesq formula for the normal surface displacement of an elastic solid subjected to a normal point load P at $x'y'$, i.e.,

$$u_z(x,y,0) = \frac{(1-\nu)P}{2\pi\mu}\left[(x-x')^2 + (y-y')^2\right]^{-\frac{1}{2}} \tag{14.158}$$

where μ is the elastic shear modulus and ν is Poisson's ratio.

Viscoelastic Solution
Generalizing (14.158) to the case of a viscoelastic half-space subjected to a time variable distributed load $P(x,y,t)$ gives under the assumption of a constant Poisson's ratio(*see* Hunter, 1967),

$$u_z(x,y,0,t) = \frac{(1-\nu)}{2\pi}G^{-1}(t) * d\int\!\!\int_{\Omega_m} P(x',y',t')\left[(x-x')^2+(y-y')^2\right]^{-\frac{1}{2}} dx'dy' \tag{14.159}$$

where $G^{-1}(t)$ is the creep function in shear and where the following notation of Gurtin and Sternberg (1962) is used

$$a*d\beta = \int_{-\infty}^{t} a(t-t')d\beta(t') = \int_{-\infty}^{t} a(t-t')\dot{\beta}(t')dt'$$

In (14.159) the double (surface) integral is taken over the maximum range Ω_m enclosing all points x',y' for which P(x',y',t') is non-zero for any time t' in the range $-\infty < t' < t$.

With reference to Eqn.(14.159), Hunter (1967) considered the following four situations:

(i) For a given load *P(x,y,t)*, Eqn. (14.159) presents the solution for the normal surface displacement and such problems may be considered within the class of the boundary value problems that can be solved by the Correspondence Principle (Subsection 14.12.4).

(ii) For the indentation problem, "***mixed-type boundary conditions***" boundary value problem, Eqn. (14.159) may be considered as an integral equation for *P* subject to the condition that for $r < a(t)$, the surface displacement u_z is given by (14.157a), while for $r > a(t)$, *P* vanishes.

(iii) For monotonously increasing *a(t)*, Ω_m is time-dependent and can be taken as $\Omega(t) = \pi a^2(t)$. In this case, the orders of space and time integration in (14.159) can be changed to give

$$u_z = \frac{(1-\nu)}{2\pi}\int\int_{\Omega(t)} dx'dy'\left[(x-x')^2+(y-y')^2\right]^{-\frac{1}{2}} G^{-1}dP \tag{14.160}$$

in which

$$G^{-1}dP = \eta(x,y,a) \tag{14.161}$$

where η is the unique solution of the corresponding elastic problem whose solution (Boussinesq, 1885) is given by

$$\eta = \frac{4}{\pi(1-\nu)R}\left[a^2-r^2\right]^{\frac{1}{2}} \tag{14.162a}$$

and,

$$\eta = 0 \quad for \quad r>a \tag{14.162b}$$

so that (14.161) leads to

$$P(x,y,t) = \frac{4}{\pi(1-\nu)R}\int_0^t G(t-t')\frac{d}{dt'}\left[(a^2(t')-r^2)^{\frac{1}{2}}\right]dt' \tag{14.163}$$

in which for a fixed r, the lower limit of the integral may be taken as t'' where t'' is the unique solution

$$a(t'') = r$$

Meantime, the total load on the indenter is given by

$$F(t) = 2\pi \int_0^{a(t)} r\, P(x,y,t)\, dr$$

which can be evaluated by interchanging the order of the space and time integrations (Hunter, 1960, 1967) to give

$$F(t) = \frac{8}{3R(1-\nu)}\int_0^t G(t-t')\frac{d}{dt'}a^3(t')dt' \tag{14.164}$$

Further, it can be shown that the depth of penetration can be expressed by

$$\delta(t) = a^2(t)/R \tag{14.165}$$

which is the same for the corresponding elastic problem.

(iv) The radius of the contact surface $a(t)$ increases monotonously to a maximum value at $t=t_m$ and then decreases to zero. In this case, the solution given above is valid for $t \le t_m$. For $t > t_m$, however, the solution fails because it is no longer permissible to replace Ω_m in (14.159) by $\Omega(t)$. To obtain the solution (14.159) for $t > t_m$, Hunter (1967) introduced the time function $t_1(t)$ defined by the relations:

For $t \le t_m$, $t_1(t) = t$

For $t \geq t_m$, $a(t_1) = a(t)$; $(t_1 < t_m)$

(14.166)

In other words, t_l is the time prior to t for which the radius of the contact circle is equal to the current value.

Further studies concerning the viscoelastic contact problem have been dealt with, for instance, by Calvit (1967) and Ting (1968). Graham (1968) and Ting (1968) have outlined restricted classes of viscoelastic contact problems which may be solved directly using the elastic-viscoelastic correspondence principle.

14.12.6. THE THERMOVISCOELASTIC BOUNDARY VALUE PROBLEM

The set of conditions that governs a thermoviscoelastic boundary value problem may be stated as follows:

(i) *Initial Conditions*
Assuming the body is initially undisturbed at a base temperature θ_o, then, the initial conditions are

$$u_i(t)=0,\ \epsilon_{ij}(t)=0,\ \sigma_{ij}(t)=0,\ \theta(t)=0,\ -\infty < t < 0$$

(14.167)

where θ denotes the temperature deviation from the base temperature θ_0.

(ii) *Boundary Conditions*
In order to account for the temperature effect, the boundary is visualized (Christensen, 1971) to be composed of two regions, i.e., B_θ is that region of the boundary upon which the temperature is prescribed and $(B\text{-}B_\theta)$ is the complimentary region over which the surface is taken to be perfectly insulated against heat flow. The thermal boundary conditions can then be stated as

$$\theta(\mathbf{x},t) = \theta(\mathbf{x},t), \qquad \mathbf{x} \text{ on } B_\theta,\ t \geq 0$$

(14.168)

$$k_{ij}(\mathbf{x},t)\theta_{,i}\, n_j = 0, \qquad \mathbf{x} \text{ on } (B\text{-}B_\theta), t \geq 0$$

(14.169)

where k_{ij} as a second-order tensor accounts for mechanical properties of the material.

Combining the thermal boundary conditions (14.168) and (14.169) with the traction and displacement boundary conditions stated earlier for the isothermal problem, (14.108), then, the set of boundary conditions for the thermoviscoelastic problem is written as

$$\sigma_{ij}(\mathbf{x},t)n_j = T_i(\mathbf{x},t) \qquad \mathbf{x} \text{ on } B_\sigma$$
$$u_i(\mathbf{x},t) = U_i(\mathbf{x},t) \qquad \mathbf{x} \text{ on } B_u$$

$$\theta(\mathbf{x},t) = \theta(\mathbf{x},t) \qquad \mathbf{x} \text{ on } B_\theta$$

$$k_{ij}(\mathbf{x},t)\theta_{,i}n_j = 0 \qquad \mathbf{x} \text{ on } (B\text{-}B\ )$$

(14.170)

(iii) *Balance of linear momentum*

The equations of (quasi-static) equilibrium are

$$\sigma_{ij,j} + \chi_i = 0 \tag{14.171}$$

or, alternatively, the equations of motion are

$$\sigma_{ij,j} + \chi_i = \rho\frac{d^2u_i}{dt^2} \tag{14.172}$$

(iv) *Strain-displacement relations, i. e.,*

$$\varepsilon_{ij}(t) = \frac{1}{2}(u_{i,j}(t) + u_{j,i}(t)) \tag{14.173}$$

(v) *Stress-strain relations*

For anisotropic materials:

The relaxation constitutive relation is expressed as

$$\sigma_{ij}(t) = \int_0^t R_{ijkl}(t-\tau)\partial\frac{\varepsilon_{kl(\tau)}}{\partial\tau}d\tau - \int_0^t \psi_{ij}(t-\tau)\frac{\partial\theta(\tau)}{\partial\tau}d\tau \tag{14.174a}$$

Meantime, the creep constitutive relation corresponding to (14.174a) is written as

$$\varepsilon_{ij}(t) = \int_0^t C_{ijkl}(t-\tau)\frac{\partial \sigma_{kl}(\tau)}{\partial \tau}d\tau - \int_0^t \alpha_{ij}(t-\tau)\frac{\partial \theta(\tau)}{\partial \tau}d\tau \qquad \textbf{(14.174b)}$$

In the case of thermorheologically simple materials, one may employ a stress-strain relation of the form (*see* Schapery, 1964)

$$\sigma_{ij}(t) = \int_0^t R_{ijkl}(\xi-\xi')\frac{\partial \varepsilon_{kl}}{\partial \tau}d\tau - \int_0^t \psi_{ij}(\xi-\xi')\frac{\partial \theta(\tau)}{\partial \tau}d\tau \qquad \textbf{(14.175a)}$$

or, equivalently,

$$\sigma_{ij}(\xi) = \int_0^\xi R_{ijkl}(\xi-\xi')\frac{\partial \varepsilon_{kl}}{\partial \xi'}d\xi' - \int_0^\xi \psi_{ij}(\xi-\xi')\frac{\partial \theta(\tau)}{\partial \tau}d\tau \qquad \textbf{(14.175b)}$$

where ξ, introduced earlier in Section 8.10 (*Chapter 8*), is the so-called "***reduced time***" defined by the relation

$$d\xi = dt/a_\theta(\theta) \qquad \textbf{(14.176)}$$

Also,

$$\xi = \int_0^t dt/a_\theta(\theta), \quad \xi' = \int_0^\tau dt/a_\theta(\theta) \qquad \textbf{(14.177)}$$

where $\tau \le t$.

The relaxation function $\psi_{ij}(\xi)$ appearing in (14.175) is assumed to have the following exponential series form

$$\psi_{ij}(\xi) = \sum_m \psi^{(m)}_{ij} \; e^{-\xi/\gamma_m} + \psi'_{ij} \qquad \textbf{(14.178)}$$

where the constants $\psi_{ij}^{(m)}$ and ψ'_{ij} define the thermal stress characteristics of the material before loading (Schapery, 1964,1967) and γ_m are appropriate exponent factors. Meantime, the relaxation moduli in (14.175) are considered (Schapery, 1964) to be given by

$$R_{ijkl}(\xi) = \sum_m R^{(m)}_{ijkl} \; e^{-t/\gamma_m} + R'_{ijkl} \tag{14.179}$$

On the other hand, when the temperature is constant, the relaxa- tion moduli may be taken as

$$R_{ijkl}(t/a_\theta) = \sum_m R^{(m)}_{ijkl} \; e^{(-t/\gamma_m)a_\theta} + R_{ijkl} \tag{14.180}$$

which reflects the effect of constant temperature on relaxation (or creep) behaviour, that is, to simply shift the time scale. Accordingly, a_θ is often referred to as "***time shift factor***").

The creep constitutive equation corresponding to (14.175b) is

$$\varepsilon_{ij}(\xi) = \int_0^\xi C_{ijkl}(\xi - \xi') \frac{\partial \sigma_{kl}}{\partial \xi'} d\xi' + \int_0^\xi \alpha_{ij}(\xi - \xi') \frac{\partial \theta}{\partial \xi'} d\xi' \tag{14.181}$$

where the function $\alpha_{ij}(\xi)$ accounts for the strain response in the absence of the stress. It is expressed (Schapery, 1964), by

$$\alpha_{ij}(\xi) = \sum_m \alpha_{ij} \quad (1 - e^{-\xi/\gamma_m}) + \alpha_{ij} \tag{14.182}$$

where $\alpha_{ij}^{(m)}$ and α_{ij} define the thermal strain characteristics of the material before loading.

For an isotropic material, the relaxation constitutive relations corresponding to (14.174a) are

$$\sigma'_{ij}(t) = \int_0^t R_1(t-\tau) \frac{\partial \varepsilon'_{ij}(\tau)}{\partial \tau} d\tau$$

$$\sigma_{kk}(t) = \int_0^t R_2(t-\tau) \frac{\partial \varepsilon_{kk}(\tau)}{\partial \tau} d\tau - 3\int_0^t \psi(t-\tau) \frac{\partial \theta(\tau)}{\partial \tau} d\tau \tag{14.183}$$

where σ'_{ij} and ϵ'_{ij} denote, respectively, the deviatoric components of the stress and strain.

The creep constitutive equations corresponding to (14.183) are expressed as

$$\varepsilon'_{ij}(t) = \int_0^t C_1(t-\tau)\frac{\partial\sigma'_{ij}(\tau)}{\partial\tau}d\tau$$

$$\varepsilon_{kk}(t) = \int_0^t C_2(t-\tau)\frac{\partial\sigma_{kk}(\tau)}{\partial\tau} - 3\int_0^t \alpha(t-\tau)\frac{\partial\theta(\tau)}{\partial\tau}d\tau \qquad \textbf{(14.184)}$$

In the case of thermorheologically simple materials, constitutive equations for isotropic materials are expressed (Schapery, 1964) for the relaxation case by

$$\sigma_{ij}(\xi) = 2\int_0^\xi R(\xi-\xi')\frac{\partial\varepsilon_{ij}}{\partial\xi'}d\xi' + \delta_{ij}\int_0^\xi\left[\lambda(\xi-\xi')\frac{\partial\varepsilon}{\partial\xi'}\right.$$

$$\left. - \psi(\xi-\xi'\frac{\partial\varepsilon}{\partial\xi'}\right]d\xi' \qquad \textbf{(14.185)}$$

where,

$$\varepsilon = \varepsilon_{11} + \varepsilon_{22} + \varepsilon_{33}$$

In (14.185), $R(\xi)$, $\lambda(\xi)$ and $\psi(\xi)$ are relaxation functions which, for thermodynamic reasons (Schapery, 1964), are considered to have the forms

$$R(\xi) = \sum_m R^{(m)} e^{-\xi/\gamma_m} + R_e$$

$$\lambda(\xi) = \sum_m \lambda^{(m)} e^{-\xi/\gamma_m} + \lambda_e \qquad \textbf{(14.186)}$$

$$\psi(\xi) = \sum_m \psi^{(m)} e^{-\xi/\gamma_m} + \psi_e$$

with constants having the properties

$$\gamma_m > 0$$

$$R^{(m)} \geq 0,\ R_e \geq 0,\ \sum_m R^{(m)} + R_e > 0$$

$$K^{(m)} = \lambda^{(m)} + \frac{2}{3} R^{(m)} \geq 0 \tag{14.187}$$

$$K_e = \lambda_e + \frac{2}{3} R_e \geq 0$$

$$\sum_m K^{(m)} + K_e > 0$$

where K_e and $K^{(m)}$ define the bulk relaxation modulus

$$K(\xi) = \lambda(\xi) + \frac{2}{3} R(\xi) = \sum_m K^{(m)} e^{-\xi/\gamma_m} + K_e \tag{14.188}$$

(vi) The heat conduction equation

For isotropic materials:

$$\frac{k_{ij}}{\theta_0} \theta_{,ij} = \frac{\partial}{\partial t} \int_0^t m(t-\tau) \frac{\partial \theta(\tau)}{\partial \tau} d\tau + \frac{\partial}{\partial t} \int_0^t \psi_{ij}(t-\tau) \frac{\partial \varepsilon_{ij}(\tau)}{\partial \tau} d\tau \tag{14.189}$$

For anisotropic materials:

$$\frac{k}{\theta_0} \theta_{,ii} = \frac{\partial}{\partial t} \int_0^t m(t-\tau) \frac{\partial \theta(\tau)}{\partial \tau} d\tau + \frac{\partial}{\partial t} \int_0^t \psi(t-\tau) \frac{\partial \varepsilon_{kk}(\tau)}{\partial \tau} d\tau \tag{14.190}$$

where k_{ij} or k, m(t), and $\psi_{ij}(t)$ or $\psi(t)$ are mechanical properties of the material.

In the general anisotropic case, the Laplace transformed governing equations for the thermoviscoelastic boundary value problem are given by:

(i) *Boundary conditions (Equations 14.170)*

$\sigma_{ij} n_j = T_i$, on B_σ

$u_i = U_i$, on B_u

$\theta = \theta$, on B_θ

$k_{ij} \theta_{,i} n_j = 0$, on $(B\text{-}B_\theta)$

(14.191)

(ii) *Balance of linear momentum*

The equations of quasi-static equilibrium (14.171)

$$\bar{\sigma}_{ij,j} + \bar{\chi}_i = 0 \quad \textbf{(14.192)}$$

or, alternatively, the equation of motion (14.172)

$$\bar{\sigma}_{ij,j} + \bar{\chi}_i = \rho s^2 \bar{u}_i \quad \textbf{(14.193)}$$

where s is the Laplace transform variable

(iii) *Strain-displacement relations (14.124)*

$$\bar{\varepsilon}_{ij} = \frac{1}{2}(\bar{u}_{i,j} + \bar{u}_{j,i}) \quad \textbf{(14.194)}$$

(iv) *The relaxation constitutive relation (14.174a)*

$$\bar{\sigma}_{ij} = s\, \bar{R}_{ijkl} \bar{\varepsilon}_{kl} - s\, \bar{\psi}_{ij} \bar{\theta} \quad \textbf{(14.195)}$$

(v) The heat conduction equation (14.189)

$$(k_{ij}/\theta_0)\bar{\theta}_{,ij} = s^2\,\bar{m}\,\bar{\theta} + s^2\,\bar{\psi}_{ij}\,\bar{\varepsilon}_{ij} \qquad \textbf{(14.196)}$$

The viscoelastic boundary value problem governed by the set of equations (14.191) to (14.196) can be solved in the same manner as in the case of coupled thermoelastic problems. Consequently, the complete solution of the viscoelastic boundary value problem under consideration is obtained by inverting the transformed solution. The procedure here is the same as in the case of treating isothermal linear viscoelastic boundary value problems discussed earlier in this section.

In problems where the coupling term involving ϵ_{ij} in (14.189) and (14.190) can be neglected, mechanical response problems and thermal response problems may be separated. Thus, after obtaining the temperature distribution, either by solving the heat conduction equation or from experimental results, the mechanical response problem would then be governed by (14.167) and (14.170) to (14.174). Integral transform methods could thus provide a useful tool in solving such problems.

14.13. Study Problems

11. Define and comment briefly on the implications of using the "Correspondence Principle" in the solution of boundary value problems in linear viscoelasticity.
12. What is meant by a nonlinear viscoelastic material? Illustrate such response in both creep and relaxation.
13. Extend your arguments, as pertaining to Problem 11 above, to the case of a nonlinear viscoelastic boundary value problem.
14. Based on the elastic solution given earlier in Chapter 6, determine the stress distribution in rotating discs if the material is considered to be isotropic, linear viscoelastic.
15. Assuming a linear thermal gradient, determine the pertaining form of the heat conduction equation in both isotropic and anisotropic linear viscoelastic materials.
16. Re-attempt the analysis of Problem 14 above with the inclusion of an assumed linear thermal gradient.

14. 14. References

Achenbach, J.D. and Chao, C.C. (1962) A three-parameter viscoelastic model particularly suited for dynamic problems, *J. Mech. Phys. Solids* **10**, 245-52.

Achenbach, J.D., Vogel, S.M. and Herrmann, G. (1966) On stress waves in viscoelastic media conducting heat, in: *Irreversible Aspects of Continuum Mechanics and Transfer of Physical Characteristics in Moving Fluids*, Eds. H. Parkus and L.I. Sedov, Springer, Berlin, pp. 1-15.

Ahrens, T.J. and Duvall, G.E. (1966) Stress relaxation behind shock waves in rocks, *J. Geophys. Res.* **71**,

4349-60.

Alfrey, T., Jr. (1948) *Mechanical Behaviour of High Polymers*, Interscience, New York.

Bailey, P. and Chen, P.J. (1971) On the local and global behaviour of acceleration Waves, *Arch. Ration. Mech. Analysis* **41**, 121. Addendum: Asymptotic Behaviour, ibid 44, 212 (1972).

Barberan, J. and Herrara, I. (1966) Uniqueness theorems and speed of propagation of signals in viscoelastic materials, *Arch. Rat. Mech. Anal.* **23**, 173-90.

Barker, L.M. (1968) The fine structure of compressive and release wave shapes in aluminium measured by the velocity interferometer technique, in: *Behaviour of Dense Media Under High Dynamic Pressures*, Gordon and Breach, pp. 483-505.

Barker, L.M. and Hollenbach, R.E. (1964) System for measuring the dynamic properties of materials, *Rev. Sci. Inst.* **35(6)**, 742-6.

Barker, L.M. and Hollenbach, R.E. (1965) Interferometer technique for measuring the dynamic mechanical properties of materials, *Rev. Sci. Inst.* **36**, 1617-20.

Barker, L.M. and Hollenbach, R.E. (1970) Shock wave studies of PMMA, Fused silica and sappire, *J. Appl. Phys.* **41**, 4208-26.

Bland, D.R. (1960) *The Theory of Linear Viscoelasticity*, Pergamon, New York.

Biot, M.A. (1958) Linear thermodynamics and the mechanics of solids, *Proceedings 3rd U.S. Natl. Congr. Appl. Mech.*, ASME, pp.1-18.

Boussinesq, M.J. (1885) In Todhunter and Pearson - *A History of Theory of Elasticity and of the Strength of Materials*, Vol. II, Part 2, Dover reprint (1960), pp. 185--357.

Bradfield, G. (1951) Internal friction of solids, *Nature* **167**, 1021-3.

Brull, M.A. (1953) A structural theory incorporating the effect of time-dependent elasticity, in *Proceedings 1st Midwestern Conf. On Solid Mechanics*, pp. 141-7.

Calvit, H.H. (1967) Numerical solution of the problem of impact of a rigid sphere onto a linear viscoelastic half-space and comparison with experiment, *Int. J. Solid Structures* **3**, 951-66.

Chao, C. and Achenbach, J.D. (1964) A Simple viscoelastic analogy for stress waves, in: *Stress Waves in Anelastic Solids*, IUTAM Symposium, Eds. H. Kolsky and W. Prager, Brown University, Providence, RI, Apr. 3-5, 1963, Springer, Berlin, pp. 222-38.

Chen, P.J. and Gurtin, M.E. (1970). On the growth of one-dimensional shock waves in materials with memory, *Arch. Ration. Mech. Analysis* **36**, 33-46.

Chen, P.J. and Gurtin, M.E. (1972a) On the use of experimental results concerning steady shock waves to predict the acceleration wave response of nonlinear viscoelastic materials, *J. Appl. Mech.* **39**, 295-6.

Chen, P.J. and Gurtin, M.E. (1972b) Thermodynamic influences on the growth of one-dimensional shock waves in materials with memory, *Z. Angew, Math. Phys.* **23**, 69-79.

Chen, P.J., Gurtin, M.E. and Walsh, E.K. (1970) Shock amplitude variation in Polymethyl Methacrylate for fixed values of the strain gradient, *J. Appl. Phys.* **41**, 3557-8.

Christensen, R.M. (1971) *Theory of Viscoelasticity*, Academic Press, New York.

Chu, B.T. (1962) *Stress Waves in Isotropic Viscoelastic Materials*, Division of Engineering, Brown University, Providence, March 1962.

Coleman, B.D. (1964) Thermodynamics, strain impulses and viscoelasticity, *Arch. Ration. Mech. Analysis* **17**, 230-254.

Coleman, B.D., Greenberg, J.M. and Gurtin, M.E. (1966) Waves in materials with memory. V. On the amplitude of acceleration waves and mild discontinuities, *Arch. Ration. Mech. Analysis* **22**, 333-54.

Coleman, B.D. and Gurtin, M.E. (1965a) Waves in materials with memory. II. On the growth and decay of one-dimensional acceleration waves, *Arch. Ration. Mech. Analysis* **19**, 239-65.

Coleman, B.D. and Gurtin, M.E. (1965b) Waves in materials with memory. III. Thermodynamic influences on the growth and decay of acceleration waves, *Arch. Ration. Mech. Analysis* **19**, 266-98

Coleman, B.D. and Gurtin, M.E. (1965c) Waves in materials with memory. IV. Thermodynamics and the velocity of general acceleration waves, *Arch. Ration. Mech. Analysis* **19(5)**, 317-38

Coleman, B.D. and Gurtin, M.E. (1966) Thermodynamics and one-dimensional shock waves in materials with memory, *Proc. R. Soc. A* **292**, 562-74.

Coleman, B.D., Gurtin, M.E. and Herrara, R.I. (1965) Waves in materials with memory. I. The velocity of one-dimensional shock and acceleration waves, *Arch. Ration. Mech. Analysis* **19**, 1-19.

Cornelliussen, A.H. et al (1961) Brown University Report NORD 18594/5.

Cunningham, J.R. and Ivey, D.G. (1956) Dynamic properties of various rubbers at high frequencies, *J. Appl. Phys.* **27**, 967-74.

Dally, J.W. and Riley, W.F. (1965) *Experimental Stress Analysis*, McGraw-Hill, New York.

Dove, R.C. and Adams, P.H. (1964) *Experimental Stress Analysis and Motion Measurement*, Charles Merril Books, Columbus, Ohio.

Dunwoody, J. (1972) One-dimensional shock waves in heat conducting materials with memory. I. Thermodynamics, *Arch. Ration. Mech. Analysis* **47**, 117-48.

Duvall, G.E. and Alverson, R.C. (1963) *Fundamental Research*, Tech. Summary Report 4, Stanford Research Inst., Menlo Park.

Edelstein, W.S. and Gurtin, M.E. (1964) Uniqueness theorems in the linear dynamic theory of anisotropic viscoelastic solids, *Arch. Ration. Mech. Anal.* **17**, 47-60.

Ferry, J.D. and Williams, M.L. (1952) II. Approximation methods for determining the relaxation time spectrum of a viscoelastic material, *J. Colloid Sci.* **7**, 347-53.

Fung, Y.C. (1965) *Foundations of Solid Mechanics*, Prentice-Hall, Englewood Cliffs, New Jersey.

Glauz, R.D. and Lee, E.H. (1954) Transient wave analysis in a linear time-dependent material, *J. Appl. Phys.* **25**, 947-53

Golden, J.M. and Graham, G.A.C. (1988) *Boundary Value Problems in Linear Viscoelasticity*, Springer-Verlag, Berlin.

Gorsky, W.S. (1936) On the transitions in the CuAu alloy. III. On the influence of strain on the equilibrium in the ordered lattice of CuAl, *Phys. Zeit Sowjet* **6**, 77-81.

Graham, G.A.C. (1965) The contact problem in the linear theory of viscoelasticity, *Int. J. Eng. Sci.* **3**, 27-46.

Graham, G.A.C. (1967) The contact problem in the linear theory of viscoelasticity when the time-dependent contact area has any number of maxima and minima, *Int. J. Eng. Sci.* **5**, 495-514.

Graham, G.A.C. (1968) The correspondence principle of linear viscoelasticity for mixed boundary value problems involving time-dependent boundary regions, *Quart. Appl. Math.* **26**, 167-74.

Graff, K.F. (1975) *Wave Motion in Elastic Solids*, Dover Publications, New York.

Green, W.A. (1964) The growth of plane discontinuities propagating into a homogeneously deformed elastic material, *Arch. Ration. Mech. Anal.* **16**, 79-89.

Greenberg, J.M. (1967) The existence of steady shock waves in nonlinear materials with memory, *Arch. Ration. Mech. Analysis* **24**, 1-21.

Greenberg, J.M. (1968) Existence of steady waves for a class of nonlinear dissipative materials, *Quart. Appl. Math.* **26**, 27-34.

Gurtin, M.E. and Herrara, I. (1964) A correspondence principle for viscoelastic wave propagation, *Quart. Appl. Math.* **22**, 360-4.

Gurtin, M.E. and Herrera, I. (1965) On dissipation inequalities and linear viscoelasticity, *Quart. Appl. Math.* **23**, 235-45.

Gurtin, M.E. and Sternberg, E. (1962) On the linear theory of viscoelasticity, *Arch. Ration. Mech. Anal.* **11** (4), 291-356.

Haddad, Y. M. (1995) *Viscoelasticity of Engineering Materials*, Kluwer, Dordrecht.

Hetenyi, M., Ed. (1950) *Handbook of Experimental Stress Analysis*, John Wiley and Sons, New York.

Hill, R. (1962) Acceleration waves in solids, *J. Mech. Phys. Solids* **10**, 1-16.

Hillier, K.W. (1949) A method of measuring some dynamic elastic constants and its application to the study of high polymers, *Proc. Phys. Soc.* **LXII, II-B**, 701-13.

Hillier, K.W. (1960) A Review of the progress in the measurement of dynamic elastic properties, *Int. Symp. on Stress Wave Propagation in Materials*, Ed. N. Davids, Inter-science Publishers, London, pp. 183-98.

Hillier, K.W. and Kolsky, H. (1949) An investigation of the dynamic elastic properties of some high polymers, *Proc. Phys. Soc.* **B62**, 111-21.

Huilgol, R.R. (1973) Growth of plane shock waves in materials with memory, *Int. J. Engg. Sci.* **11**, 75-86.

Hunter, S.C. (1960a) Viscoelastic waves, in: *Progress in Solid Mechanics*, Vol.I, Eds. I.N. Sneddon and R.Hill, North- Holland Pub. Co., Amsterdam, pp. 1-57.

Hunter, S.C. (1960b) The Hertz problem for a rigid spherical indenter and a viscoelastic half-space, *J. Mech. Phys. Solids* **8**, 219-34.

Hunter, S.C. (1961) The rolling contact of a rigid cylinder with a viscoelastic half-space, *J. Appl. Mech.* **28**, 611-17.

Hunter, S.C. (1967) *The solution of boundary value problems in linear viscoelasticity*, Proc. 4th *1965 Symposium on Naval Structural Mechanics*, Eds. A.C. Eringen, H. Liebowitz, S.L. Koh and J.M. Crowley, Pergamon, Oxford, pp. 257-95.

Ivey, D.G., Mrowea, B.A. and Guth, E. (1949) Propagation of ultrasonic bulk waves in high polymers, *J. Appl. Phys.* **20**, 486-92.

Keast, D.N. (1967) *Measurements in Mechanical Dynamics*, McGraw-Hill, New York.

Kolsky, H. (1954a) Attenuation of short mechanical pulses by high polymers, *Proc. 2nd Int. Congr. on Rheology*, Butterworths, London, pp. 79-84.

Kolsky, H. (1954b) The propagation of longitudinal elastic waves along cylindrical bars, *Phil. Mag.* **45**, 712-26.

Kolsky, H. (1956) The propagation of stress pulses in viscoelastic solids, *Philosophical Magazine* **8(1)**, 693-710.

Kolsky, H. (1958) The propagation of stress waves in viscoelastic solids, *Appl. Mech. Rev.* **11**, 465-8.

Kolsky, H. (1960) Viscoelastic waves, *Int. Symp. on Stress Wave Propagation in Materials*, Ed. N. Davids, Interscience Publishers, London, pp. 59-90.

Kolsky, H. (1963) *Stress Waves in Solids*, Dover Publications, New York.

Lee, E.H. (1955) Stress analysis in viscoelastic bodies, *Quart. Appl. Math.* **13**, 183-90.

Lee, E.H. (1960) Viscoelastic stress analysis, in: *First Symposium on Naval Structural Mechanics*, Eds. J.N. Goodier and N.J. Hoff, Pergamon Press, New York, pp. 456-82.

Lee, E.H., (1966) Some recent developments in linear viscoelastic stress analysis, in: *Proceedings, Eleventh Int. Cong. of Applied Mechanics*, Ed. H. Gortler, Springer-Verlag, Berlin, pp. 396-402.

Lee, E.H. and Kanter, I. (1953) Wave propagation in finite rods of viscoelastic materials, *J. Appl. Phys.* **24(9)**, 1115-22.

Lee, E.H. and Morrison, J.A. (1956) A comparison of the propagation of longitudinal waves in rods of viscoelastic materials, *J. Polymer Sci.* **XIX**, 93-110.

Lee, E.H. and Radok, J.R.M. (1960) The contact problem for viscoelastic bodies, *J. Appl. Mech., Trans. ASME* **27**, 438-44.

Lee, E.H., Radok, J.R.M. and Woodward, W.B. (1959) Stress analysis for linear viscoelastic materials, *Trans. Soc. Rheol.* **3**, 41-59.

Lockett, F.J. (1962) The reflection and refraction of waves at an interface between viscoelastic materials, *J. Mech. Phys. Solids* **10**, 53-64.

Love, A.E.H. (1944) *A Treatise on the Mathematical Theory of Elasticity*, Cambridge University Press.

Lubliner, J. and Sackman, J.L. (1967) On uniqueness in general linear viscoelasticity, *Q. Appl. Math.* **25**, 129-38.

Magrab, E.B. and Blomquist, D.S. (1971) *The Measurement of Time-Varying Phenomena*, Wiley-Interscience, New York.

Malvern, L.E. (1951) Plastic wave propagation in a bar of material exhibiting a strain rate effect, *Quart. Appl. Math.* **8**, 405-11.

Mason, W.P. and McSkimin, H.J. (1947) Attenuation and scattering of high frequency sound waves in metals and glasses, *J. Acoust. Soc. Amer.* **19**, 464-73.

Morland, L.W. (1962) A plane problem of rolling contact in linear viscoelasticity theory, *J. Appl. Mech.* **29**,

345-58.
Morland, L.W. (1967) Exact solution for rolling contact between viscoelastic cylinders, Quart. *J. Appl. Math.* **20**, 73-106.
Morland L.W. and Lee, E.H. (1960) Stress analysis for linear viscoelastic materials with temperature variation, *Trans. Soc. Rheol.* **4**, 233-63.
Morrison, J.A. (1956) Wave propagation in rods of Voigt material and viscoelastic materials with three-parameter models, *Quart. Appl. Math.* **14**, 153-69.
Muki, R. and Sternberg, E. (1961) On transient thermal stresses in viscoelastic materials with temperature dependent properties, *J. Appl. Mech.* **28**, 193-207.
Nashif, A.D., Jones, D.I.G. and Henderson, J.P. (1965) *Vibration Damping*, John Wiley & Sons, New York.
Nunziato, J.W. and Sutherland, H.J. (1973) Acoustical determination of stress relaxation functions in Polymers, *J. Appl. Phys.* **44(1)**, 184-7.
Nunziato, J.W. and Walsh, E.K. (1973) Propagation of steady shock waves in nonlinear thermoviscoelastic solids, *J. Mech. Phys. Solids* **21**, 317-35.
Odeh, F. and Tadjbakhsh, I. (1965) Uniqueness in the linear theory of viscoelasticity, *Arch. Ration. Mech. Anal.* **18**, 244-50.
Onat, E.T. and Breuer, S. (1963) On uniqueness in linear viscoelasticity, in: *Progress in Applied Mechanics*, The Prager Anniversary Volume, McMillan, New York, pp. 349-53.
Orowan, E. (1934) Zür kristall plastizität.III. Über den mechanismus des gleitvorganges, *Zeits f. Phys.* **89**, 634-59.
Polanyi, M. (1934) Über eine art gitterstörung, die einen kristall plastisch machen könnte, *Zeits f. Phys.* **89**, 660-4.
Predeleanu, M. (1965) Stress analysis in bodies with time-dependent properties, *Bull. Math. Soc. Sci. Math., R.S. de Roumanie* **9**, 115-27.
Rayleigh, J.W.S. (1894) *Theory of Sound*, Dover reprint, Dover Publications Inc., New York.
Read, W.T. (1950) Stress analysis for compressible viscoelastic materials, *J. Appl. Phys.* **21**, 671-4.
Rogers, T.G. (1965) Viscoelastic stress analysis, in: *Proceedings, Princeton University Conference on Solid Mechanics*, Princeton, New Jersey, pp. 49-74.
Rogers, T.G. and Lee, E.H. (1962) Brown Univ. Report NORD 18594/6.
Schapery, R.A. (1964) Application of thermodynamics to thermomechanical, fracture, and birefringant phenomena in viscoelastic media, *J. Appl. Phys.* **35(5)**, 1451-65.
Schapery, R.A. (1967) Stress analysis of viscoelastic composite materials, *J. Composite Materials* **1**, 228-66.
Schapery, R.A. (1974) Viscoelastic behaviour and analysis of composite materials, Ed. G. Sendeckj, Vol. 2, Academic Press, New York, pp. 86-168.
Schuler, K.W. (1970) Propagation of steady shock waves in Polymethyl Methacrylate, *J. Mech. Phys. Solids* **18**, 277-93.
Schuler, K.W., Nunziato, J.W. and Walsh, E.K. (1973) Recent results in nonlinear viscoelastic wave propagation, *Int. J. Solid Structures* **91**, 1237-81.
Schuler, K.W. and Walsh, E.K. (1971) Critical-induced acceleration for shock propagation in Polymethyl Methacrylate, *J. Appl. Mech.* **38**, 641-45.
Schwarzl, F. and Staverman, A.J. (1952) Time-temperature dependence of linear viscoelastic behaviour, *J. Appl. Phys.* **23(8)**, 838-43.
Sips, R. (1951) General theory of deformation of viscoelastic substances, *J. Polymer Sci.* **9**, 191-205.
Snoek, J.E. (1941) Effect of small quantities of carbon and nitrogen on the elastic and plastic properties of Iron, *Physica* **8**, 711-33.
Sternberg, E. (1964) On the analysis of thermal stresses in viscoelastic solids, in: *High Temperature Structures and Materials, Proc. of the 3rd Symp. on Naval Structural Mechanics*, Eds. A.M. Freudenthal, B.A. Boles and H. Liebowitz, Pergamon, Oxford, pp. 348-82.
Sternberg, E. and Gurtin, M.E. (1963) Uniqueness in the theory of thermorheologically simple ablating viscoelastic solids, in: *Progress in Applied Mechanics*, The Prager Anniversary

Volume, Ed. D.C. Drucker, MacMillan, New York, pp. 373-84.
Sternberg, E. and Gurtin, M.E. (1964) Further study of thermal stresses in viscoelastic materials with temperature dependent properties, in: *Proceedings, IUTAM Symposium on Second Order Effects in Elasticity, Plasticity and Fluid Mechanics*, Haifa, pp. 51-76.
Thomas, T.Y. (1957) The growth and decay of sonic discontinuities in ideal gases, *J. Math. Mech.* **6**, 455-69.
Thomas, T.Y. (1961) *Plastic Flow and Fracture in Solids*, Academic Press, New York.
Ting, T.C.T. (1966) The contact stresses between a rigid indenter and a viscoelastic half-space, *J. Appl. Mech.* **33**, 845-54.
Ting, T.C.T. (1968) Contact problems in the linear theory of viscoelasticity, *J. Appl. Mech.* **35**, 248-54.
Tobolsky, A., Powell, R.E. and Eyring, H. (1943) *The Chemistry of Large Molecules*, Interscience, New York.
Truesdell, C. and Toupin, R.A. (1960) The classical field theories, *Handbuk der Physik* **III/1**, Ed. S.Flügge, Springer, Berlin.
Valanis, K.C. (1965) Propagation and attenuation of waves in linear viscoelastic solids, *J. of Mathematics & Physics* **44(3)**, 227-39.
Varley, E. (1965) Acceleration fronts in viscoelastic materials, *Arch. Ration. Mech. Analysis* **19**, 215-25.
Varley, E. and Cumberbatch, E. (1965) Nonlinear theory of wave-front propagation, *J. Ins. Math. and Appl.* **1**, June 1965, 1, 101-112.
Volterra, V. (1909) Sulle equazioni integro-differenziali Della teoria dell' elasticita, *R. Accademia dei Lincei*, **18(1)**, 167. Also, Ibid **18(2)**, 295.
Volterra, V. (1931) *Theory of Functional*, Dover Reprint, Dover Publications, New York.
Worely, W.J. (Ed.) (1962) *Experimental Techniques in Shock and Vibration*, ASME, New York.

14.15. Further Reading

Abbott, B.W. and Cornish, R.H. (1965) A stress wave technique for determining the tensile strength of brittle materials, *Exp. Mech.* **22**, 148-53.
Aboudi, J. (1979) The dynamic indentation and impact of a viscoelastic half-space by an axisymmetric rigid body, *Compu. Math. Appl. Mech. Eng.* **20**, 135-50.
Achenbach, J.D. and Reddy, D.P. (1967) Note on wave propagation in linearly viscoelastic media, *Z. Angew. Math. Phys.* **18**, 141-4.
Alblas, J.B. and Kuipers, M. (1970) The contact problem of a rigid cylinder rolling on a thin viscoelastic layer, *Int. J. Eng. Sci.* **8**, 363-80.
Alfrey, T. (1944) Nonhomogeneous stresses in viscoelastic media, *Quart. Appl. Math.* **2**, 113- 19.
Arenz, R.J. (1964) Uniaxial wave propagation in realistic viscoelastic materials, *J. Appl. Mech., Trans. ASME* **86(E)**, Mar. 1964, 17-21.
Arenz, R.J. (1965) Two-dimensional wave propagation in realistic viscoelastic materials, *J. Appl. Mech., Trans. ASME* **32(2)**, June 1965, 303-14.
Asay, J.R., Lamberson, D.L. and Guenther, A.H. (1969) Pressure and temperature dependence of velocities in Polymethyl Methacrylate, *J. Appl. Phys.* **40**, 1768-83.
Atkinson, C. and Coleman, C.J. (1977) On some steady-state moving boundary problems in the linear theory of viscoelasticity, *J. Inst. Maths. Appl.* **20**, 85-106.
Baker, W.E. and Dove, R.C. (1962) Measurements of internal strains in a bar subjected to longitudinal impact, *Exp. Mech.* **19**, 307-11.
Ballon, J.W. and Smith, J.C. (1949) Dynamic measurements of polymer physical properties, *J. Appl. Phys.* **20**, 493-502.
Barton, C.S., Volterra, E.G. and Citron, S.J. (1958) On elastic impacts of spheres on long rods, *Proc. 3rd U.S. Nat. Cong. Appl. Mech.*, pp. 89-94.

Battiato, G., Ronca, G. and Varga, C. (1977) Moving loads on a viscoelastic double layer: Prediction of recoverable and permanent deformations, in: *Proceedings, Fourth International Conf. on Structural Design of Asphalt Pavements*, The University of Michigan, Ann Arbor, Michigan, USA, pp. 459-60.

Becker, E.C.H. and Carl, H. (1962) Transient-loading technique for mechanical impedance measurement, in *Experimental Techniques in Shock and Vibration*, Ed. W.J. Worley, ASME, New York, pp.1-10.

Berry, D.S. (1958) A note on stress pulses in viscoelastic rods, *Phil. Mag.* **8**, 100-2.

Berry, D.S. and Hunter, S.C. (1956) The propagation of dynamic stresses in viscoelastic rods, *J. Mech. Phys. Solids* **4**, 72-95.

Calvit, H.H. (1967) Experiments on rebound of steel spheres from blocks of polymers, *J. Mech. Phys. Solids* **15**, 141-50.

Chou, P.C. (1968) Introduction to wave propagation in composite materials, in *Composite Materials Workshop*, Eds. S.W. Tsai, J.C. Halpin and N.J. Pagano, Technomic, Stamford, Conn., pp. 193-216.

Chref, C. (1889) The equations of an isotropic elastic solid in polar and cylindrical coordinates, their solutions and applications, *Trans. Camb. Phil. Soc. Math. Phys. Sci.* **6**, 115-17.

Chu, B.T. (1962) Stress waves in isotropic linear viscoelastic materials (Part one), *J. Mécanique* **1(1)**, 439-62.

Chu, B.T. (1965) Response of various material media to high velocity loadings. I. Linear elastic and viscoelastic materials, *J. Mech. Phys. Solids* **13**, 165-87.

Comninou, M. (1976) Contact between viscoelastic bodies, *J. Appl. Mech.* **43**, 630-2.

Dally, J.W. (1968) A dynamic photoelastic study of a doubly loaded half-plane, *Develop. Mech.* **4**, 649-64.

Dally, J.W., Durelli, A.J. and Riley, W.F. (1960) Photoelastic study of stress wave propagation in large plates, *Proc. Soc. Exp. Stress Analysis* **17**, 33-50.

Dally, J.W. and Lewis, D. (1968) A photoelastic analysis of propagation of Rayleigh waves past a step change in elevation, *Bull. Seism. Soc. Am.* **58**, 539-63.

Dally, J.W. and Riley, W.F. (1967) Initial studies in three-dimensional dynamic photoelasticity, *J. Appl. Mech.* **34**, 405-10.

Dally, J.W. and Thau, S.A. (1967) Observations of stress wave propagation in a half-plane with boundary Loading, *Int. J. Solids Struct.* **3**, 293-7.

Davies, R.M. (1948) A critical study of the Hopkinson pressure bar, *Phil. Trans. R. Soc.* **A240**, 375-457.

Dohrenwend, C.O., Drucker, D.C. and Moore, P. (1944) Transverse impact transients, *Exp. Stress Analysis* **1**, 1-10.

Dunwoody, J. (1966) Longitudinal wave propagation in a rate dependent material, *Int. J. of Engineering Sci.* **4**, 277-87.

Dziecielak, R. (1985) The effect of temperature on the propagation of discontinuity waves in a porous medium with a viscoelastic skeleton, *Studia Geotechnica et Mechanica*, Vol. **VII (2)**, 17-34.

Edelstein, W.S. (1969) The cylinder problem in thermoviscoelasticity, *Res. Natl. Bur. Standards* **73B**, 31-40.

Engelbrecht, J. (1979) One-dimensional deformation waves in nonlinear viscoelastic materials, *Wave Motion* **1**, 65-74.

Evans, J.F., Hadley, C.F., Eisler, J.D. and Silverman, D. (1954) A three-dimensional seismic wave model with both electrical and visual observation of waves, *Geophysics* **19**, 120-36.

Ferry, J.D. (1961) *Viscoelastic Properties of Polymers*, Wiley, New York.

Fisher, H.C. (1954) Stress pulse in bar with neck or swell, *Appl. Scient. Res.* **A4**, 317-28.

Fisher, G.M.C. and Gurtin, M.E. (1965) Wave propagation in the linear theory of viscoelasticity, *Q. Appl. Math.* **23**, 257-63.

Fichera, G. (1972) Boundary value problems of elasticity with unilateral constraints, in: *Encyclopedia of Physics*, Vol.VI a/2: Mechanics of Solids II, Ed. C. Truesdell, Springer-Verlag, Berlin, pp. 391-423

Frederick, J.R. (1965) *Ultrasonic Engineering*, John Wiley and Sons, New York.

Frydrychowich, W. and Singh, M.C. (1986) Similarity representation of wave propagation in a nonlinear viscoelastic rod on a group theoretic basis, *Appl. Math. Modeling*. **10 (8)**, 284-93.

Gakhof, F.D. (1966) *Boundary Value Problems*, Pergamon, Oxford.

Gaul, L. (1992) Substructure behaviour of resilient support mounts for single and double stage mounting systems, *Computers & Structures*, Vol. **44**, No. **1/2**, pp. 273-78.

Gaul, L., Klein, P. and Kemple, S. (1991) Damping description involving fractional operators, *Mechanical Systems and Signal Processing* **5(2)**, 81-82.

Gaul, L., Schanz, M. and Fiedler, C. (1992) Viscoelastic formulations of BEM in time and frequency domain, *Engineering Analysis with Boundary Elements* **10**, 137-41.

Gladwwell, G.M.L. (1980) *Contact Problems in the Classical Theory of Elasticity*, Sijthoff and Noordhoff, Alphen aan den Riju.

Golden, J.M. and Graham, G.A.C. (1988) *Boundary Value Problems in Linear Viscoelasticity*, Springer-Verlag, Berlin.

Goldsmith, W., Polivka, M. and Yang, T. (1966) Dynamic behaviour of concrete, *Exp. Mech.* **23**, 65-79.

Goodier, J.N., Jahsman, W.E. and Ripperger, E.A. (1959) An experimental surface-wave method for recording force-time curves in elastic impacts, *J. Appl. Mech.* **26**, 3-7.

Gopalsamy, K. and Aggarwala, B.D. (1972) Propagation of disturbances from randomly moving sources, *ZAMM* **52**, 31-35.

Graffi, D. (1952) Sulla teoria dei materiali elastico-viscosi, *Atti Accod. Ligure Sci. Lett.* **9**, 1-10.

Graffi, D. (1982) Mathematical models and waves in linear viscoelasticity, in *Wave Propagation in Viscoelastic Media*, V52, Ed. F. Mainardi, Pitman, Boston, 1-27.

Graham, G.A.C. (1965) On the use of stress functions for solving problems in linear viscoelasticity theory that involve moving boundaries, *Proc. R. Soc.* (Edin.) **A67**, 1-8.

Graham, G.A.C. (1969) The solution of mixed boundary value problems that involve time-dependent boundary regions, for viscoelastic materials with one relaxation function, *Acta Mech.* **8**, 188-204.

Graham, G.A.C. and Golden, J.M. (1988) The generalized partial correspondence principle in linear viscoelasticity, *Q. Appl. Math.* **56(3)**, 527-38.

Graham, G.A.C. and Sabin, G.C.W. (1973) The correspondence principle of linear viscoelasticity for problems that involve time-dependent regions, *Int. J. Engg. Sci.* **11**, 123-40.

Graham, G.A.C. and Sabin, G.C.W. (1978) The opening and closing of a growing crack in a linear viscoelastic body that is subject to alternating tensile and compressive loads, *Int. J. Fracture* **14**, 639-49.

Graham, G.A.C. and Sabin, G.C.W. (1981) Steady-state solutions for a cracked standard linear viscoelastic body, *Mech. Res. Commun.* **8**, 361-68.

Graham, G.A.C. and Williams, F.M. (1972) Boundary value problems for time-dependent regions in aging viscoelasticity, *Utilitas Mathematica* **2**, 291-03.

Green, W.A. (1960) Dispersion relations for elastic waves in bars, in *Progress in Solid Mechanics*, Vol.I, Eds. I.N. Sneddon and R. Hill, Chapter 5, North Holland Publishing Co., Amsterdam.

Gurtin, M.E. and Herrara, I. (1964) A Correspondance principle for viscoelastic wave propagation, *Quart. Appl. Math.* **22**, 360-4.

Harris, C.M. and Crede, E. (1961) *Shock and Vibration Handbook*, Vols. I, II and III, McGraw-Hill, New York.

Harvey, R.B. (1975) On the deformation of a viscoelastic cylinder rolling without slipping, *Q.J. Mech. Appl. Math.* **28**, 1-24.

Hatfield, P. (1950) Propagation of low frequency ultrasonic waves in rubbers and rubber-like polymers, *Brit. J. Appl. Phys.* **1**, 252-56.

Hrusa, W.J. and Renardy, M. (1985) On wave propagation in linear viscoelasticity, *Quart. of Appl. Math.* **XLIII (2)**, July 1985, 237-53.

Hsieh, D.Y. and Kolsky, H. (1958) An experimental study of pulse propagation in elastic cylinders, *Proc. Phys. Soc.* **71**, 608-12.

Hudson, G.E. (1943) Dispersion of elastic waves in solid circular cylinders, *Phys. Rev.* **63**, 46-51.

Hunter, S.C. (1961) Tentative equations for the propagation of stress, strain and temperature fields in viscoelastic solids, *J. Mech. Phys. Solids* **9**, 39-51.

Hunter, S.C. (1967) The transient temperature distribution in a semi-infinite viscoelastic rod subject to longitudinal oscillations, *Int. J. Eng. Sci.* **5**, 119-43.

Hunter, S.C. (1968) The motion of a rigid sphere embedded in an adhering elastic or viscoelastic medium, in: *Proceedings, Edinburgh Mathematical Society* **16** (Series II), Part I, pp. 55-69.

Jeffrey, A. (1978) Nonlinear wave propagation, *ZAMM* **58**, T38-T56.

Jeffrey, A. and Taniuti, T. (1964) *Nonlinear Wave Propagation*, Academic Press, New York.

Kalker, J.J. (1975) Aspects of contact mechanics, in: *The Mechanics of the Contact Between Deformable Media*, Ed. A.D. de Pater and J.J. Kalker, Delft University Press, pp. 1-25.

Kalker, J.J. (1977) A survey of the mechanics of contact between solid bodies, *J. Appl. Math. Phys. (ZAMP)* **57**, 13-17.

Knauss, W.G. (1968) Uniaxial wave propagation in a viscoelastic material using measured material properties, *J. Appl. Mech., ASME*, Sept. 1968, 449-53.

Koeller, R.C. (1984) Application of fractional calculus to the theory of viscoelasticity, *J. Appl. Mech.* **51**, 299-307.

Kolsky, H. (1960) Experimental wave-propagation in solids, in *Structural Mechanics*, Eds. J. N. Goodier and N. Hoff, Pergamon Press, Oxford, pp. 233-62.

Kolsky, H. (1965) Experimental studies in stress wave propagation, in *Proc. V^{th} U.S. Natn. Congr. Appl. Mech.*, pp. 21-36.

Kolsky, H. and Prager, W., Eds. (1964) Stress waves in anelastic solids, *IUTAM Symposium*, Brown University, Providence, R.I., April 3-5, 1963, Springer-Verlag, Berlin, pp. 1-341.

Lamb, H. (1904) On the propagation of tremors over the surface of an elastic solid, *Phil. Trans. R. Soc.* **A203**, 1-42.

Lamb, H. (1917) On waves in an elastic plate, *Proc. Roy. Soc.* **A93**, 114-28.

Langhaar, H.L. (1962) *Energy Methods in Applied Mechanics*, John Wiley and Sons, New York.

Lee, E.H. (1966) Some recent developments in linear viscoelastic stress analysis, in *Proceedings of the 11th Congress of Applied Mechanics*, Ed. H. Gortler, Springer-Verlag, Berlin, pp. 396-402.

Lifshitz, J.M. and Kolsky, H. (1964) Some experiments on anelastic rebound, *J. Mech. Phys. Solids* **12**, 35-43.

Lifshitz, J.M. and Kolsky, H. (1965) The propagation of spherical divergent stress pulses in linear viscoelastic solids, *J. Mech. Phys. Solids* **13**, 361-76.

Lindholm, U.S. (1964) Some experiments with the split Hopkinson pressure bar, J. *Mech. Phys. Solids* **12**, 317-35.

Lindsay, R.B. (1960) *Mechanical Radiation*, McGraw-Hill, New York.

Lockett, F.J. (1961) Interpretation of mathematical solutions in viscoelasticity theory illustrated by a dynamic spherical cavity problem, *J. Mech. Phys. Solids* **9**, 215-29.

Lockett, F.J. (1962) The reflection and refraction of waves at an interface between viscoelastic materials, *J. Mech. Phys. Solids* **10**, 53-64.

Lockett, F.J. and Morland, L.W. (1967) Thermal stresses in a viscoelastic thin-walled tube with temperature-dependent properties, *Int. J. Engg. Sci.* **5**, 879-98.

Love, A.E.H. (1944) *A Treatise on the Mathematical Theory of Elasticity*, Dover Publications, New York.

Mahalanabis, R.K. and Mandal, B. (1986) Propagation of thermomagneto-viscoelastic waves in a half-space of Voigt-type material, *Indian Journal of Technology* **24**, Sept. 1986, 365-567.

Mainardi, F. and Nervosi, R. (1980) Transient-waves in finite viscoelastic rods, *Lett. Nuovo Cim* **29**, 443-7.

Mainardi, F. and Turchetti, G. (1975) Wave front expansions for transient viscoelastic waves, *Mech. Res. Comm.* **2**, 107-12.

Mainardi, F. and Turchetti, G. (1979) Positive constraints and approximation methods in linear viscoelasticity, *Lett. Nuovo Cim.* **26**, 38-40.

Margeston, J. (1971) Rolling contact of a smooth viscoelastic strip between rotating rigid cylinders, *Int. J. Mech. Sci.* **13**, 207-15.

Margeston, J. (1972) Rolling contact of a rigid cylinder over a smooth elastic or viscoelastic layer, *Acta Mech.* **13**, 1-9.

McCartney, L.N. (1978) Crack propagation in linear viscoelastic solids: Some new results, *Int. J. Fract.* **14**, 547-54.

Medick, M.A. (1961) On classical plate theory and wave propagation, *J. Appl. Mech.* **28**, 223-8.

Meyer, M.L. (1964) On spherical near fields and far fields in elastic and viscoelastic solids, *J. Mech. Phys. Solids* **12**, 77-111.

Miklowjiz, J. (1964) Pulse propagation in a viscoelastic solid with geometric dispersion, in *Stress Waves in Anelastic Solids*, Springer-Verlag, Berlin, pp. 255-76.

Morland, L.W. (1963) Dynamic stress analysis for a viscoelastic half-plane subject to moving surface traction,

Proc. London Math. Soc. **13**, 471-92.
Morland, L.W. (1968) Rolling contact between dissimilar viscoelastic cylinders, *Q. Appl. Math.* **25**, 363-76.
Morse, P. and Feshbach, H. (1953) *Methods of Theoretical Physics*, Vols. I and II, McGraw-Hill, New York.
Nachman A. and Walton, J.R. (1978) The sliding of a rigid indenter over a power law viscoelastic layer, *J. Appl. Mech.* **45**, 111-13.
Norris, J.M. (1967) Propagation of a stress pulse in a viscoelastic rod, *Experimental Mechanics* 7(7), 297-301.
Nunziato, J.W. and Walsh, E.K. (1973) Amplitude behavior of shock waves in a thermoviscoelastic solid, *Int. J. Solids Structures* **9**, pp. 1373-83.
Oliver, J. (1957) Elastic wave dispersion in a cylindrical rod by a wide-band short duration pulse technique, *J. Acoustic Soc. Am.* **29**, 189-94.
Pao, Y.H. (1955) Extension of the Hertz theory of impact to the viscoelastic case, *J. Appl. Phys.* **26**, 1083-8.
Petrof, R.C. and Gratch, S. (1964) Wave propagation in a viscoelastic material with temperature-dependent properties and thermomechanical coupling, *J. of Applied Mechanics, ASME*, Sept. 1964, 423-9.
Pindera, J.T. (1986) New research perspectives opened by isodyne and strain gradient photoelasticity, in *Proceedings of the International Symposium on Photoelasticity*, Tokyo, pp. 193-202.
Reinhardt, H.W. and Dally, J.W. (1970) Some characteristics of Rayleigh wave interaction with surface flaws, *Mater. Eval.* **28**, 213-20.
Renardy, M. (1982) Some remarks on the propagation and non-propagation of discontinuities in linearly viscoelastic liquids, *Rheologica Acta* **21**, 251-4.
Ricker, N.H. (1977) *Transient Waves in Viscoelastic Media*, Elsevier, Amsterdam.
Riley, W.F. and Dally, J.W. (1966) A photoelastic analysis of stress wave propagation in a layered model, *Geophysics* **31**, 881-89.
Ripperger, E.A. (1953) The propagation of pulses in cylindrical bars. An experimental study, *Proc. 1st Midwest Conf. Solid Mech.*, 29-39.
Rogers, T.G. (1965) Viscoelastic stress analysis, in: *Proceedings of the Princeton Univ. Conf. on Solid Mechanics*, Princeton, N.J., USA, pp. 49-74.
Rogers, C.A., Barker, D.K. and Jaeger, L.A. (1988) Introduction to smart materials and structures, in *Proceedings, Smart Materials, Structures and Mathematical Issues Workshop*, Virginia Polytech. Inst. & State University, Blacksburg, VA, Sept. 15-16, 1992, pp. 17-28.
Rubin, J.R. (1954) Propagation of longitudinal deformation waves in a prestressed rod of material exhibiting a strain-rate effect, *J. Appl. Phys.* **25**, 528-36.
Sabin, G.C.W. (1975) *Some Dynamic Mixed Boundary Value Problems in Linear Viscoelasticity*, Ph.D. Thesis, Univ. of Windsor, Windsor, Ont., Canada.
Sabin, G.C.W. (1987) The impact of a rigid axisymmetric indenter on a viscoelastic half-space, *Int. J. Eng. Sci.* **25**, 235-51.
Sackman, J.L. and Kaya, I. (1968) On the propagation of transient pulses in linearly viscoelastic media, *J. Mech. Phys. Solids* **16**, 349-56.
Schapery, R.A. (1955) A method of viscoelastic stress analysis using elastic solutions, *J. Franklin Institute* **279(4)**, 268-89.
Schapery, R.A. (1962) Approximate method of transform inversion for viscoelastic stress analysis, *Proc. 4th U.S. National Cong. Appl. Mech., ASME*, New York, pp. 1075-85.
Schapery, R.A. (1974) Viscoelastic behaviour and analysis of composite materials, *Mechanics of Composite Materials*, Vol.2, Ed. G.P. Sendeckj, Academic Press, New York, pp. 85-168.
Schapery, R.A. (1978) A method for predicting crack growth in nonhomogeneous viscoelastic media, *Int. J. Fract.* **14**, 293-309.
Schapery, R.A. (1979) On the analysis of crack initiation and growth in nonhomogenous viscoelastic media, in: *Fracture Mechanics, Proceedings of the Symposium in Applied Mathematics of the A.M.S. and S.I.A.M.*, Vol. XII, Ed. R. Burridge, American Math. Soc., Providence, pp. 137-52.
Sips, R. (1951) Propagation phenomena in elastic viscous media, *J. Polymer Sci.* **6**, 285-93.
Skalak, R. (1957) Longitudinal impact of a semi-infinite circular elastic bar, *J. Appl. Mech.* **34**, 59-64.
Sokolnikoff, I.S. (1956) *Mathematical Theory of Elasticity*, 2nd ed., McGraw Hill, New York.
Stackgold, I. (1967) *Boundary Value Problems of Mathematical Physics*, Vol.1, McMillan, New York.

Stoneley, R. (1924) Elastic waves at the surface of separation of two Solids, *Proc. R. Soc.* **A106**, 416-28.

Sultanov, K.S. (1984) Longitudinal wave propagation in a viscoelastic semi-space including an absorbing layer, *J. Appl. Mech. Tech. Phys.* **25(5)**, Sept.-Oct. 1984, 790-5.

Sutherland, H.J. and Lingle, R. (1972) An acoustic characterization of Polymethyl Methacrylate and three epoxy formulations, *J. Appl. Phys.* **43(10)**, 4022-6.

Tanagi, T. (1990) A Concept of intelligent material, *U.S.-Japan Workshop on Smart/Intelligent Materials and Systems*, Eds. C.A. Rogers, *C. Andrew and A. Masuo*, March 19-23, 1990, Honolulu, Hawaii, pp. 3-10.

Tanary, S. and Haddad, Y.M. (1988) *Characterization of Adhesively Bonded Joints Using Acousto-Ultrasonics*, Final Report prepared under contract serial number 31946-6-0012/01-ST, Department of Mechanical Engineering, University of Ottawa, Ottawa, Canada.

Tatel, H.E. (1954) Note on the nature of a seismogram II., *J. Geophys. Res.* **59**, 289-94.

Thau, S.A. and Dally, J.W. (1969). Subsurface characteristics of the Rayleigh wave, *Int. J. Eng. Sci.* **7**, 37-52.

Timoshenko, S.P. (1921) On the correction for shear of the differential equation for transverse vibrations of prismatic bars, *Phil. Mag.* **6 (41)**, 744-6.

Timoshenko, S.P. (1928) *Vibration Problems in Engineering*, Van Nostrand, New Jersey.

Ting, T.C.T. (1969) A mixed boundary value problem in viscoelasticity with time-dependent boundary regions, in: *Proceedings of the 11th Midwestern Mechanics Conf.*, Eds. H.J. Weiss, D.F. Young, W.F. Riley and T.R. Rogge, Iowa Univ. Press, pp. 591-8.

Ting, E.C. (1970) Stress analysis for a nonlinear viscoelastic cylinder with ablating inner surface, *Trans. ASME, J. Appl. Mech.* **37E**, 44-7.

Tsai, Y.M. and Kolsky, H. (1968) Surface wave propagation for linear viscoelastic solids, *J. Mech. Phys. Solids* **16**, 99-109.

Tschoegl, N.W. (1989) *The Phenomenological Theory of Linear Viscoelastic Behaviour*, Springer, Berlin.

Viktorov, I.A. (1967) *Rayleigh and Lamb Waves: Physical Theory and Applications*, Plenum Press, New York.

Walsh, E.K. (1971) The decay of stress waves in one-dimensional polymer rods, *Trans. Soc. Rheology* **15:2**, 345-53.

Watson, G.N. (1960). *A Treatise on the Theory of Bessel Functions*, Cambridge University Press, New York.

Whitham, G.B. (1974) *Linear and Nonlinear Waves*, J. Wiley & Sons, New York.

Willis, J.R. (1967) Crack propagation in viscoelastic media, *J. Mech. Phys. Solids* **15**, 229-40.

Zemanek, J. (Jr.) and Rudnick, I. (1961) Attenuation and dispersion of elastic waves in a cylindrical bar, *J. Acoust. Soc. Am.* **33**, 1283-8.

Zener, C. (1948) *Elasticity and Anelasticity of Metals*, Univ. Press, Chicago.

Zukas, J.A. (1982) Stress Waves in Solids, in *Impact Dynamics*, Eds. J.A. Zukas et al., John Wiley & Sons, New York, Chapter 1, pp. 1-27.

CHAPTER 15

TRANSITION TO THE DYNAMIC BEHAVIOUR OF STRUCTURED AND HETEROGENEOUS MATERIALS

15.1. Introduction

The current technology of the design and manufacturing of laminated and fibre-reinforced composites is faced with problems essentially related to the inherent nature of the mechanical response of the different constituents of the microstructure, the formation of interfaces between such constituents and the evolution of the associated deformation processes under loading. Optimal design of such material systems is becoming a very progressive and challenging domain in both applied mechanics and material science.

Thus, the increasing use of such materials is inciting new developments to be made within the context of macro- and micro-mechanical constitutive modelling, applications of such materials under variable boundary conditions, experimental testing methods, computational methods of analysis and optimization. A new dimension of optimal design is being realized by building new composite systems through direct tailoring of the microstructure, e. g., by judicious reinforcement and mixing (hybridization) of the constituents of the microstructure within a specific topological frame of reference and to satisfy the boundary conditions involved.

In this context, theoretical and experimental studies of the dynamic stress-strain relations of hybrid composites have become significantly important. The increased interest in the subject matter has been motivated recently by the increasing number of engineering applications and, as well, by the contributions provided by such studies to a better understanding of the mechanisms of deformation of such material systems when subjected to a dynamic loading environment. This chapter reviews some research efforts pertaining to the microdynamics of polymeric composite systems. For other classes of composite systems, the reader is referred to the bibliography cited at the end of the chapter.

15.2. Influences of Material Properties on Dynamic Behaviour

The dynamic mechanical behaviour of fibre-reinforced composite materials is governed primarily by their stiffness and damping properties. One of the goals of composite micromechanics has been to predict these macromechanical properties by using information on constituent microstructural properties and the interaction between constituent elements of the microstructure; e. g., Hashin and Rosen (1965), Hashin (1970 a&b) and Jones (1975). Several authors report values for elastic moduli as deduced from vibration tests of beams and rods; for instance, Schultz and Tsai (1968 &1969), Adams and Bacon (1973 a&b) and Paxson (1975). Gibson and Plunkett (1976) presented a critical review of the literature on dynamic properties of fibre-reinforced composites. Polymeric composite systems exhibit in general viscoelastic behaviour (*see Chapter 8*). The viscoelastic nature of glass fibre/unsaturated polyester, for instance, has been studied, e. g., by Suzuki and Miyano (1976) and that of carbon fibre/epoxy composites was investigated by Miyano *et al.* (1986), among others. Meantime, some aspects of the dynamic behaviour of different classes of polymeric composites have been investigated by, e.g., Cavaille *et al.* (1987), Chua (1987), Kodama (1976), Reed (1979) and Kimoto (1990).

Kimoto (1990) investigated the influence of the reinforcement surface treatment on the viscoelastic response characteristics of glass fibre (GF)/epoxy composites. In this, time and temperature dependence of flexural fracture properties and dynamical properties of different composites, within the mentioned class, were investigated. The epoxy resin used was a mixture of *diglycidyl ether* of *bisphenol-A* and a class of *aminopropyl* curing agent. These were used in a stoichiometric ratio of 2:1. Type E of GF woven cloths were used as reinforcement. The cloths were treated with two kinds of silane coupling reagents, namely, *γ-glycidoxypropyltrimethoxysilane* (ES) and *vinyltris* (*β-methoxyethoxy*)-silane (VS). Glass fibre woven cloths with only heat cleaning treatment (HC) were also employed in the measurement of dynamic mechanical properties. GF composite plates were prepared by a hand-lay-up method using two sheets of GF woven cloths for the purpose of the dynamic test. The volume ratio of GF content was about 25 vol%. Epoxy resin and composite plates were cured at room temperature for 24h, and then at 80°C for 3h. GF composite plates containing surface treated GF are denoted, according to the above terms, by ES-P, VS-P and HC-P. Flexural load was applied by a three-point bending method. A span of 50 mm was maintained in all measurements, and the loading bar has a fixed diameter of 10 mm. Strain rates were calculated from the cross-head speed. The flexural stress (σ) and the flexural strain (ϵ) were calculated from the maximum load and the corresponding displacement, respectively. Dynamic properties were measured using a viscoelastic spectrometer at a frequency of 10 Hz.

Figures 15.1 and 15.2, due to Kimoto (1990), show, respectively, values of flexural stress (σ) and flexural strain (ϵ) obtained for ES-P as a function of strain rate $\dot{\gamma}$ at various constant temperatures. Similar results were obtained for the epoxy matrix and the VS-P composite. As shown in Figure 15.2, the flexural strain ϵ varies with temperature and $\dot{\gamma}$ in a complex manner.

When the curves in Figure 15.1 were joined smoothly to the curve at 60°C, by applying horizontal shift on a logarithmic scale of $\dot{\gamma}$, a master curve of the flexural stress σ for ES-P was produced as shown in Figure 15.3. In a similar manner, master curves of σ for epoxy and VS-P were produced. As shown in Figure 15.3, the flexural stress σ decreases monotonously with decreasing γ (or increasing temperature), and the shapes of the master curves, for the ES-P and VS-P composites, are similar to each other. Master curves for the flexural strain ϵ were also obtained by Kimoto (1990) and are shown in Figure 15.4. The latter curves have a minimum (for epoxy resin) or a maximum value for (GF composites). With reference to Figure 15.2, the flexural strain ϵ , as mentioned earlier, varies with temperature and $\dot{\gamma}$ in a complex fashion which, in turn, is affecting the pertaining master curves of Figure 15.4.

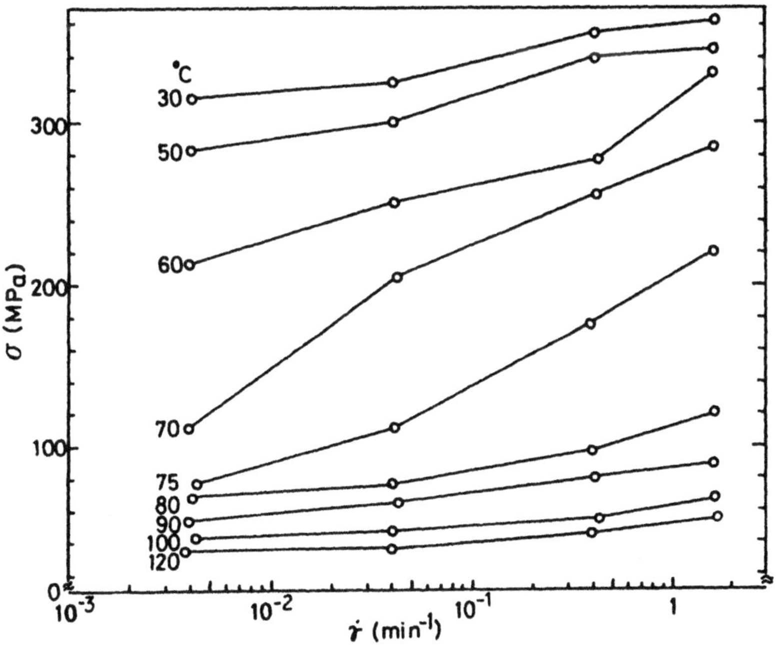

Figure 15.1. Flexural strength for ES-P (glass fibre-epoxy composite; reinforcement is surface treated with *γ- glycidoxpropyltrimethoxy silane* (ES)) as a function of strain rate at various constant temperatures. "Reprinted from *Journal of Materials Science* **25** (1990) 3327-32, Kimoto, M., Flexural properties and dynamic mechanical properties of glass fibre-epoxy composites, with kind Permission from Chapman and Hall Ltd.".

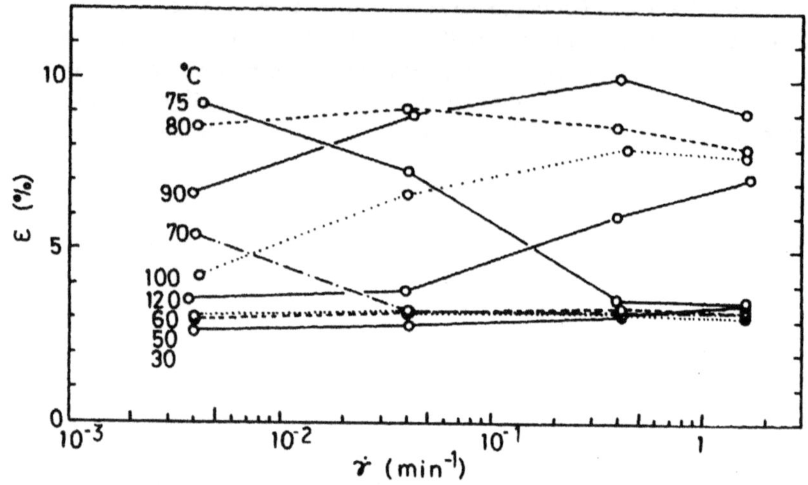

Figure 15.2. Flexural strain for ES-P (glass fibre-epoxy composite; reinforcement is surface treated with *γ- glycidoxpropyltrimethoxy* silane (ES)) as a function of strain rate at various constant temperatures. "Reprinted from *Journal of Materials Science* **25** (1990) 3327-32, Kimoto, M., Flexural properties and dynamic mechanical properties of glass fibre-epoxy composites, with kind Permission from Chapman and Hall Ltd.".

Figure 15.5 shows the temperature dependence of the storage moduli (E') obtained for the matrix epoxy, ES-P, and VS-P. This figure shows also the temperature dependence of E' for HC-P as a reference. As seen from this figure, E' values were larger for GF composites by comparison with those pertaining to the epoxy resin over the considered temperature range, and the difference in the values of the referred-to moduli is more remarkable in the rubbery region than in the glassy region. Such an increase in E' for composites in the rubbery region has been, also, observed by Souma (1982) and Lewis and Nielsen (1970). The magnitude of E' for GF composites was greater for ES-P than for VS-P over the entire temperature range.

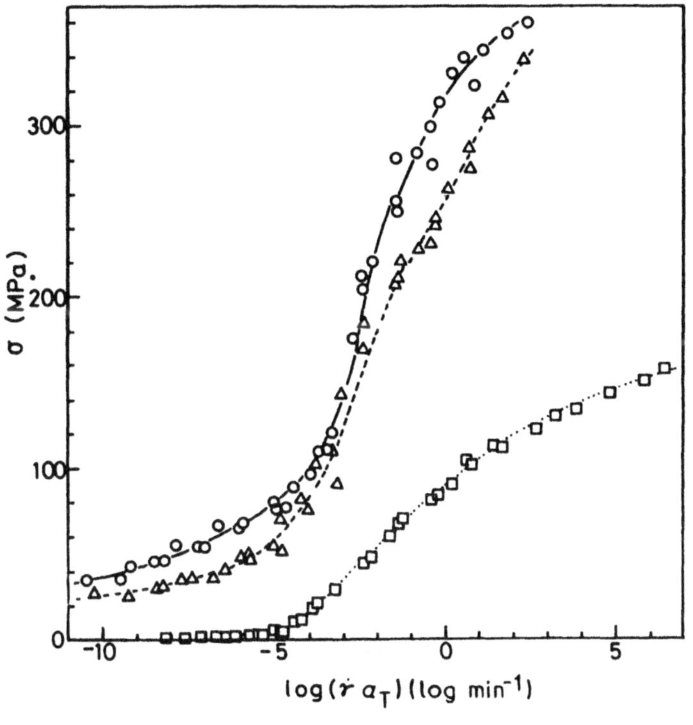

Figure 15.3. Master curves of flexural strength σ (□) Matrix epoxy, (O) ES-P, (Δ) VS-P (glass fibre-epoxy composite; reinforcement is surface treated with *β- methoxyethoxy silane* (VS)). "Reprinted from *Journal of Materials Science* **25** (1990) 3327-32, Kimoto, M., Flexural properties and dynamic mechanical properties of glass fibre-epoxy composites, with kind Permission from Chapman and Hall Ltd.".

Figure 15.6 shows the temperature dependence of the loss modulus E", for matrix epoxy, ES-P, VS-P and HC-P. As shown in this figure, temperature dependence of E" for epoxy has a peak maximum at ~ 75°C. The temperature dependence of E" of GF composites also shows a shoulder at ~75°C, and in addition, other peaks appear on the higher temperature side (*see* Kodama (1976) and Reed (1979).

Agbossou *et al.* (1993) studied a series of polystyrene/glass bead- composites by using dynamic spectrometry. The composite specimens were based on polystyrene ($\overline{M}_n$ = 99 500, $\overline{M}_w$ = 306 800) reinforced by 6%-50% volume fraction of glass beads. Two different size distributions of particles were used: The first is within the range of 1-45 μm and the second is within the range of 70-110 μm. The glass beads were dried at 100°C, but no particular treatment was performed on them. The composite was extruded at 200°C. The extruded samples, of varying volume fraction, were moulded at 200°C under high pressure (200 bar) and cooled at room temperature. In order to give the same thermal history to each sample, specimens were heated at temperatures higher than their glass temperature and then cooled to room temperature at the same cooling rate. The moulded samples were finally cut to the following dimensions: 20 mm x 4 mm x 5 mm. For dynamic analysis, frequency scans were performed, using a Viscoanalyser, by increasing the temperature from 30°C to 200°C at several frequencies over the range of 5-100 Hz. Several measurements were repeated for both frequency and temperature scans in order to verify that no physical ageing occurred in the material during the experiment.

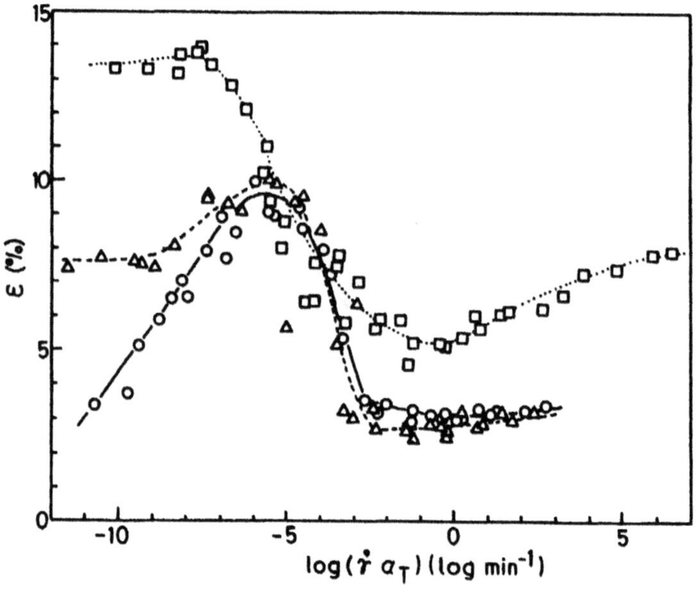

Figure 15.4. Master curves of flexural strain ε (□) Matrix epoxy, (O) ES-P, (Δ) VS-P. "Reprinted from *Journal of Materials Science* **25** (1990) 3327-32, Kimoto, M., Flexural properties and dynamic mechanical properties of glass fibre-epoxy composites, with kind Permission from Chapman and Hall Ltd.".

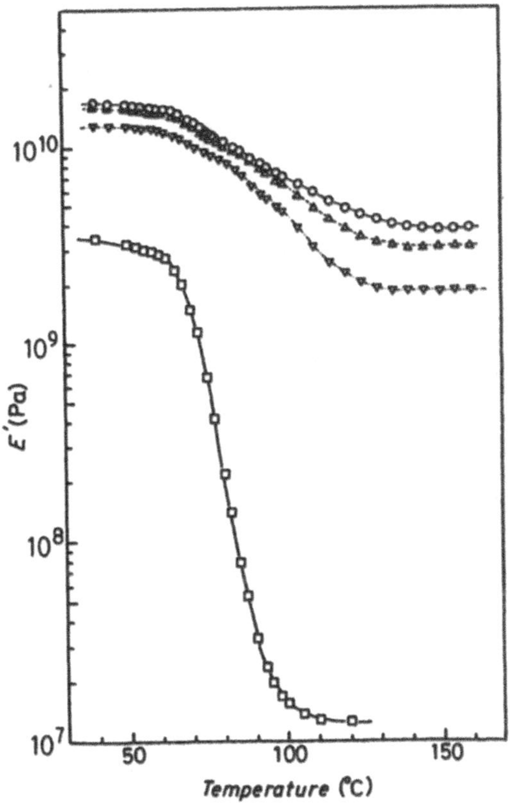

Figure 15.5. Temperature dependence of the storage modulus E': (□) Matrix epoxy, (O) ES-P, (Δ) VS-P, (∇) HC-P (glass fibre-epoxy matrix; reinforcement is with only heat cleaning treatment). "Reprinted from *Journal of Materials Science* **25** (1990) 3327-32, Kimoto, M., Flexural properties and dynamic mechanical properties of glass fibre-epoxy composites, with kind Permission from Chapman and Hall Ltd.".

Plots of log E' and the loss tangent tan δ of a polystyrene matrix at five frequencies, i.e. 5, 10.5, 22.3, 47.2 and 100 Hz, versus temperature are shown in Figure 15.7. The angle δ is of particular interest as it represents the phase angle by which the strain lags behind the stress in a viscoelastic material (e.g., Haddad, 1995). Accordingly, the loss tangent tan δ is simply the ratio between the loss modulus E" and the storage modulus E'. As demonstrated in Figure 15.7, the tan δ maxima and log E' plots, as related to the temperature, show a frequency dependence.

Figure 15.8 shows log E', log E" and tan δ spectra recorded at 5 Hz for composites reinforced by 6%, 15%, 21%, 35% and 50% volume fraction of fillers with a size distribution within the range of 70-110 μm. As illustrated in Figure 15.8, with increasing volume fraction of fillers, the magnitude of the mechanical relaxation is decreased and the tan δ maximum is shifted towards higher temperatures.

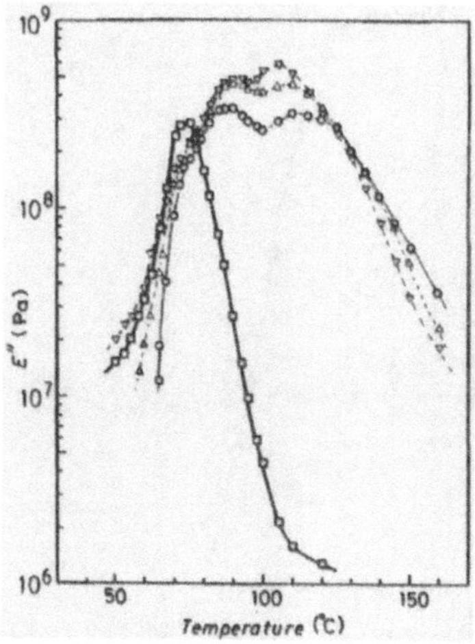

Figure 15.6. Temperature dependence of the loss modulus E" (□) Matrix epoxy, (O) ES-P, (Δ) VS-P, (∇) HC-P. "Reprinted from *Journal of Materials Science* **25** (1990) 3327-32, Kimoto, M., Flexural properties and dynamic mechanical properties of glass fibre-epoxy composites, with kind Permission from Chapman and Hall Ltd.'.

Figure 15.9, due to Agbossou *et al.* (1993), shows log E', log E" and tan δ spectra recorded at 5 Hz for composites reinforced by 50% volume fraction of glass beads with two different size distributions, i.e., within the ranges of 1-45 μm and 70-110 μm. For 50% volume fraction of glass beads, the composite reinforced with the largest glass beads shows a higher magnitude of relaxation than that exhibited by the composite reinforced by the smallest ones. Thus, for similar volume fraction of fillers, it can be observed that the reinforcement-effect increases with decreasing average size of glass beads. Then, for similar volume fraction of fillers, the specific surface of the glass beads increases as their average size decreases. Thus, it can be concluded that the interface related to the specific surface of the glass beads could influence the dynamic response behaviour of such composite materials. In this context, it may be suggested, following Agbossou *et al.* (1993) , that the interface in such composite materials could tend to decrease the molecular motion ability of the matrix and the interface contribution appears to be greater in composites reinforced with the smallest glass beads.

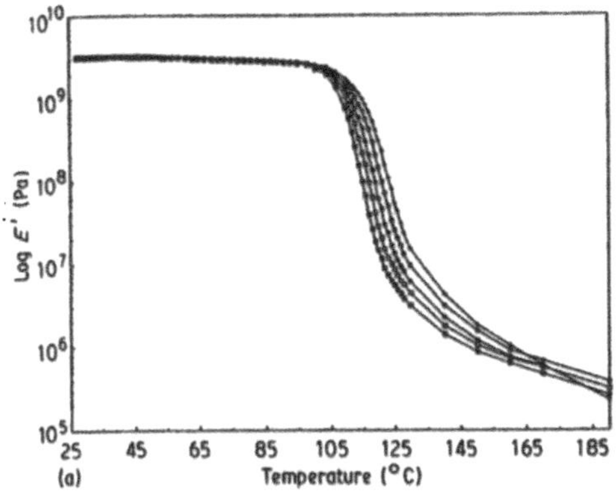

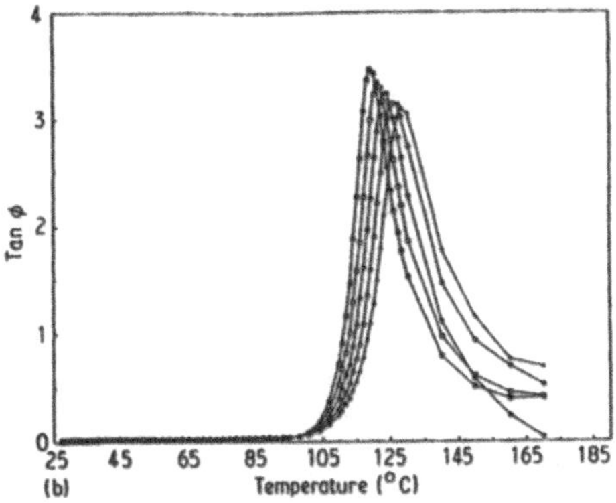

Figure 15.7. Plots of log E' and tan δ versus temperature for polystyrene matrix for various frequencies: (■) 5 Hz, (□) 10.5 Hz, (●) 22.3 Hz, (○) 47.2 Hz and (+) 100 Hz. "Reprinted from *Journal of Materials Science* **28** (1993) 1963-72, Agbossou, A., Bergeret, A., Benzarti, K. and Alberola, N., Modelling of the viscoelastic behaviour of amorphous thermoplastic/glass beads composites based on the evaluation of the complex Poisson's ratio of the polymer matrix, with kind Permission from Chapman and Hall Ltd.".

15.3. "Discontinuous" *vs.* "Continuous" Fibre-Reinforcement

In the study of the dynamic behaviour of polymeric material systems, loss modulus is as important as storage modulus, as the former measures sound and vibration damping capacity. Mclean and Read (1975) showed both experimentally and analytically that discontinuous reinforcement of a rubber-like viscoelastic matrix can produce a large increase in both moduli in the axial direction. On the contrary to the case of continuous fibre-composite systems, where the ratio of compliance to breaking strength is invariant for a given fibre material; with discontinuous fibre-systems, however, this ratio can be varied. A variable ratio of the compliance to breaking strength would give more latitude in the design of various structural and mechanical components using composite materials.

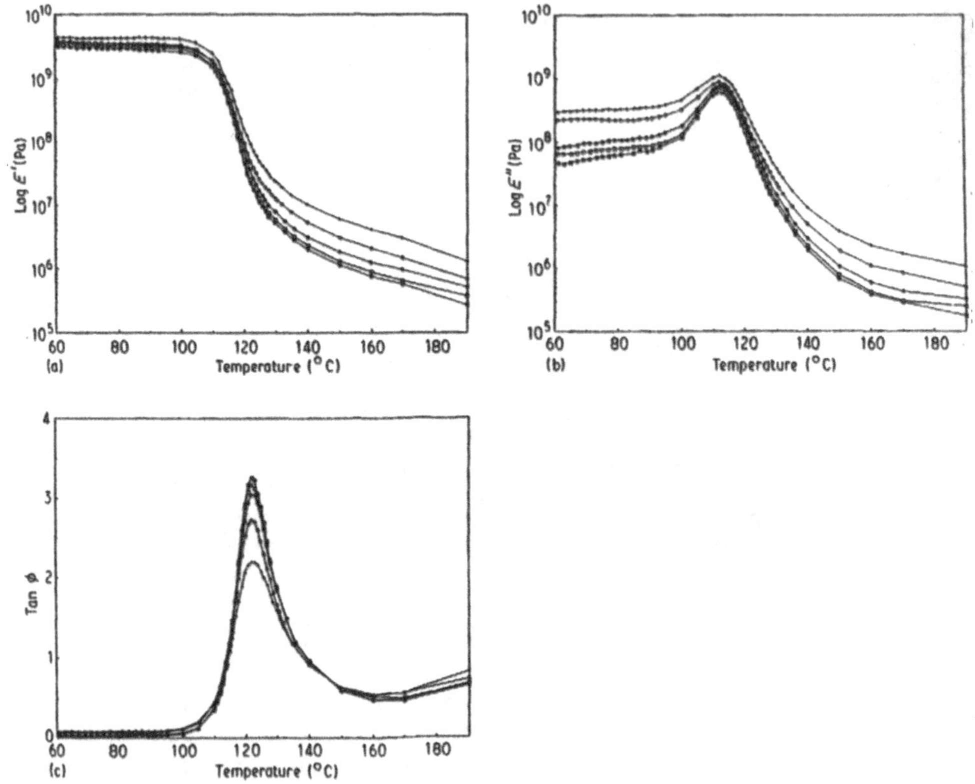

Figure 15.8. Plots of log E', log E" and tan δ versus temperature at 5 Hz for polystyrene/glass beads 70-110 μm composite reinforced by: (■) 6%, (□) 15% (●) 21%, (○) 35%, and (+) 50% volume fraction of fillers. "Reprinted from *Journal of Materials Science* **28** (1993) 1963-72, Agbossou, A., Bergeret, A., Benzarti, K. and Alberola, N., Modelling of the viscoelastic behaviour of amorphous thermoplastic/glass beads composites based on the evaluation of the complex Poisson's ratio of the polymer matrix, with kind Permission from Chapman and Hall Ltd.'.

Based on a strain energy model, McLean and Read (1975) assumed that the sum of the strain energy in the matrix and that in the fibres would give the magnitude of the strain energy in the composite, i.e.,

$$W_c = W_m + W_f \quad \textbf{(15.1)}$$

and

$$W_c = \frac{1}{2}\sigma\left(\varepsilon_m + \varepsilon_f\right) \tag{15.2}$$

where σ and ϵ indicate, respectively, the longitudinal stress and strain.

From their model, McLean and Read arrived at the following expression of the composite longitudinal modulus E_c in terms of that of the matrix E_m and of the fibre E_f.

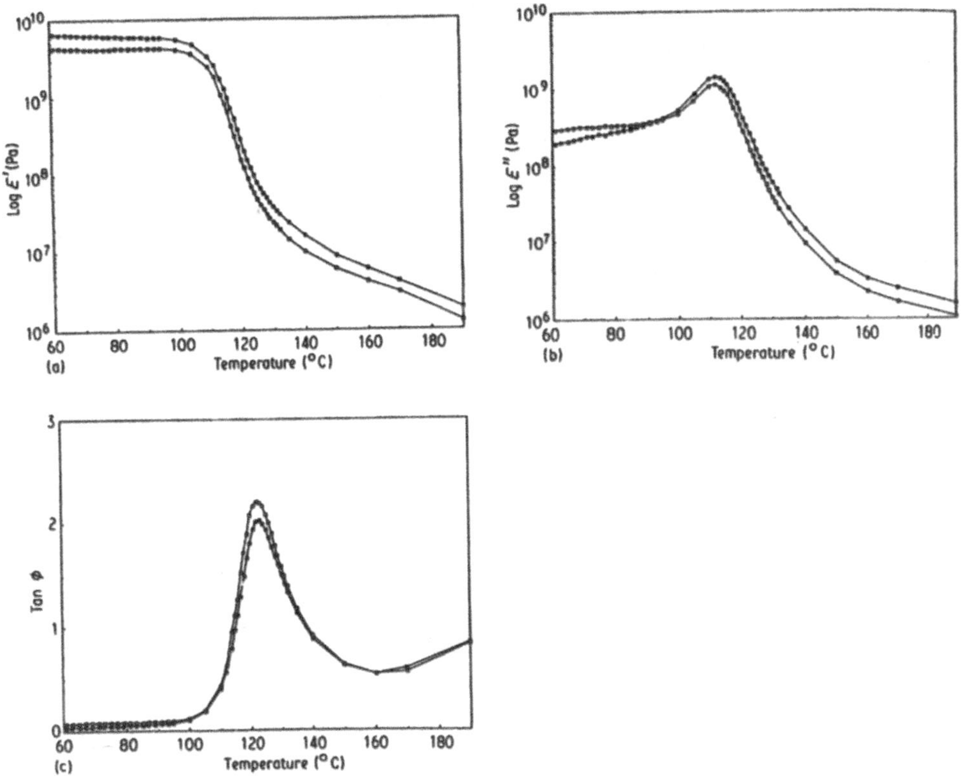

Figure 15.9. Plots of log E', log E" and tan δ versus temperature at 5 Hz for: (■) polystyrene/glass beads 1-45 μm composites reinforced by 50% volume fraction of fillers, and for *(□) polystyrene/glass* beads 70-100 μm composites reinforced by 50% volume fraction of fillers. "Reprinted from *Journal of Materials Science* **28** (1993) 1963-72, Agbossou, A., Bergeret, A., Benzarti, K. and Alberola, N., Modelling of the viscoelastic behaviour of amorphous thermoplastic/glass beads composites based on the evaluation of the complex Poisson's ratio of the polymer matrix, with kind Permission from Chapman and Hall Ltd.".

$$\frac{1}{E_c} = \frac{16}{E_m V_m (\ell/s)^2} + \frac{8}{15 E_f V_f} \tag{15.3}$$

in which V_m and V_f indicate, respectively, the volume fraction of the matrix and the fibre, ℓ is the fibre length and s is the transverse spacing between the fibres.

Using Eqn. (15.3), the composite modulus E_c is plotted versus the fibre volume fraction V_f in Fig. 15.10 as the solid line for the polymer-carbon fibre composite: $E_m = 0.03$ GN m^{-2}, $E_f = 400$ GN m^{-2}, and the value of ℓ/s assuming a square array of fibres on a transverse section is calculated as

$$\ell/s = \frac{2r V_f}{\sqrt{(\pi V_f)} - 2 V_f} \tag{15.4}$$

where the fibre aspect ratio r is considered to be equal to 112. Equation (15.3), however, does not apply when $V_f \to 0$ as a term pertaining to the effect of an unreinforced matrix has not been included.

In Fig. 15.10, the lower curve shows the calculated influence of the fibre volume fraction on the longitudinal composite modulus E_c in the case of discontinuous reinforcement. The upper curve, however, relates to continuous reinforcement when E_f is much greater than E_m.

In the mentioned paper, the composite loss modulus E_c' is defined as

$$\frac{E_c''}{E_c} = \frac{\text{energy dissipated per cycle}}{\text{peak energy stored per cycle}} = \frac{h W_m}{W_c} \tag{15.5}$$

in which the symbol h expresses some appropriate function.

Equation (10.5) was further expressed in terms of the constituent parameters as

$$\frac{E_c''}{E_m''} = \frac{V_m (\ell/4s)^2}{\left[1 + \frac{8 E_m V_m (\ell/4s)^2}{15 E_f V_f}\right]^2} \tag{15.6}$$

Equation (15.6) also follows from the elastic-viscoelastic correspondence principle which enabled the authors to replace E_c and E_m in Eqn. (15.3) by complex moduli $E_c^* = E_c^{\prime} + i\,E_c^{\prime\prime}$ and $E_m^* = E_m^{\prime} + i\,E_m^{\prime\prime}$. The curve of $E_{fl}^{\prime\prime}$ versus V_f given by Eqn. (15.6) is drawn in Fig. 15.11 using an experimental value for $E_m^{\prime\prime}$ of 0.041 GNm^{-2}.

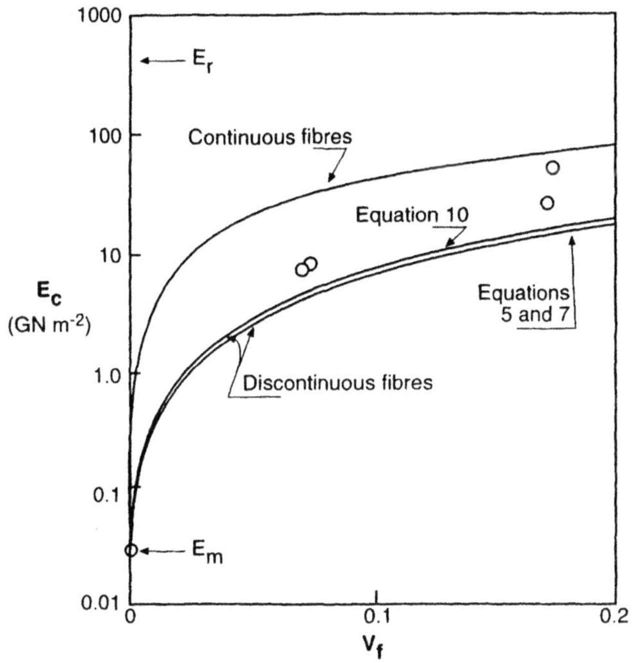

Figure 15.10. The lower curves show the calculated influence of V_f on longitudinal composite modulus E_c in the case of discontinuous reinforcement. The upper curve relates to continuous reinforcement when $E_f > E_m$. The circles are experimental measurements. Carbon fibres in soft polymer. "Reprinted from *Journal of Materials Science* **10** (1975) 481-92, Mclean, D. and Read, B. E., Storage and loss moduli in discontinuous composites, with kind Permission from Chapman and Hall Ltd.". The equation numbers shown in the figure pertain to the original paper.

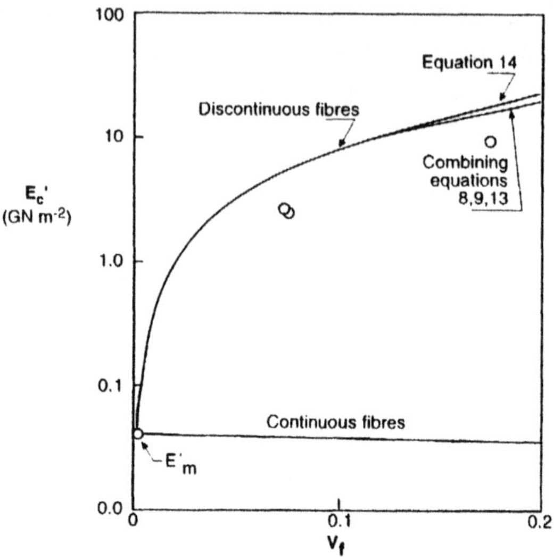

Figure 15.11. Influence of V_f on longitudinal loss modulus E_c''. The two upper lines are theoretical values for discontinuous fibres, and the lower line is the theoretical result for continuous fibres when $E_f'' << E_m''$. The circles are experimental measurements. Carbon fibres in soft polymer. "Reprinted from *Journal of Materials Science* **10** (1975) 481-92, Mclean, D. and Read, B. E., Storage and loss moduli in discontinuous composites, with kind Permission from Chapman and Hall Ltd.". The equation numbers shown in the figure pertain to the original paper.

15.3.1. DESIGN FLEXIBILITY

In some industrial applications, rubber is reinforced with continuous steel wires or with continuous polymer fibres. The reinforcement is added to raise the longitudinal tensile strength. At the same time, it significantly reduces the longitudinal compliance, thus, casting away one of the favourable merits of rubber. Moreover, little variation is possible in the ratio of longitudinal modulus to longitudinal strength. To show that with discontinuous reinforcement much greater latitude in the ratio is possible, McLean and Read (1975) used, as presented below, the equations for compliance and tensile strength.

Continuous Reinforcement

With continuous reinforcement, the compliance C_{cc} is expressed by

$$C_{cc} = \frac{1}{E_{cc}} = \frac{1}{E_f V_f} \tag{15.7}$$

where the subscript *cc* refers to continuously reinforced composite and the elastic modulus of the matrix has been neglected, which, in view of the authors, is a reasonable approximation for the case, for instance, of steel fibres in rubber. The breaking strength, of the composite, σ_b is calculated by

$$\sigma_b = \sigma_{fb} V_f \tag{15.8}$$

where σ_{fb} is the fibre breaking strength and, again, the matrix contribution has been neglected.

Combining equations (15.7) and (15.8), the ratio of compliance/strength, for a composite with continuous reinforcement, is

$$\frac{C_{cc}}{\sigma_b} = \frac{C_f}{\sigma_{fb} V_f^2} \tag{15.9}$$

in which C_f is the fibre compliance. The ratio (15.9) is invariant with any given matrix and fibre for a given composite ultimate strength since the fibre-volume fraction V_f is then fixed.

Discontinuous Reinforcement

In this case, Mclean and Read (1975) calculated the composite ultimate strength by

$$\sigma_b = \frac{2}{3} \sigma_{fb} V_f \tag{15.10a}$$

where it is assumed that one third of the applied load is carried by the matrix. This is under the condition that $\sigma_b < 1/3\ \sigma_{mb} V_m$, σ_{mb} being the breaking strength of the matrix. In the case of rubber reinforced with steel, for instance, σ_{fb} is approximately equal to 10 σ_{mb} so that $V_f \leq 0.05$. A second condition is that the stress transfer between the matrix and the fibre must be adequate. When these two conditions are met, C_{cd} / σ_b is obtained from equations (15.3) and (15.10), and it can be varied considerably by altering *l/s*, i.e. by varying the aspect ratio *r*.

$$C_{cd} = \frac{1}{E_{cd}} \approx \frac{1}{E_f V_f}$$

(15.10b)

Here, C_{cd} is the compliance for the composite with discontinuous reinforcement. A measure of the extra latitude offered by discontinuous reinforcement is given by the range which C_{cd} / C_{cc} can take for a given breaking strength σ_b.

From equations (15.9) and (15.10), it follows that

$$\frac{C_{cd}}{C_{cc}} = \frac{2 E_f V_f}{3 E_m V_m (\ell / 4 s)^2} + \frac{16}{4 s} \qquad (15.11)$$

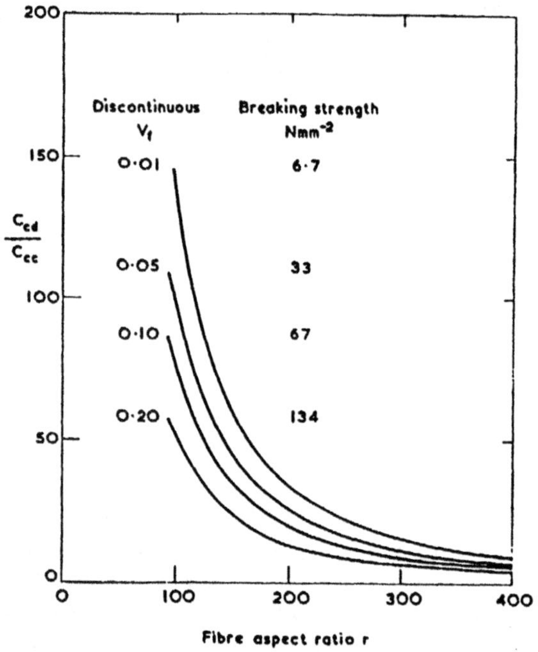

Figure 15.12. Showing the extra design flexibility offered by discontinuous versus continuous reinforcement. C_{cd} is the compliance of rubber reinforced with steel wires of aspect ratio *r*. C_{cc} is the compliance of rubber reinforced with continuous steel wires to have the same breaking strength. The ratio C_{cc}/C_{cd} can be very large. "Reprinted from *Journal of Materials Science* **10** (1975) 481-92, Mclean, D. and Read, B. E., Storage and loss moduli in discontinuous composites, with kind Permission from Chapman and Hall Ltd.".

The value of C_{cd} / C_{cc} for different values of r and V_f has been calculated by McLean and Read (1975) with $E_f = 2 x 10^5$ and $E_m = 1 \, N mm^{-2}$, which are considered to be appropriate values for steel and rubber, respectively. The results are shown on Fig 15.12 due to McLean and Read (1975). The tensile strength indicated on Fig. 15.12 is derived from Eqn. (15.10) assuming $\sigma_{fb} = 1000$ Nmm^{-2}. Fig. 15.12 shows, e.g., that for a sample breaking strength of 33 N mm^{-2}, the compliance can be increased 100-fold if continuous wire is replaced by a discontinuous one. Fig.15.12, also, illustrates the range of variation of C_{cd} for a given tensile strength. The ordinate axis shows the wide range of C_{cd} available by varying the aspect ratio r.

Dynamic mechanical measurements over a range of temperature provide valuable insight into the structure, morphology and viscoelastic behaviour of polymeric materials. These measurements form an important part of the approach for establishing relaxation transitions. Akay (1993) cautioned, however, that care must be taken in the interpretation of dynamic mechanical analysis (DMA) spectra of complex material systems such as advanced polymer composites, particularly when determining the glass-transition temperature T_g. The latter is commonly defined as the temperature corresponding to the maximum value of the loss tangent (*tan* δ). The glass transition temperature T_g is also sometimes defined as the temperature corresponding to the maximum value of the loss modulus E'', or, alternatively, as the temperature of the maximum change in real modulus E . DMA data of continuous carbon-fibre (CF) reinforced epoxy laminates have been studied by Akay (1993) and represented here in Figures 15.13 and 15.14. These data indicate that glass transition, in the two considered materials, occurred over different temperature intervals depending on the mode of testing. As shown in the indicated figures, a sharp transition of low intensity is indicated in the longitudinal mode where the properties are fibre dominated. The fibres are stiffer and carry, in this mode, more load than the matrix and, thus, the observable properties are not appreciably sensitive to variations in the mechanical properties of the matrix. In the transverse mode, however, both the fibres and the matrix experience the applied stress and, by consequence, the properties of the composite show much greater sensitivity to variations in the properties of the matrix. The data clearly show that glass transitions occurred over a wider temperature range when the tests were conducted in the transverse mode. This was attributed by the authors to a combination of the following reasons:

(i) Non-uniform compliance in the matrix due to various fibre-resin interactions.

(ii) Non-uniformity of the temperature distributions within the 90° specimens as compared withe the 0° specimens.

With reference to Figures 15.13 and 15.14, different values of T_g, depending on its definition, can be obtained and this can be misleading:

- T_g expressed as the temperature corresponding to the maximum value of *tan* δ produced a higher value in the transverse mode compared with the value obtained in the longitudinal mode by as much as 25° C without any fundamental reason other than the broadening of the transition region.

- Alternatively, the glass transition temperature T_g as corresponding to the maximum value of E'' revealed increases from 2°C to 14°C in the longitudinal mode compared to the transverse mode. This is more realistic since in the longitudinal mode most of the load is carried by the fibres and thus the matrix experiences a delayed transition. Further, the definition of T_g by the maximum value of E''also indicates more precisely the temperature at which stiffness (as expressed by E') suffers significant deterioration.

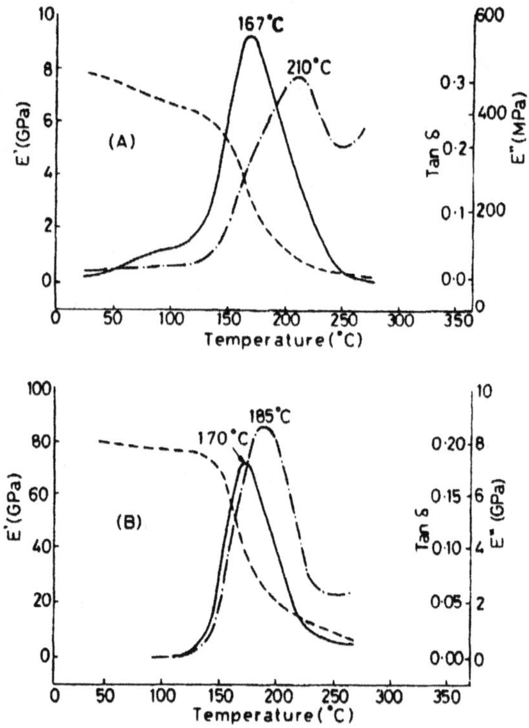

Figure 15.13. Dynamic mechanical properties for CF/epoxy-a at (A) transverse, (B) longitudinal modes of testing. "Reprinted from *Composite Science and Technology* 47, Akay, M., Aspects of dynamic mechanical analysis in polymeric composites, 419-23, 1993, with kind permission from Elsevier Science Ltd, The Boulevard, Langford Lane, Kidlington 0X5 1GB, UK".

Thus, in the class of composite materials considered, T_g , as defined by the temperature corresponding to E''-maximum value, is considered to be a more consistent and an appropriate index than the one based on the maximum value of the tangent modulus *tan δ* (Akay, 1993).

Gerard, Perret and Chabert (1990) studied the dynamic mechanical properties of unidirectional carbon fibre/epoxy matrix composites in order to determine the influence of the presence of carbon fibre and the effect of surface treatment (untreated, oxidized and sized) on the behaviour of epoxy matrix. The study shows that when carbon fibres are introduced at different volume fractions (from 40% to 70%), the mobility of the macromolecular chains of the epoxy matrix is reduced at the fibre/matrix interface. An oxidization treatment, leading to a larger number of functional groups present at the interface, increases this effect by creating additional interactions. However, an epoxy sizing induces a higher mobility by creating less crosslinked interphases. By studying the viscoelastic properties in a large range

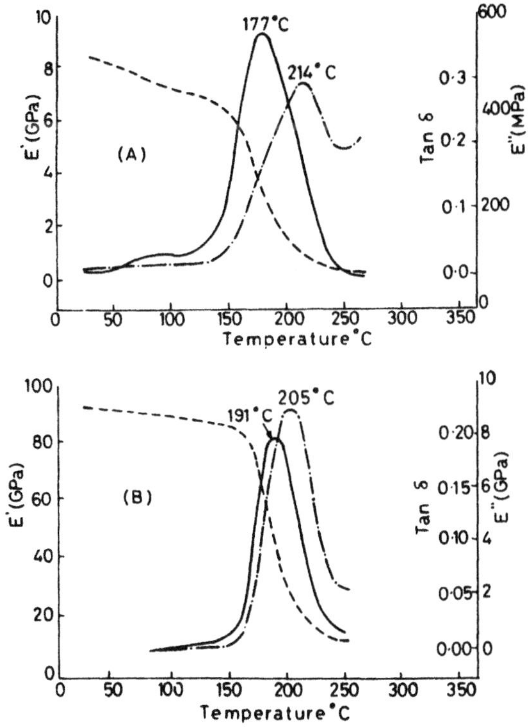

Figure 15.14. Dynamic mechanical properties for CF/epoxy-b at (A) transverse, (B) longitudinal modes of testing. "Reprinted from *Composite Science and Technology* **47**, Akay, M., Aspects of dynamic mechanical analysis in polymeric composites, 419-23, 1993, with kind permission from Elsevier Science Ltd, The Boulevard, Langford Lane, Kidlington 0X5 1GB, UK".

of temperatures and frequencies, the main relaxation of a component of the sizing, displaying a phase separation in the interphase, could be observed as a shoulder of the secondary relaxation peak of the epoxy matrix. In this context, dynamic mechanical measurements were performed by the authors with a polymer Lab. DMTA apparatus in a temperature range from -120 to 250° C and a frequency range from 0.033 to 10 Hz. As shown in Figure 15.15, the viscoelastic spectra of the epoxy matrix in unidirectional composites display three relaxations:

i) *A major relaxation α*. It is associated with the glass transition of the DGEBA-MDA matrix and occurs at a high temperature of about 180° C.
ii) *A secondary relaxation β*: It occurs at a low temperature (near - 65°C at 0.1 Hz). It is associated with motions of small parts of the macromolecular chains (hydroxyether groups and diphanylpropane units)
iii) *A third relaxation γ*: It occurs at about 50 °C and corresponds to unreacted molecular segments and/or crosslink inhomogeneities

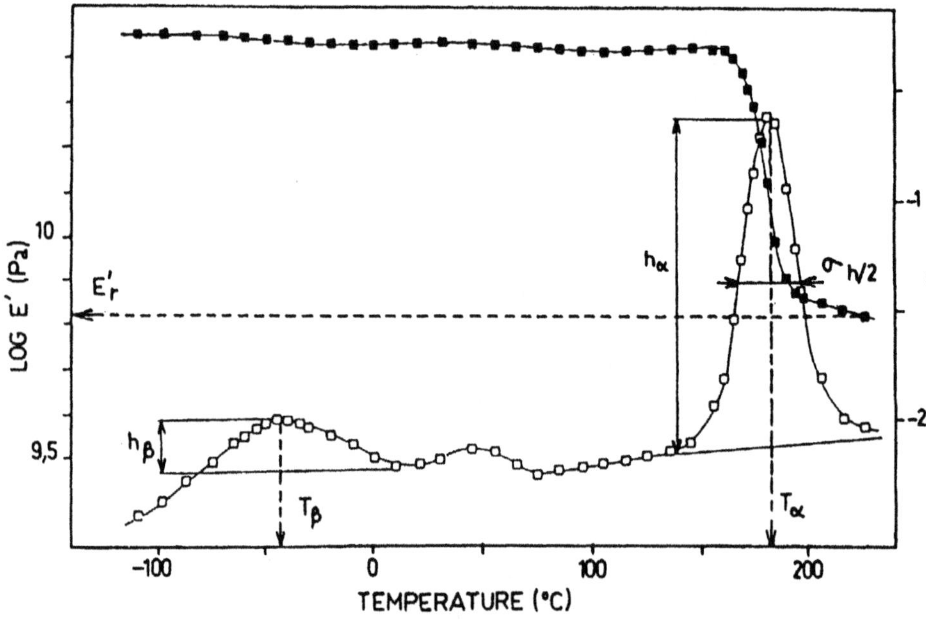

Figure 15.15. Dynamic mechanical spectrum of unidirectional composite material based on DGEBA-MDA matrix and AS4 (oxidized carbon fibre), 51% vol. fraction, at 1 Hz. "Reprinted from Controlled Interfaces in Composite Materials, Proc. Third International Conference on Composite Interfaces, H. Ishida (ed.), May 21-24, 1990, Cleveland, Ohio, pp. 449-55, Gerard, J. F., Perret, P. and Chabert, B., Study of carbon/epoxy (or interface): Effect of surface treatment of carbon fibres on the dynamic mechanical behaviour of carbon/epoxy unidirectional composites, with kind Permission from Elsevier Science Ltd., The Boulevard, Langford Lane, Kidlington OX5 1GB, UK".

The major relaxation, α, of the epoxy matrix in the unidirectional composites was found to be greatly influenced by the presence of carbon fibres, Fig. 15.16. The increase of T_α with fiber volume fraction was attributed by the authors to the restriction of the mobility of the macromolecular chains at the interface due to interactions with the fibre surface. For the β relaxation, Fig. 15.17, an additional peak could be observed as a shoulder of the β peak at high temperature and for low frequency measurements. In this, the authors postulate that the carbon surface restricts the mobility of macromelocular chains by creating interactions. Surface treatments such as oxidization which introduce additional functional groups on the fibre surface, induce an added rigidity at the interfacial zones. However, epoxy sizing leads to a larger mobilty by creating less crosslinked interphases.

Gibson and Plunkett (1976) considered the determination, both analytically and experimentally, the elastic stiffness and internal damping of E-glass fiber-reinforced epoxy beams under flexural vibration. A mathematical model for predicting the effective complex moduli of unidirectional and (0/90) crossply laminated glass-epoxy beams to flexural vibration was presented. The model, which is an extension of previous work by Hashin and Rosen (1965), employs the complex moduli of the matrix and of the fibres, together with geometric information pertaining to the microstructure, to predict the effective complex moduli of the various components. Upper and lower bounds on the moduli, due to Hashin (1965), were used to show the effects of fibre packing geometry. Comparison of measured and predicted values of the complex moduli with predicted bounds on the moduli shows that, for small vibration amplitudes, the predicted values are reasonably accurate. In this context, it was concluded by Gibson and Plunkett (1976) that damping and stiffness are independent of the vibration amplitude as long as the maximum strain value does not exceed the threshold strain for material damage. However, once this threshold strain is exceeded, permanent changes occur in damping and stiffness. The resulting increase in damping is much more significant than the corresponding reduction in stiffness.

In the work of Gibson and Plunkett (1976), the composite material selected for the forced vibration experiments was 3M Scotchply 1002, an E-glass reinforced epoxy, in both unidirectional and (0/90) crossply configurations. Specimens of pure epoxy, transverse unidirectional, longitudinal unidirectional, and (0/90) crossply material were tested in order to find the pertaining storage and loss moduli. Specimens of double-cantilever type were machined from pre-cured 51 ply thick panels with each ply having a nominal thickness of 0.25mm (0.01"). Epoxy matrix specimens, of the same double-cantilever type, were cut from a 3.3 mm (0.13") thick sheet of cured epoxy.

Specimens, after being clamped, were excited in a steady-state, resonant flexural vibration mode by an electromagnetic shaker while measurements of specimen resonant frequency, base acceleration, and bending strain were made. Electrical resistance strain gages, bonded to the surface of the specimen, were used to measure bending strain, while the base acceleration was measured by a piezoelectric accelerometer mounted on the specimen support clamp. An electronic frequency counter was used to measure the frequency of the accelerometer signal.

Storage and loss moduli found for each series of resonant dwell tests are shown in 10.18 to 10.22. Static storage moduli determined by tensile tests (for the matrix) or 3-point flexture tests (for the composites) are also presented for comparison with the presented dynamic values. Figures 15.19 to 10.22 show the predicted moduli and predicted bounds on the composite moduli.

As shown in Fig. 15.18, the matrix loss modulus, E_m'', increased by nearly a factor of two over the range of test frequencies, while the storage modulus, E_m', increased by only 9%. Fig. 15.19 shows very good agreement between measured and predicted values of the transverse modulus.

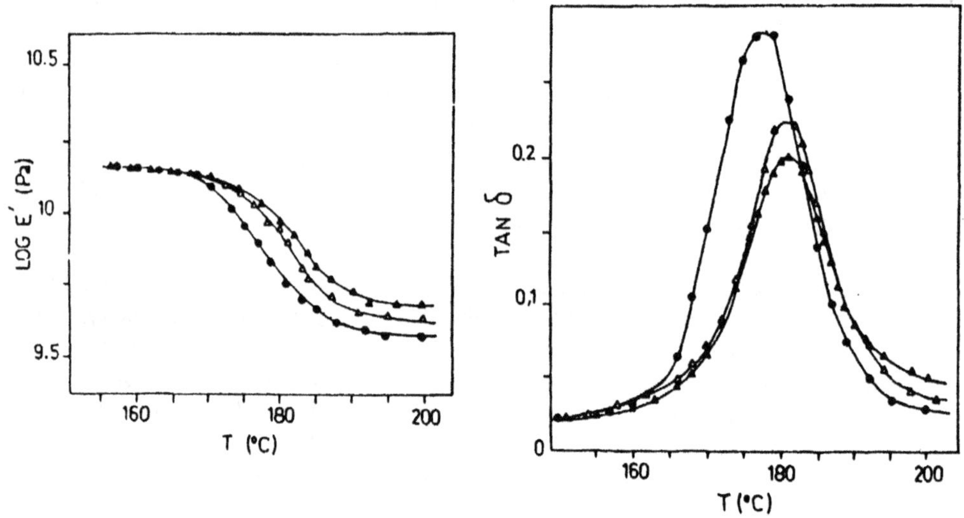

Figure 15.16. Dynamic mechanical spectra (at 0.1 Hz) for unidirectional composites based on different volume fractions of carbon fibres: (●) 60%, (Δ) 64.4%, (Δ) 68 %. "Reprinted from Controlled Interfaces in Composite Materials, Proc. Third International Conference on Composite Interfaces, H. Ishida (ed.), May 21-24, 1990, Cleveland, Ohio, pp. 449-55, Gerard, J. F., Perret, P. and Chabert, B., Study of carbon/epoxy (or interface): Effect of surface treatment of carbon fibres on the dynamic mechanical behaviour of carbon/epoxy unidirectional composites, with kind Permission from Elsevier Science Ltd., The Boulevard, Langford Lane, Kidlington OX5 1GB, UK".

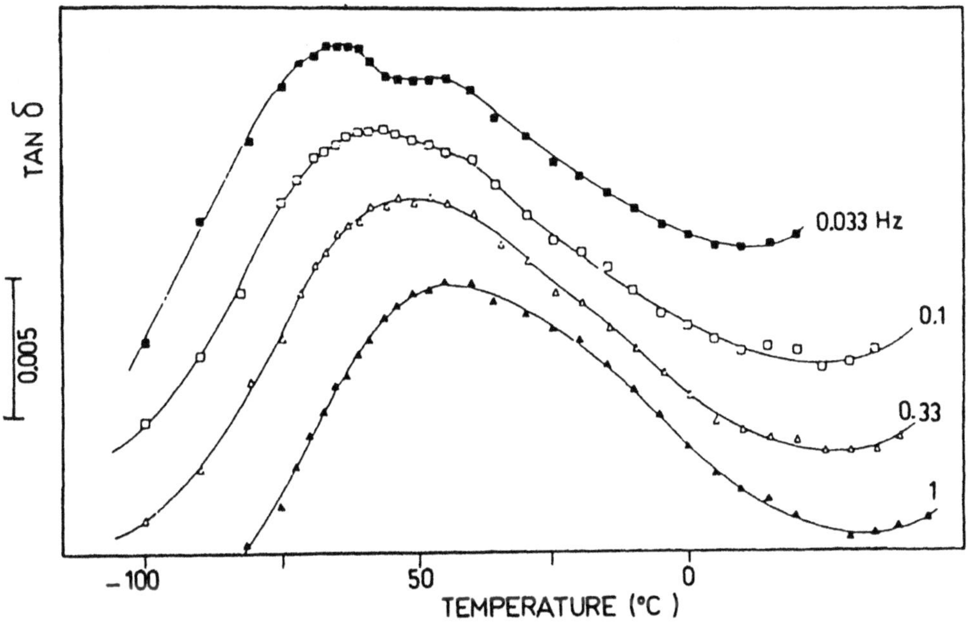

Figure 15.17. β relaxation (↓) of the epoxy network in UD carbon/epoxy composite as a function of a frequency of measurement, (48% vol. Fraction of T300 carbon fibres) (additional relaxation (↓) due to the sizing) (a vertical shifting has been used to clarify the figure). "Reprinted from Controlled Interfaces in Composite Materials, Proc. Third International Conference on Composite Interfaces, H. Ishida (ed.) , May 21-24, 1990, Cleveland, Ohio, pp. 449-55, Gerard, J. F., Perret, P. and Chabert, B., Study of carbon/ epoxy (or interface): Effect of surface treatment of carbon fibres on the dynamic mechanical behaviour of carbon/epoxy unidirectional composites, with kind Permission from Elsevier Science Ltd., The Boulevard, Langford Lane, Kidlington OX5 1GB, UK".

Comparison of the magnitudes of E_T'' for the transverse ply (Figure 10.19) with those of E_L' for the longitudinal ply (Figure 15.20) indicates that, in the crossply laminate, most of the dissipation occurs in the transverse plies. Since the matrix is far more dissipative than the fibers are, one would expect that the configuration in which the matrix is subjected to the greatest strain (the transverse ply) would be more dissipative than the configuration in which the matrix carries the least strain (the longitudinal ply). As shown in Fig. 15.20, the value of E_L' determined by static flexure shows good agreement with measured and predicted dynamic values. Although, as shown in Fig. 15.20, the relative differences between predicted and experimentally determined values of E_L'' are significant, the corresponding relative differences are not as great for the 0/90° crossply laminate, as shown in Figure 15.21. In the latter case, the experimentally determined values of E_C'' are slightly greater than the predicted values, but are generally below the upper bound. The differences are attributed by the Gibson and Plunkett (1976) to the longitudinal plies, as the resulting calculated values of the crossply loss modulus are much closer to the experimentally determined values than are the calculated values based on the properties of the fibers and the matrix (Figure 15.22).

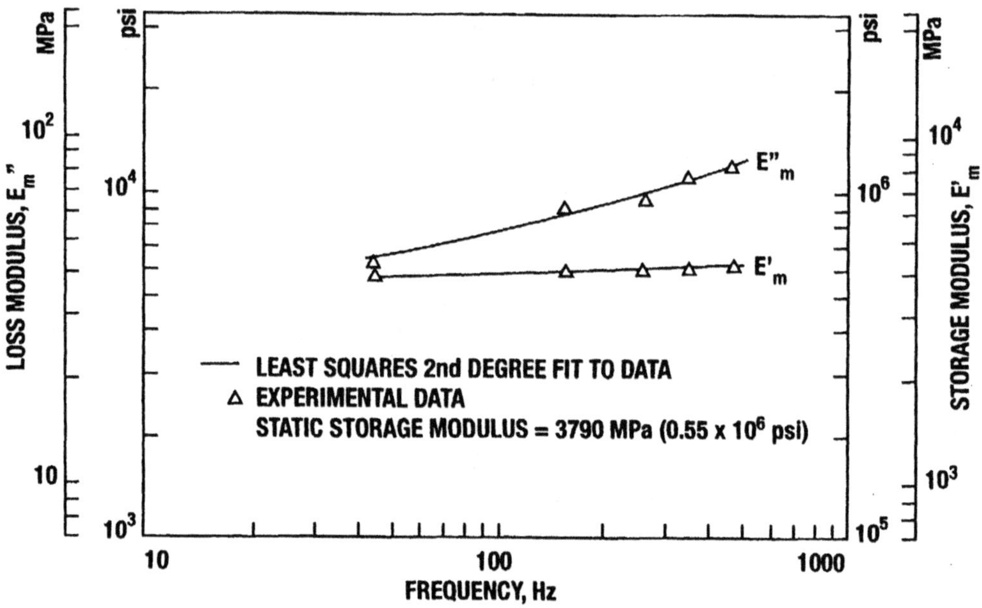

Figure 15.18. Matrix storage and loss moduli versus frequency. "Reprinted from *J. Composite Materials* **10**, Gibson, R. F. and Plunkett, R., Dynamic mechanical behaviour of fibre-reinforced composites: Measurement and analysis, 325-41, 1976, with kind permission from Technomic Publishing Co. Inc., Lancaster, PA".

Among other conclusions arrived at by Gibson and Plunkett (1976), one may emphasise the followings:

1. Measured and predicted storage moduli of the glass-epoxy composites tested are practically independent of vibration frequency over the nominal range from static to 500 Hz.
2. Measured and predicted loss moduli of the composites all increased with increasing frequency in the range from static to 500 Hz. Frequency dependence of the composite loss moduli is governed by the viscoelastic behaviour of the epoxy matrix.

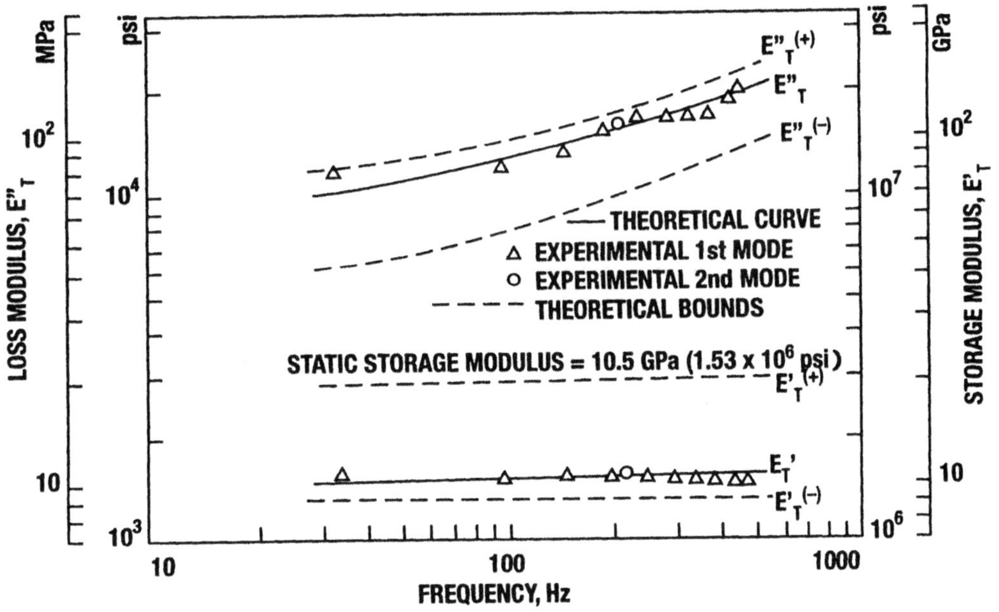

Figure 15.19. Storage and loss moduli versus frequency for transverse ply material. "Reprinted from *J. Composite Materials* **10**, Gibson, R. F. and Plunkett, R., Dynamic mechanical behaviour of fibre-reinforced composites: Measurement and analysis, 325-41, 1976, with kind permission from Technomic Publishing Co. Inc., Lancaster, PA".

3. The predicted and measured values of the complex transverse and longitudinal extensional moduli are within the predicted bounds on the transverse moduli; actual storage moduli are close to the lower bound on the storage modulus, while actual loss moduli are close to the upper bound on the loss modulus.
4. Damping and stiffness of the crossply laminate tested in flexure are independent of amplitude as long as the maximum strain value does not exceed the magnitude of the fracture strain of the transverse plies, otherwise, the material damage causes increased damping and decreased stiffness.
5. Damping is far more sensitive to microstructural damage in the composite than is stiffness.

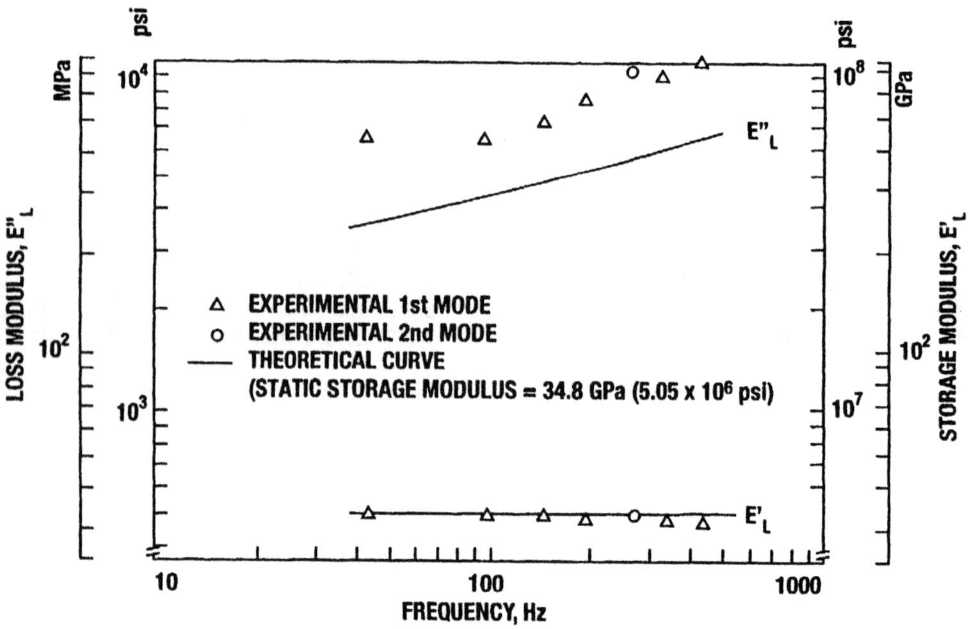

Figure 15.20. Storage and loss moduli versus frequency for longitudinal ply material. "Reprinted from *J. Composite Materials* **10**, Gibson, R. F. and Plunkett, R., Dynamic mechanical behaviour of fibre-reinforced composites: Measurement and analysis, 325-41, 1976, with kind permission from Technomic Publishing Co. Inc., Lancaster, PA".

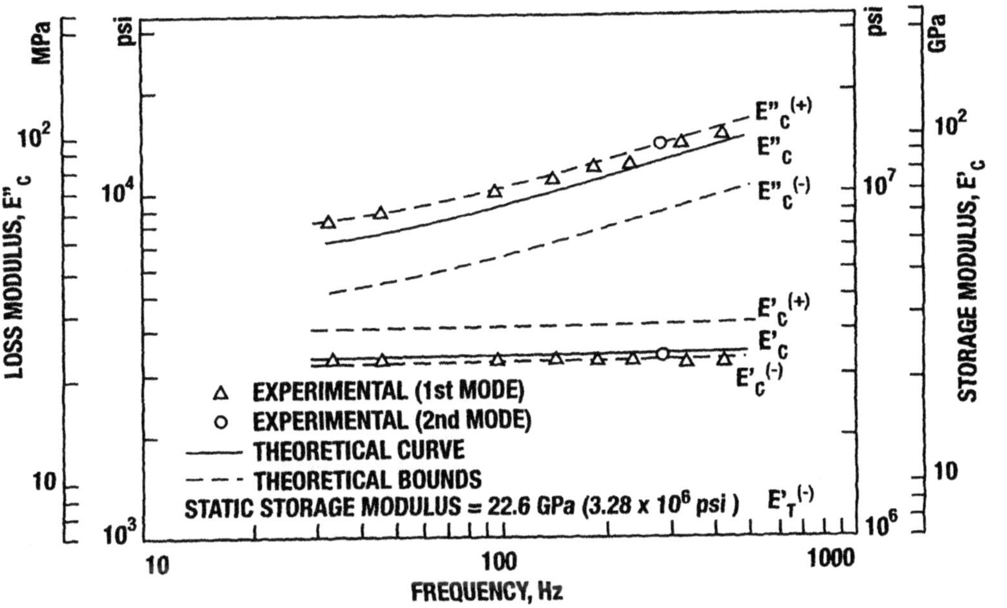

Figure 15.21. Storage and loss moduli versus frequency for 0/90 crossply laminate. Theoretical values found from micromechanical analysis of plies. "Reprinted from *J. Composite Materials* **10**, Gibson, R. F. and Plunkett, R., Dynamic mechanical behaviour of fibre-reinforced composites: Measurement and analysis, 325-41, 1976, with kind permission from Technomic Publishing Co. Inc., Lancaster, PA".

15.4. Sheet Molding Compounds (SMC)

Recent activity in composite materials research in the automobile industry has led to the development of inexpensive choped fibre reinforced plastics, often referred to as sheet molding compounds (SMC). These materials are generally made up of 25.4 mm (1 in) long glass fibres randomly dispersed in a polyster resin matrix. Static mechanical properties of SMC are reasonably well characterized (e. g., Helmbuch and Sanders, 1978 and Jutte, 1978), but scarce results have been reported on dynamic response properties of such material.

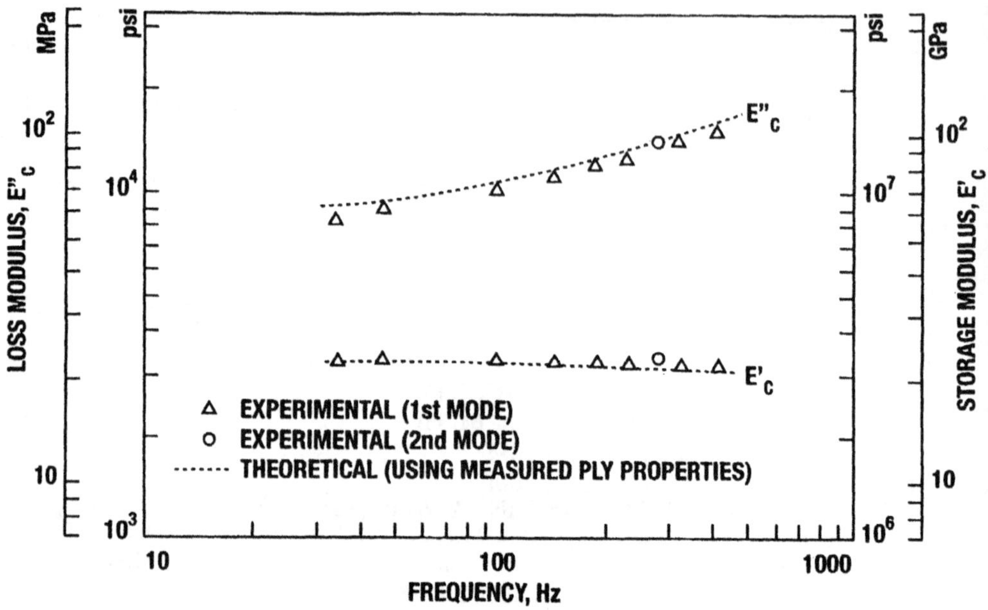

Figure 15.22. Storage and loss moduli versus frequency for 0/90 crossply laminate. Theoretical values found from measured ply properties show better agreement with measured crossply values than do those shown in Fig. 10.21. "Reprinted from *J. Composite Materials* **10**, Gibson, R. F. and Plunkett, R., Dynamic mechanical behaviour of fibre-reinforced composites: Measurement and analysis, 325-41, 1976, with kind permission from Technomic Publishing Co. Inc., Lancaster, PA".

Gibson and Yau (1980) presented measurements of complex moduli of SMC-25 (25% by weight random E-glass fibers in a polyster matrix), of SMC-R65 (65% by weight random E-glass fibers in a polyster matrix), and of XM-3 (25% by weight random E-glass fibers, 50% by weight continuous E-glass fibers @ ± 7.5° in a polyster matrix). Comparison of the obtained results were compared with previously obtained data on continuous fiber composites. The latter are unidirectional and crossply configurations of 3M Scotchply (approximately 50% by volume continuous aligned E-glass fibers in an epoxy matrix). In the referred-to paper, Gibson and Yau (1980) have shown that the use of complex modulus to describe the small-amplitude dynamic behaviour of chopped fiber and of continuous fiber reinforced plastics is appropriate since both stiffness and damping are independent of amplitude.

The approach taken by Gibson and Yau (1980) is to see if the complex moduli fall within the so-called "*elementary bounds*". For this purpose, bounds on the elastic moduli of composite materials have been derived using a variety of methods. Paul (1960), for instance, used the principles of minimum potential energy and minimum complementary energy to show that the bounds on the Young's modulus of a macroscopically isotropic, two-phase composite with an arbitrary phase geometry are

$$\frac{1}{\frac{V_f}{E_f} + \frac{V_m}{E_m}} \le E_c \le E_f\, V_f + E_m\, V_m \tag{15.12}$$

where the upper bound is theoretically valid only if the two Poisson's ratios, i.e., of the fibre and matrix are equal. On the other hand, Hill (1964) showed that, when the transverse strain mismatch due to differences in Poisson's ratios is accounted for, the upper bound may be taken as

$$E_c \le E_f V_f + E_m V_m + 0(\nu_f - \nu_m)^2 \tag{15.13}$$

which reduces to the same equation (15.12) when the Poisson's ratios of the fibre and matrix are equal.

Due to the random orientation of fibers in SMC, the properties should be *nearly* isotropic, and the assumptions leading to (15.12) may be applicable. Wolf and Carne (1979), for instance, showed that the anisotropy of SMC stiffness is less than 10%. In a transversely isotropic continuous fibre composite, these bounds may be applied to the transverse modulus (i.e., perpendicular to the fibres).

To establish the bounds on the complex viscoelastic moduli, Gibson and Yau (1980) used the Correspondence Principle, following Hashin (1970 a&b), and replaced the elastic moduli in (15.12) with the corresponding complex moduli. The resulting bounds on the storage modulus of the composite are the same as those in (15.12). That is.

$$E_c^{'(+)} = \mathrm{Re}\left[E_f^* V_f + E_m^* V_m \right] \tag{15.14}$$

and

$$E_c^{'(-)} = \mathrm{Re}\left[\frac{1}{\frac{V_f}{E_f^*} + \frac{V_m}{E_m^*}} \right] \tag{15.15}$$

Meantime, the corresponding "*bounds*" on the loss moduli of the composite are

$$E_c''^{(+)} = \mathrm{Im}\left[\frac{1}{\dfrac{V_f}{E_f^*} + \dfrac{V_m}{E_m^*}}\right] \tag{15.16}$$

$$E_c''^{(-)} = \mathrm{Im}\,[\,E_f^*\,V_f + E_m^*\,V_m\,] \tag{15.17}$$

as it can be shown that

$$\mathrm{Im}\left[\frac{1}{\dfrac{V_f}{E_f^*} + \dfrac{V_m}{E_m^*}}\right] > \mathrm{Im}\,[\,E_f^*\,V_f + E_m^*\,C_m\,] \tag{15.18}$$

The use of equations (15.14) - (15.17) requires knowledge of the complex moduli and volume fractions of the fibres and the matrix resin. As a first approximation, the fibres are assumed to be elastic , i.e. $E_f'' = 0$; since data on glass fiber damping might not be easily accessible.

The measured complex moduli and the estimated bounds on the moduli of the SMC materials are shown in Figures 15.23 to 15.25. The validity of the data here is supported by the overlap between the data for different modes of vibration and between specimens. Static modulus data provide an additional check on the storage modulus. As shown in the figures, while the measured storage moduli generally fall within the bounds, the loss moduli are significantly greater than the upper bounds for SMC-R25 and SMC-R65. The measured loss moduli for XMC-3 fall on both sides of the upper bound, but most of the data fall above the bound. As mentioned earlier, the bounds are based on the assumption that no dissipation occurs in glass fibres. Comparing the results of Fig. 15.23 with those of Fig. 15.24, one sees that the relative separation between measured loss moduli and the corresponding upper bound is greater for SMC-R65 than for SMC-R25. Since the SMC-R65 composite has more fibres per unit volume than the SMC-R25, it follows that the total matrix-fibre interfacial area of SMC-R65 would be greater than that of the SMC-R25. Thus, any dissipation mechanism operating at the interface would have more effect on the damping of SMC-R65 than on that of SMC-R25. Figure 10.25 shows that the storage modulus for the XMC-3 longitudinal specimens fall near the upper bounds, while the storage modulus for the XMC-transverse specimen falls near the lower bound. The loss moduli for both specimens fall near the upper bound, however, it is also likely that additional dissipation occurs by virtue of fibre discontinuity and resulting shear stress concentrations at the fibre ends.

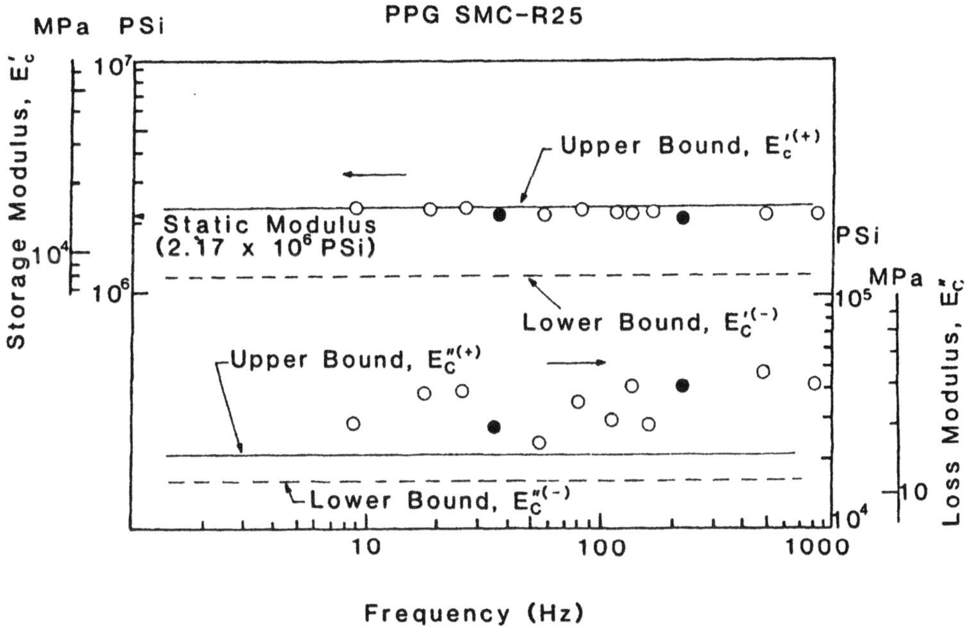

Figure 15.23. Measured complex moduli and estimated bounds for SMC-R25, ___ Upper Bound, - - - Lower Bound, o Specimen A, • Specimen B. "Reprinted from *J. Composite Materials* 14, Gibson, R. F. and Yau, A., Complex moduli of chopped and continuous fibre-composites: Comparison of measurements with estimated bounds, 155-67, 1980, with kind permission from Technomic Publishing Co. Inc., Lancaster, PA".

Figures 15.26 and 15.27 present the corresponding data for the case of a continuous fibre-composite 3M Scotchply. There is less scatter in the loss modulus data here than in Figures 15.23 to 15.25, because only first and second modes were used (the data in Figures 15.23 to 15.25 were obtained by testing up through the fifth mode, and it was postulated by the authors that the proximity of strain gages to nodal points in the higher modes produced more errors in the data). The storage and loss moduli generally fall within the bounds for all three configurations (longitudinal, transverse, and cross ply).

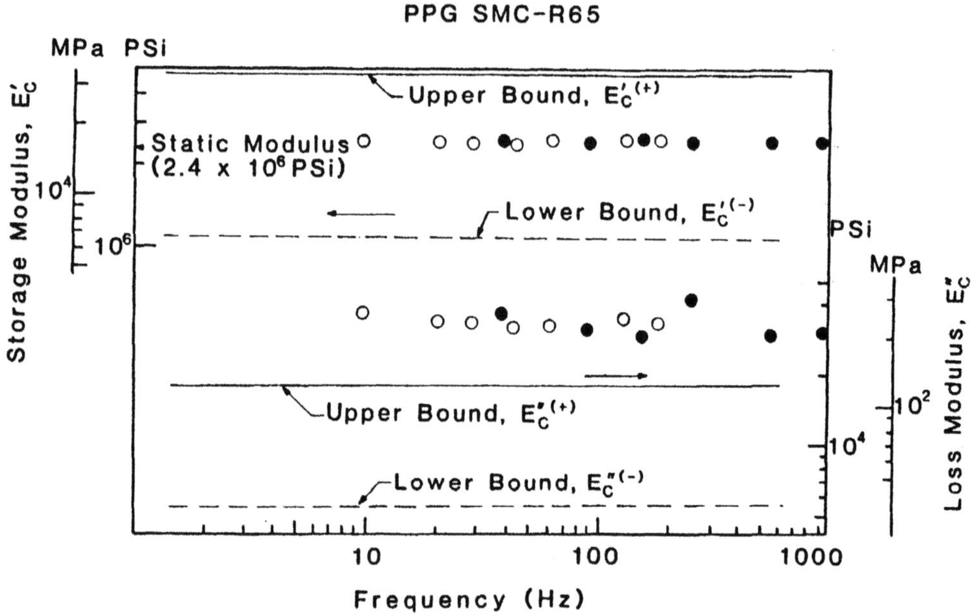

Figure 15.24. Measured complex moduli and estimated bounds for SMC-R65, ___ Upper Bound, - - - Lower Bound, o Specimen A, • Specimen B. "Reprinted from *J. Composite Materials* **14**, Gibson, R. F. and Yau, A., Complex moduli of chopped and continuous fibre-composites: Comparison of measurements with estimated bounds, 155-67, 1980, with kind permission from Technomic Publishing Co. Inc., Lancaster, PA".

Along with the continuous competing requirements for improving the weight, interdisciplinary performance, and reliability of composite components, the development of real time non-destructive "health-monitoring" techniques based on the global dynamic characteristics of the composite structure is receiving growing attention (e.g., Lee *et al.*, 1987, Tracy and Pardoen, 1989, Grady and Meyn, 1989, Raju *et al.*, 1992). In this realm, one approach is concerned with developing the capability to detect delamination by monitoring changes in the dynamic response characteristics.

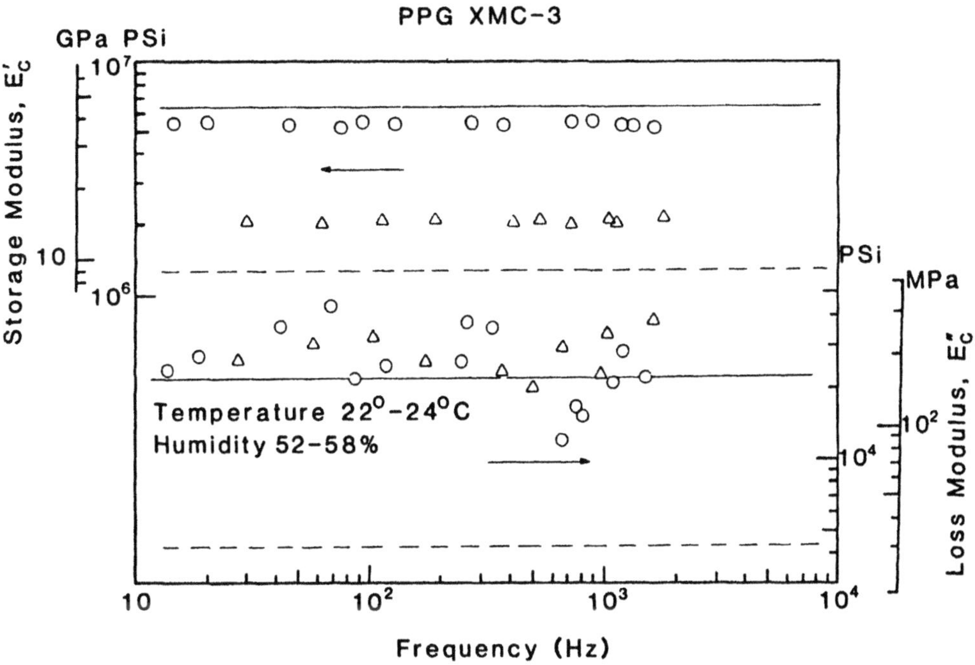

Figure 15.25. Measured complex moduli and estimated bounds for XMC-3, ___ Upper Bound, - - - Lower Bound, o Measured Longitudinal, Δ Measured Transverse. "Reprinted from *J. Composite Materials* 14, Gibson, R. F. and Yau, A., Complex moduli of chopped and continuous fibre-composites: Comparison of measurements with estimated bounds, 155-67, 1980, with kind permission from Technomic Publishing Co. Inc., Lancaster, PA".

Although significant work has been reported in the general area of delamination prediction and growth, limited research efforts have been reported on structural dynamic characteristics. Tracy and Pardoen (1989), and Hanagud *et al.* (1992), for instance, have considered the effects of a single delamination on the natural frequencies and modes by applying the classical beam theory on a delaminated beam of four longitudinal distinct regions. Saravanos (1993) considered the development of a discrete laminate mechanical approach for predicting the effects of delamination on the dynamic characteristics of composite laminates including damping. An analytical procedure was introduced for the prediction of natural frequencies, modes and modal damping in composite beams with an interlaminar delamination. The predicted effects of delamination vary based on crack size, laminate configuration, and mode order. The results also indicate that delamination effects could be more profound in angle-ply laminates due to resultant changes in the extension-flexure and flexure-twisting stiffness/damping coupling in the delaminated sections.

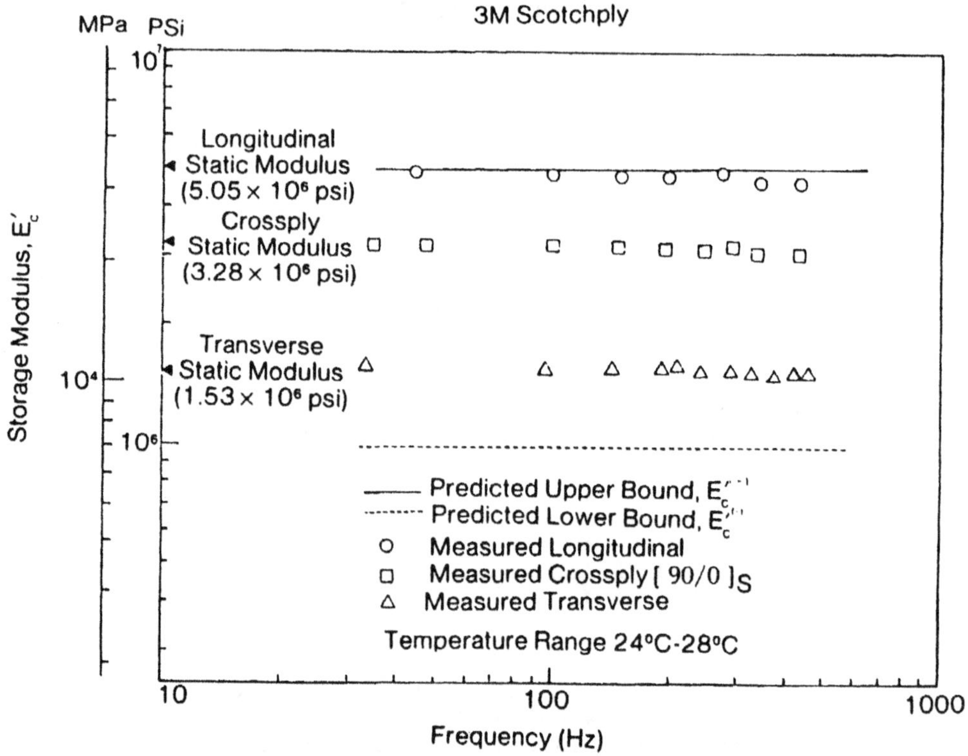

Figure 15.26. Measured storage moduli and estimated bounds for *3M* Scotchply. "Reprinted from *J. Composite Materials* **14**, Gibson, R. F. and Yau, A., Complex moduli of chopped and continuous fibre-composites: Comparison of measurements with estimated bounds, 155-67, 1980, with kind permission from Technomic Publishing Co. Inc., Lancaster, PA".

The developed method was applied to predict the dynamic characteristics of cantilever composite beams with a single delamination. Predicted natural frequencies were correlated with reported measured data of a $[0/90]_{2s}$ T300/934 graphite/epoxy cantilever beam and a $[90/45/-45/0]_{2s}$ simply-supported AS4/3501-6 graphite/epoxy beam. In all experimental results, the delamination was artificially induced during the lay-up of the composite using a teflon tape. Subsequently the effects of a central delamination on the first three modal frequencies, damping and shapes of cantilever beams were investigated. The composite material in this case was either 0.60 FVR T300/934 epoxy or 0.50 FVR HM-S Graphite/Epoxy. Three types of beam configurations were considered with ply thickness of 0.127 mm (0.005 in) each: cross-ply $[0/90]_{2s}$, [0/90/45/-45], and $[45/-45/90/0]_s$ laminates. All beams were assumed to have a delamination at their mid-plane. The delamination was also assumed to be symmetrically located about the centre of the beam (50% span).

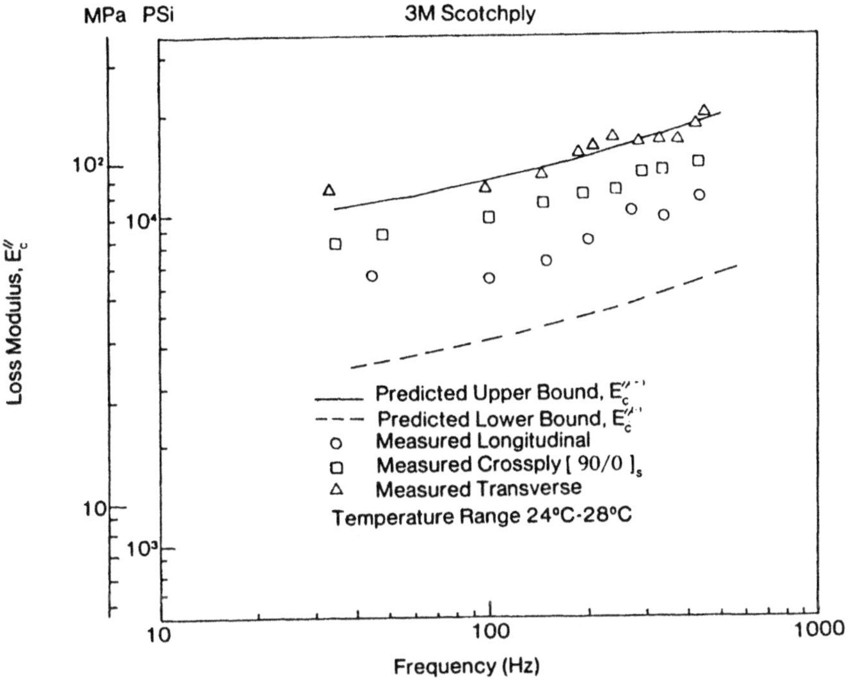

Figure 15.27. Measured loss moduli and estimated bounds for *3M* Scotchply. "Reprinted from *J. Composite Materials* **14**, Gibson, R. F. and Yau, A., Complex moduli of chopped and continuous fibre-composites: Comparison of measurements with estimated bounds, 155-67, 1980, with kind permission from Technomic Publishing Co. Inc., Lancaster, PA".

$[0/90]_{2s}$ Beam:

The predicted effects of delamination crack length on the modal characteristics of the $[0/90]_{2s}$ beam are shown in Fig. 15.28. The effects of crack length on the first three bending mode shapes of the beam are shown in Fig. 15.29. Local opening modes were also observed for large delamination lengths, but they are not presented herein (Saravanos, 1993). As a general trend, the delamination reduced the natural frequency and increased the modal damping of the structure, even with friction effects being neglected. The effects of delamination were more obvious in the characteristics and shapes of the higher modes. The delamination did not change significantly the modal frequency, damping, and shape of the fundamental mode. Small delaminations (less than 20%) produced little change in most modal characteristics. The damping of the third mode appears to be a promising "early" damage indicator for this type of laminate (Saravanos, 1993). As presented in the same article (Saravanos, 1993), the presence of a central delamination drastically reduces the flexural rigidity of the delaminated portion, but on the other hand is increasing both strain energy and laminate damping in the delaminated sub-laminates. This explains the reductions in modal frequencies and the increase in modal damping.

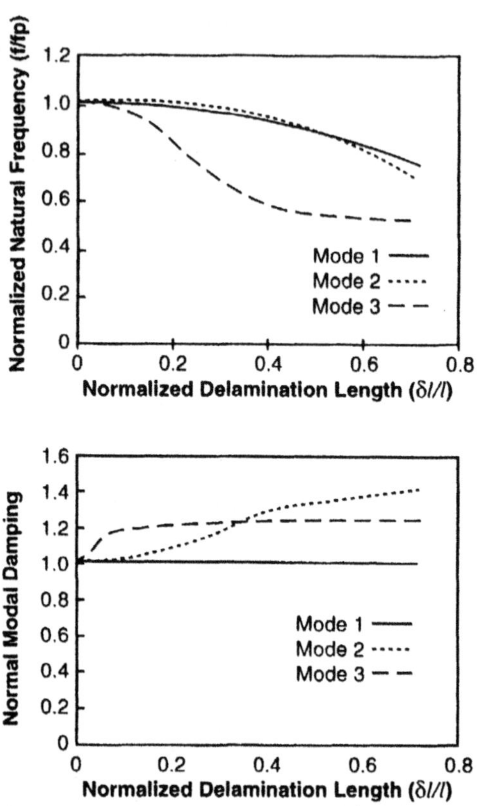

Figure 15.28. Effect of delamination size on modal frequencies and damping of the $[0/90]_{2s}$ beam. "Reprinted from NCA-Vol. 16/AMD-Vol. 172, Saravanos, D. A., Mechanics for the effects of delamination on the dynamic characteristics of composite laminates, ASME 1993, pp. 11-21, with kind permission from ASME International, New York, NY".

$[0/90/45/\text{-}45]_s$ Beam:

A more general laminate configuration was investigated by Saravanos (1993) in this case. The modal characteristics of the beam are shown in Table 15.1 (pristine beam) and in Fig. 15.30 for various crack length. The effects of crack length on the first three mode shapes of the beam are shown in Fig. 15.31. Most of the trends described in the previous case are also observed here, although they are more profound than in the previous case. The delamination effects are more observable in the higher modes and the characteristics of the fundamental mode provide minimal damage indication.

TABLE 15.1. Modal frequencies and damping of pristine cantilever beams (Saravanos, 1993)

Lamination	*[0/90]$_{2s}$*	*[0/90/45/-45]$_s$*	*[45/-45/90/0]$_s$*
Modal SDCs, %			
Mode 1	0.95	0.572	1.964
Mode 2	0.95	0.572	1.964
Mode 3	0.95	0.572	1.964
Natural Frequencies, Hz			
Mode 1	78.6	87.3	39.8
Mode 2	492.6	547.2	249.5
Mode 3	1379.2	1532.3	698.4

$[45/-45/90/0]_s$ Beam:
The natural frequencies and the corresponding modal damping and mode shapes of the beam are shown in Table 15.1 and in Figures 15.32 and 15.33, respectively for various crack sizes. Contrary to the former cases, the effect of delamination on the mode shapes and natural frequencies is modest. The damping is more sensitive at the presence of the delamination, but depending on the mode order, the damping may either increase or decrease. The effects of various thermal treatments on the dynamic properties of PEEK (Polyether ether ketone) and carbon fibre reinforced PEEK (APC2) have been studied by Folkes, Kalay and Ankara (1993). PEEK is one of a new generation of engineering polymers having good high temperature properties. As such, it has received much attention as a likely contender in replacing more traditional thermosetting resins. Continuous carbon fibre reinforced PEEK has also been developed and can offer favourable physical properties. This type of composite is referred to as "aromatic polymer composite (APC2). In addition, these materials have comparatively short processing cycles by virtue of their thermoplastic nature.

A Rheometrics RSA2 Solids Analyzer was used by Folkes *et al.* (1993) to record the dynamic mechanical spectra of the various specimens. Unclamped, three-point bend testing at a frequency of 62.8 rad s^{-1} was used. The specimen dimensions were approximately 1mm x 6.5 mm x 48 mm and 3.25 mm x 6.5 mm x 48 mm for APC2 and PEEK samples, respectively. A strain of 0.01 was applied to the samples, and the heating rate was 2.5°C min^{-1}.

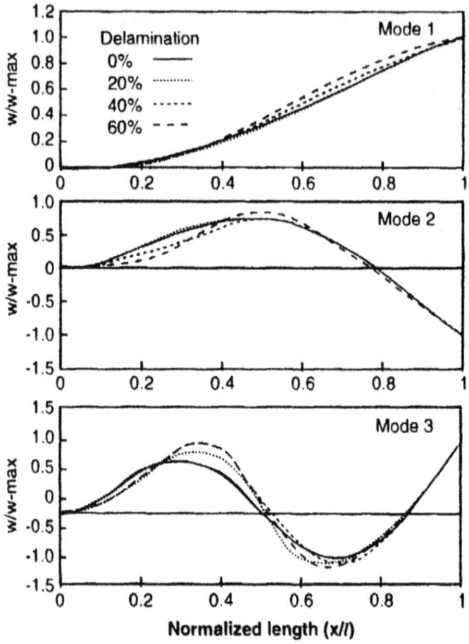

Figure 15.29. Effect of delamination on the mode shapes of the $[0/90]_{2s}$ beam. "Reprinted from NCA-Vol. 16/AMD-Vol. 172, Saravanos, D. A., Mechanics for the effects of delamination on the dynamic characteristics of composite laminates, ASME 1993, pp. 11-21, with kind permission from ASME International, New York, NY".

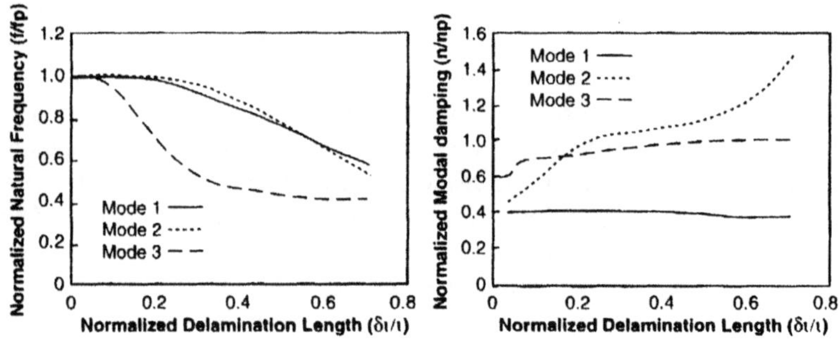

Figure 15.30. Effect of delamination size on modal frequencies and damping of the $[0/90/45/-45]_s$ beam. "Reprinted from NCA-Vol. 16/AMD-Vol. 172, Saravanos, D. A., Mechanics for the effects of delamination on the dynamic characteristics of composite laminates, ASME 1993, pp. 11-21, with kind permission from ASME International, New York, NY".

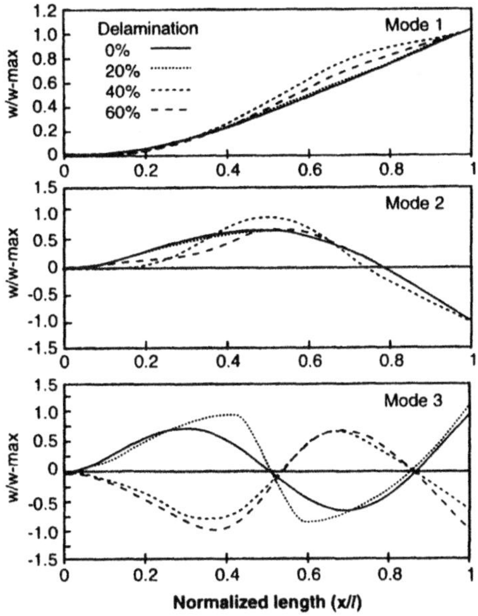

Figure 15.31. Effect of delamination on the mode shapes of the [0/90/45/-45]$_s$ beam. "Reprinted from NCA-Vol. 16/AMD-Vol. 172, Saravanos, D. A., Mechanics for the effects of delamination on the dynamic characteristics of composite laminates, ASME 1993, pp. 11-21, with kind permission from ASME International, New York, NY".

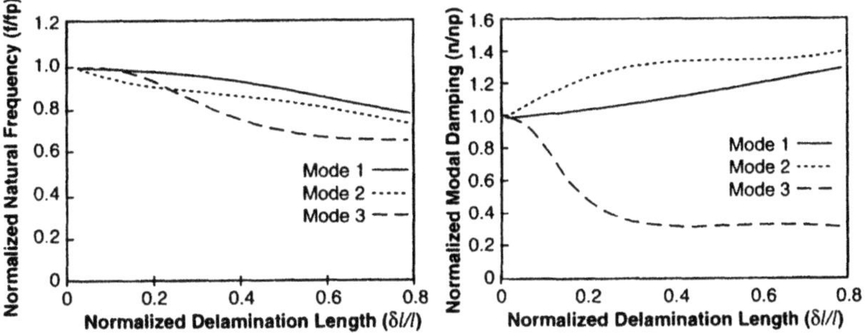

Figure 15.32. Effect of delamination size on modal frequencies and damping of the [45/-45/90/0]$_s$ beam. "Reprinted from NCA-Vol. 16/AMD-Vol. 172, Saravanos, D. A., Mechanics for the effects of delamination on the dynamic characteristics of composite laminates, ASME 1993, pp. 11-21, with kind permission from ASME International, New York, NY".

Rotem (1993) studied the fatigue behaviour of a [0/± 45/90] graphite/epoxy laminate under reverse loading at different frequencies. It was found that for this laminate, the interlaminar stresses at the free edges dominate the fatigue life by introducing interlaminar cracks which cause the laminate to collapse under the compression portion of the load cycle. This failure occurred at all frequencies that were studied, namely, 0.1, 1, 2.8, 10 and 28 Hz. It was reported by Rotem (1993) that the fatigue life decreases considerably as the frequency rises from 2.8 Hz to 10 Hz, while the changes as frequency increases from 0.1 Hz to 2.8 Hz and from 10 Hz to 28 Hz were more moderate. It was also found that the axial modulus hardly changes in this frequency range and therefore, as was advanced by Rotem (1993), it cannot be the cause of the fatigue life decrease. It was suggested, however, that the reason for the reduction of fatigue life could be due the heat generated at the free edge location by the hysteresis of the stresses amplitude. More heat is generated on the higher frequency loading which cause higher temperature at these locations. The higher temperature reduces the local strength and causes earlier crack initiation which results in shorter fatigue life.

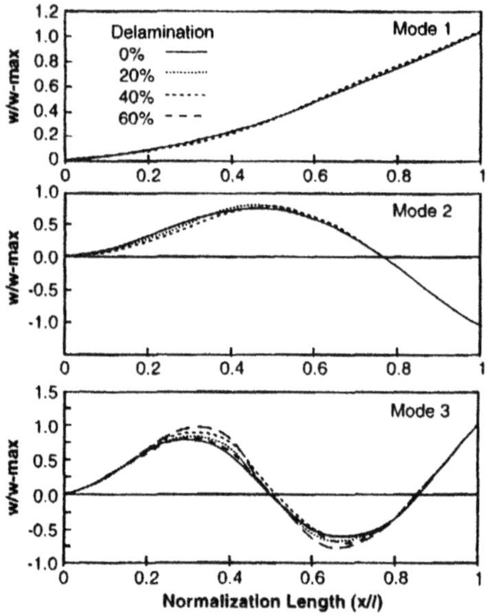

Figure 15.33. Effect of delamination on the mode shapes of the [45/-45/90/0]$_s$ beam. "Reprinted from NCA-Vol. 16/AMD-Vol. 172, Saravanos, D. A., Mechanics for the effects of delamination on the dynamic characteristics of composite laminates, ASME 1993, pp. 11-21, with kind permission from ASME International, New York, NY".

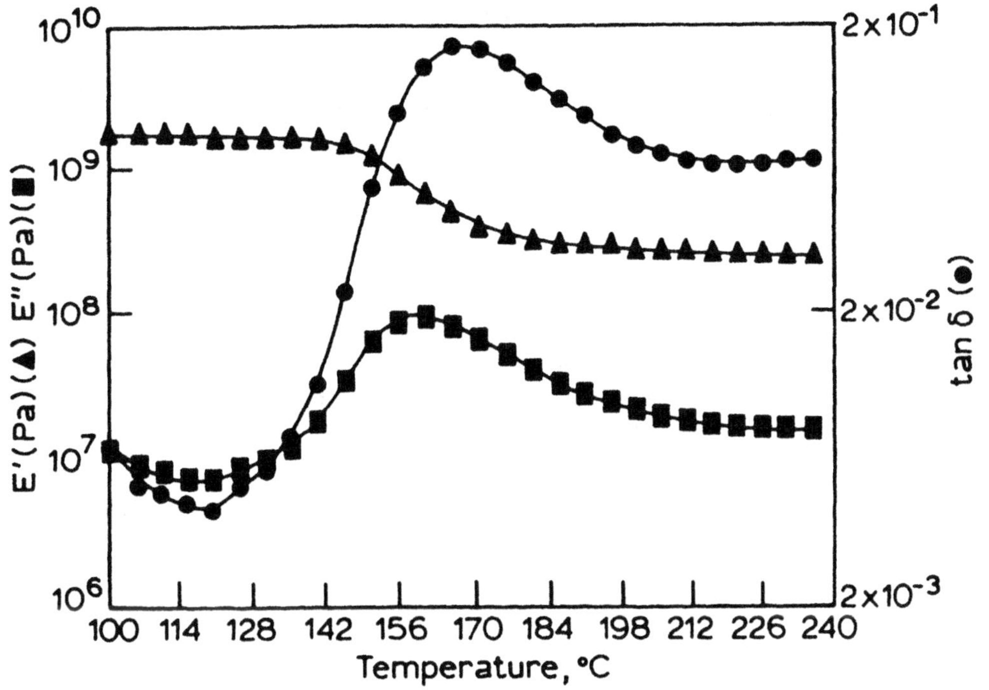

Figure 15.34. The dynamic mechanical spectrum of PEEK. "Reprinted from *Composite Science and Technology* **46**, Folkes, M. J., Kalay, G. and Ankara, A., The effect of heat treatment on the properties of PEEK and APC2, 77-83, 1993, with kind permission from Elsevier Science Ltd, The Boulevard, Langford Lane, Kidlington 0X5 1GB, UK".

The results of the fatigue tests of the isotropic laminate for two loading frequencies are shown in Fig. 15.36, as well as the S-N curves fitted to these data (Rotem, 1993). The fit was done by the least squares regression technique and it can be seen that the fatigue lifes at the higher frequency (10Hz) are much shorter than those at the lower frequency (2.8 Hz). In effect, the S-N curve for the 10Hz tests indicates a fatigue life of about one tenth of the fatigue life of the 2.8 Hz S-N curve. Since fatigue loading was tension-compression with zero mean, there was no creep effect and the entire degradation process took place within the interlaminar zones.

Adding the test results for the other frequencies shows a very interesting phenomenon, as shown in Fig. 15.37. The lower frequency results tend to crowd at the 2.8 Hz location while the higher frequency results tend to crowd at the 10 Hz location. It seems that for this material, therefore, the strongest effect of frequency is between 2.8 Hz and 10 Hz. It is clear from the results on Fig. 10.37 that the frequency effect is not linear.

There is a good correlation between the failure processes and the stiffness behaviour. Figure 15.38 is a plot of the modulus change for fatigue loading frequency of 10 Hz. The failure process showed the beginning of inter-laminar crack initiation at about 99% of the fatigue life, meantime the accelerated modulus degradation starts at about 90% of the fatigue life as seen in the two examples on Figure 15.38. On the other specimens, Fig. 15.39, for which the loading frequency was 1 Hz, the accelerated modulus degradation also starts at about 96% of the fatigue life, for different load levels which gave different fatigue life.

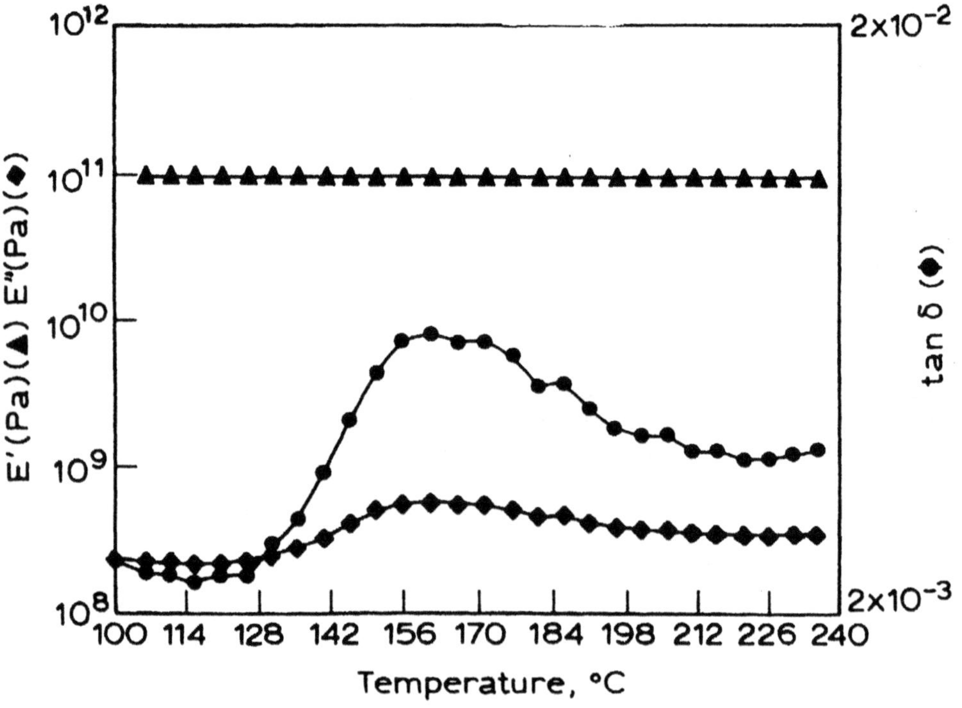

Figure 15.35. The dynamic mechanical spectrum of APC2. "Reprinted from *Composite Science and Technology* **46**, Folkes, M. J., Kalay, G. and Ankara, A., The effect of heat treatment on the properties of PEEK and APC2, 77-83, 1993, with kind permission from Elsevier Science Ltd, The Boulevard, Langford Lane, Kidlington 0X5 1GB, UK".

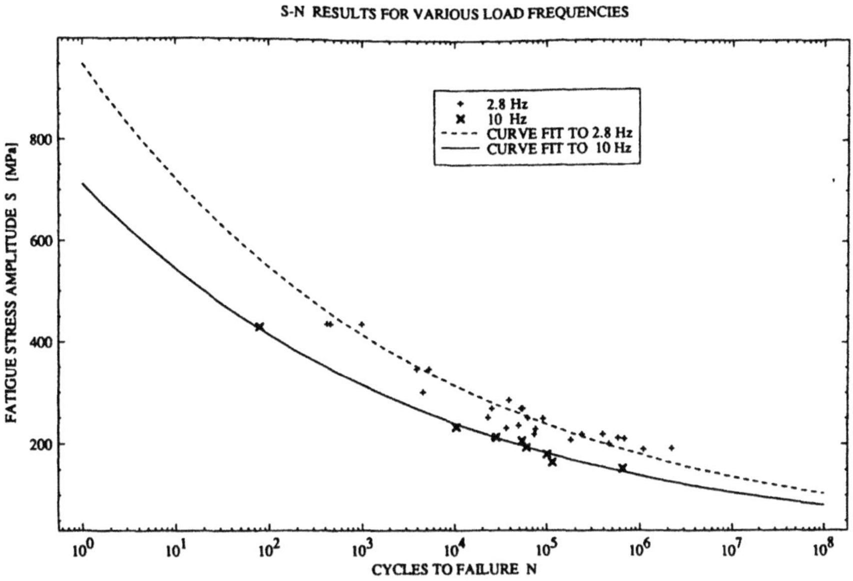

Figure 15.36. Fatigue results of two loading frequencies at R= - 1. "Reprinted from *Composite Science and Technology* **46**, Rotem, A., Load frequency effect on the fatigue strength of isotropic laminates, 129-38, 1993, with kind permission from Elsevier Science Ltd, The Boulevard, Langford Lane, Kidlington 0X5 1GB, UK".

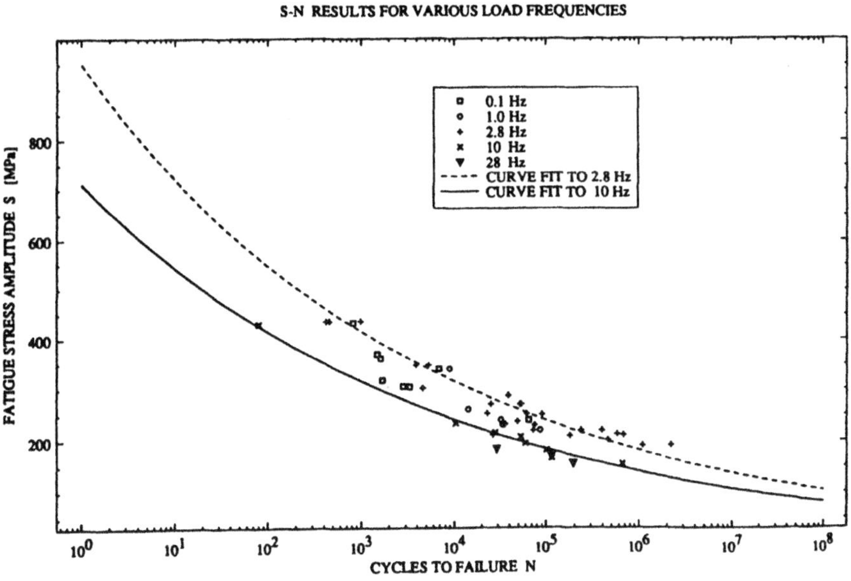

Figure 15.37. Fatigue results of five loading frequencies at R= - 1. "Reprinted from *Composite Science and Technology* **46**, Rotem, A., Load frequency effect on the fatigue strength of isotropic laminates, 129-38, 1993, with kind permission from Elsevier Science Ltd, The Boulevard, Langford Lane, Kidlington 0X5 1GB, UK".

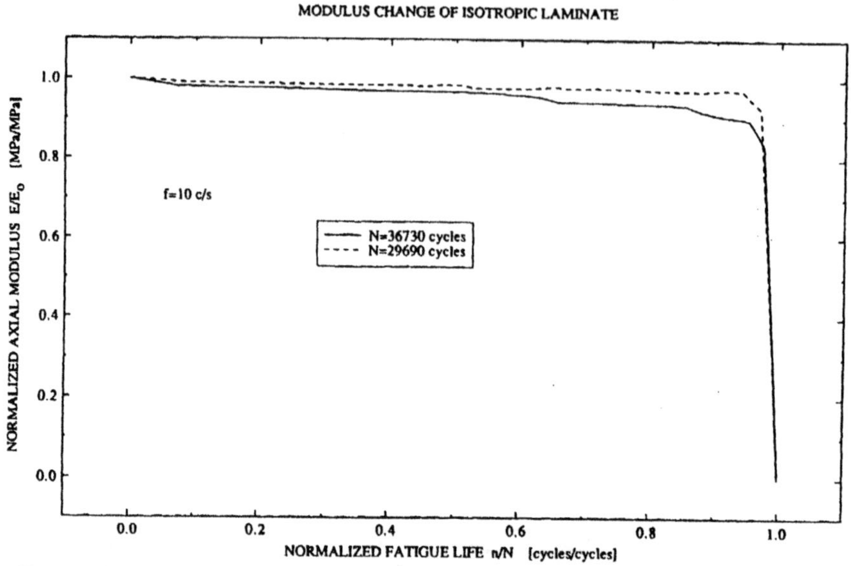

Figure 15.38. Modulus change under fatigue loading of 10 Hz and R = - 1. "Reprinted from *Composite Science and Technology* **46**, Rotem, A., Load frequency effect on the fatigue strength of isotropic laminates, 129-38, 1993, with kind permission from Elsevier Science Ltd, The Boulevard, Langford Lane, Kidlington 0X5 1GB, UK".

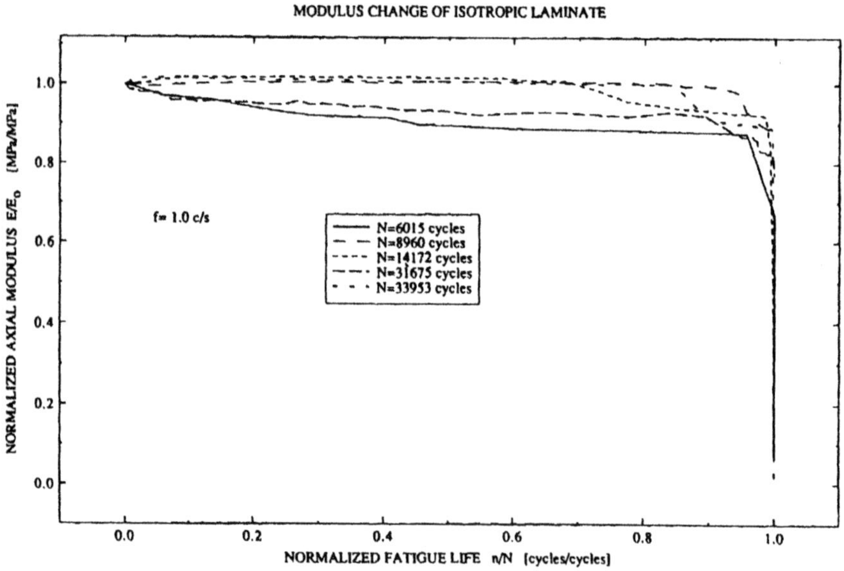

Figure 15.39. Modulus change under fatigue loading of 1 Hz and R = - 1. "Reprinted from *Composite Science and Technology* **46**, Rotem, A., Load frequency effect on the fatigue strength of isotropic laminates, 129-38, 1993, with kind permission from Elsevier Science Ltd, The Boulevard, Langford Lane, Kidlington 0X5 1GB, UK".

15.5. The Trade-off between Damping and Stiffness in the Design of Discontinuous Fibre-Reinforced Composites

It is well-known that lightweight fibre-reinforced polymer composite materials have higher specific strength and stiffness when compared with conventional structural materials such as metals. Much effort has been devoted to the improvement and optimization of these properties in various composite structures. Good vibration damping properties are also particularly important for composite structures when they are used under dynamic loading, such as in aerospace structures and rotor blades. Due in part to the extensive accumulated experience with conventional structural materials which in general have poor internal damping, the potential for the improvement and optimization of damping in fibre- reinforced composites has not been yet fully realized. Meanwhile, the full use of discontinuous fibre-reinforcement has not been yet fulfilled in composite materials research. This is due, on the other hand, to the direct accomplishment of higher specific strength and stiffness in the more familiar continuous fibre composites.

In the conventional damping theory, the loss tangent tan δ, often referred to as the damping factor, is assumed to vary comparatively little with frequency for a large class of polymers, particularly at temperatures near the polymer glass-transition temperature (e.g., Nashif *et al.*, 1965). Thus, a large number of researchers, by following this assumption, considered in their models that the damping factor to be constant. However, in the case of fibre-reinforced composites, and, in particular, discontinuous fibre composites, the damping factor is a frequency dependant, as it relates strongly with the particulars of the microstructure, e.g., fibre-aspect-ratio, fibre volume faction and fibre off-axis angle; e.g., Gibson and Yau (1980), Gibson *et al.* (1982), Sun *et al.* (1985), and Suarez *et al.* (1986).

The damping properties of continuous fibre composites have been studied by a number of researchers; e.g., Bert and Clary (1974) and Bert (1980). There are relatively few research publications on the damping of discontinuous fibre composites. However, studies reported by, for instance, McLean and Read (1975) and Gibson *et* al.(1982) indicate that vibration damping of fibre-reinforced composites, of polymeric matrix, may be significantly improved, and possibly can be readily optimized by using, as a reinforcement, discontinuous fibres rather than continuous ones.

A possible explanation of the above mentioned advantages concerning the damping of discontinuous fibre composites is the presence of shear stress concentration at the fibre-segment ends, and, thus, the shear loading transfer mechanism that occurs between the reinforcement and the matrix material. In this context, it is often argued in the literature that shear deformation is primarily responsible for the vibrational energy dissipation in viscoelastic materials such as polymers. An approximate stress distribution along a short fibre embedded in a continuous matrix was reported, for instance, by Cox (1952).

The research work of Gibson and Yau (1980) and Gibson *et al.* (1982) indicates that by varying the fibre-aspect-ratio and fibre orientation, highest damping and maximum stiffness could be achieved separately. This observation implies that the optimum conditions (in terms of microstructural parameters such as fibre-aspect-ratio and orientation) for damping may not be necessarily the same for stiffness. Consequently, it is important to study the influence of the various governing microstructural parameters as pertaining to both damping and stiffness. The optimization, in terms of the microstructure, of this trade-off between damping and stiffness is the main intention of this section.

It is obvious that the most ideal situation for designing a discontinuous fibre-reinforced polymer composite structure is to optimize the damping and stiffness simultaneously with respect to the microstructure controlling parameters. In this context, the general procedure of the *"Force-Balance Approach"*, e.g., Sun *et al.* (1985), is used below in this section to formulate an analytical model pertaining to the optimization of the damping and stiffness of a class of discontinuous fibre-reinforced composite materials. In this context, a multi-objective optimization functional is established to optimize these two properties simultaneously. In this context, a particular application, i.e., concerning a discontinuous E-glass/epoxy composite system, is dealt with in this section (*see* Feng, 1999, and Haddad and Feng, 1999).

15.5.1. INFLUENCE OF SELECTED MICROSTRUCTURAL PARAMETERS

There appear to be two primary sources of damping in fibre-reinforced composites:

(i) the extent of the viscousc nature of the bulk matrix, and
(ii) the friction mechanism at the interface as caused by the relative motion between the matrix and the fibre.

Both of these effects may prove to be particularly significant in polymeric base composites that are reinforced with discontinuous fibres, whereas high shear stresses are often developed at the fibre-matrix interface. When a short-fibre composite is subjected to a cyclic loading, the matrix at regions surrounding the fibre-segment and adjacent to its ends undergoes high cyclic shear strains, thus, producing significant viscous energy loss. The shear stress concentration at these regions may also induce partial debonding at the fibre/matrix interface that might eventually result in a slip between the fibre and the matrix and, thus, in accompanying frictional losses. Such a fibre/matrix debonding would, however, affect adversely the stiffness of the composite and, by consequence, its ultimate strength. It is, however, often argued in the literature that it is often desirable to have a strong interfacial bond so that slip at the interface could be avoided. If this is accomplished, then, the most viable source of possible enhanced dissipation would appear to be the occurring shear deformation in the matrix as a result of the shear stress concentration adjacent to the fibre ends. Based on the particular mechanism of stress transfer between the fibre and the matrix, it is obvious that there are several microstructural parameters (e.g. fibre-aspect-ratio, fibre volume fraction, fibre/matrix modulus ratio, etc.) that could influence the shear stress distribution surrounding the fibre-segment. The situation becomes further complicated when the interaction between neighboring fibre-segments, in the composite laminate, is taken into account.

Force-Balance Approach
As mentioned earlier, the *"Force-Balance Approach"* is used below in this section to predict the damping and stiffness for this class of materials. The *"Force-Balance Approach"* is often regarded as a combination of elastic mechanics-of-materials analysis, which could reasonably predict the stress transfer at the fibre/matrix interface, in conjunction with the well-known elastic-viscoelastic correspondence principle of linear viscoelasticity; e.g., Hashin (1970) and Haddad (1995).

The basic assumptions for the *"Force Balance Approach"* are:

- A round fibre is surrounded by a cylindrical matrix under the effect of an extensional load. This is illustrated in Figure 15.40 below.
- Both the fibre and matrix are isotropic.
- The mechanical response of the matrix is linear viscoelastic.
- The fibre contributes, to a certain extent, to the overall energy dissipation in the composite specimen.
- The bonding between the fibre and the matrix is assumed to be perfect. Further, the fibre/matrix interface is assumed to have the same viscoelastic properties of the bulk matrix.
- The load transfer between the matrix and the fibre depends upon the difference between the actual displacement at a point on the interface and the displacement that would exist if the fibre were absent.

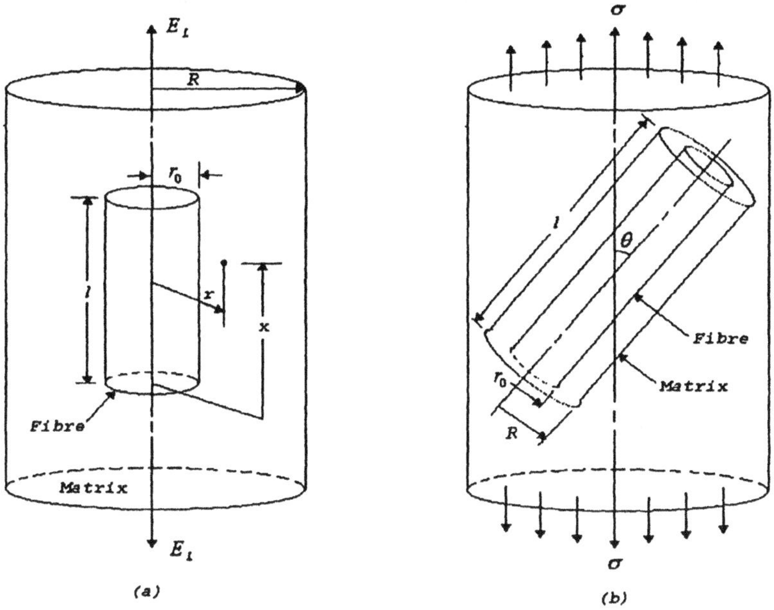

Figure 15.40. Representative volume element. (a) Aligned case. (b) Off-axis case. (Adapted after Sun, C. T., Gibson, R. F. and Chaturvedi, S. K. (1985), *Journal of Materials Science* **20**, 2575-85, with kind permission from Kluwer academic publishers, Dordrecht).

In the force-balance approach, the expression for the elastic stiffness of a discontinuous-fibre composite is derived from the average fibre stress as based upon using Cox's analytical model concerning fibre stress distribution (Cox, 1952). Subsequently, the elastic-viscoelastic correspondence principle is employed to obtain the expression for the complex modulus of the assumed linear viscoelastic composite laminate. This involves the replacement of the static elastic moduli of the fibre, matrix and composite in the expressions resulting from the linear elastic analysis, with the corresponding viscoelastic moduli. In the case of sinusoidal loading, the expression for the arrived at complex modulus would involve both the storage and loss moduli.

For a typical representative volume element, Figure 15.40, the expression of the longitudinal modulus E_x of the composite along the loading axis may be expressed (*see* Agarwal and Broutman, 1980 and Feng, 1999) by:

$$\frac{1}{E_x} = \frac{\cos^4\theta}{E_L} + \frac{\sin^4\theta}{E_T} + \left(\frac{1}{G_{LT}}\right)\sin^2\theta\cos^2\theta \tag{15.19}$$

where E_L, E_T and G_{LT} are the longitudinal modulus, transverse modulus, and in-plane shear modulus, respectively, and can be expressed in term of both the fibre and matrix material parameters, i. e., E_f, E_m, G_f, G_m, etc, and the fibre volume fraction V_f, by using, for instance, the rule-of-mixtures if one deals with a continuous fibre composite. For a short fibre composite, however, one cannot use the rule-of mixtures to represent the longitudinal modulus E_L. For the latter case, i.e., a short fibre composites, the longitudinal modulus would depend also on the fibre aspect ratio, *l/d*. Based upon the shear-lag model (Cox, 1952), E_L may be expressed for the case of short fibre composite by (Feng, 1999)

$$E_L = E_f\left(1 - \frac{\tanh(\chi/2)}{(\chi/2)}\right)V_f + E_m(1 - V_f) \tag{15.20}$$

where,

$$\chi 2 = 4\frac{G_m}{E_f}\,\frac{(\ell/d)^2}{\ln(2R/d)} \tag{15.21}$$

The ratio *R/d*, as illustrated in Figure 15.40, is related to the fibre volume fraction V_f, for the particular packing array under consideration. For instance,

$$\text{For a square array:}\quad \left(\frac{R}{d}\right)^2 = \frac{\pi}{16V_f} \tag{15.22}$$

$$\text{For a hexagonal array:}\quad \left(\frac{R}{d}\right) = \frac{\pi}{8(3V_f)^{1/2}} \tag{15.23}$$

Based upon the work of Gibson, *et al.* (1982), the packing geometry has an insignificant effect on the magnitude of damping. Thus, we adopt, in the following analysis, expression (15.22) corresponding to the square packing array. Combining equations (15.21) and (15.22), it follows that

$$\chi^2 = 8\,\frac{G_m}{E_f}\,\frac{(\ell/d)^2}{\ln\left(\dfrac{\pi}{4V_f}\right)^{1/2}} \tag{15.24}$$

Equation (15.24) above shows that the parameter χ is essentially a function of the fibre matrix stiffness ratio E_f/G_m, fibre-aspect-ratio ℓ/d, and the fibre volume fraction V_f.

The transverse modulus E_T and the transverse in-plane shear modulus G_{LT}, of the short fibre composite, are almost independent of fibre-aspect-ratio ℓ/d. In this context, Feng (1999) used the following prediction equations adopted by Halpin-Tsai in their model concerning continuous fibre composites (e.g., Agarwal *et al.*, 1980),

$$E_T = E_m(1 + 2\eta_1 V_f)/(1 - \eta_1 V_f) \tag{15.25}$$

$$G_{LT} = G_m(1 + \eta_2 V_f)/(1 - \eta_2 V_f) \tag{15.26}$$

where the coefficients η_1 and η_2 of above equations can be expressed, respectively, as

$$\eta_1 = [(E_f/E_m) - 1]/[(E_f/E_m) + 2] \tag{15.27}$$

$$\eta_2 = [(G_f/G_m) - 1]/[(G_f/G_m) + 1] \tag{15.28}$$

The Poisson's ratio ν_{LT} of the short fibre composite, which is assumed to be insensitive to fibre length, may be expressed, using the '*rule-of-mixture*' form, as

$$\nu_{LT} = \nu_f V_f + \nu_m(1 - V_f) \tag{15.29}$$

According to the previous assumptions, both the fibre and matrix are considered to behave in a linear viscoelastic manner. This permits us to use the elastic-viscoelastic correspondence principle (e.g., Haddad, 1995) in order to redefine the basic material properties as,

$$\begin{aligned} E_x^* &= E_x' + i\,E_x'' \\ E_f^* &= E_f' + i\,E_f'' \\ E_m^* &= E_m' + i\,E_m'' \\ G_m^* &= G_m' + i\,G_m'' \end{aligned} \tag{15.30}$$

where, as introduced earlier, the over-prime designates the storage modulus and the double over-prime designates the loss modulus. Meantime, the damping (or loss) factor is defined as the ratio between the loss modulus to the storage modulus, i.e.,

$$\begin{aligned} \eta_c &= E_x'' / E_x' \\ \eta_f &= E_f'' / E_f' \\ \eta_m &= E_m'' / E_m' \\ \eta_{Gm} &= G_m'' / G_m' \end{aligned} \tag{15.31}$$

Upon using equation (15.30), equation (15.20) and equations (15.24) to (15.28) may be written, respectively, as

$$E_L^* = (E_f' + i\,E_f'')\left(1 - \frac{\tanh\left(\dfrac{\chi^*}{2}\right)}{\left(\dfrac{\chi^*}{2}\right)}\right) + (E_m' + i\,E_m'')\,(1 - V_f) \tag{15.32}$$

$$\chi^{*2} = 8\,\frac{(G_m' + i\,G_m'')\,(l/d)^2}{(E_f' + i\,E_m'')\,\ln\left(\dfrac{\pi}{4V_f}\right)^{1/2}} \tag{15.33}$$

$$E_T^* = (E_m' + i\,E_m'')\,\frac{1 + 2\eta_1^*\,V_f}{1 - \eta_1^*\,V_f} \tag{15.34}$$

$$G_{LT}^* = (G_m' + i\,G_m'')\,\frac{1 + \eta_2^*\,V_f}{1 - \eta_2^*\,V_f} \tag{15.35}$$

$$\eta_1^* = \frac{[(E_f^{'} + i\,E_f^{''})/(E_m^{'} + i\,E_m^{''})] - 1}{[(E_f^{'} + i\,E_f^{''})/(E_m^{'} + i\,E_m^{''})] + 2} \tag{15.36}$$

$$\eta_2^* = \frac{[(G_f^{'} + i\,G_f^{''})/(G_m^{'} + i\,G_m^{''})] - 1}{[(G_f^{'} + i\,G_f^{''})/(G_m^{'} + i\,G_m^{''})] + 1} \tag{15.37}$$

Substituting E_L^*, E_T^* and G_{LT}^* from equations (15.32), (13.4) and (15.35) for E_L, E_T and G_{LT}, respectively, and E_x^* for E_x into equation (15.19), one obtains

$$\frac{1}{E_x^{'} + i\,E_x^{''}} = \frac{\cos^4\theta}{E_L^*} + \frac{\sin^4\theta}{E_T^*} + \left(\frac{1}{G_{LT}^*} - \frac{2\nu_{LT}}{E_L^*}\right)\sin^2\theta\cos^2\theta \tag{15.38}$$

This equation can be used to determine $E_x^{'}$ and $E_x^{''}$ for the composite by separating its real and imaginary parts.

Examining equation (15.32), one finds that E_L^* is an exponential function of χ^* which depends upon the complex stiffness ratio G_M^*/E_f^*.

Since the loss moduli are generally small, one may neglect the higher order terms of loss factors such as η_f^2 and $\eta_{Gm}\eta_f$. Following this approach, and by careful manipulation, one can obtain the following expression from equation (15.33) as,

$$\frac{\chi^*}{2} = \frac{\chi}{2}\left[1 + \frac{1}{2}i\,(\eta_{Gm} - \eta_f)\right] \tag{15.39}$$

Then, one may use a Taylor's series approximation and similarly neglect any resulting higher order terms in the loss factors to obtain

$$\tanh\frac{\chi^*}{2} = \tanh\frac{\chi}{2} + i\frac{\chi}{4}\,\frac{(\eta_{Gm} - \eta_f)}{\cosh^2\frac{\chi}{2}} \tag{15.40}$$

Finally, by combining equations (15.32 to 15.40), one determines analytical representations of E'_x, E''_x and η_x as dealt with by Feng (1999) and Haddad and Feng (1999).

The analytical expressions reached out by Feng (1999) are represented here symbolically as,

$$E'_x = \psi_1 (E'_m, E'_f, G'_m, (\ell/d), \theta, V_f, \eta_f, \eta_m, \eta_{Gm}, \nu_m, \nu_f) \tag{15.41}$$

$$E''_x = \psi_2 (E'_m, E'_f, G'_m, (\ell/d), \theta, V_f, \eta_f, \eta_m, \eta_{Gm}, \nu_m, \nu_f) \tag{15.42}$$

$$\eta_x = \psi_3 (E'_m, E'_f, G'_m, (\ell/d), \theta, V_f, \eta_f, \eta_m, \eta_{Gm}, \nu_m, \nu_f) \tag{15.43}$$

In the equation (15.43), η_x is defined as the ratio of E''_x/E'_x and will not have the same changing pattern as E'_x. Therefore, two dependent variables, the non-dimensional ratio E'_x/E'_m and the non-dimensional ratio η_x/η_m, are used in the following numerical presentation. In general, from the expressions for E'_x and η_x, e.g., equations (15.41) and (15.43), we have eleven variables, namely $E_f, E_m, \theta, V_f, G_m, \eta_f, \eta_m, \eta_{Gm}, \nu_f, \nu_m$ and the fibre-aspect-ratio ℓ/d. An optimization that would include all possible variations of these eleven variables are almost impossible and unnecessary. Therefore, in the following optimization analysis, we choose, as a particular example, the matrix material to be Scotchply epoxy and the fibre-reinforcement to be of E-glass in the dealt with composite layer. The corresponding material properties of these two selected materials are presented in Table 15.2. Consequently, we narrow down the optimization variables to three, namely, θ, V_f and ℓ/d.

For a large class of composite materials used in aerospace and automotive industries, the fibre volume fraction varies within the range of 50 to 70%. Meanwhile, the observation made by Cox (1952) shows that, for short fibre reinforced composites, the reduction of the effective longitudinal modulus due to the load transfer from fibre to fibre is considered significant only for fibre aspect ratios, ℓ/d, less than 100 Therefore, one could set that the fibre volume fraction to range from 50% to 70% and the fibre aspect ratio to range from 1 to 100.

Figures 15.41 to 15.45 are obtained by setting, respectively, the fibre of-axis angle θ as 0°, 40°, 60°, 80° and 90°, and plotting the non-dimensional ratios η_x/η_m and E'_x/E'_m against the fibre volume fraction V_f and the fibre aspect ratio ℓ/d. In these figures, it is clear that with the increase of fibre off-axis angle θ, the values of the non-dimensional ratio η_x/η_m are increasing and those for E'_x/E'_m are decreasing. When the fibre off-axis angle reaches a value between 40° to 60°, both curves pertaining to η_x/η_m and E'_x/E'_m change their directions, which demonstrate that for a fibre off-axis angle θ within the range of 40° to 60°, both the ratios η_x/η_m and E'_x/E'_m reach their extreme values (maximum and minimum, respectively) almost simultaneously.

TABLE 15.2 Selected material properties of Scotchply 1002 matrix epoxy and E-glass fibres at room temperature (Adapted after Gibson and Plunkett 1976).

Material Properties	Expoxy	E-glass
Young's modulus, GPa	3.79	72.4
Shear modulus, GPa	1.38	30.3
Damping factor	0.015	0.0014
Shear damping factor	0.018	0.0014
Poisson's ratio, υ	0.36	0.2
Specific Gravity, *g*	1.23	2.54

Figures 15.46 to 15.50, which are plotted by setting the fibre-aspect-ratio ℓ/d at 5, 20, 40, 80 and 100, respectively, present the non-dimensional ratios η_x/η_m and E'_x/E'_m against the fibre volume fraction V_f and the fibre off-axis angle θ. One could note that the value η_x/η_m decreases monotonously as the fibre-aspect-ratio ℓ/d increases and for $\ell/d>15$, the rate of decrease in the value of η_x/η_m slows down until the fibre-aspect-ratio ℓ/d reaches 20, whereby the value of η_x/η_m maintains a constant value afterwards. The non-dimensional ratio E'_x/E'_m also increases sharply until the fibre-aspect-ratio ℓ/d reaches a value of about 20. With the fibre-aspect-ratio ranging from 20 to 60, the non-dimensional ratio E'_x/E'_m increases slowly with the increase of the fibre-aspect-ratio ℓ/d and it appears to have a constant value from $\ell/d=60$ upwards.

By setting the fibre volume fraction V_f at 50%, 60% and 70%, the non-dimensional ratios η_x/η_m and E'_x/E'_m versus the fibre aspect ratio ℓ/d and the fibre off-axis angle θ could be plotted as shown in Figures 15.51 to 15.53. In these figures, one could identify that with the increase of the fibre volume fraction V_f, both the non-dimensional ratios η_x/η_m and E'_x/E'_m change almost linearly, with the values of η_x/η_m are monotonously decreasing and those for E'_x/E'_m are monotonously increasing.

It is apparent from the above mentioned results that among the three considered independent variables, the fibre off-axis angle θ has the most significant influence on the damping and stiffness of fibre reinforced composites, e.g., Sun *et al.* (1985). The numerical results obtained appear to have good agreement with the observations made by Gibson *et al.* (1982), Sun *et al.* (1985), and Suarez *et al.* (1986).

15.5.2. OPTIMIZATION

From the above numerical results, one can note that in order to increase the damping property of a discontinuous fibre reinforced composite, it is necessary to sacrifice the stiffness of such material. The analysis of the trade-off between damping and stiffness has recently gain remarkable attention from researchers and design engineers, particularly in aerospace and automotive industries, due to the high-volume use of fibre composite materials in such industries. Therefore, the simultaneous optimization of these two properties for the design of high performance fibre-reinforced composite structures is becoming remarkably important in the realm of the development of composite materials for applications involving both quasi-static and dynamic loading.

In addition, as it is well recognized, one of the most important advantages of a fibre reinforced composite material over its metallic counterparts is its light specific weight. The latter property is particularly attractive to aerospace and automotive industries. Thus, it is necessary to include this property in the dealt with optimization problem. Thus, three objective functions become involved in the optimization problem of interest to the present work, i.e., the maximization of both damping and stiffness and, on the other hand, minimization of the specific weight. Thus, the so-called "*inverted utility function method*", e.g., Rao (1984), appears to be suitable to deal with this multi objective optimization problem.

The Inverted Utility Function Method

In this method, a utility function U_i (f_i) is defined for each objective function as

$$U_i = -w_i f_i(X) \tag{15.44}$$

where f_i (X), the ith objective function, with a weighting factor w_i (i = 1, 2, ...k), is to be minimized. In the process of optimization, one inverts each utility function and attempt to minimize or reduce the total undesirability. Thus, it follows that

$$U^{-1} = \sum_{i=1}^{k} U_i^{-1} = \sum_{i=1}^{k} \left(\frac{1}{U_i}\right) = \sum_{i=1}^{k} -\frac{1}{w_i}\frac{1}{f_i(X)} = \sum_{i=1}^{k} -a_i\left(\frac{1}{f_i(X)}\right) \tag{15.45}$$

where the scalar weighting factor a_i is defined by

$$a_i = \frac{1}{w_i}; \quad \sum_{i=1}^{k} a_i = 1$$

The solution of the optimization problem is established by minimizing the function U_i^{-1}, as expressed by (15.45), subject to the imposed constraints. The selection of scalar weighting factor a_i would depend on the extent of importance of each objective function.

Multivariable Non-linear Optimization
The utility functions of the dealt with objective functions are set as

$$U_1 = w_1\left(\frac{\eta_x}{\eta_m}\right) \quad (15.46)$$

$$U_2 = w_2\left(\frac{E_x'}{E_m'}\right) \quad (15.47)$$

$$U_3 = -w_3\left(\frac{\overline{W}}{\overline{W}_m}\right) \quad (15.48)$$

where $\overline{W}$ is the specific weight for the dealt with discontinuous fibre-reinforced composite, which is defined in term of fibre volume fraction V_f as

$$\overline{W} = \overline{W}_f V_f + \overline{W}_m (1 - V_f) \quad (15.49)$$

where E_x' and η_x are defined, respectively, by equations (15.41) and (15.43). $\overline{W}_f$, $\overline{W}_m$, E_m' and η_m are set corresponding to the material properties of considered E-glass/epoxy discontinuous fibre-reinforced composite as shown in Table 15.2.

Substituting equation (15.46) to (15.48) into equation (15.45), the total undesirability of this design problem can be expressed as

$$U^{-1} = a_1\left(\frac{\eta_m}{\eta_x}\right) + a_2\left(\frac{E_m'}{E_x'}\right) - a_3\left(\frac{\overline{W}_m}{\overline{W}}\right) \quad (15.50)$$

In the presently dealt with optimization problem, both damping and stiffness are considered to be approximately of the same importance to the optimization problem in hand, but, each is of a more significance than the specific weight. Accordingly, one may set the associated-with weighting factors selectively as $a_1 = 0.513$, $a_2 = 0.387$, $a_3 = 0.1$. For the reasons stated early in this section, the constraints for this optimization problem may be set as

$$\begin{aligned} 0.5 \le\ & V_f \le 0.7 \\ 1 \le\ & \ell/d \le 100 \\ 0^\circ \le\ & \theta \le 90^\circ \end{aligned} \quad (15.51)$$

Implementation of Non-Linear Programming

Following the above presentation, the optimization of this design work becomes

$$\text{Minimize} \quad U^{-1}$$

$$\text{Subject to} \quad \begin{aligned} 0.5 \le V_f \le 0.7 \\ 1 \le \ell/d \le 100 \\ 0^\circ \le \theta \le 90^\circ \end{aligned} \tag{15.52}$$

This is a typical constrained non-linear optimization problem. In order to simplify this problem, one can adopt the so-called "*mapping technique*", often referred-to as the "*variable transformation technique*" to deal with the above parametric constraints (e.g., Rao, 1984). By using this technique, the constrained optimization problem could be solved by a non-constrained optimization technique.

In the mapping technique, one assumes that there is a minimization problem f(X) whereby $X^T = [x_1, x_2, \ldots, x_j]$ has the parametric constraints

$$\ell_i \le x_i \le u_i \tag{15.53}$$

where j = 1, 2, 3, ..., n; i = 1, 2, 3, ..., m.

In this context, one can use the general mapping procedure (Rao, 1984) as

$$x_i = \ell_i + (u_i - \ell_i)\sin^2 y_i \tag{15.54}$$

Therefore, the objective function f(X) changes to $f(X^*, Y)$, whereby $Y^T = [y_1, y_2, \ldots, y_i]$ and X^* represents all the components of one variable vector X with the exception of x_i. Meanwhile if m = n, $f(X^*, Y)$ becomes f(Y); *see* Feng (1999).

Accordingly, the mapping procedure can be utilized on the above mentioned parametric constraints, equation (15.52), as

$$\begin{aligned} \theta &= 90^\circ \sin^2 y_1 \\ \ell/d &= 1 + 99 \sin^2 y_2 \\ V_f &= 0.5 + 0.2 \sin^2 y_3 \end{aligned} \tag{15.55}$$

where $Y^T = [y_1, y_2, y_3]$ is the mapping vector in the above procedure. Therefore, one can convert this constrained non-linear optimization problem $U^{-1}(\theta, \ell/d, V_f)$, Eqn. (15.52), to a non-constrained non-linear optimization $U^{-1}(y_1, y_2, y_3)$ and solve this problem by using a non-constrained optimization techniques for the purpose of the simplification of the analysis.

The "*Simplex Method*", *see* Rao (1984), seems to be suitable for this non-constrained non-linear optimization problem with a relatively small number of variables. In the referred-to method, the movement of the "*Simplex*" of n + 1 points in an n-dimensional space towards an optimal point is achieved by using three operations known as "*Reflection*", "*Contraction*" and "*Expansion*" techniques (e.g., Rao, 1984).

Following the algorithm of the "*Simplex Method*", one can implement (Feng, 1999) a numerical scheme to solve this optimization problem (*see* Fig. 15.54 for the pertaining flowchart).

With reference to the flowchart of Fig. 15.54, one is to set first the values of a number of involved parameters, namely, the desired starting point, accuracy of the problem and probe length. Such parameters are used to construct the initial "*Simplex*", as well as the "*Reflection*", "*Expansion*" and "*Contraction*" coefficients. In this context, one sets the starting point as $Y^T = [0, 0, 0]$, the accuracy of the problem as ACCUR = 0.001, the probe length as PLE = 0.1, the "*Reflection*" coefficient as A = 1.0, the "*Expansion*" coefficient as Y = 2.0, the "*Contraction*" coefficient as B= 0.5 (e.g., Rao, 1984). The final result is given out as: the optimal fibre off-axis angle $\theta = 44.4°$, the optimal fibre volume fraction V_f = 60.605 and the fibre aspect ratio $\ell/d = 1.49$.
These results have good agreement with the observations made by Sun et al. (1985) that for small off-axis angles θ (say $\theta \leq 45°$), η_x becomes maximum in the whisker or micro-fibre composites range (i.e., for very small ℓ /d, say $\ell/d \leq 5$) and the stiffness E_x' for micro-fibre and whisker composites is also relatively high. Therefore, in order to achieve high stiffness E_x' and damping η_x, micro-fibre and whisker composites appear to be the ideal candidates.

If Y^T are set as various starting points within a useful interval from [0, 0, 0] to [95, 95, 95], with the increments 0.5 or 5.0, and the same input parameters are used as in the above case, one arrives at outputs as shown in Table 15.3. In this case, it is obvious that one obtains multiple local minima, such as, for instance, the local minimum: 0.5827. For each case of these local minima, the off-axis angle remains almost the same, i.e., approximately 43.75°, and the fibre volume fraction and the fibre-aspect-ratio change in opposite directions and could be catalogued into two groups, i.e., ($\ell/d \approx 1.38$, $V_f \approx 62\%$) and ($\ell/d \approx 85.09$, $V_f \approx 54\%$). It is obvious that this interesting observation gives more flexibility to the design of high performance fibre-reinforced composites by using, for instance, fibre-reinforcement with either a relatively low fibre aspect ratio, i.e., $\ell/d \approx 1.38$ and a relatively high fibre volume fraction, i.e., $V_f \approx 62\%$, or of a higher fibre aspect ratio, i.e., $\ell/d \approx 85.09$ and a lower fibre volume fraction, i.e., $V_f \approx 54\%$. Thus, in the presented case, one may conclude that concluded from the above results that, when at the small fibre off-axis angle $\theta \approx 43.75°$ and by approximately setting the fibre-aspect-ratio at $\ell/d \approx 1.38$ or ℓ /d ≈ 85.09, the corresponding fibre volume fraction V_f reaches about 62%, or 54%, we could get the maximum damping η_x, relatively high stiffness E_x' and relatively low weight $\overline{W}$.

Summary of Section 15.5

Analytical predictions which were determined by the "*Force-Balance Method*" show that damping and stiffness are functions of fibre off-axis angle, fibre volume fraction and fibre-aspect-ratio. In order to increase the damping of the fibre-composite, it may be necessary to sacrifice the stiffness, and vice versa.

The "*Inverted Utility Function Method*" and "*Simplex Method*" were found to be suitable to deal with the multi objective optimisation problem with relatively small number of variables. The use of the "*Variable Transformation Technique*", Rao (1984), to convert the constrained non-linear optimisation problem to a non-constrained one, makes such an optimisation much easier to handle.

For a given E-glass/epoxy composite material, the results of optimization of damping, stiffness and specific weight show that, approximately at fibre off-axis angle $\theta \approx 43.75°$, by setting fibre aspect ratio $\ell / d \approx 1.38$ or $\ell / d \approx 85.09$, the corresponding fibre volume fraction V_f reaches 62% or 54%, and one could obtain maximum damping, relatively high stiffness and relatively low specific weight for this class of materials.

The existence of multiple local minima gives more flexibility to the design of high performance discontinuous fibre-reinforced composites. That is, in the presented case, for instance, both the micro-fibre or whisker composites ($\ell/d \approx 1.38$) and the discontinuous fibre reinforced composites with longer fibre ($\ell \approx d$ 85.09) can be selected to satisfy the above mentioned design criteria.

15.6. Study Problems

1. Comment, using an analytical proof, on the validity of the following statement: "*On the contrary to the case of continuous fibre-composite systems, where the ratio of compliance to breaking strength is invariant for a given fiber material; with discontinuous fibre-systems, however, this ratio can be varied*".

2. Following your answer to Problem 1 above, clarify, with the support of a mathematical model the statement: "*A variable ratio of the compliance to breaking strength would give more latitude in the design of various structural and mechanical components using composite materials*".

3. What is meant by the "*glass transition temperature T_g*" ? How one would determine it for a particular polymeric composite system. Discuss the effect of reinforcement-discontinuity in a composite laminate on the pertaining value of T_g.

15.7. References

Adams, R. D. and Bacon, D. G. C. (1973a) The dynamic properties of unidirectional fibre-reinforced composites in flexure and torsion, *J. Composite Materials* **7**, 53-67.

Adams, R. D. and Bacon, D. G. C. (1973b) Effects of fibre orientation and laminate geometry on the dynamic properties of CFRP, *J. Composite Materials* **7**, 402-28.

Agarwal, B.D. and Broutman, L. J. (1980) *Analysis and Performance of Fibre Composites*, Wiley Interscience, New York.

Agbossou, A., Bergeret, A., Benzarti, K. and Alberola. N. (1993) Modelling of the viscoelastic behaviour of amorphous thermoplastic/glass beads composites based on the evaluation of the complex Poisson's ratio of the polymer matrix, *J.* Materials Science **28**, 1963-72.

Akay, M. (1993) Aspects of dynamic mechanical analysis in polymeric Composites, Composite Science and Technology **47**, 419-23.

Bert, C.M. (1980) Damping applications for vibrations control, ASME AMD-38, *The American Society of Mechanical Engineers*, New York, pp. 53-63.

Bert, C.M. and Clary, R.R. (1974) Composite materials: Testing and design, ASTM STP 546 - *The American Society for Testing and Materials*, Philadelphia, pp. 250-65.

Cavaille, J.Y., Johari,G.P. and Mikalajczak, G. (1987) Dynamic mechanical properties of structural glass-fibre epoxy composites, *Polymer* **28**, 1841-46 .

Chua, P. S. (1987), 42th Annual Conference, Composites Institute, The Society of the Plastics Industry, February 1987, Session 21-A, Ohio, USA.

Cox, H.L. (1952) The elasticity and strength of paper and other fibrous materials, *British Journal of Applied Physics* **3**, 72-84.

Feng, J. (1999) *On the Viscoelastic Response of Laminated Composites*, Masters Thesis, University of Ottawa, Canada.

Folkes, M. J., Kalay, G. and Ankara, A. (1993) The effect of heat treatment on the properties of PEEK and APC2, *Composites Science and Technology* **46**, 77-83.

Gerard, J. F., P. Perret and Chabert, B. (1990) Study of carbon/epoxy interface (or Interphase): Effect of surface treatment of carbon fibres on the dynamic mechanical behaviour of carbon/epoxy unidirectional composites, in: *Controlled Interphases in Composite Materials*, H. Ishida (ed.), Proceedings of the Third International Conference on Composite Interfaces (ICCI- III), May 21-24, 1990, Cleveland, Ohio, pp. 449-56.

Gibson, R.F., Chaturvedi, S.K. and Sun, C.T. (1982) Complex moduli of aligned discontinuous fibre-reinforced polymer composites, *Journal of Materials Science* **17**, 3499-509.

Gibson, R. F. and Plunkett, R. (1976) Dynamic mechanical behaviour of fibre-reinforced composites: Measurement and analysis, *J. Composite Materials.* **10** (October 1976), 325-41.

Gibson, R. F. and Yau, A. (1980) Complex moduli of chopped fibre and continuous fibre composites: Comparison of measurements with estimated bounds, *J. Composite Materials* **14** (April 1980), 155-67.

Grady, J. E., and Meyn, E. H. (1989) Vibration testing of impact damaged composite laminates, Proceedings, 30th AIAA/ASME/ASCE/AHS/ASC SDM Conference, Mobile, AL, pp.2186-93.

Haddad, Y.M. (1995) *Viscoelasticity of Engineering Materials*, Kluwer, Dordrecht.

Haddad, Y.M. and Feng, J. (1999) On the optimization of the mechanical behavior of a class of composite systems under both quasi-static and dynamic loading, AMPT'99, Dublin, Ireland.

Hanagud, S., Nagesh Babu, G. L., Roglin, R., L. and Savanur, S. G. (1992) Active control of delamination in composite Structures, Proceedings, 33rd AIAA/ASME/ASCE/ASC SDM Conference, Dallas, TX, pp. 1819-29.

Hashin, Z. (1965) On elastic behaviour of fibre reinforced materials of arbitrary transverse phase geometry, *J. Mechanics and Physics of Solids* **13**, 119.

Hashin, Z. (1970a) Complex moduli of viscoelastic composites. I. General theory and application to particulate composites, *Int. J. Solids Structures* **6**, 539-52.

Hashin, Z. (1970b) Complex moduli of viscoelastic composites. II. Fibre-reinforced materials, *Int. J. Solids and Structures* **6**, 797-807.

Hashin, Z. and Rosen, B. W. (1965) The elastic moduli of fibre-reinforced materials, *J. Applied Mechanics* **32**, 219.

Helmbuch, R. A. and Sanders, B. A. (1978) Mechanical properties of automotive fibre reinforced plastics, in: *Composite Materials in the Automotive Industry*, ASME, Kulkarni, S. V., Zweben, C. H., and Pipes, R.B. (Eds.), 111-39.

Hill, R. (1964) Theory of mechanical properties of fibre-strengthen materials. I. Elastic behaviour, *J. Mech. Phys. Solids* **12**, 199-212.

Jones, R. M. (1975) *Mechanics of Composite Materials*, McGraw-Hill,

Jutte, R. B. (1978) Structural SMC-material, Process and performance review, SAE Paper No. 780355, 1978 SAE Congress, Detroit.

Kimoto, M. (1990) Flexural properties and dynamic mechanical properties of glass fibre-epoxy composites, *J. Materials Science* **25**, 3327-32.

Kodama, M. (1976) Mechanical dispersion of cross-link polymers reinforced with randomly distrubted short fibre, *J. Appl. Polym.* Sci. **20**, 2165- 84.

Lee, B. T., Sun, C. T. and Liu, D. (1987) An assessment of damping measurement in the evaluation of integrity of composite beams, *J. Reinforced Plastics and Composites* **6**, 114-25.

Lewis, T. B. and Nielsen, L. E. (1970) *J. Appl. Polym. Sci.* **14**, 1449.

Mclean, D. and Read, B. E.(1975) Storage and loss moduli in discontinuous composites, *J. Materials Science* **10**, 481-92.

Mitoh, M. and Nakao, I. (1983) *J. Adhesion Soc. Jpn*, **19**, 485.

Miyano, Y., Kanemitsu, M. , Kunio, T. and Kuhn, H. (1986) Role of matrix resin on fracture strength of unidirectional C. F. R. P., *J. Compos. Mater.* **20**, 520-38.

Nashif, A.D., Jones, D.I.G. and Henderson, J.P. (1965) *Vibration Damping*, Wiley, New York.

Nemes, J. A. and Randles, P. W. (1994) Constitutive modeling of high strain-rate deformation and spall fracture of graphite/peek composites, *Mechanics of Materials* **19**, 1-14.

Paul, B. (1960) Prediction of elastic constants of multiphase materials, *Trans. of the Metallurgical Society of AIME* **218** (February 1960), pp. 36-41.

Paxson, E. B., Jr. (1975) Real and imaginary parts of the complex viscoelastic modulus for boron fibre reinforced plastics, *J. Acoustical Society of America* **57**, 891.

Raju, P. K., Vaidya, U. K. and Crocker, M. J. (1992) Characterization of carbon-carbon (C/C) composites using vibration measurements, in: *Vibro-Acoustic Characterization of Materials and Structures*, P. K. Raju (ed.), NCA-Vol. 14, ASME, New York, N. Y., 177-81.

Rao, S.S. (1984) *Optimization: Theory and Applications*, Wiley Eastern Limited, Second Edition, pp. 649-51.

Reed, K. E. (1979) 34th Annual Technical Conference, *Reinforced Plastics/Composites Institute*, The Society of the Plastics Industry, Jan. 1979, Session 22G.

Rotem, A. (1993) Load frequency effect on the fatigue strength of isotropic laminates, *Composites Science and Technology* **46**, 129-38.

Roylance, D. (1980) Stress-wave damage in graphite/epoxy laminates, *J. Compos. Mater.* **13**, 14111-119.

Saravanos, D. A. (1993) Mechanics for the effects of delamination on the dynamic characteristics of composite laminates, NCA-Vol. 16/AMD-Vol. 172, *Dynamic Characterization of Advanced Materials*, ASME 1993, P. K. Raju and R. F. Gibson (editors), pp. 11-21.

Schultz, A. B. and Tsai, S. W. (1968) Dynamic moduli and damping ratios in fibre-reinforced composites, *J. Composite Materials* **2**, 368.

Schultz, A. B. and Tsai, S. W. (1969) Measurements of complex dynamic moduli for laminated fibre-reinforced composites, *J. Composite Materials* **3**, 434.

Souma, I, (1982) Dynamic mechanical properties of polyvinylchloride composite system, *J. Appl. Polym. Sci.*

27, 1523-32.

Suarez, S.A., Gibson, R.F., Sun, C.T. and Chaturvedi, S.K. (1986) The influence of fibre length and fibre orientation on damping and stiffness of polymer composite materials, *Experimental Mechanics* **6**, 175-84.

Sun, C.T., Gibson, R.F. and Chaturvedi, S.K. (1985) Internal materials damping of polymer matrix composites under off-axis loading, *Journal of Materials Science* **20**, 2575-85.

Suzuki, K. and Miyano, Y. (1976) Time-temperature dependence of flexural strength of GRP laminates, *J. Soc. Mater. Sci. Jpn.* **25**, 302.-8.

Tracy, J. J. and Pardoen, G. C. (1989) Effect of delamination on the natural frequencies of composite laminates, *J. Composite Materials* **23**, 1200-15.

Wolfe, J. A. and Carne, T. G. (1979) Identification of elastic constants for composites using modal analysis, *Society for Experimental Stress Analysis*, May 1979 Spring Meeting, San Francisco.

15.8. Further Reading

Akay, M. (1971) *Dynamic Mechanical Properties of Polymeric Paint Films*, Ph. D. Thesis, University of Manchester Institute of Science and Technology, UK.

Barker, L. M. (1971) A model for stress-wave propagation in composite materials, *J. Compos. Mater.* **5**, 141-62.

Bennouna, M. M. and White, R. G. (1984) The effects of large vibration amplitudes on the dynamic strain response of a clamped-clamped beam with consideration of fatigue life, *J. Sound Vib.* **96(3)**, 281-308.

Bert, C. W. (1977) Optimal design of a composite material plate to maximise its fundamental frequency, *J. Sound Vib.* **50**, 229-39.

Brinson, L. C. and Knauss, W. G. (1991) Thermorheologically complex behaviour of multiphase viscoelastic materials, *J. Mechanics and Physics of Solids* **39(7)**, 859-80.

Broutma, L. J. and Krock, R. H. (1967) *Modern Composite Materials*, Addison-Wesley.

Caprino, G., Crivelli-Visconti, T. And Di-Illo, A. (1984) Composite materials response under low velocity impact, *Comp. Struct.* **2(3)**, 261-71.

Clarkson, B. L. (1968) Acoustic fatigue test facilities, in: *Noise and Acoustic Fatigue in Aeronautics*, E. J. Richards and D. J. Mead (Eds.), Chap. 19, John Wiley, New York.

Curran, D. R., Seamen, L. and Shockey, D. A. (1987) Dynamic failure of solids, *Phys. Rep.* **147**, 253.

Davison, L. and Graham, R. J. (1979) Shock compression of solids, *Phys. Rep.* **55**(4), 255-379.

Dobynes, A. L. and Porter, T. R. (1981) A study of structural integrity of graphite composite structures subjected to low velocity impact, *Poly. Eng. Sci*, **21(8)**, 493-8.

Folk, R., Fox, G., Shook, C. A. and Curtis, C. W. (1958) Elastic strain produced by sudden application of pressure to one end of a cylindrical bar. I. Theory, *J. Acousto. Soc. Am.* **30**, 552-8.

Fujii, T. (1993) Dynamic response of sandwich beams with an adhesive damping layer, *Int. J. Adhesion and Adhesives* **13(3)**, July 1993, 201-9.

Gibson, R. F. (1975) *Elastic and Dissipative Properties of Fibre-Reinforced Composite Materials in Flexural Vibration*, Ph.D. Dissertation, University of Minnesota, Minneapolis, USA.

Gibson, R. F. (1989) Dynamic mechanical properties of advanced composite materials and structures: A review of recent research, *Shock and Vibration Digest* **9**, 9-17.

Goldsmith, W. (1960) *Impact*, Edward Arnod, London.

Greszczuk, L. B. (1982) Damage in composite materials due to low velocity impact, in: *Impact Dynamics*, John Wiley, New York, 55-94.

Haddad, Y. M. (1986) A microstructural approach to the mechanical response of composite systems with randomly oriented short fibers, *J. Mat. Sci* **21**, 3767-76.

Haddad, Y. M. (1992) On the deformation theory of a class of randomly structured composite systems,

ASME, J. Energy Resources Technology **114**, 110-16.

Haddad, Y. M. (Editor) (1998) *Advanced Multilayered and Fibre-Reinforced Composites*, Kluwer, Dordrecht.

Haddad, Y. M. and Tanary, S. (1989) On the micromechanical characterization of the creep response of a class of composite systems, *ASME, J. Pressure Vessel Technology* **111**, 177-82.

Hagen, R., Salmen, L. and de Ruvo, A. (1993) Dynamic mechanical studies of a highly filled composite structure: A lightweight coated paper, *J. Applied Polymer Science* **48**, 603-10.

Hashin, Z. (1970c) Dynamic behaviour of viscoelastic composites, *Int. J. Solid Struct.* **5**, 539-54.

Herrman, W. (1969) Nonlinear stress waves in metals, in: *Wave propagation in solids*, ASME, New York.

Holehouse, I. (1984) *Sonic Fatigue Design Techniques for Advanced Composite Airplane Structures*, Ph. D. Thesis, University of Southhampton.

Holmes, J. W., Wu, X. And Ramakrishnau, V. (1992) High-frequency fatigue of fibre-reinforced ceramics, 16th Annual Conference and Exposition on Ceramics and Advanced Composites, Cocoa Beach, Fl., January 1992.

Hooker, R. J. (1975) *High Damping Metals*, Ph.D. Thesis, University of Sothampton.

Khalil, A. A. and Bayoumsi, M. R. (1991) Effect of loading rate on fracture toughness of bonded joints, *Int. J. Adhesion and Adhesives* **11(1)**, January 1991, 25-29.

Kinra, V. K., Wren, G. G., Rawal, S. P. and Misra, M. S. (1991) On the influence of ply-angle on damping and modulus of elasticity of a metal matrix composite, *Metallurgical Transactions A* **22a**, 641-51.

Kinslow, R. (1967) Stress waves in laminated materials, AIAA paper No. 67-140, AIAA-5th Aerospace Sciences Meeting, January 1967.

Lai, K. M. (1983) Low velocity transverse impact behaviour of 8 ply graphite-epoxy laminates, *J. Rein. Plast. Comp.* **2**, 216.

Lai, K. M. (1983) Residual strength assessment of low velocity impact damage of graphite-epoxy laminates, *J. Rein. Plast. Comp.* **2**, 226-39.

Lazan, B. J. (1968) *Damping of Materials and Members in Structural Mechanics*, Pergamon Press, Oxford.

Lesientre, G. A. (1994) Modelling frequency-dependent longitudinal dynamic behaviour of linear viscoelastic long fibre composites, *J. Composite Materials* **28 (18)**, 1770-82.

Lesientre, G. A., Yarlagadda, D., Christensen, D. and Whatley, W. (1993) Enhanced flexural damping of composite plates using intercalated graphite fibre, *AIAA Journal* **31(4)**, 746-50.

Lin, D. X., Ni, R. G. and Adams, R. D. (1984) Prediction and measurement of the vibrational damping parameters of carbon and glass fibre-reinforced plastic plates, *J. Composite Materials* **18**, 132-52.

Murayama, T. (1978) *Dynamic Mechanical Analysis of Polymeric Material*, Elsevier, Amsterdam.

Ni, R. G. and Adams, R. D. (1984) The damping and dynamic moduli of symmetric laminated composite beams - Theoretical and experimental results, *J. Composite Materials* **18**, 104-21.

Nielsen, L. E. (1974) *Mechanical Properties of Polymers and Composites*, Marcel Dekker, New York, Vol. 2, Ch. 7.

Palmer, T. A. and White, R. G. (1984) Development of carbon fibre reinforced plastics for use in high performance structures, in Proceedings, *Conf. on Fibre Reinforced Plastics*, University of Liverpool, April 1984.

Peck, J. C. (1971) Pulse attenuation in composites, in: *Shock Waves and the Mechanical Properties of Solids*, J. J. Burke and V. Weiss (Eds.), Proc. 17th Army Materials Research Conf., Syracuse Univ. Press, pp. 155-84.

Peck, J. C. and Gurtman, G. A. (1969) Dispersive pulse propagation parallel to the interfaces of a laminated composite, *J. Appl. Mech.* **36**, 479-84.

Saravanos, D. A. and Chamis, C. C. (1990) Mechanics of damping for fibre composite laminates including hygro-thermal effects, *AIAA Journal* **28 (10)**, 1813-19.

Saravanos, D. A. and Pereira, J. M. (1992) Effects of interply daming layers on the dynamic characteristics of composite plates, *AIAA Journal* **30(12)**, 2906-13.

Schier, J. F. and Juergens, R. J. (1983) Design impact of composites on fighter aircraft. Pt. 1. They force a

fresh look at the design process, Aeronautics and *Astronautics*, **21** (9), 44-9.
Schuler, K. W. and Nunziato, J. W. (1974) The dynamic mechanical behaviour of polymethyl methacrylate, *Rheol. Acta* **13**, 265-73.
Shalak, R. (1957) Longitudinal impact of a circular elastic rod, *J. Appl. Mech.* **24**, 59-64.
Shen, M. H. H. and Grady J. E. (1992) Free vibrations of delaminated beams, NASA TM 105582.
Shuler, S. A., Holmes, J. W. and Wu, X. (1993) Influence of loading frequency on the room-temperature fatigue of a carbon-fibre/Si C- matrix composite, *J. Am. Ceram. Soc.* **76** (9), 2327-36.
Souvere, J. (1984) *Dynamic Response of Acoustically Excited Stiffened Composite Honeycomb Panels*, Ph. D. Thesis, University of Southhampton. (3), 309-31.
Suarez, S. A., Gibson, R. F., Sun, C. T. and Chaturvedi, S. K. (1986). The influence of fibre length and fibre orientation on damping and stiffness of polymer composite materials, *Experimental Mechanics* **26**, 175-84.
Tauchert, T. R. and Hsu, N. N. (1973) Influence of stress upon internal damping in a fibre-reinforced composite material, *J. Composite Materials* **7**, 546.
Teh, C. E., and White, R. G. (1980) Dynamic response of isotropic and anisotropic panels under simulated flight loading conditions, *Proceedings of the 1st International Conference on Recent Advances in Structural Dynamics*, ISVR
Thomson, J. L. (1990) Investigation of composite interphase using dynamic mechanical analysis, artifacts and reality, *Polymer Composites* **11**, 105-13.
Ting, T. C. T. (1980) Dynamic response of composites, *Appl. Mech. Rev.* **33**, 1629-35.
Tsai, G. C., Doyle, J. F.and Sun, C. T. (1987) Frequency effects on the fatigue life and damage of graphite/epoxy composites, *J. Comp. Mater.* **21(1)**, 2-13.
Tyler, F. R. and Butcher, R. M. (1968) A criterion for the time-dependence of dynamic fracture, *Int. J. Fract. Mech.* **4**, 431.
Velupillai, D. (1978) British Aerospace goes for composites, *Flight International*, November 1978.
Ward, I. M. (1985). *Mechanical Properties of Solid Polymers*, 2nd Edn, Wiley, New York.
White, R. G. (1975) Some measurements of the dynamic properties of mixed carbon fibre reinforced plastic beam and plates, *The Aeronautical Journal of the Royal Aeronautical Society*, 318-25.
Whittler, J. S. and Peck, J. C.(1969) Experiments on dispersive pulse propagation in laminated composites and comparison with theory, *J. Appl. Mech.* **36**, 485-90.

TABLE 15.3. **Optimization results for various starting points Y^T (Feng, 1999)**

Starting point	Fibre off-axis angle θ°	Fibre aspect ratio ℓ/d	Fibre volume fraction V_f %	Local minima
[0, 0, 0]	44.4	1.49	60.6	0.5833
[0.5, 0.5, 0.5]	43.9	83.6	54.2	0.5827
[1.0, 1.0, 1.0]	43.9	66.7	54.6	0.5828
[2.5, 2.5, 2.5]	44.1	1.46	60.8	0.5832
[5.0, 5.0, 5.0]	43.9	97.4	54.3	0.5827
[7.5, 7.5, 7.5]	44.3	38.2	55.2	0.583
[10, 10, 10]	43.7	49.7	54.3	0.5829
[15, 15, 15]	44	1.47	60.8	0.5832
[20, 20, 20]	43.7	100	54.1	0.5827
[25, 25, 25]	43	1.05	65	0.5837
[30, 30, 30]	43.7	99.8	54.2	0.5827
[35, 35, 35]	43.5	97	54	0.5827
[40, 40, 40]	44.2	1.44	60.6	0.5832
[45, 45, 45]	43.8	99.1	54	0.5827
[50, 50, 50]	43.7	57.9	54.8	0.5828
[55, 55, 55]	43.7	72.6	53.8	0.5828
[60, 60, 60]	43.3	97.2	53.6	0.5827
[65, 65, 65]	43.4	95.3	53.8	0.5827
[70, 70, 70]	43.7	96.4	54.2	0.5827
[75, 75, 75]	43.4	99.1	53.9	0.5827
[80, 80, 80]	44.8	88	54.2	0.5827
[85, 85, 85]	43.8	75.1	54.1	0.5827
[90, 90, 90]	43.5	84.8	54.1	0.5827
[95, 95, 95]	43.2	89.4	53.7	0.5827

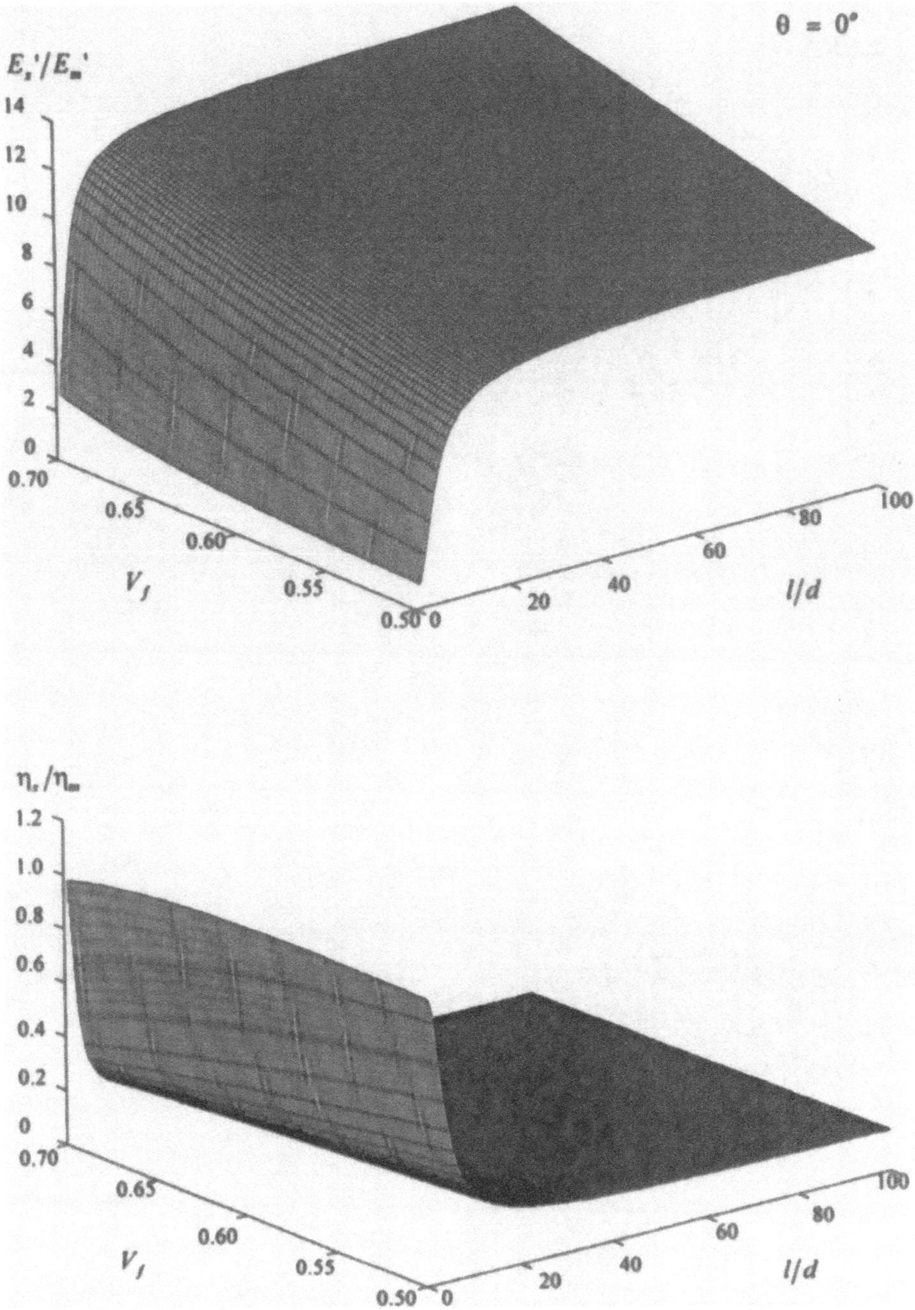

Figure 15.41. Non-dimensional ratios η_x/η_m and E_x'/E_m' *vs*. fibre volume fraction V_f and fibre-aspect-ratio l/d. Off-axis angle θ is set set to be 0°. Reprinted with permission from Feng (1999); *see* Haddad and Feng (1999).

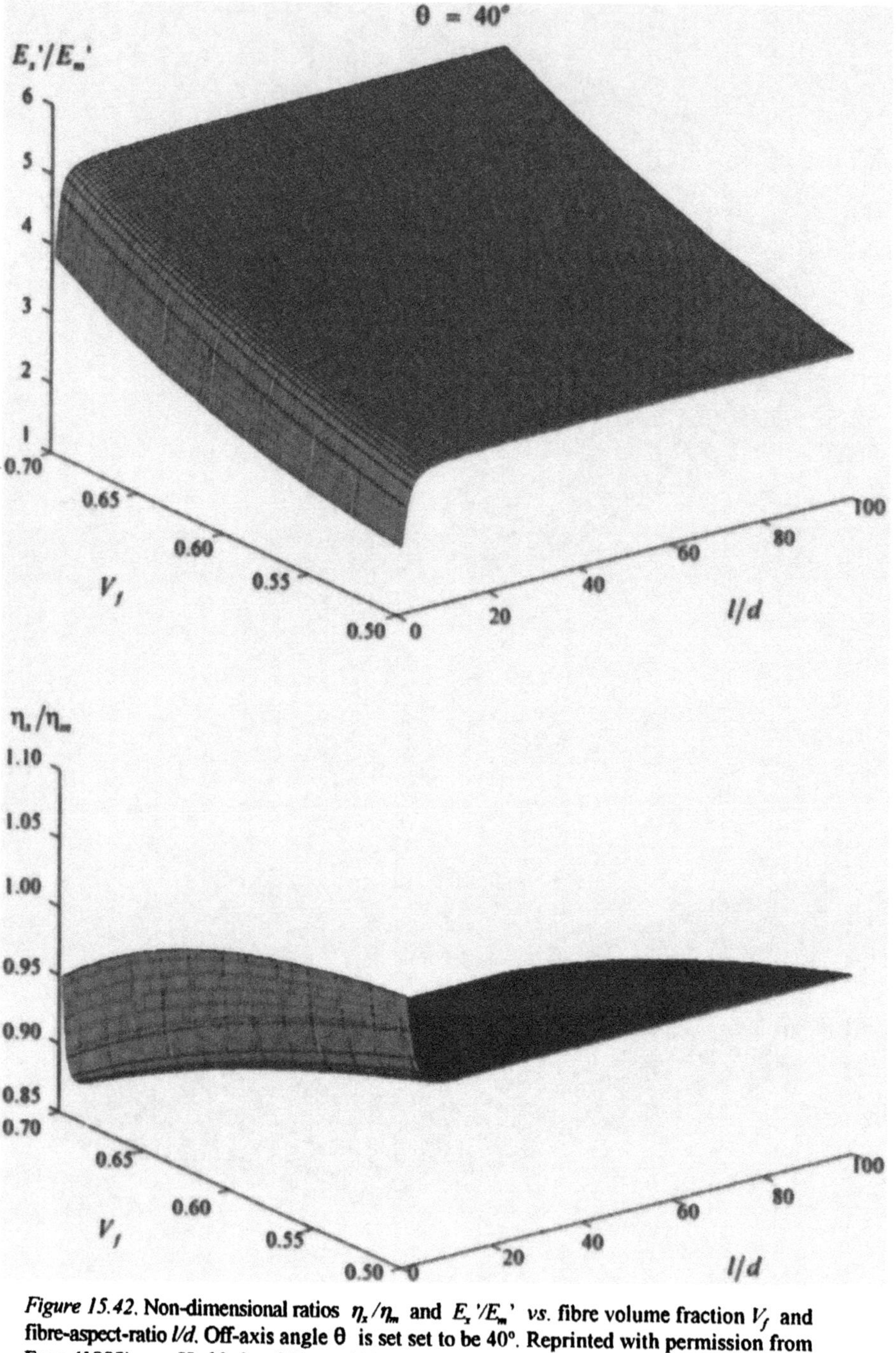

Figure 15.42. Non-dimensional ratios η_x/η_m and E_x'/E_m' *vs.* fibre volume fraction V_f and fibre-aspect-ratio l/d. Off-axis angle θ is set set to be 40°. Reprinted with permission from Feng (1999); *see* Haddad and Feng (1999).

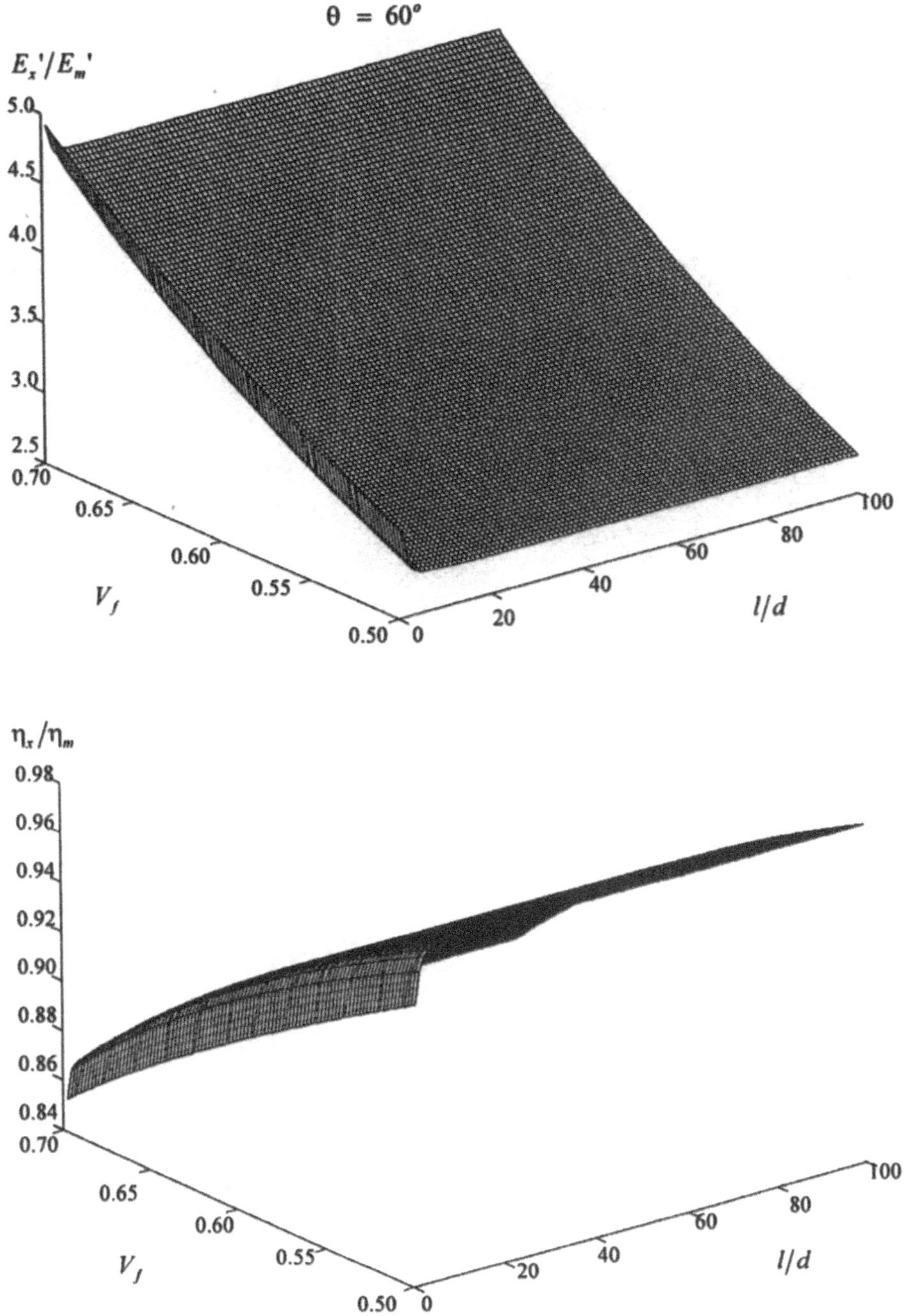

Figure 15.43. Non-dimensional ratios η_x/η_m and E_x'/E_m' *vs*. fibre volume fraction V_f and fibre-aspect-ratio l/d. Off-axis angle θ is set set to be 60°. Reprinted with permission from Feng (1999); *see* Haddad and Feng (1999).

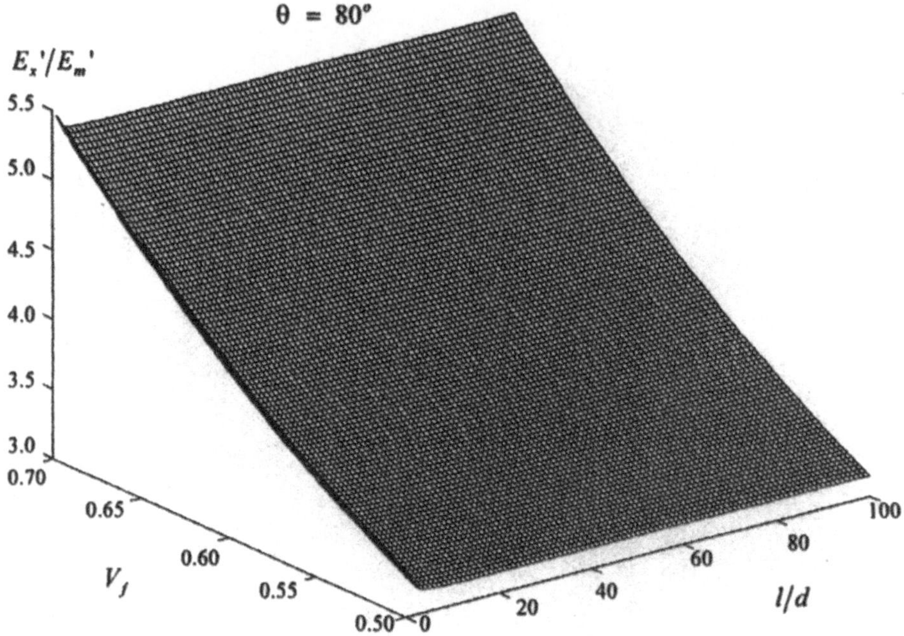

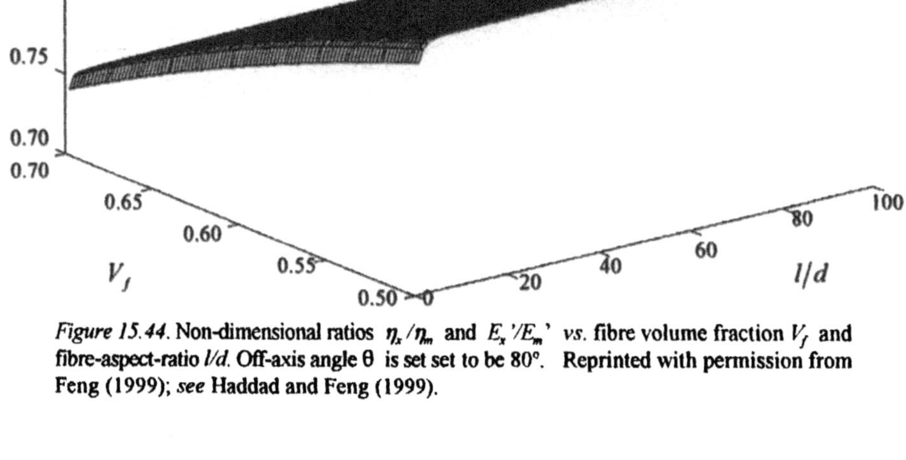

Figure 15.44. Non-dimensional ratios η_x/η_m and E_x'/E_m' *vs.* fibre volume fraction V_f and fibre-aspect-ratio l/d. Off-axis angle θ is set set to be 80°. Reprinted with permission from Feng (1999); *see* Haddad and Feng (1999).

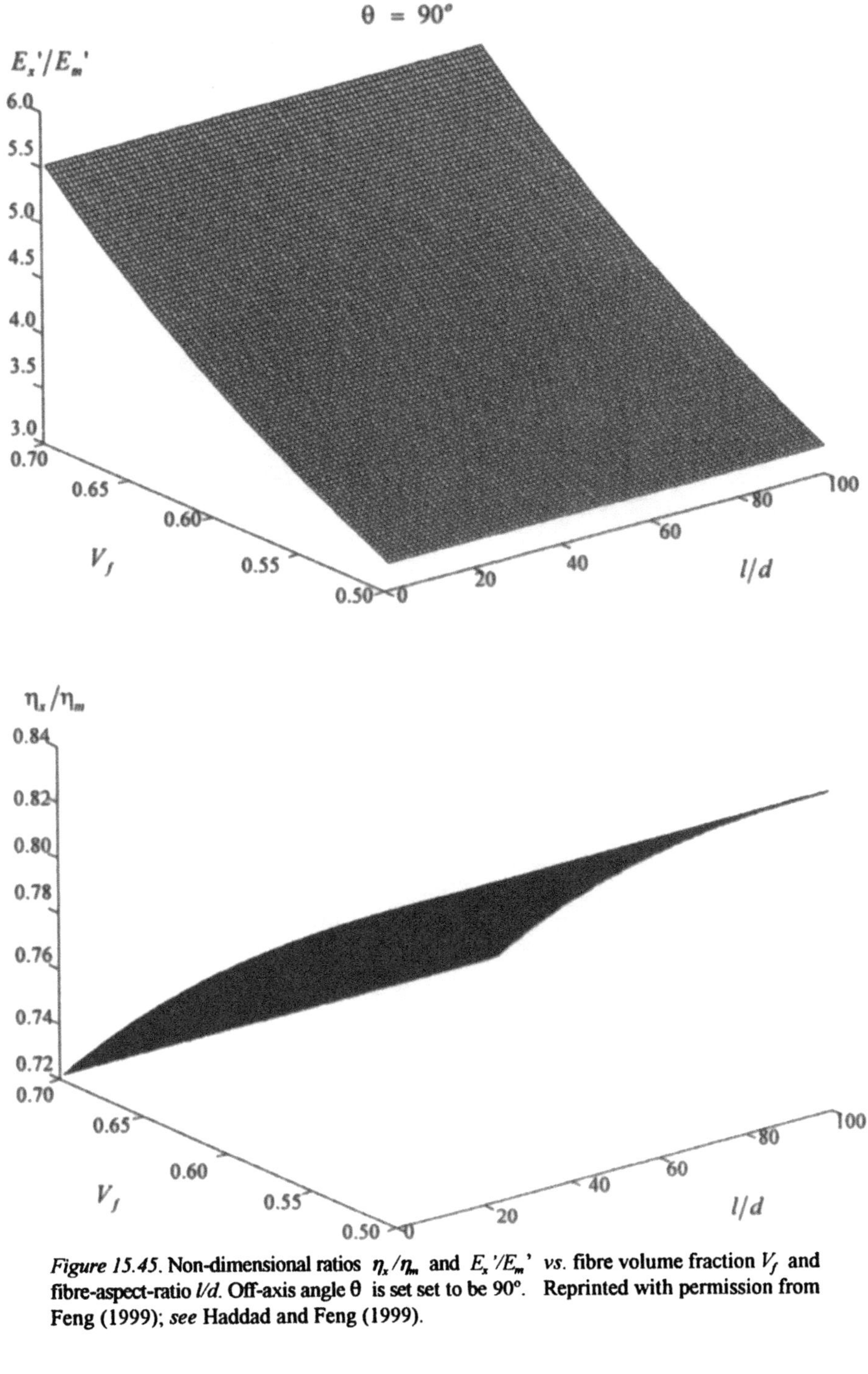

Figure 15.45. Non-dimensional ratios η_x/η_m and E_x'/E_m' *vs*. fibre volume fraction V_f and fibre-aspect-ratio l/d. Off-axis angle θ is set set to be 90°. Reprinted with permission from Feng (1999); *see* Haddad and Feng (1999).

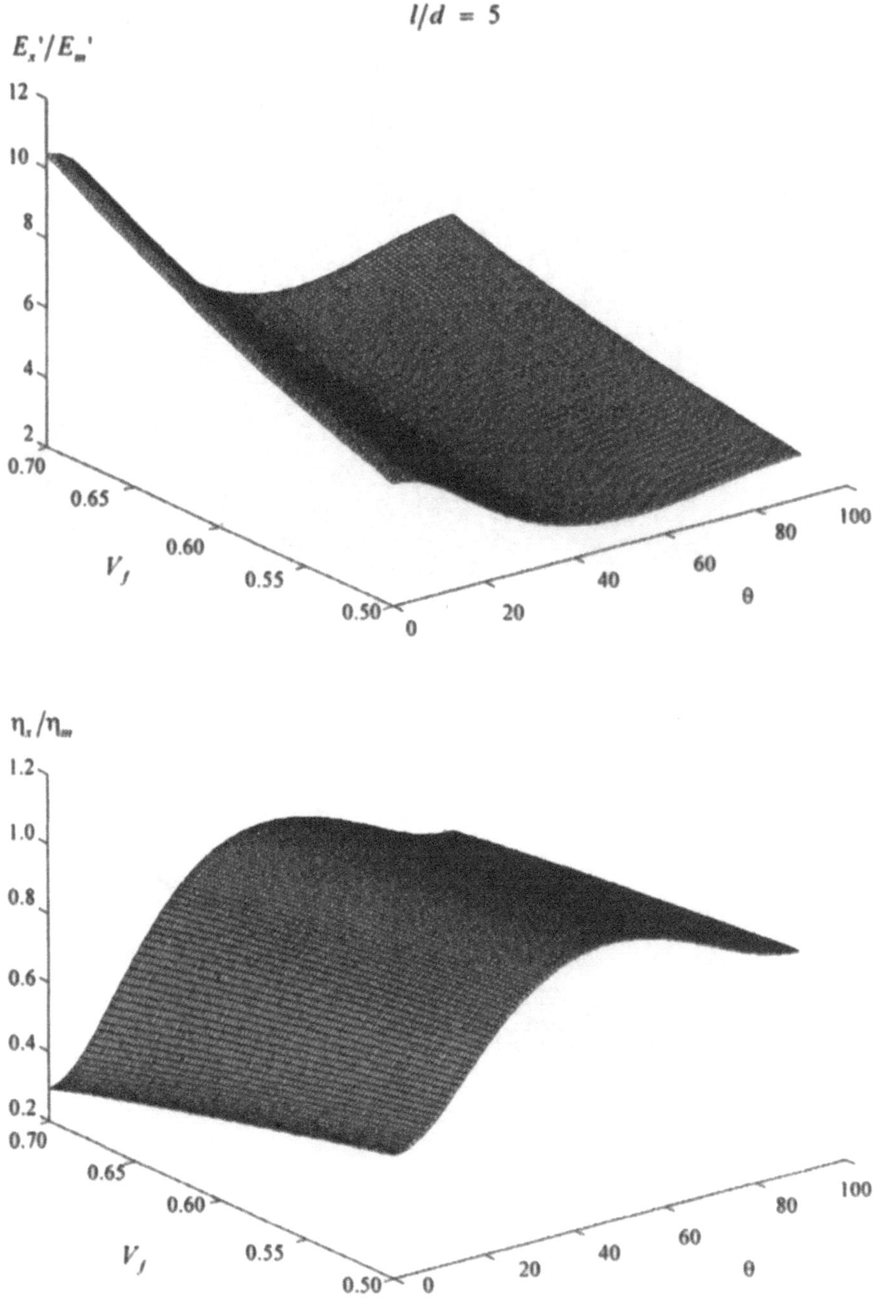

Figure 15.46. Non-dimensional ratios η_x/η_m and E_x'/E_m' *vs.* fibre volume fraction V_f and off-axis angle θ. Fibre-aspect-ratio l/d is set to be 5. Reprinted with permission from Feng (1999); *see* Haddad and Feng (1999).

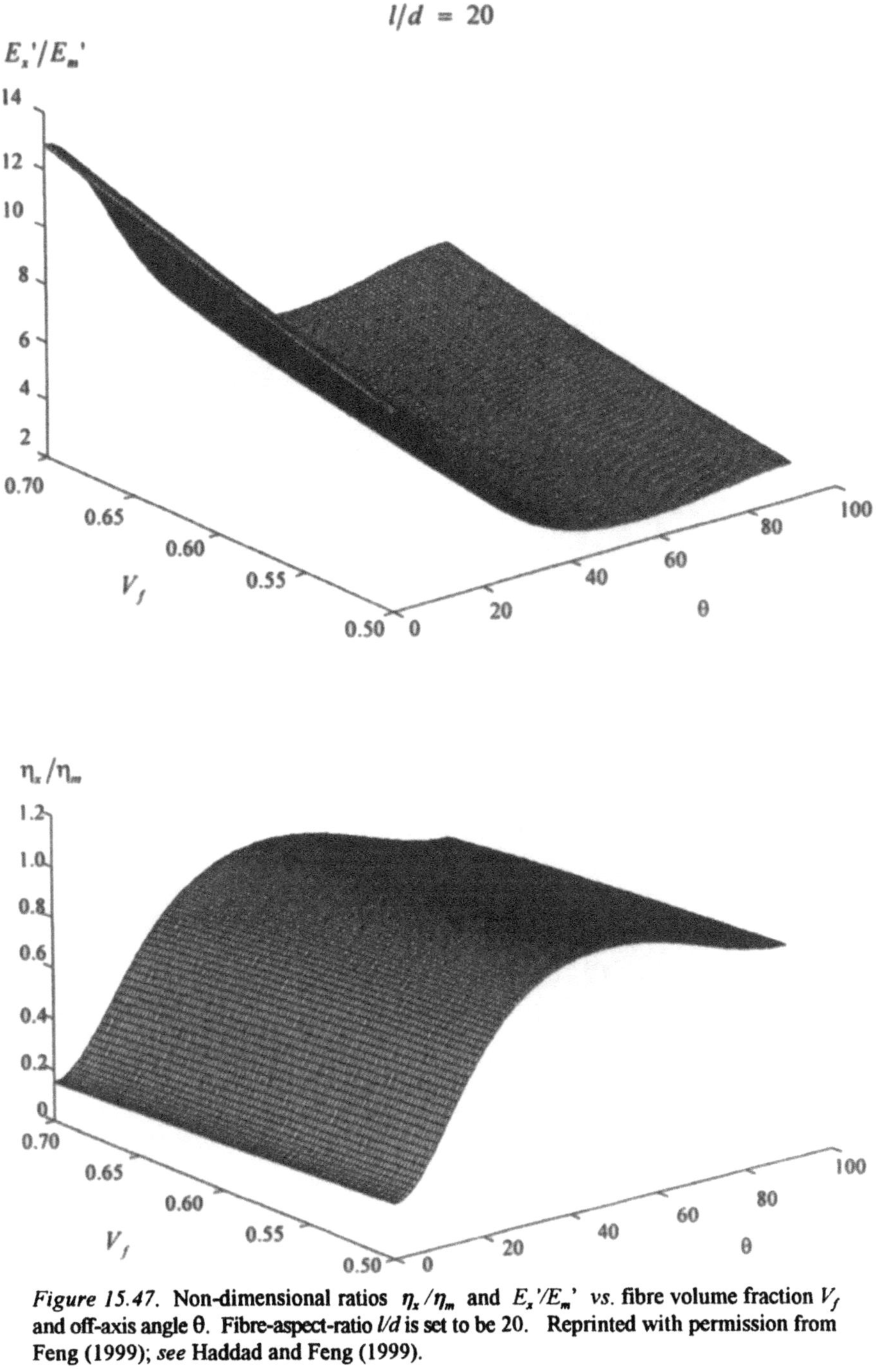

Figure 15.47. Non-dimensional ratios η_x/η_m and E_x'/E_m' *vs.* fibre volume fraction V_f and off-axis angle θ. Fibre-aspect-ratio l/d is set to be 20. Reprinted with permission from Feng (1999); *see* Haddad and Feng (1999).

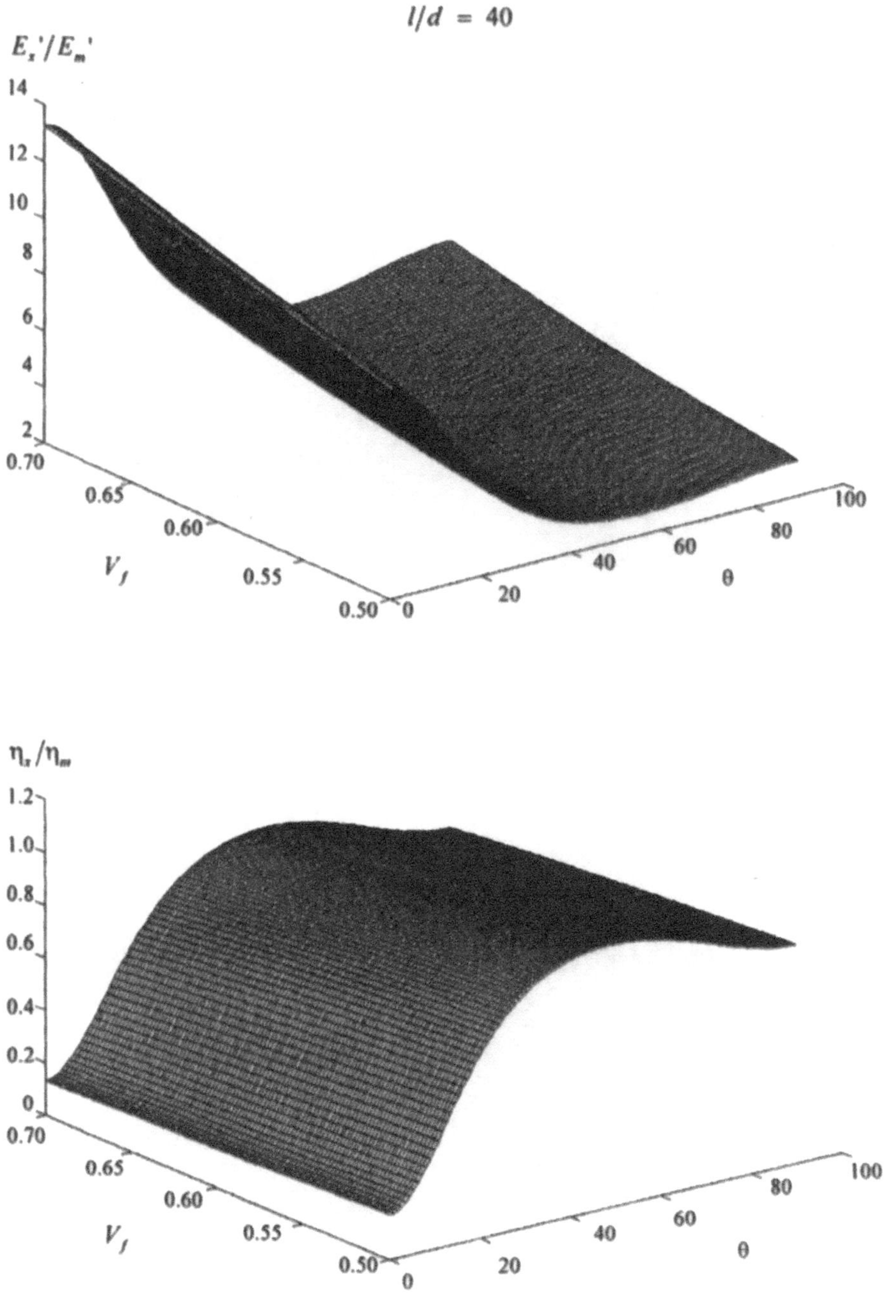

Figure 15.48. Non-dimensional ratios η_x/η_m and E_x'/E_m' *vs.* fibre volume fraction V_f and off-axis angle θ. Fibre-aspect-ratio l/d is set to be 40. Reprinted with permission from Feng (1999); *see* Haddad and Feng (1999).

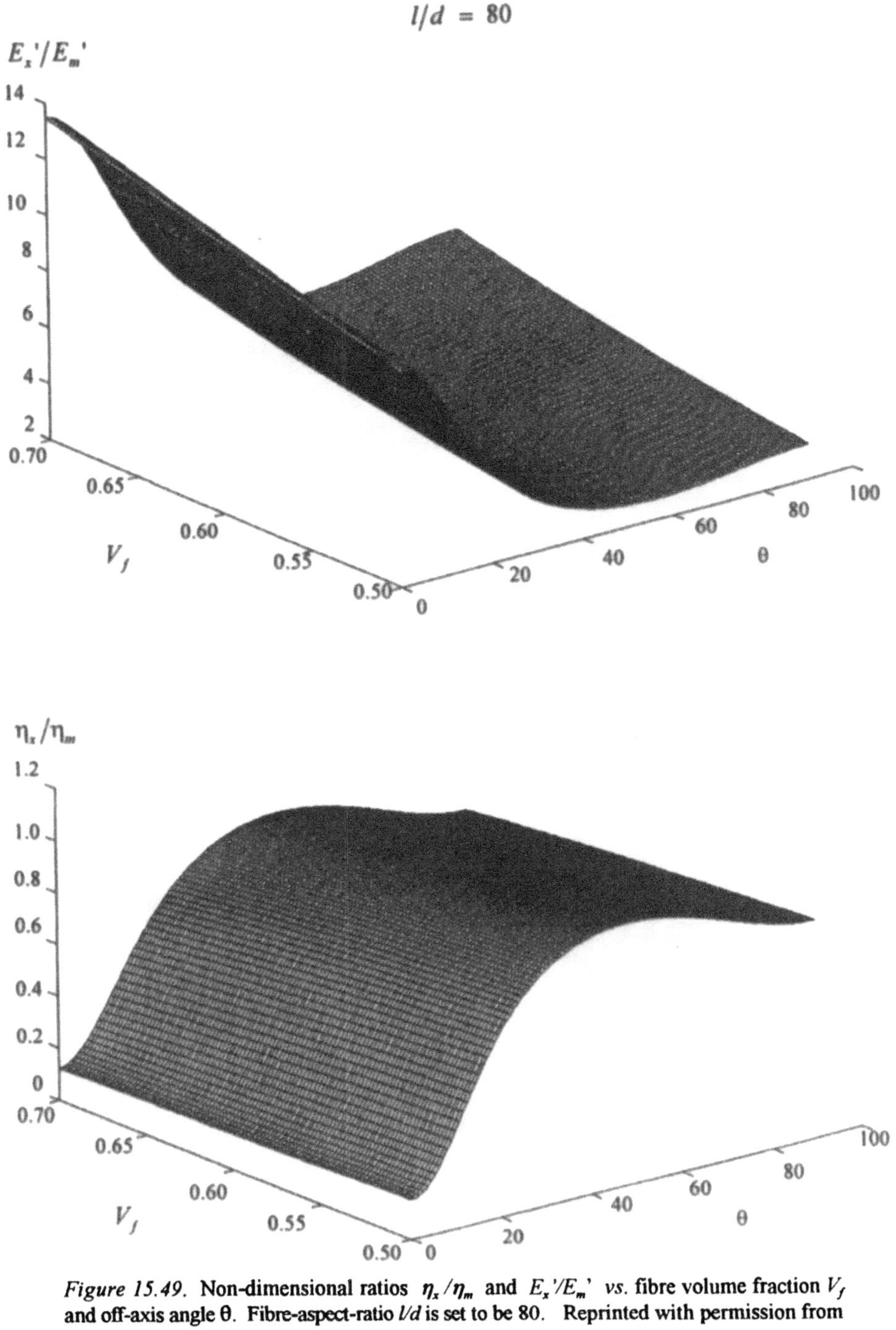

Figure 15.49. Non-dimensional ratios η_x/η_m and E_x'/E_m' *vs.* fibre volume fraction V_f and off-axis angle θ. Fibre-aspect-ratio l/d is set to be 80. Reprinted with permission from Feng (1999); *see* Haddad and Feng (1999).

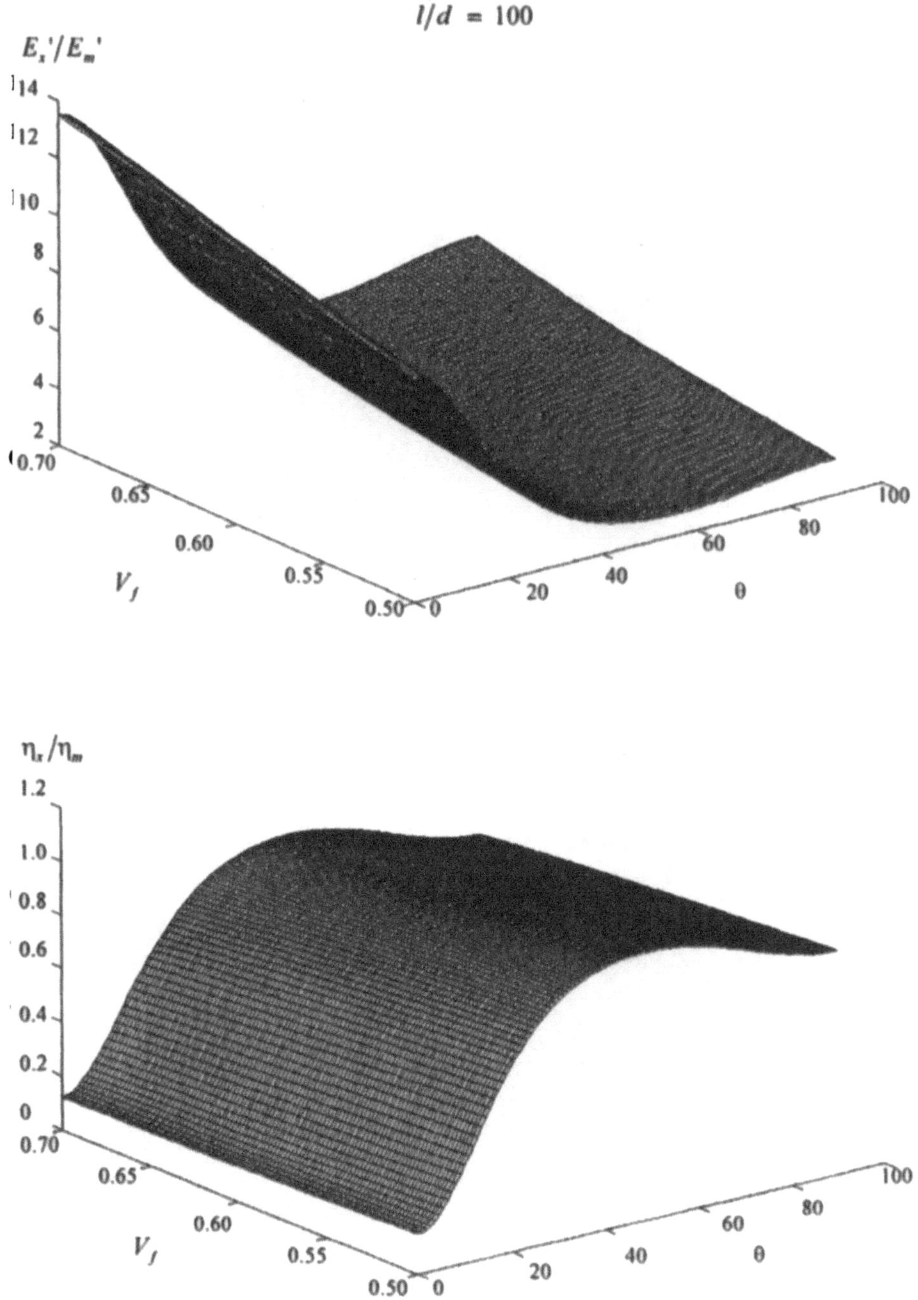

Figure 15.50. Non-dimensional ratios η_x/η_m and E_x'/E_m' *vs.* fibre volume fraction V_f and off-axis angle θ. Fibre-aspect-ratio l/d is set to be 100. Reprinted with permission from Feng (1999); *see* Haddad and Feng (1999).

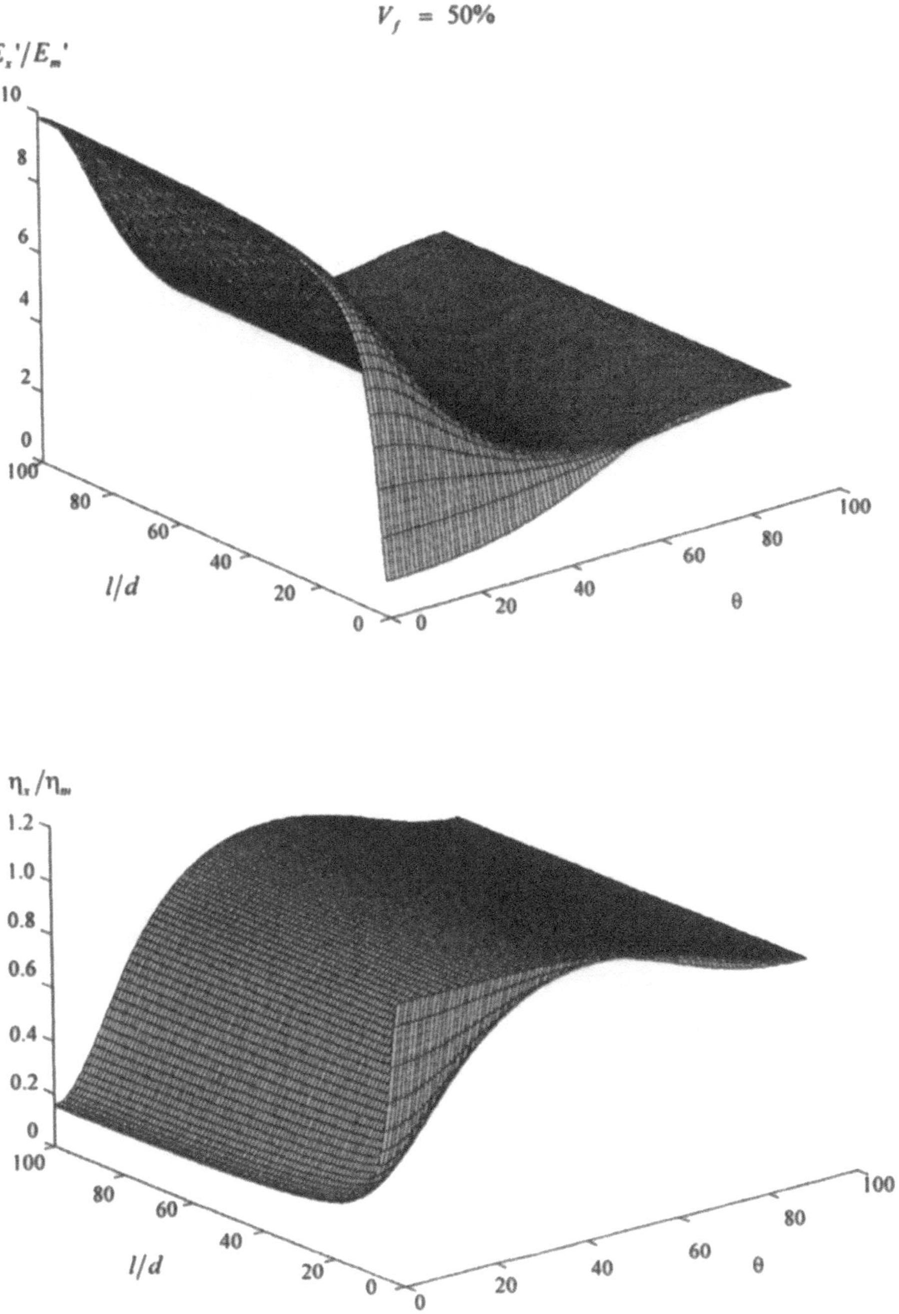

Figure 15.51. Non-dimensional ratios η_x/η_m and E_x'/E_m' *vs*. fibre-aspect-ratio l/d and off-axis angle θ. Fibre volume fraction V_f is set to be 50%. Reprinted with permission from Feng (1999); *see* Haddad and Feng (1999).

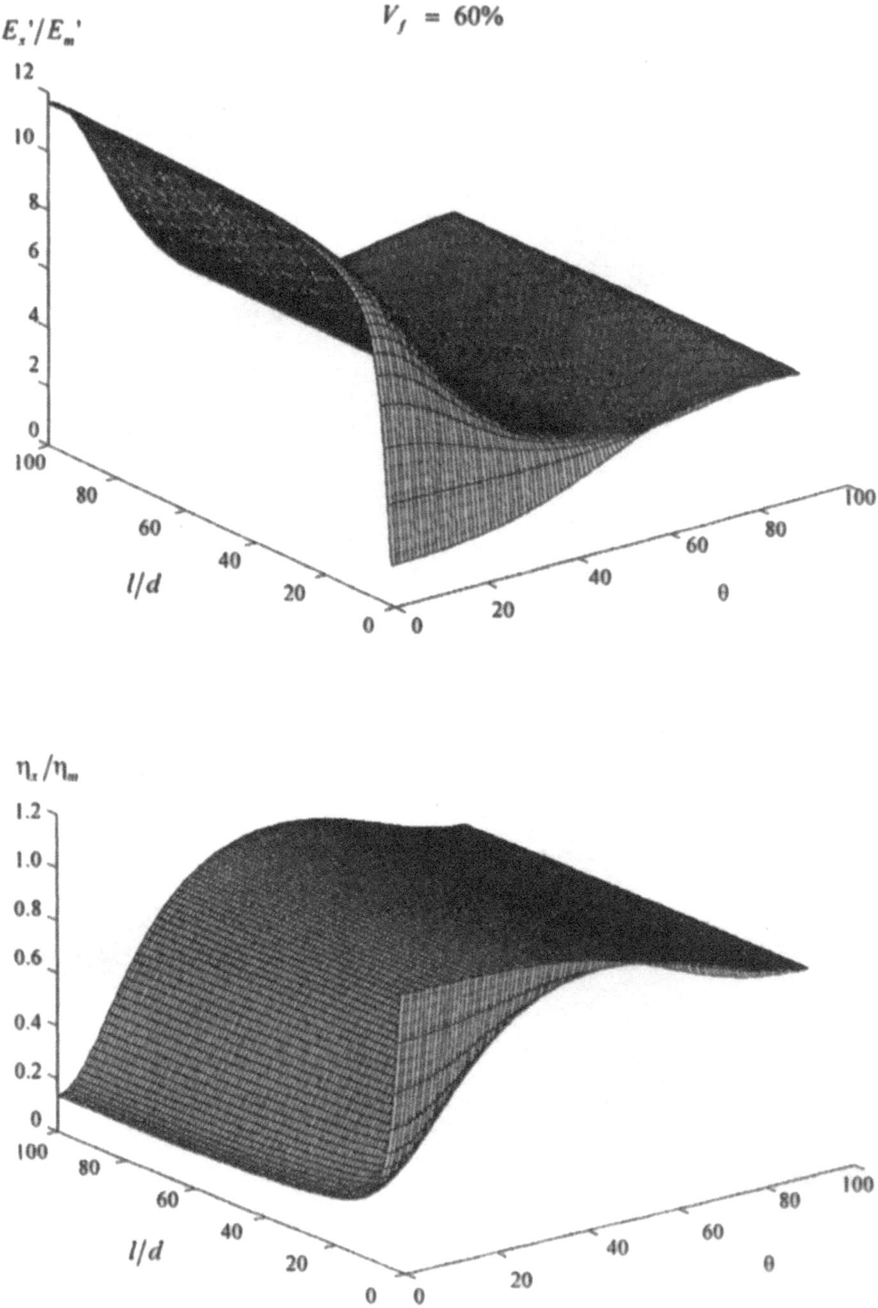

Figure 15.52. Non-dimensional ratios η_x/η_m and E_x'/E_m' *vs.* fibre-aspect-ratio l/d and off-axis angle θ. Fibre volume fraction V_f is set to be 60%. Reprinted with permission from Feng (1999); *see* Haddad and Feng (1999).

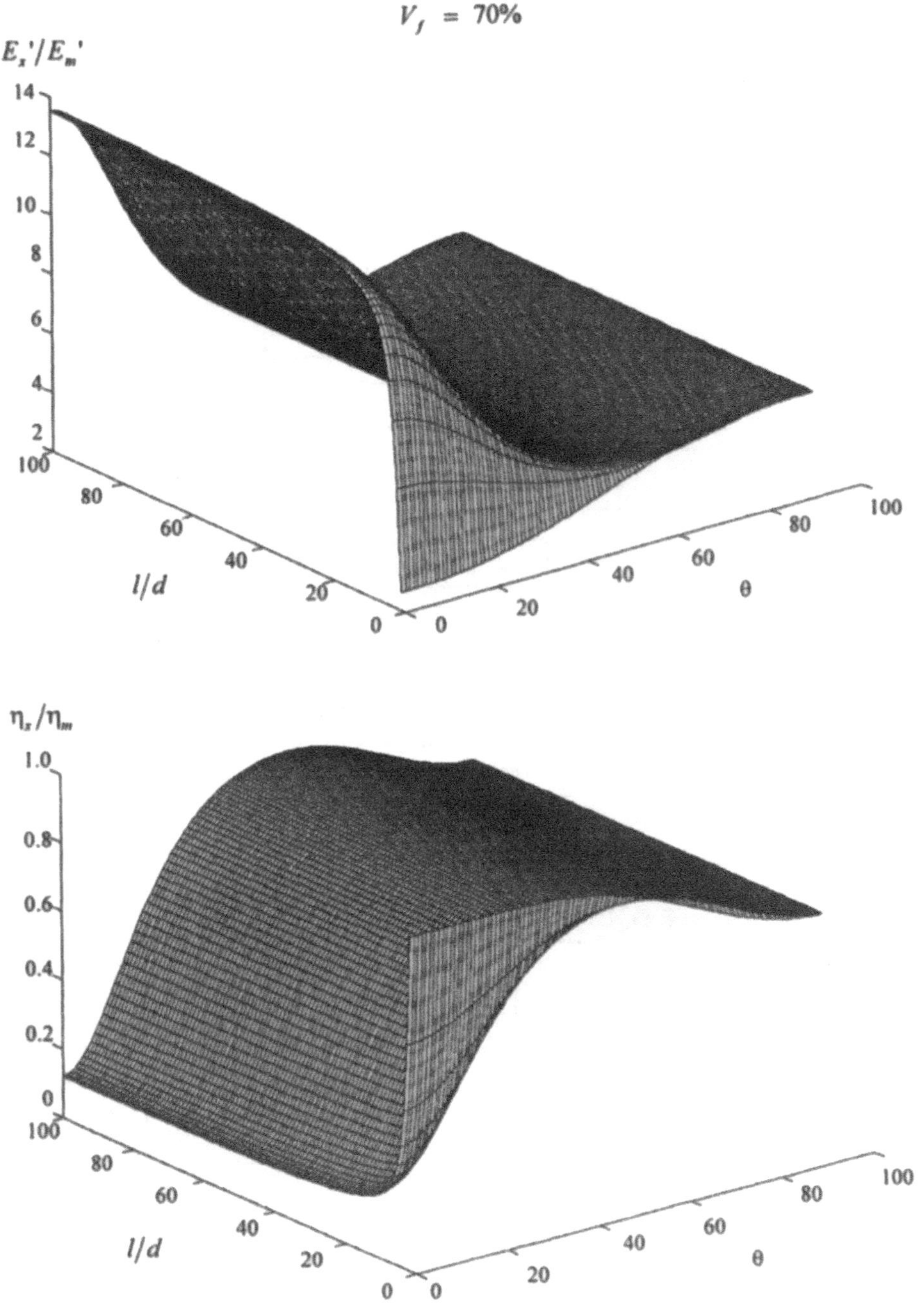

Figure 15.53. Non-dimensional ratios η_x/η_m and E_x'/E_m' *vs.* fibre-aspect-ratio l/d and off-axis angle θ. Fibre volume fraction V_f is set to be 70%. Reprinted with permission from Feng (1999); *see* Haddad and Feng (1999).

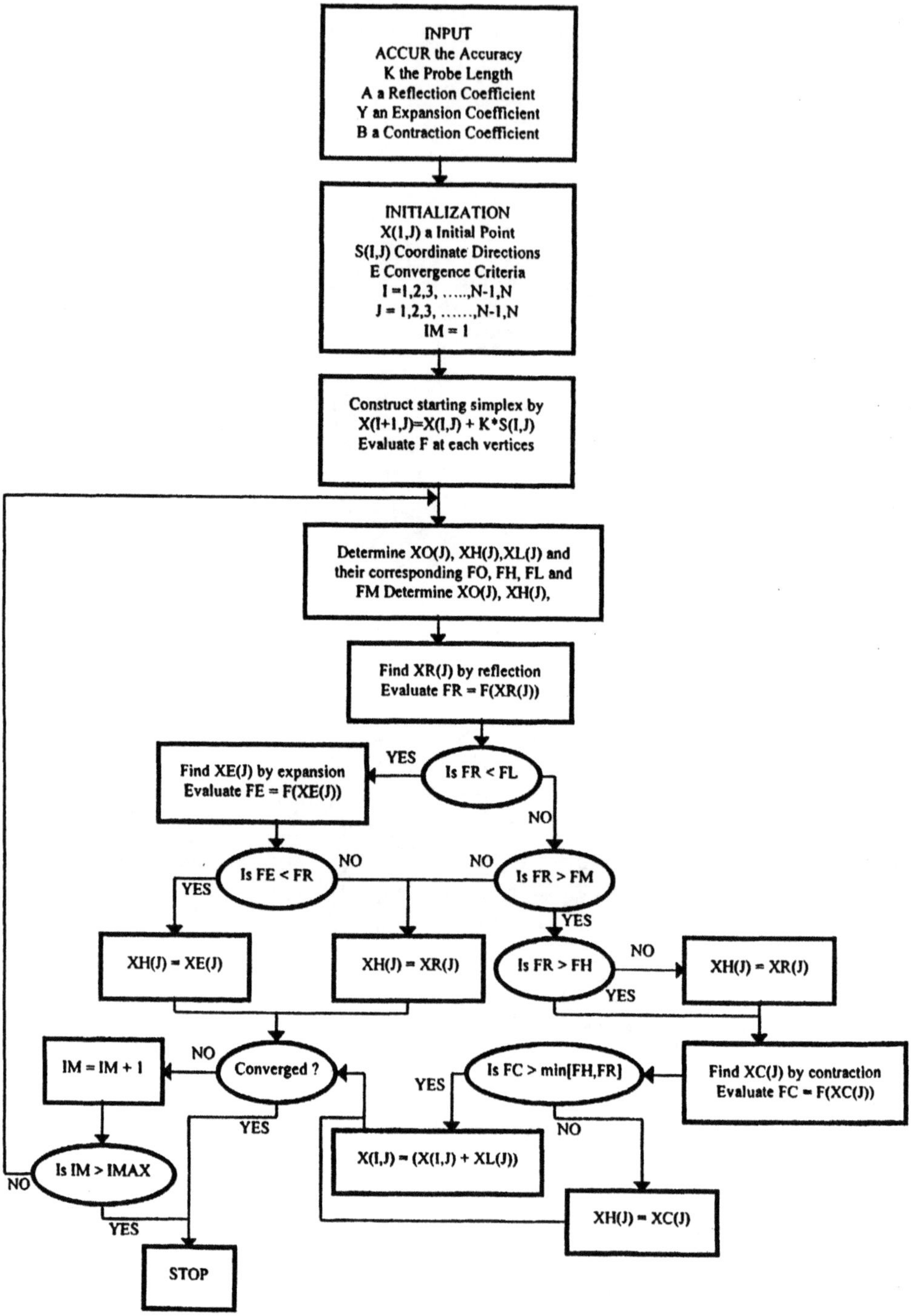

Figure 15.54. Computer program flowchart of "*Simplex Optimization Method*". Reprinted with permission from Feng (1999); *see* Haddad and Feng (1999).

CHAPTER 16

THE STOCHASTIC MICROMECHANICAL APPROACH TO THE RESPONSE BEHAVIOUR OF ENGINEERING MATERIALS

16.1. Introduction

In the past several decades, the micromechanical approach has been recognized as a promising tool for the description of the response behaviour of engineering materials with the inclusion of the so-called "*local*" or "*microstructural*" effects. The microstructure of a class of such materials, however, is discrete in the sense of being heterogeneous and/or discontinuous. In view of this fact, the so-called "deterministic micromechanics", that are based on the concepts of continuum mechanics, could no longer be accepted for the interpretation of the experimental results concerning the behaviour of discrete materials. It has been, therefore, increasingly appreciated that a more appropriate representation of discrete materials would only be achieved by including the random characteristics of the real microstructure. Further, the response behaviour of such microstructure is often both time- and loading history-dependent. Thus, the pertaining deformation process and its space- and time-evolutions are expected to be stochastic in character. In this context, the establishment of the connection between the response behaviour of the individual elements of the microstructure, their interactions, and the observable macroscopic behaviour would be an essential requirement. The fulfilment of the latter seems possible (Axelrad, 1993, and Axelrad and Haddad, 1998) by the introduction of the principles of set theory, together with the concepts of measure theory and topology. Thus, in the stochastic micromechanical formulations, continuum mechanics concepts are generally replaced by considerations of microstructural response variables in the form of discrete statistical functions. The latter are established within well-defined "*measuring scales*" defining the levels of observation into the material system.

In order to describe the mechanical response of the nonhomogeneous material system from a microstructural point of view, it is necessary to consider the response of the individual structural elements which on a local scale could differ considerably from an average response, which would be arrived at by the phenomenological continuum formulations. Such local deviations in the response behaviour, which are usually neglected within the continuum mechanics approach, are directly related to basic properties of the nonhomogeneous material system. Accordingly, the stochastic microstructural analysis begins with a definition of the "*structural element*" of the particular material system under consideration and deals with the formulation of its response behaviour in a probabilistic sense.

In order to extend the formulation, pertaining to the response behaviour of the structural element, to the practical case of a macroscopic material system, the stochastic micromechanical approach makes use of "*intermediate quantities*" arising from the consideration of the existence of a statistical ensemble of structural elements within an intermediate domain of the material specimen. Further, it is equally important to find a connection between the microscopic and the macroscopic response formulations. Thus, the analysis aims at the formulation of a set of "*governing response equations*" for the structured material system that, in contrast to the classical continuum mechanics formulations, are based on the concepts of statistical theory and probabilistic micromechanics; *see* Axelrad (1984, 1993), Haddad (1990, 1995), and Axelrad and Haddad (1998). In this context, it has been found useful to employ operational representation of the various relations. Hence, the notion of a "*Material Operator*" characteristic of the response behaviour of an intermediate domain of the material is introduced. This material operator provides the connection between the stress field and the occurring deformations within the intermediate domain, of the material system, under consideration. The "*Material Operator*" would generally contain in its argument those stochastic variables or functions of such variables distinctive of the response behaviour of the microstructure within the intermediate domain. In a very reduced and simplified form, such an operator may be expressed as

$$\Gamma(\epsilon, t) = \Gamma({}^{\alpha}\Gamma(\epsilon,t), {}^{\alpha\beta}\Gamma(\epsilon,t), {}^{\alpha}K, p_1, p_2, \dots)$$

where ${}^{\alpha}\Gamma$ and ${}^{\alpha\beta}\Gamma$ are random material operators expressing the response characteristics of elements of the microstructure, ${}^{\alpha}K$ is a function of one or more geometrical parameters, p_1 and p_2 are geometrical probabilities, ϵ is the microstrain and t is the time parameter. Other variables that may be included in the argument of the material operator $\Gamma(\epsilon, t)$ above could include, for instance, the temperature T and relative humidity ϕ, among others. An approximate classification of structured solids that could be treated within the context of the presented stochastic micromechanical approach is presented in Table 16.1. Micrographs of the microstructures of a number of materials representative of some of the classes given in Table 16.1 are shown , with various magnifications, in Figures 16.1 to 16.6. Meanwhile, a comparison between some of the basic concepts of the stochastic micromechanical approach adopted here and the corresponding postulates of the conventional continuum mechanics approach is shown in Table 16.2.

16.2. Probabilistic Micromechanical Response

One of the main concepts of "*stochastic micromechanics*" (Axelrad, 1993) is the use of "*three measuring scales*". The smallest scale is identifiable with a "*structural element*" of the actual microstructure. The next scale is intermediate between the level of the structural element and the macroscopic scale of the material system. It is termed "*meso*". The third and largest scale is identifiable with the macroscopic material body. It is referred-to as "*macroscopic*". The concept of "*three measuring scales*" in stochastic micromechanics is dealt with in more details below.

TABLE 16.1. Some classes of structured engineering materials that could be treated within the context of the stochastic micromechanical approach (after Axelrad, 1978)

A	*B*	*C*	*D*	*E*
Polycrysalline solids	Composite Materials	Fibrous systems	Polymeric systems	Particulate materials
High temperature solids	Two-phase materials	Paper	Synthetic fibrous structures	Soils
Directional solidified metals		Textiles		Dispersed particle systems

Figure 16.1. TEM-micrograph (X5700) of Silicon-Steel. Reprinted from Axelrad, D. R. (1993) *Stochastic Mechanics of Discrete Media*, Springer-Verlag Berlin Heidelberg, pp. 64, with kind permission from Springer-Verlag Berlin Heidelberg.

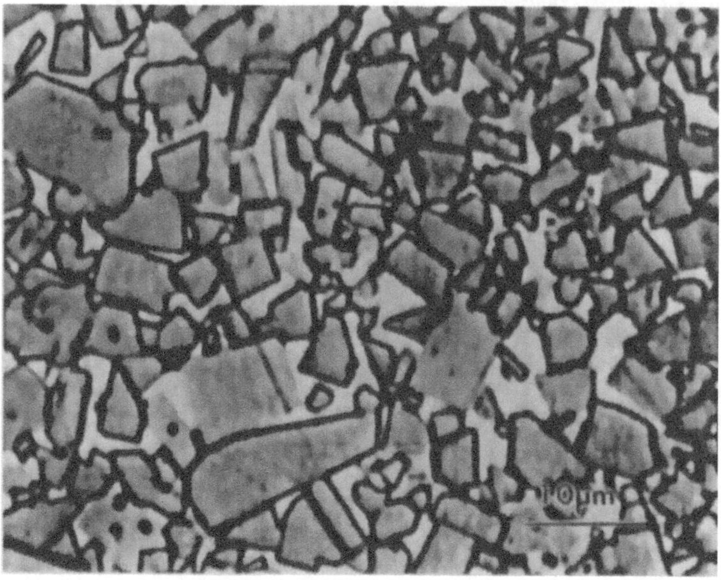

Figure 16.2. TEM-micrograph (X1500) of Tungsten-Cobalt compound. Reprinted from Axelrad, D. R. (1993) *Stochastic Mechanics of Discrete Media*, Springer-Verlag Berlin Heidelberg, pp. 64, with kind permission from Springer-Verlag Berlin Heidelberg.

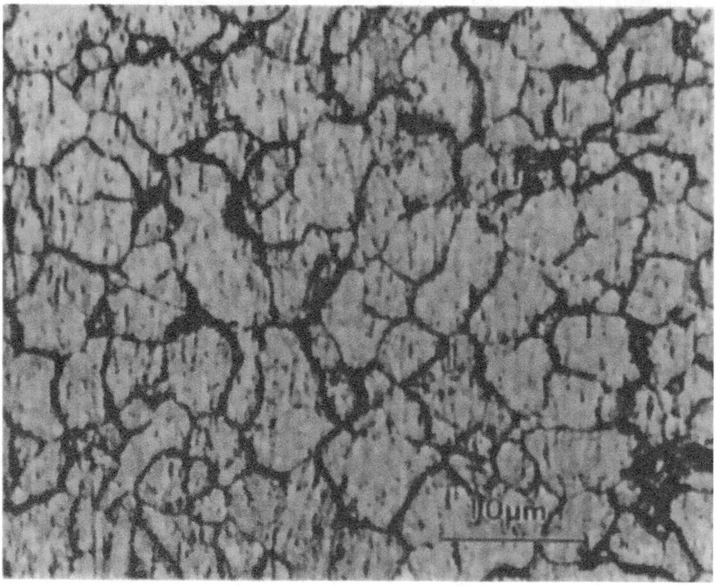

Figure 16.3. SEM-micrograph (X1700) of Zirconium-Alloy. Reprinted from Axelrad, D. R. (1993) *Stochastic Mechanics of Discrete Media*, Springer-Verlag Berlin Heidelberg, pp. 65, with kind permission from Springer-Verlag Berlin Heidelberg.

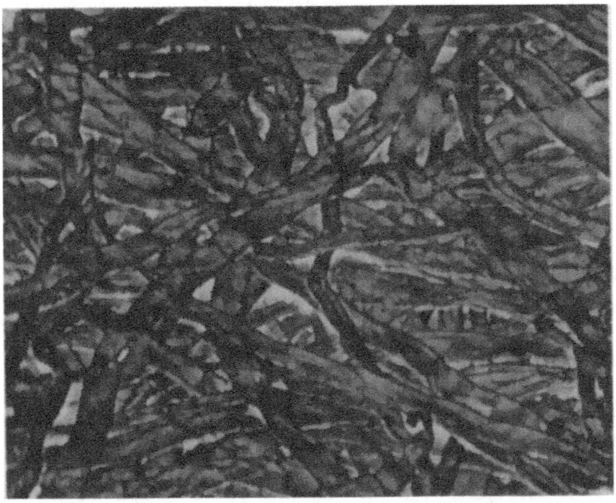

Figure 16.4. SEM-micrograph (X170) of Sulphite-paper (Fibrous structure). Reprinted from Axelrad, D. R. (1993) *Stochastic Mechanics of Discrete Media*, Springer-Verlag Berlin Heidelberg, pp. 65, with kind permission from Springer-Verlag Berlin Heidelberg.

Figure 16.5. SEM-micrograph (X4800) of a Kaolin-Water compound (Soil). Reprinted from Axelrad, D. R. (1993) *Stochastic Mechanics of Discrete Media*, Springer-Verlag Berlin Heidelberg, pp. 66, with kind permission from Springer-Verlag Berlin Heidelberg.

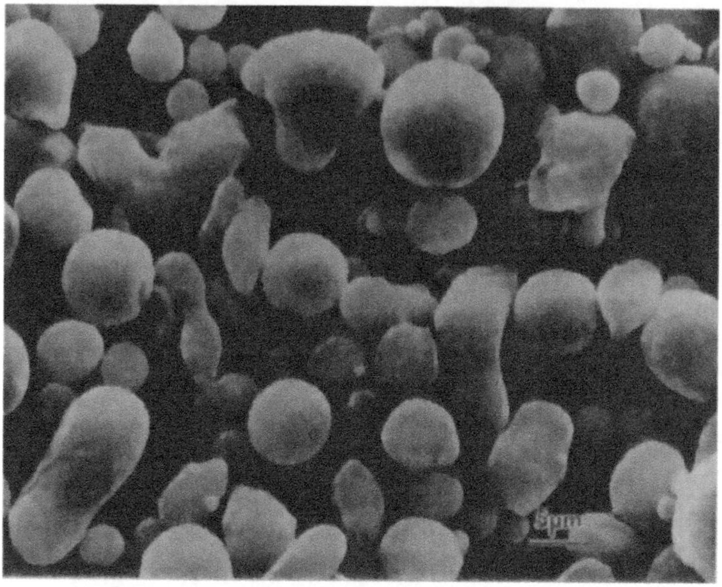

Figure 16.6. SEM-micrograph (X2000) of Al-Glycol compound. Reprinted from Axelrad, D. R. (1993) Stochastic Mechanics of Discrete Media, Springer-Verlag Berlin Heidelberg, pp. 66, with kind permission from Springer-Verlag Berlin Heidelberg.

TABLE 16.2. A comparison between basic concepts of the probabilistic micromechanical approach and the corresponding concepts of classical continuum mechanics.

	Classical Continuum Mechanics	Stochastic Micromechanics
Material System	Continuous	Discrete
Local description	Mathematical point	Structural element
Stress and deformation	Continuous	Discontinuous
Analytical approach	- Deterministic - Constitutive theory	- Stochastic - Operational formalism of the response of a structured material system

16.2.1. A STRUCTURAL ELEMENT

A structural element (κ) is defined as the smallest part of the medium that represents the mechanical and physical characteristics of the microstructure at the "*micro*" level. In a large class of structured material systems, this element is chosen arbitrarily to represent an individual micrelement as well as the binding interaction within the boundary between two matching microelements. Figure 16.7 illustrates the concept of a "structural element" for different classes of structured solids.

Throughout this Chapter, a superscipt (κ) to the left of the symbol will refer, in general, to a structural element. The quantities referring to an individual microelement is denoted by a superscript "α" while those referring to the bonding interaction within the boundary between two matching adjoining microelements α and β are designated by superscripts $\alpha\beta$.

Microelement "α"

In the stochastic micromechanical approach, the continuum approach is maintained for the formulation of the response behaviour of a single microelement. Thus, it is understood that the effects of the microelement's substructural mechanisms, such as dislocations and other lattice imperfections, are not considered at this stage of presentation. Further development of the analysis, however, may include the effects of such mechanisms. Hence, it is considered in the present analysis that the overall response of the microelement is of greater significance to the overall response of the macroscopic material system.

16.2.2. AN INTERMEDIATE SCALE "MESODOMAIN"

The next scale is a "*meso*" of the material body and is associated with a countable set (finite) of structural elements κ; ($\kappa = 1, \ldots N$) where N is large enough to comply with the low of large numbers of probability theory. The "*meso*" scale is of utmost significance since it defines a set of κ; ($\kappa = 1, \ldots N$) where all the statistics of the physical, geometric and field quantities governing the behaviour of the elements of the microstructure are assumed to be independent of position or index number. The various dependence relations and limiting procedures available to distinguish between "independent" and "dependent" random variables have been discussed by Axelrad (1993).

6.2.3. THE MACROSCOPIC SCALE

The third and largest scale is identifiable with the macroscopic material body and is defined as the union of disjoint mesodomains. It is the mathematical manifold representing the macroscopic body of the medium. It has to be recognized, however, that the considerations also involve the notion of "mean values" for an ensemble of structural elements for which experimental observations can be easily made.

The scope of the stochastic micormechanical approach to the mechanical response of a randomly structured material system (of mutually interacting microelements) is demonstrated in Figure 16.8.

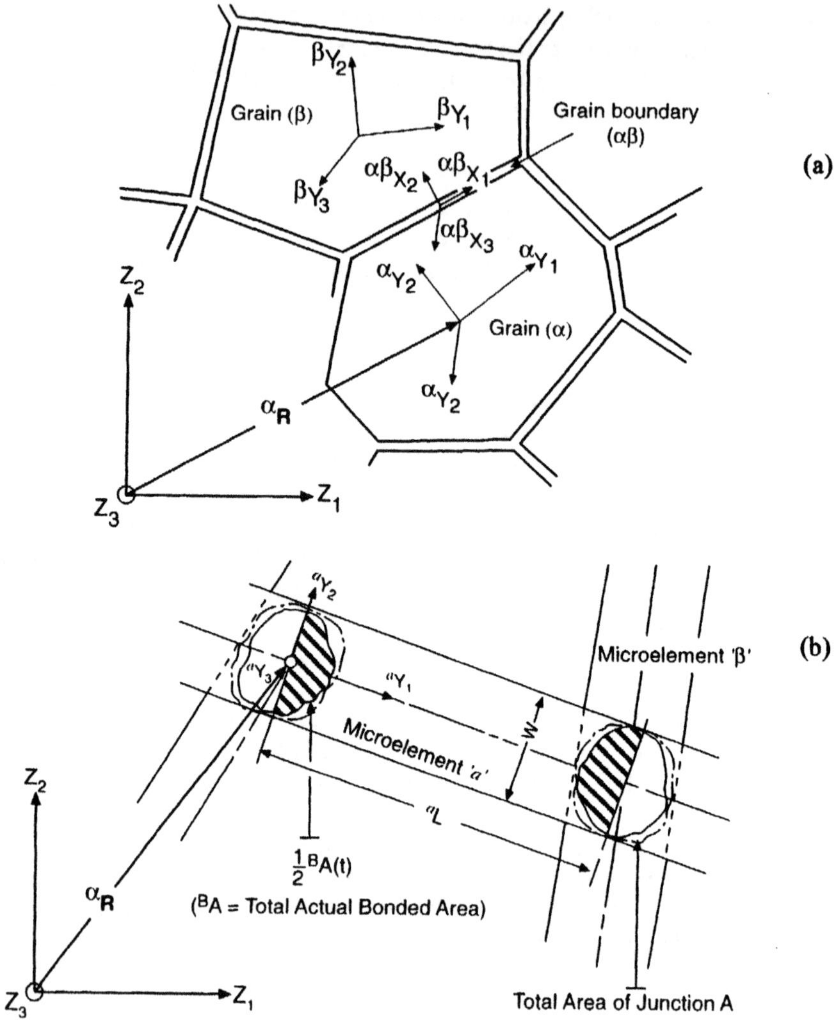

Figure 16.7. **The concept of a "*structural element*" for different classes of structured solids: (a) Polycrystalline solid, and (b) Fibrous structure.**

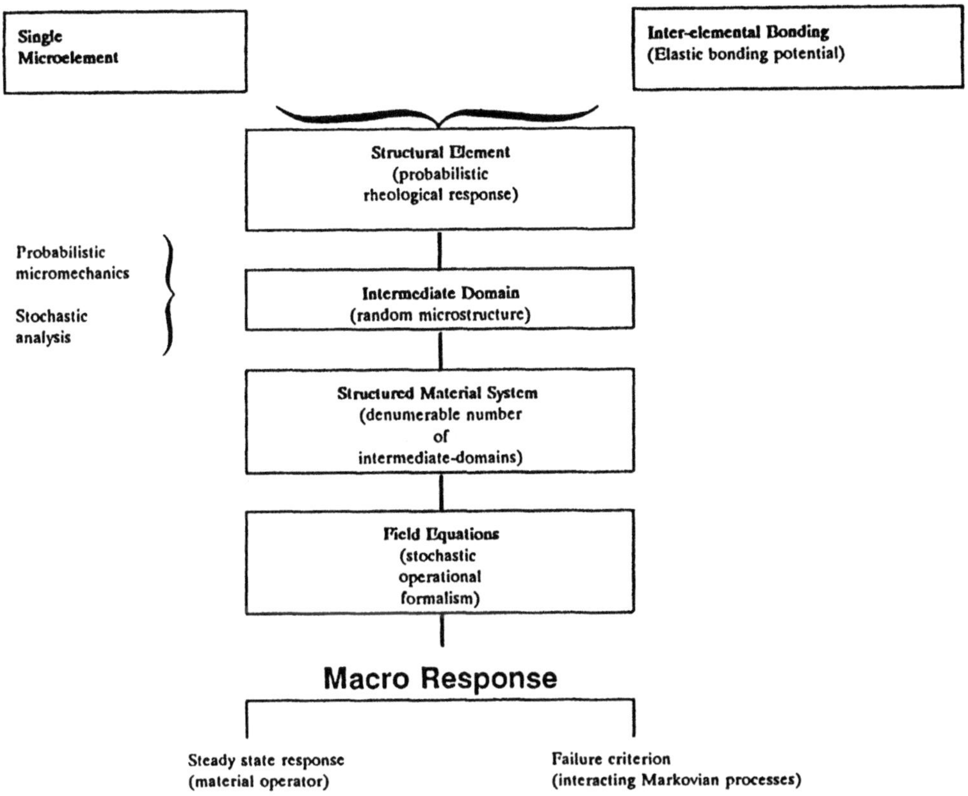

Figure 16.8. Scope of the stochastic micromechanical approach to the response behaviour of a randomly structured material system.

In the following, the stochastic micromechanical approach will be demonstrated for the case of polycrystalline solid. Case studies concerning the application of this approach to fibrous structures can be found in Haddad (1995). The apptoach has been also demonstrated by the author for the case of composite systems (*see* Haddad, 1986b, and Haddad and Tanary, 1989).

16. 3. The Stochastic Micromechanical Approach to the Response Behaviour of Polycrystalline Solids

16.3.1. STEADY-STATE RESPONSE

Structural Element

Microelement (α). The strain increment, in a continuous "*elasoplastic*" microelement, may be expressed as the sum of elastic and plastic increments as

$$\Delta^{\alpha}\varepsilon_{ij} = \Delta^{\alpha}\varepsilon_{ij}^{(e)} + \Delta^{\alpha}\varepsilon_{ij}^{(p)} \tag{16.1}$$

Introducing, with reference to the microelement's local coordinate frame, Fig. 16.9, a micro-stress ${}^{\alpha}\xi_{ij}$ in the Cauchy sense, one can write the elastic response equation in an incremental form as

$$\Delta^{\alpha}\xi_{ij} = {}^{\alpha}E_{ijkl}\,\Delta^{\alpha}\varepsilon_{kl}^{(e)} \tag{16.2}$$

where ${}^{\alpha}\varepsilon_{ijkl}$ is the elastic tensor modulus of the "*continuous*" microelement (α).

The plastic strain increment is assumed to be given by the flow rule as

$$\Delta\varepsilon_{ij}^{(p)} = T_{ij}\,\Delta\lambda \qquad T_{ij} = \partial f / \partial\xi_{ij} \tag{16.3}$$

in which f is the yield function and $\Delta\lambda$ is a scalar function. Assuming that the yield function is to be given by the plastic work $W^{(p)}$, i.e.

$$f(\xi_{ij}, \varepsilon_{ij}^{(p)}) = F(W^{(p)}) \tag{16.4}$$

then, the scalar function is calculated by (Kitagawa and Tomita, 1973)

$$\Delta \lambda = (T_{mn} \Delta \xi_{mn} \Delta \varepsilon^{(p)}_{mn}) / (F' \xi'_{ij} T_{ij})$$

where

$$R_{mn} = \frac{\partial f}{\partial \varepsilon^{(p)}_{mn}}, \qquad F' = \frac{dF}{dW^{(p)}} \quad \text{and} \quad \xi'_{ij} = \xi_{ij} - \frac{1}{3} \delta_{ij} \delta_{rs} \tag{16.5}$$

From equations (16.1), (16.2) and (16.5), it follows that

$$\Delta \lambda = T_{ij} E_{ijkl} \Delta \varepsilon_{kl} / \{F' \xi_{mn} + E_{mnrs} T_{rs} - R_{mn}\} T_{mn} \tag{16.6}$$

Assume the linear incremental constitutive equation

$$\Delta^{\alpha} \xi_{ij} = {}^{\alpha} A_{ijkl} \Delta^{\alpha} \varepsilon_{kl} \tag{16.7}$$

whereby the material operator A_{ijkl} can be written, in view of equations (16.1) - (16.7), as

$$\mathbf{A}_{jkl} = E_{ijkl} - T_{\upsilon w} T_{pq} E_{pqkl} E_{ij\upsilon w} / (F' \xi_{mn} + E_{mnrs} T_{rs} - R_{mn}) T_{mn} \tag{16.8}$$

From the above, Eqn. (16.8) is valid for any arbitrary yield function f. Assuming, for instance, the von Mises yield function

$$f = \frac{1}{2} \xi'_{ij} \xi'_{ij} = \frac{1}{3} \overline{\xi^2}$$

then, Eqn. (16.8) reduces to

$$A_{ijkl} = E_{ijkl} - 2 \mu \xi'_{ij} \xi'_{ij} \xi'_{kl} / \frac{2}{3} \overline{\xi^2} (F' / (2 \mu + 1)) \tag{16.9}$$

in which μ is the shear modulus. Let the operator L_{ijkl} represent the second term on the right-hand side of Eqn (16.9); hence, in terms of this operator, one can write, with reference to Eqn (16.7), that

$$A_{ijkl} = E_{ijkl} - L_{ijkl} \tag{16.10}$$

where L_{ijkl} is seen as a material operator representing the plastic response of the microelement. Meantime, the response behaviour of the microelement is expressed in terms of the incremental microdeformation as

$$\Delta^{\alpha} \xi_{ij} = {}^{\alpha}\Gamma_{ijs} \Delta^{\alpha} u_s \tag{16.11}$$

where the material operator ${}^{\alpha}\Gamma_{ijs}$ takes, in view of (16.10), the form

$${}^{\alpha}\Gamma_{ijs} = {}^{\alpha}A_{ijkl} \, {}^{\alpha}\nabla_{kls} = [\, {}^{\alpha}E_{ijkl} - {}^{\alpha}L_{ijkl} \,] \, {}^{\alpha}\nabla_{kls} \tag{16.12}$$

Inter-elemental Boundary "$\alpha\beta$"

In any mathematical approach to the response behaviour of material systems that would be based explicity on microstructural considerations, it is of utmost importance to include in the formulation the internal surface effects caused by existing inter-elemental boundaries. In this regard, several models have been proposed in the literature to assess the intercrystal energies as associated with the possible inter-elemental boundary topology.

In the case of polycrystals, for instance, Bollmann (1970) defined grain boundaries in terms of '*coincidence lattices*' obtained from the interpretation of two adjoining grains. This led to Bollmann's 'low misfit angle O_1-lattice' and '*high misfit angle O_2-lattice*' theories. The two latter concepts, introduced as the sum of all positions of '*best fit*', represent a description of the possible boundaries between two idealized crystals of given structure and crystallographic orientation.

Within the present analysis, one seeks an expression incorporating the mechanical response of the inter-elemental boundary separating two neighbouring microelements α and β. In this context, the interaction between a pair of atoms $^{\alpha}q$ and $^{\beta}q$, on the matching surfaces of α and β, separated by a distance vector δ is assumed to be given by a 'pair potential function' defined by (Morse, 1929)

$$D(|d|) = D_0[\exp\{-2b(|\delta - \Delta|)\} - 2\exp\{-b(|\delta - \Delta|)\}]; \qquad (d = \delta - \Delta) \tag{16.13}$$

in which D_0 is the equilibrium value of the potential, Δ is the equilibrium separation vector corresponding to D_0 and *b* is a material constant. The values of the above parameters are given in Table 16.3 for a number of material systems.

TABLE 16.3. Potential function parameters

	Copper	Aluminium	Gold
D_0 (eV)	0-216 67	0-140 0	0-180 0
b (Å^{-1})	2-233 49	2-277 75	2-969 98
$\lvert\Delta\rvert$ (Å)	2-547 56	2-847 80	2-874 13

In a manner similar to the operational formulation of the response of an individual microelement, Eqn. (16.11), one can express the bonding response in an operational form. For this reason, a transform operator $^{\alpha\beta}\Gamma$ for the bonding interaction is introduced such that

$$^{\alpha\beta}\zeta_{ij} = {}^{\alpha\beta}\Gamma_{ijk}\, {}^{\alpha\beta}d_k \tag{16.14}$$

where $^{\alpha\beta}d_k(t)$ is now the generalized relative displacement between the two matching microelements *α* and *β*. This relative displacement may be expressed, following Gel'fand and Vilenkin's generalized functions concepts (1964), as

$$^{\alpha\beta}d = \delta\,({}^{\alpha\beta}d - {}^{i}\hat{d})\,{}^{i}\hat{d} \tag{16.15}$$

in which the symbol ^ indicates a discrete value of the parameter and δ is the three-dimensional '*Dirac-delta*' function.

With reference to (16.13) and (16.14), the expression for the operator ${}^{\alpha\beta}\Gamma_{ijk}$ of (16.14) may be approximated by

$$ {}^{\alpha\beta}\Gamma_{ijk}(t) = \frac{-2b^2 D_0}{{}^{\alpha\beta}a} \, {}^{\alpha\beta}g_i \, {}^{\alpha\beta}v_j \, {}^{\alpha\beta}g_k^{-1} \qquad \textbf{(16.16)} $$

where ${}^{\alpha\beta}a$ is the area per bond, and ${}^{\alpha\beta}v_j$ are the components of the unit normal to the grain boundary $(\alpha\beta)$.

Transition to the Macroscopic Response Behaviour

Following the concepts of the micromechanical theory of structured media (Axelrad, 1984, 1993, and Axelrad and Haddad, 1998), all microscopic field quantities within the intermediate domain are considered to be stochastic functions of primitive random variables. Thus, the components of the microstress, for instance, are seen as stochastic functions ${}^{\kappa}\xi(\mathbf{r},t)$ that can be regarded as a family of random variables ${}^{\kappa}\xi_t(\mathbf{r})$ within the intermediate domain depending on the time parameter t, or a family of curves ${}^{\kappa}\xi_r(t)$ depending on the structural element position vector ${}^{\kappa}\mathbf{r}$.

The basic kinematic quantities pertaining to the deformation of the material microstructure are considered as follows:

The microelement deformation vector,

$$ {}^{\alpha}\mathbf{d} : {}^{\alpha}d_i \; ; \; i = 1,2,3 $$

and the interfacial bonding deformation within the inter-elemental boundary,

$$ {}^{\alpha\beta}\mathbf{d} : {}^{\alpha\beta}d_j \; ; \; j = 1,2,3 $$

Within an intermediate domain of the medium, referred to as a "*mesodomain*" (Axelrad, 1984, 1993), the above kinematic quantities are considered to be stochastic functions of primitive random variables.

The basic kinematic parameters, at any particular time, which describe the changes that have taken place in the structural element, may be seen as the outcome of (κ) due to the deformation process and is designated by

$$^{\kappa}\eta : {}^{\kappa}\eta\zeta \quad (\zeta = 1,2,\ldots,m)$$

where $\eta = 6$ represents the number of basic kinematic parameters above. The entire set of possible outcomes define the sample space $\sum$, i. e. $^{\kappa}\eta \in \sum$.

It is understood, however, that due to experimental limitations, $^{\kappa}\eta$ cannot be determined in an exact fashion. This, then, calls for a "*parameter cell*" type of formulation which is common in statistical mechanics.

Thus, the event Ξ is taken to be the experimentally specified parameter cell in Σ, such that

$$\Xi : (\eta \,, \eta + \Delta\, \eta)\,; \qquad \Xi \in \Sigma$$

where $\Delta\eta$ is the experimental range of the measurement of the kinematic parameter. Thus, during the deformation process, the probability of the kinematic parameter being in the event Ξ is a probability measure that changes with time and may be designated by $\{\Xi\}$. Thus, one could identify this probability measure by arbitrarily setting

$$\{^{\alpha}\eta \in \Xi\} \equiv \{\eta = \Omega\}$$

where Ω indicates a particular value in the event Ξ.

Now, consider the basic kinematic random variable (vector) d_t (s) for some fixed time (s); the most convenient definition of such variable may be provided by the choice of the image set

$$d_t(\eta \,, s) = \eta$$

i.e., the value of the random variable at this particular time s is the outcome η. Furthermore, the probabilistic distribution for the random variable is established by the condition that the set

$$\{d_t(s) \le \Omega\} = \{\eta \in \Sigma : d_t(\eta, s) \le \Omega\}$$

is an event for all values of $\Omega \in \Sigma$.

Now, the basic kinematic stochastic process d_t can be considered, as an extension of the foregoing, as a family of random variables $\{d_t(s);\ s, t > 0\}$, where its probability distribution function can be read as

$$\Pi\{d_t(\Omega, t)\} = \Pi\{d_t \le \Omega\}$$

16.3.2. STOCHASTIC APPROACH TO THE INTERNAL DAMAGE IN A STRUCTURED SOLID

One of the main objectives of the mechanics of deformable solids is the formulation of the ductile fracture process in materials. In metals, for instance, the ductile fracture phenomenon is generally associated with the nucleation of micro voids, their growth and interlinking in the material specimen (Rosenfield, 1968, and McClintock, 1968). Although significant research efforts have been made in recent years towards understanding the controlling factors involved, no comprehensive ductile fracture criterion has yet been reached. This is due primarily to the great theoretical complexity of the problem and the difficulty in carrying out definitive experiments (Rosenfield, 1968, and Sih, 1983).

Microscopic voids usually form at sites of second phase particles such as inclusions, precipitates and dispersions. There is, also, the possibility that such voids may form at highly strained regions in the specimen regardless of the presence of second phase particles (Rogers, 1971). With the evolution of the fracture process, such voids grow and join together into larger voids or cracks. The latter becomes, then, the source of localized microstresses and microdeformation bands which spread in the microstructure in directions determined by the boundary conditions of the macroscopic specimen. As a result, additional micro voids may also form and they, too, would grow and coalesce leading to additional cracks in the specimen. Such a process would, then, repeat itself pending on the rate of energy transfer in the microstructure until the final collapse of the material.

Void nucleation at the sites of included particles is attributed to a large extent to the strength of the particle, the particle-matrix interfacial bonding, as well as the mechanism of load transfer between the matrix and the particle during the deformation process. When the particles are weak or brittle, void nucleation occurs by the shattering of the inclusions at very small strains (e.g., Nemat-Nasser, 1977). On the other hand, when the particles are strong, but weakly bounded to the matrix, nucleation would occur by particle-matrix bond decohesion. Equivalently, a case when the particle is not bonded to the matrix may demonstrate a pre-existing void. When the metal contains, however, strong particles which are strongly bonded to the matrix, void formation is retarded and the material demonstrates improved ductility (e.g., Nemat-Nasser, 1977, and Argon and Safoglu, 1975).

While void formation is an essential part of the ductile fracture process, the most influential factors leading to the final failure of the macroscopic specimen are void growth and void coalescence within the highly strained matrix between the voids. In effect, such events occur in a rather cooperative manner: the nucleation of voids due to particle-matrix bond decohesion concentrates the strain in narrow bands emanating from the voids (e.g., Rosenfield, 1968), while bands impinging on particles cause holes to form around these particles (*see* Rosenfield, 1968, and Ashby, 1966). Either of such events may occur first commencing the process of ductile fracture. Further, during the fracture process, the particles may block the path of the deformation lines within the matrix, hence, resulting in large stress concentration. On the other hand, these stresses could be partially relieved by void formation. However, if the metal is of high ductility, the void may not grow immediately into a crack, but is blunted by local plastic flow (Rosenfield, 1968). In this regard, Bluhm and Morrissey (1966) pointed out that hole formation and growth are usually gradual, but, hole coalescence is rapid and catastrophic. Once holes begin to join together, they are very rapidly converted into a crack which may transverse the cross-section of the material specimen in short time.

The conventional approach to the formulation of the fracture process in ductile solids is based on continuum and '*modified*' continuum models which usually ignore the nondeterministic influence of microscopic events such as those referred to above. Examples of such models are due to McClintock (1968), Hult and McClintock (1956), Hult (1957), Rice and Tracey (1969), among others. These models imply, within the restrictions of continuum mechanics, that all the basic quantities involved in the fracture process are continuous variables or functions of such variables. Due to the randomness of events leading to ductile fracture, the field quantities fail to be continuous particularly at the evolved boundaries within the specimen. Hence, it becomes necessary to consider such local events as an integral part of the problem. In this context, the need for a probabilistic approach to the problem, that accounts for the random nature of the phenomenon, has been discussed frequently in the literature (e.g., Haddad, 1985a).

In the following presentation, a probabilistic, microstructural approach to the formulation of the fracture process in a ductile solid is introduced. Following the concepts of the stochastic micromechanical theory, as introduced earlier in this Chapter, the material system is regarded as heterogeneous medium of actual microstructural elements. These elements may exhibit random geometric and physical characteristics and are further disturbed by a random distribution of second phase particles, Fig. 16.9. The latter are, in general, irregular in size, shape, orientation as well as interspacing. Hence, in the stochastic micromechanical approach, the mechanics of the discrete microstructure introduce the relevant field quantities as random variables or functions of such variables and their corresponding distribution functions. Further, the evolution of the internal events in the microstructure and their interaction effects are considered in this approach to be time-dependent. Hence, as introduced earlier, it appears appropriate to consider the accompanying evolution process as a stochastic process (e.g., Axelrad, 1984, Provan, 1971 and Haddad, 1983).

For the simplification of the analysis, it is assumed in the present model that the included particles are of sufficient strength so that they would not break during the deformation process. Accordingly, void formation at the particle-matrix interface is seen, rather, to play the prominent role in the initiation of the fracture process in the specimen. Here, the transfer of microstress between the matrix and the particle is considered to be carried out through an interfacial binding mechanism. Hence, from a micromechanical point of view, the nucleation of a microvoid at the particle-matrix interface is considered to occur when the associated binding stress satisfies the criterion of a maximum binding displacement value corresponding to the cut-off of the binding potential. The growth of a resulting microvoid is, then, assumed to follow a '*transgranular*' random walk, of the discrete Markov type (*see* Bharucha Reid, 1960). The latter is associated with the build-up of strain in the matrix surrounding the void (*see* McClintock, 1958, and Haddad and Sowerby, 1977). Two probabilities of absorption are involved here, i.e., at the structural element boundary and at infinity. As the crack reaches the boundary between two neighboring elements (grains), an inter-elemental (*intergranular*) fracture process may set up. Thus, a time-dependent, non-homogeneous intergranular fracture process is examined in relation to the intensities of transformations within the grain boundary.

A Structural Element

As defined in the foregoing, a structural element of the medium is defined as the smallest region of the microstructure that represents the mechanical and physical properties at the microlevel. In the case of a polycrystalline material, this element is chosen to represent an individual grain (α), as well as the grain boundary between two matching grains (α&β). To model the ductile fracture process involved, one assumes that there exists the probability that the microelement is disturbed by the presence of second phase particles within the grain. Following our presentation above, a superscript "α" on the left of the symbol will refer, in general, to the individual grain within the structural element (κ). A distinction is made, however, between the matrix within the grain and the particle that may be present in the grain. Hence, the quantities referring to the former will be designated by "m", while those identifying the particle will be designated by "i". The grain boundary between two matching grains α and β is referred to by "αβ", as introduced earlier.

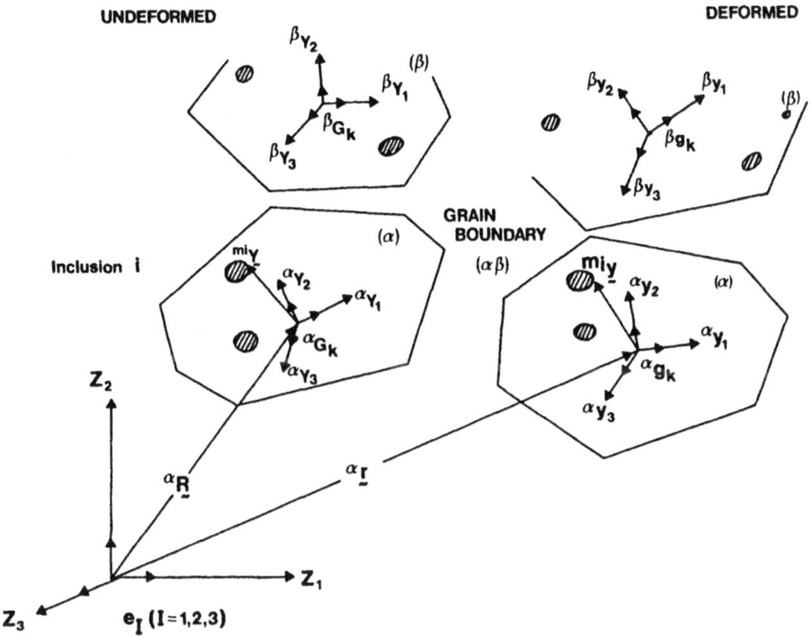

Figure 16.9. A structural element (κ). Reprinted from Haddad, Y. M. (1986) A stochastic approach to the internal damage in a structured solid, *Theoretical and Applied Fracture Mechanics* 6, 175-85, with kind permission from Elsevier Science Publishers B.V. (North-Holand).

For the description of the deformation kinematics of a structural element (κ), it is convenient, as shown in Fig. 16.9, to use two local Cartesian frames of reference, i.e., $^{\alpha}Y_k\,(k = 1,2,3)$ attached to the center of the grain and $^{i}X_k\,(i = 1,2,3)$ that can be used to describe the orientation of the particle "i" of "α". These coordinate frames are to express the local motion of the microstructure relative to an external Cartesian frame $Z_I\,(I = 1,2,3)$.

Particle-Matrix Interface. In any microstructural approach to the formulation of the fracture process in a ductile solid, it is of utmost importance to include in the analysis the effect of interfacial bonding between the second phase particles and the matrix material within the specimen. Due to the complexity of the interfacial conditions of the two components, there could be significant physical difficulties that would make any direct attempt on the problem rather impossible. One may, however, advance the argument that two types of bonding may be responsible for the strength of the particle-matrix interface, i.e., chemical and the so-called physical (or frictional) bonding. The first is essentially determined by the compatibility of the atomic structure of the two materials to form a particular type of bonding. Such compatibility may be translated in terms of the type of matching atoms that might be available to form the bond, interfacial energy and topology, among other factors. On the other hand, frictional bonding is primarily due to the contact forces that may develop between the two materials during metal forming and subsequent heat treatment operations. It is also possible that the products of chemical reaction between second phase particles and the matrix may enhance or impair the bonding between the two components.

With reference to Fig. 16.10, the distance vector $\boldsymbol{\Delta}$ between, for example, two matching points ${}^{m}q$ and ${}^{i}q$ of the matrix and the particle i, respectively, is considered to be the basic kinematic parameter of an interfacial bonding. The counterpart of this vector in the deformed state is denoted by $\boldsymbol{\delta}$ and the microdeformation in the bond, at time t, can thus be read as

$$d(t) = \delta(t) - \Delta \qquad \textbf{(16.17)}$$

In case of chemical bonding, for instance, between matching points of the particle and the matrix, a '*pair potential*' form appears to be most suitable for the description of the binding interaction. One of the usual forms of such a potential is represented by 'Morse function', presented earlier by (16.13).

Based on the bonding potential form of (16.13), an operational response relation for the particle matrix bonding interaction can be expressed, as demonstrated earlier in this Chapter (see, also, Haddad, 1985b) as

$$^{B}\xi_{IJ}(t) = {}^{B}\Gamma_{IJK}\,{}^{B}d_{K}(t) \qquad \textbf{(16.18)}$$

In the response equation ${}^{B}\xi_{IJ}(t)$ is the matrix-particle bonding microstress at time t, ${}^{B}\Gamma_{IJK}$ is the pertaining material operator which takes a form similar to (16.16), i.e.,

$$ {}^{B}\Gamma_{IJK} = \frac{-2D_0 b^2}{{}^{mi}a(t)} \, {}^{mi}n_J \, e_K \, e_I^{-1}, \tag{16.19} $$

where ${}^{mi}a$ is the interfacial bond area, ${}^{mi}n_J$ is the unit normal to the interface at the point in question and e_K is a unit base vector associated with the external coordinate frame, Fig. 16.10.

The particle-matrix interfacial stress ${}^{B}\xi(t)$, Eqn. (16.18), may be also related to the macroscopic stress on the specimen as established by Haddad (1986). Hence, in view of the latter reference, one can write that

$$ {}^{B}d_K(t) = {}^{B}\Gamma^{-1}_{IJK} \, \zeta(t) \, \sigma_{IJ}(t) \tag{16.20} $$

in which $\zeta(t)$ is a probabilistic, time-dependent function expressed in terms of the geometrical characteristics and the orientation of the local microstructure (*see* Haddad,1986), and σ_{IJ} is the macroscopic stress. Equation (16.20) establishes the "*criterion of microvoid initiation*" at the particle-matrix interface due to an interfacial bond failure.The latter is seen to correspond to the value of $|\,{}^{B}d(t)\,| \geq |\,{}^{B}d\,|_{max}$, i.e., at the cut-off of the binding potential. In view of equations (16.19) and (16.20), the position of the interfacial bond under consideration is determined by the unit normal ${}^{mi}n_J$ to the surface of the particle i whose orientation is determined by $({}^{i}g_k \cdot e_I)$, Fig. 16.10.

Following the above, a debonding process along the particle-matrix interface would, in turn, create a debonded (*free*) zone at the site of the particle. In the present model, it is assumed that the free zone is one which would initiate a "*transgranular*" crack propagation within the grain.

Growth of a Transgranular Crack

In the dealt with model, the growth of a transgranular crack is thought to be consisting of a series of steps (McClintock, 1958 and Haddad and Sowerby, 1977). First, the local stress is increased until a point of incipient fracture is reached in the matrix material surrounding the void. Fracture then occurs for an incremental distance during which time the redistribution of stress in the neighborhood of the crack will cause a further increase in the total accumulated strain there. If this increased strain due to the redistribution of the stress is not enough to leave the local matrix in a condition to satisfy the fracture criterion, further growth will not occur until the local stress is increased. The process of increasing the applied stress and then cracking with associated straining and redistribution of stress is repeated stochastically until a stage is reached in which there is enough strain to satisfy the fracture instability criterion. Under these conditions, no further increase in the local stress might be required and the crack will become unstable in growth until it reaches an absorbing barrier. The latter is assumed, in the present analysis, to be the grain boundary ahead the direction of crack growth.

In view of the above, the growth of a nucleated fissure within the grain is assumed to follow a random walk of a finite set of states $\chi = 0, 1, \ldots, i, \ldots, \phi$ where the states 0 and ϕ are seen as absorbing barriers representing, respectively, the nucleation site of the fissure and

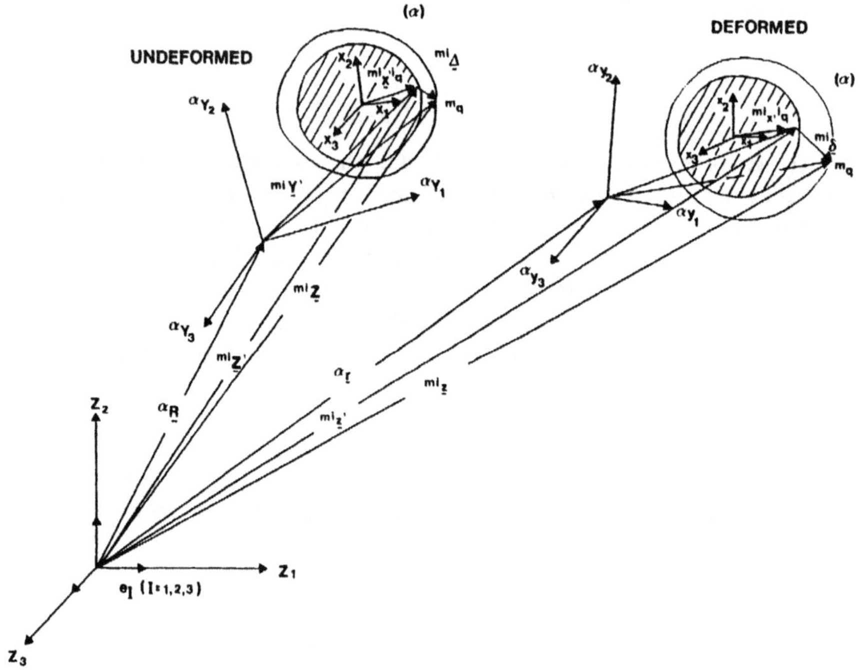

Figure 16.10. Particle-matrix interface. Reprinted from Haddad, Y. M. (1986) A stochastic approach to the internal damage in a structured solid, *Theoretical and Applied Fracture Mechanics* 6, 175-85, with kind permission from Elsevier Science Publishers B.V. (North-Holand).

the grain boundary ahead of the crack tip as shown in Fig. 16.11. In this figure, the state i is identified by the site of the crack tip as designated by the position vector r (i, t). In this context, two probabilities p and q are introduced:

- the probability, independent of the position, that the crack tip will move one step from position i to i + 1.
- the probability that the crack tip will move, independent of the position from position i to i - 1.

With the understanding that p + q = 1.

With reference to (McClintock, 1958, and Haddad and Sowerby, 1977), one may express the probability of growth, p, as follows

$$p = p\left\{\left(\frac{\partial \epsilon}{\partial y}\right)_{c,\xi} dc - \left(\frac{\partial \epsilon}{\partial c}\right)_{\xi} dc \leq \left(\frac{\partial \epsilon}{\partial \xi}\right)_{c,y} d\xi\right\} \tag{16.21}$$

The first term on the right-hand side of (16.21) represents the plastic strain gradient at a unit distance in front of the current tip of the crack, for an instantaneous crack length c and at the current grain stress level ξ. The second term in this equation represents the plastic strain gradient occurring during the previous growth step. The last term, however, designates the increase in the strain level that would be necessary to provide for the difference between the first two terms.

Let $\pi_{i,n}$ denote the probability that the random walk of the crack tip will terminate with the nth increment of time at the barrier (ϕ) between the two joining elements α and β when the initial position of the crack tip is i.

After the first step, the position is either i + 1 or i - 1 with probabilities p and q = 1 - p, respectively, as indicated above. Thus, one can write for $0 < i < \phi - 1$ and n = 1 that

$$\pi_{i,n+1} = p\pi_{i-1,n} + q\pi_{i+1,n}, \tag{16.22}$$

subject to the boundary conditions:

$$\pi_{0,n} = \pi_{\phi,n} = 0 \quad \text{for } n > 1, \qquad \pi_{\phi,0} = 1 \text{ and } \pi_{i,0} = 0 \quad \text{for } i > 1 \tag{16.23}$$

The boundary conditions (16.22) are valid for all i with $0<i<\phi$ and $n\geq 0$. The solution of (16.21) subject to (16.22) is given by

$$\pi_{i,n} = \phi^{-1}\, 2^n\, p^{(n+i)/2}\, q^{(n-i)/2} \times \sum_{k=1}^{n-1} \left\{ \cos^{n-1}\left(\frac{\pi k}{\phi}\right) \sin\left(\frac{\pi i k}{\phi}\right) \right\} \tag{16.24}$$

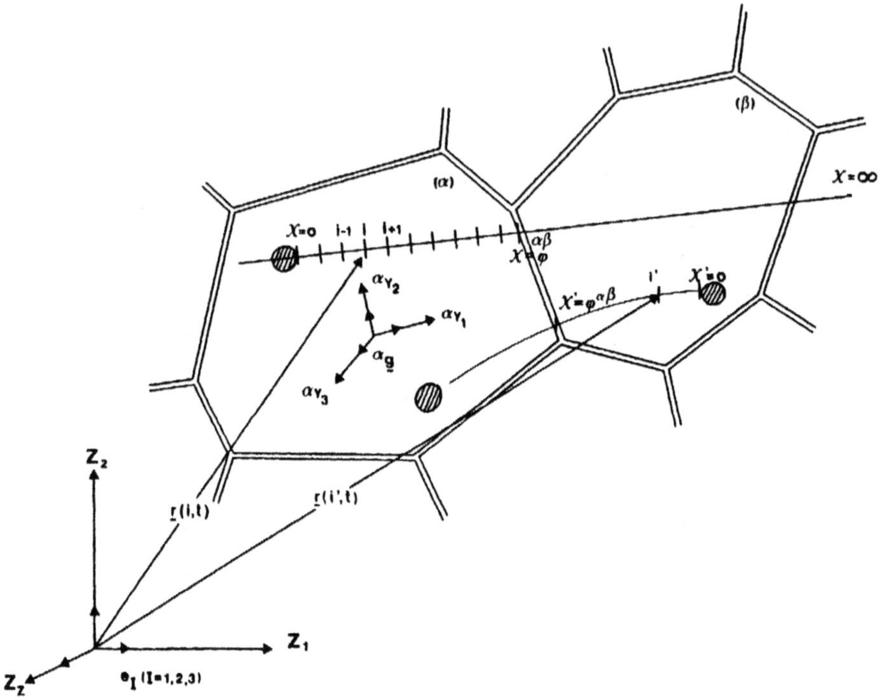

Figure 16. 11. Growth of a transgranular crack. Reprinted from Haddad, Y. M. (1986) A stochastic approach to the internal damage in a structured solid, *Theoretical and Applied Fracture Mechanics* 6, 175-85, with kind permission from Elsevier Science Publishers B.V. (North-Holland).

Considering, now, the case in which $\chi = 0$ is the only absorbing barrier, i.e., when $\phi = \infty$. In this case, by adjusting the boundary conditions (16.22) accordingly and letting $\phi \to \infty$ in (16.23), the latter becomes

$$\pi_{i,n} = 2^n p^{(n+i)/2} q^{(n-i)/2} \times \int_0^1 \left\{ \cos^{n-1}(\pi\chi) \sin(\pi\chi) \sin(\pi\chi i) \right\} d\chi \tag{16.25}$$

Equation (16.24) leads to the probability of eventual absorption at the barrier ϕ, i.e., when the transgranular crack reaches the grain boundary ahead. Denoting the latter by P, then

$$P = \sum_{n=0} \pi_{i,n} = \left(\frac{1 - \sqrt{1 - 4pq}}{2q} \right)^i \tag{16.26}$$

which can be written as

$$P = \begin{cases} \left(\dfrac{p}{q} \right)^i & \text{for } q \geq p, \\ 1 & \text{for } q \leq p \end{cases} \tag{16.27}$$

Two Propagating Fissures. In Fig. 16.11, the local positions of two neighboring particles are identified by the two local coordinate frames ${}^iX_1, {}^iX_2, {}^iX_3$ and ${}^{i'}X_1\, {}^{i'}X_2\, {}^{i'}X_3$. The two particles may be sited within the same grain α or within two adjoining grains α and β as shown in the figure.

Thus, during the fracture process, the tip of the nucleated fissure of particle i may follow a random walk characterized by a finite set of states ${}^iX = 0, 1, \ldots, j, \ldots, {}^{\alpha\beta}\phi$ with probabilities ${}^ip$, ${}^iq$ and ${}^i\pi_{j,n}$ as identified previously in the section above pertaining to the growth of a transgranular crack. The second fissure, nucleated of particle i', may also follow in its growth a random walk defined by the finite set of states ${}^{i'}X = 0', 1, \ldots, j', \ldots, {}^{\alpha\beta}\phi$ with probabilities ${}^{i'}p$, ${}^{i'}q$ and ${}^{i'}\pi_{j',n'}$, where ${}^{i'}p = {}^{i'}p({}^ip)$. Accordingly, a simultaneous absorption of the two propagating fissures may occur at the grain boundary, ${}^{\alpha\beta}\phi$, if

$${}^i\pi_{j,n} = {}^{i'}\pi_{j',n'}, \tag{16.28}$$

subject to the conditions (16.23).

Further, the probability of eventual, simultaneous absorption of the two fissures at the grain boundary could be expressed by the condition

$$\sum_{n=0}^{\infty} {}^{i}\pi_{j,n} = \sum_{n'=0}^{\infty} {}^{i'}\pi_{j',n'}, \tag{16.29}$$

which can be further expressed in terms of the associated probabilities of growth, p and p', by utilizing equations (16.26) and (16.27).

As soon as the crack tip strikes in it growth, the grain boundary ahead, a debonding effect might take place between the two adjoining grains.

Intergranular Facture

In the present model, a fracture zone between two adjoining grains is seen to consist of two parts (Axelrad, 1984). A cohesive zone, in which the neighbouring grains act as completely bonded and a debonded (*free*) zone in which bonding has ceased to exist. The existence of such a free zone could have been initiated by a perfect debonding process due to the increase of local stress, or as a result of ductile fracture by void formation intercepting the grain boundary as dealt with in the previous section. From this point of view, the free zone is one which will initiate an intergranular crack propagation or debonding process towards the cohesive zone. In this context, following Axelrad (1984), we assume that the intergranular fracture process occurs in a rather cooperative manner, i.e., bonds can dissociate and reform within the same mechanical state. Thus, it may be visualized that the breaking of intergranular bonds will occur in such a manner that energy is released activating bond formation within the same sites in the specimen. Hence, we consider a process such that the number of intergranular bonds can experience positive as well as negative jumps. Thus in general, a time-dependent nonhomogeneous birth-and-death model (Kendall, 1948) is seen to be applicable.

If at time t, the material system is in the state $\sum$ ($\sum = 1, 2, \ldots$) corresponding to a number of existing intergranular bonds n(t): n_{Σ}, one considers that both the intensities of positive and negative transitions to be time-dependent. The latter are designated, respectively, in the following analysis by $\lambda(t)$ and $\mu(t)$. Accordingly:

(i) The probability of transition from the state $\sum$ to ($\sum$ + 1) in the interval (t, t + Δt) is λ(t) Δt + O(Δt);

(ii) The probability of transition from the state $\sum$ to ($\sum$ - 1) in the interval (t, t + Δt) is μ (t) Δt + O(Δt);

(iii) The probability of a transition to a state other than a neighbouring state is O(Δt);

(iv) The probability of no change is 1 - (λ (t) + μ (t)) + O(Δt);

(v) The state $\sum = 0$ is an absorbing state corresponding to the breakage of all inter-granular bonds within the material specimen.

The above assumptions lead to the relation

$$P_{\Sigma}(1+\Delta t) = \lambda(t)P_{\Sigma-1}(t)\Delta t + [1-\{\lambda(t)+\mu(t)\}\Delta t]P_{\Sigma}(t) + \mu(t)P_{\Sigma+1}(t)\Delta t + O(\Delta t), \tag{16.30}$$

where P_{Σ} is the probability that the material system is in the state $\sum$ as defined above. Equation (16.30) leads in the limit to the following differential equation

$$\frac{dP_{\Sigma}(t)}{dt} = \lambda(t)P_{\Sigma-1}(t) - [\lambda(t)+\mu(t)]P_{\Sigma}(t) + \mu(t)P_{\Sigma+1}(t), \tag{16.31a}$$

which holds for $\sum = 1, 2, ...$ for $\sum = 0$, however, one has

$$\frac{dP_0(t)}{dt} = \mu(t)P_1(t) \tag{16.31b}$$

The solution of Eqn. (16.31) can be obtained with the aid of generating functions. Hence;

$$P_\Sigma(t) = [1-\zeta(t)][1-Y(t)][Y(t)]^{\Sigma-1},$$
$$\Sigma = 1,2,\ldots, \tag{16.32}$$
$$P_0(t) = \zeta(t),$$

where:

$$\zeta(t) = 1 - \frac{e^{-\gamma(t)}}{\Omega(t)}$$
$$Y(t) = 1 - \frac{1}{\Omega(t)}$$
$$\gamma(t) = \int_0^t [\mu(\tau) - \lambda(t)]\,dt, \text{ and} \tag{16.33}$$
$$\Omega(t) = e^{-\gamma(t)}\left[1 + \int_0^t \mu(\tau)\,e^{\gamma(t)}\,d\tau\right].$$

The probability of total intergranular bond dissociation is, then, given with reference to (16.32) and (16.33) by

$$P_0(t) = \frac{\int_0^t \mu(\tau)\,e^{\gamma(t)}\,dt}{1 + \int_0^t \mu(\tau)\,e^{\gamma(t)}\,dt} \tag{16.34}$$

16.4. References

Argon, A.S. and Safoglu, R. (1975) Cavity formation from inclusions in ductile fracture, *Metallurg. Trans. A*, *6A*, 825-37.

Ashby, M.F. (1966) Work hardening of dispersion-hardened crystals, *Phil. Mag. 14*, 1157-78.

Axelrad, D. R. (1984) *Foundations of the Probabilistic Mechanicsof Discrete Media*, Pergamon Press, Oxford.

Axelrad, D. R. (1993) *Stochastic Mechanics of Discrete Media*, Monograph, Springer Verlag, Berlin

Heidelberg.
Axelrad, D. R. and Haddad, Y. M. (1998) On the behavior of materials with binary microstructures, in Proceedings NATO ARW, Kiev, Ukraine, *Advanced Multilayered and Fibre-Reinforced Composites*, Y. M. Haddad (editor), Kluwer, Dordrecht, pp. 163-72.
Bharucha Reid, A. T. (1960) *Elements of the Theory of Markov Processes and their Applications*, McGraw-Hill, NewYork.
Bluhm, J.I. and Morrissey, R.J. (1966) Fracture in a tensile specimen, in: T. Yokobori et al., eds., *Proc. 1st Internat. Conf Fracture*, Japanese Society for Strength and Fracture of Materials, Sendai, Japan, III, pp. 1739- 80.
Bollmann, W. (1970) *Crystal Defects and Crystalline Interfaces*, Springer-Verlag, New York.
Gel'fand, I. M. And Vilenkin, N. Ya. (1964) *Generalized Functions*, Academic Press, New York.
Haddad, Y. (1983) A stochastic approach to the growth and coalescence of microvoids in a ductile solid, in *Proc. 10th Canad. Fracture Conf. Modelling Problems in Crack Tip Mechanics,* University of Waterloo, Waterloo, Canada, pp. 239-47.
Haddad, Y.M. (1985a) A Stochastic approach to the internal damage in a structural solid: in *Proc. Amer. Physical Soc. Topical Conf. Shock Waves in Condensed Matter*, Spokane, WA., USA .
Haddad, Y.M. (1985b) A Stochastic approach to the rheology of randomly structured networks, *Mater. Sci. Engrg. 72,* 135-47.
Haddad, Y. M. (1986a) A stochastic approach to the internal damage in a structured solid, *Theoretical and Applied Fracture Mechanics* **6**, 175-85.
Haddad, Y. M. (1986b) A microstructural approach to the mechanical response of composite systems with randomly oriented, short fibres, *J. Materials Science* **21**, 3767-76.
Haddad, Y. M. (1990) A microstructural approach to the mechanical response of a class of polycrystalline systems, *Res Mechanica* **28**, 177-96.
Haddad, Y.M. (1995) *Viscoelasticity of Engineering Materials*, Kluwer, Dordrecht.
Haddad, Y.M. and Sowerby, R. (1977) A Micro-probabilistic approach to the ductile deformation and fracture of metals, in *Proc. Fracture 1977,* ICF4, Univ. of Waterloo, Waterloo, Canada, 2, pp. 457-66.
Haddad, Y. M. and Tanary, S. (1989) On the micromechanical characterization of the creep response of a class of composite systems, *J. Pressure Vessel Technology* **111**, 177-82.
Kendall, D.G. (1948) On the generalized birth-and-death process, *Ann. Math. Statist. 19,* 1-15 .
Hult, J.A. (1957) Fatigue crack propagation in torsion, *J. Mech. Phys. Solids 6,* 47-52.
Hult, J.A. and McClintock, F.A. (1956) Elastic-plastic stress and strain distributions around sharp notches under repeated shear, in: *Proc. 9th International Congress Appl. Mech.*, Brussels, **8**, pp. 51-58.
Kitagawa, H. and Tomita, Y. (1973) An incremental finite element analysis of two-dimensional large strain and large displacement problems for elasto-plastic material, Proc. 21st Japan Nat. Congr. Appl. Mech., 1971, University of Tokyo Press, 243-55.
Kolmogorov, A.N. (1950) *Foundations of the Theory of Probability*, Chelsea, N.Y.
McClintock, F. A. (1958) Ductile fracture instability in shear. *Trans. ASME* **12**, 582-8.
McClintock, F.A. (1968) A criterion for ductile fracture by the growth of holes, *J. Appl. Mech. 6,* 363-71.
Morse, P. M. (1929) Diatomic molecules according to the wave mechanics, II. Vibrational levels, *Physics Review* **34**, 57-64.
Nemat-Nasser, S. (1977) Overview of the basic progress in ductile fracture, *Trans. 4th Internat. Conf. Structural Mechanics in Reactor Technology,* San Francisco, CA, **L2/1**, pp.1-11.
Provan, J.W. (1971) Deformation of arbitrary oriented media, *Arch. Mech. 23(3),* 339-52.

Rice, J.R. and Tracey, D.M. (1969) On the ductile enlargement of voids in triaxial stress fields, *J. Mech. Phys. Solids* ***17***, 201-17.

Rogers, H.C. (1971) *Metal Forming: Interrelation between Theory and Practice*, Plenum Press, New York.

Rosenfield, A.R. (1968) Criterion for ductile fracture of two-phase alloys, *Metallurgical Reviews* ***121***, 29-40.

Sih, G.C. (1983) The state of affairs near the crack tip, in: J.J. Pindera and B.R. Krasnowski, Eds., *Proc. 10th Canad. Fracture Conf. Modeling Problems in Crack Tip Mechanics*, University of Waterloo, Waterloo, Canada, 65-90.

16.5. Further Reading

Axelrad, D.R. (1986) On the transient behaviour of structured solids, in: *Trends in Application of Pure Mathematics to Mechanics*, Eds. Kröner and Kirchgässner. Lecture Notes in Physics No. 249, Springer Verlag, Berlin-Heidelberg.

Axelrad, D.R. and Basu, S. (1973a) Operational approach to the deformation of structured media. in *Stochastic Problems in Mechanics*, Eds. S.T. Ariaratnam and Leipholz, H.H.E., Study No. 10, University of Waterloo, Canada, pp. 61-77.

Axelrad, D.R. and Basu, S. (1973b) Mechanical relaxation of crystalline solids, *Advances in Molecular Relaxation Processes* **6**, 185-99.

Axelrad, D.R., Basu, S. and Haddad, Y.M. (1976) Microrheology of cellulosic systems, *Proc. 7th Int. Congress on Rheology*, Gothenburg, Sweden.

Axelrad, D.R., Haddad, Y.M. and Atack, D. (1975) Stochastic deformation theory of a two-dimensional fibrous network, *Proc. 11th Annual Meeting, Soc. Engg. Sci.*, Dvorak, G.J., Ed., Duke University, North Carolina, 166-167.

Axelrad, D.R. and Frydrychowicz, W. (1992) Stochastic theory of the inelastic behaviour of multi-component high-temperature materials, *Math. Models and Methods in Appl. Sciences*, **Vol. 2**, No. 3, Ed. N. Bellomo, World Scientific Publ.

Axelrad, D.R. and Frydrychowicz, W. (1995) Stochastic analysis of the fracture of solids with microcracks. *Z. angew. Math. Phys.* (ZAMP), **46**, Birkhäuser Verlag, Basel.

Basu, S. (1975) *On a General Deformation Theory of Structural Solids*, Ph.D. Thesis, McGill University, Montreal, Canada.

Bernal, J.D. (1958) General introduction: Structure arrangements of macromolecules, *Discussions of Faraday Soc.* **25**, 7-18.

Blumenthal, R.C. and Getoor, R.K. (1963) *Markov Processes and Potential Theory*, Academic Press, New York.

Bourbaki, N. (1951) *Topologie Génèrale*, Hermann, Paris.

Cox, T.B. and Low Jr., J.R. (1974) An investigation of the plastic fracture of AISI 4340 and 18 nickel-200 grade (A. 10) merging steels, *Metallurg. Trans.* **5**, 1427-70.

Domb, C. and Green, M..S., (eds) (1972) *Phase Transitions and Critical Phenomena*, Vol. **1-6**, Academic Press, N.Y..

Doob, J.L. (1953) *Stochastic Processes*, John Wiley and Sons Inc., New York.

Dynkin, E.P. (1965) *Markov Processes*, Volumes 1 & 2, Academic Press, New York.

Ebeling, W. (1993) Entropy, predictability and historicity of non-linear process, in: Statistical Physics and Thermodynamics of Non-Linear Non-Equilibrium Systems, eds. W. Ebeling and W. Muschik, World

Scientific. Publ. London.
Ehrenfest, P. and Ehrenfest, T. (1959) *The Conceptual Foundations of the Statistical Approach in Mechanics*, Cornell University Press, Ithaca, New York.
Foguel, S.R. (1969) *The Ergodic Theory of Markov Processes*, Van Nostrand Reinhold Co., New York.
Fung, Y.C. (1965) *Foundations of Solid Mechanics*, PrenticeHall, Englewood Cliffs, N.J..
Gikhman, I.I. and Shorohod, A.V. (1969) *Introduction to the Theory of Random Processes*, W.B. Saunders and Co., Philadelphia.
Goffman, C. and Pedrick, G. (1965) *First Course in Functional Analysis*, Prentice-Hall, New Jersey.
Guiasu, S. (1977) *Information Theory with Applications*, McGraw Hill, N.Y.
Halmos, P. R. (1950) *Measure Theory*, van Nostrand, N.Y.
Högfors, C. (1987) History-dependent systems, *Rheol. Acta*, **26**, 317-21.
Hope, E. (1954) The general temporally discrete Markoff process, *J. Rat. Mech. Anal.* **3**, 12-45.
Hsu, G.S. (1988) Cell-to-cell mapping., *Appl. Math. Sciences* **64**, Springer Verlag, N.Y.
Jongschaap, R.J.J. (1987) On the derivation of some fundamental expressions for the average stress tensor in systems of interacting particles, *Rheol. Acta* **26**, 328-35.
Kakutani, S. (1940) Ergodic theorems and the Markoff processes with a stable distribution, *Proc. Imp. Acad. of Japan, Tokyo* **16**, 49-54.
Kampe de Fereit, J. (1962) Statistical mechanics of continuous media, *Proc. Symp. Appl. Math. on Hydrodynamic Instability* **13**, 165-98.
Kappos, D.A. (1969) *Probability Algebras and Stochastic Spaces*, Academic Press, New York.
Kendall, D.G. (1955) Some analytical properties of continuous stationary markov transition functions, *Trans. Am. Math. Soc.* **78**, 529-40.
Khinchine, A.I. (1949) *Mathematical Foundations of Statistical Mechanics*, Dover, New York.
Kolmogorov, A.N. (1956) *Foundations of Probability Theory*, Chelsea, New York.
Mende, W. and Peschel, M. (1981) Structure-building phenomena in systems with power produced force, in: *Chaos and Order in Nature*, ed. H. Haken, Springer-Verlag, Berlin.
Meyer, S. .P. and Tweedie, R.L. (1993) *Markov Chains and Stochastic Stability*, Springer Verlag, London.
Moreau, J.J. (1971) Sur l'evolution d'un système elasto-viscoplastique, *C.R. Acad. Sci., Series A* **273**, 118-21.
Pugachev, V.S. (1965) *Theory of Random Functions*, Addison-Wesley, Mass.
Renyi, A. (1970) *Foundations of Probability*, Holden-Day Inc., San Francisco.
Rosenblatt, M. (1971) *Markov Processes, Structure and Asymptotic Behaviour*, Springer-Verlag, New York.
Simmons, G.F. (1963) *Introduction to Topology and Modern Analysis*, McGraw-Hill, New York.
Treves, F. (1967) *Locally Convex Spaces and Partial Differential Equations*, Springer-Verlag, New York.
Yaglom, A.M. (1962) *Theory of Stationary Random Functions*, Prentice-Hall, New York.
Yosida, K. (1965) *Functional Analysis*, Springer-Verlag, Berlin.

CHAPTER 17

INTELLIGENT MATERIALS - AN OVERVIEW

17.1. Introduction

Engineering materials are used either for their inherent structural strength or for their functional properties. Often a feed back control loop is designed so that the mechanical response of the material is monitored and the environment that is causing such a response can be controlled. The evolution of a new kind of material termed "*Intelligent*", "*Smart*", or "*Adaptive*" by various researchers, e.g., Rogers (1988) and Ahmad (1988), witnesses a significant development in materials science whereby the referred-to smart material adapts itself to suit the environment rather than necessitating to control the same. In this context, development in the area of materials research aims at incorporating intelligence into engineering materials, enabling them to sense the external stimuli and alter their own properties to adapt to the changes in the environment.

This chapter discusses possible forms of intelligence that may be incorporated in these materials. Three basic mechanisms of intelligent materials, namely, the sensor, processor and actuator functions are described. Implementation of these in the microstructure of various materials, as well as associated algorithms and techniques are illustrated. Different models, control algorithms and analyses developed by various researchers are reviewed and their potential applications in engineering materials are presented.

17.2. Definition of an Intelligent Material

"*Intelligent*" or "*Smart*" materials may be defined as "Those materials which sense any environmental change and respond to it in an optimal manner", e.g., Rogers *et al.* (1988). From this definition and the analogy of the *bionic* system of humans and animals, it can be seen that the following mechanisms may be essential for any material to be made intelligent.

(i) A sensing device to perceive the external stimuli (e.g., skin which senses thermal gradients, an eye that senses optical signals, etc.), termed as "*sensor*" function.

(ii) A communication network by which the sensed signal would be transmitted to a decision-making mechanism (e.g., the nervous system in humans and animals), termed as "*memory*" function.
(iii) A decision-making device which has the capability of reasoning (e.g., the brain), termed as "*processor*" function.
(iv) An actuating device, which could be inherent in the material or externally coupled with it (e.g., stiffening of muscles in humans and animals to resist deformation due to external loading), termed as "*actuator*" function.

All of the above mechanisms need to be active in real time applications, for the material to respond intelligently. Another important factor in the overall process is the time of response. This is the interval between the instant when the sensor senses the stimulus and that of the actuator response. An optimum time interval is crucial in the design of intelligent materials and is dependent on the type of application.

17.3. The Concept of Intelligence in Engineering Materials

As mentioned earlier, designing a material system which incorporates sensor, processor and actuator functions is the fundamental step in the evolution of an intelligent material for achieving a desired response adaptable to the environment. This concept is illustrated in Fig. 17.1.

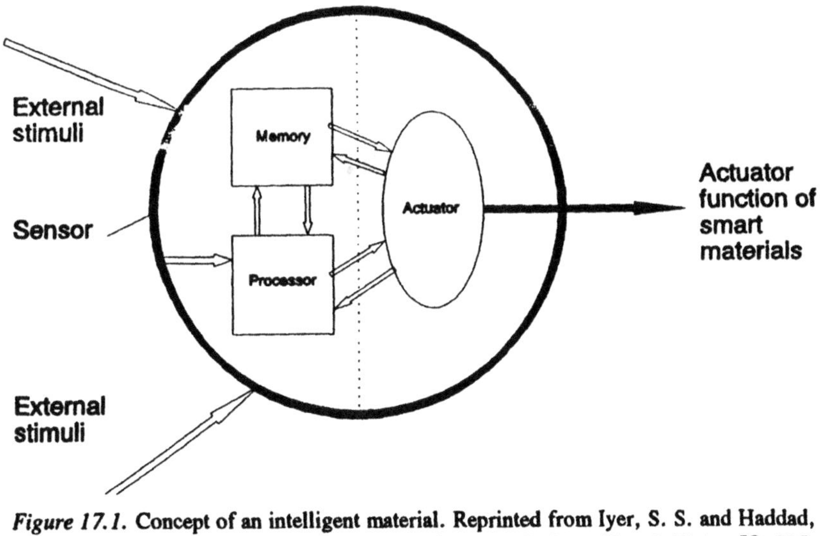

Figure 17.1. Concept of an intelligent material. Reprinted from Iyer, S. S. and Haddad, Y. M. (1994) Intelligent materials - An overview, *Int. J. Pres. Ves. & Piping* **58**, 335-44 with kind permission of Elsevier Science Limited.

17.3.1. SENSOR FUNCTION

The concept of a sensor function in a smart material is defined as the ability of the material to sense the response characteristics of self with respect to environmental factors such as mechanical loading, temperature, humidity and electrical inputs. An example of this function is that of a piezoelectric sensor embedded in a composite material. Such sensor diagnoses the mechanical disturbance imposed on the material by generating a voltage which can be further measured and analysed.

17.3.2. MEMORY AND PROCESSOR FUNCTION

This mechanism stores the signals which are sensed and transmitted earlier by the sensor function. The characteristics of these signals are then compared with pre-stored acceptable values acquired during the '*training*' process of the processor (see *Chapter 18*). The training process may be carried out using an artificial intelligence technique, e.g., pattern recognition method (*Chapter 18*). Typically, this function is in the form of an executable artificial intelligence software that could produce a logical output in the form of an electrical voltage that could further be amplified and used to activate an actuator mechanism.

17.3.3 ACTUATOR FUNCTION

This mechanism is coupled with the material. It produces an output corresponding to the signal received from the processor function. This output is usually in the form of a

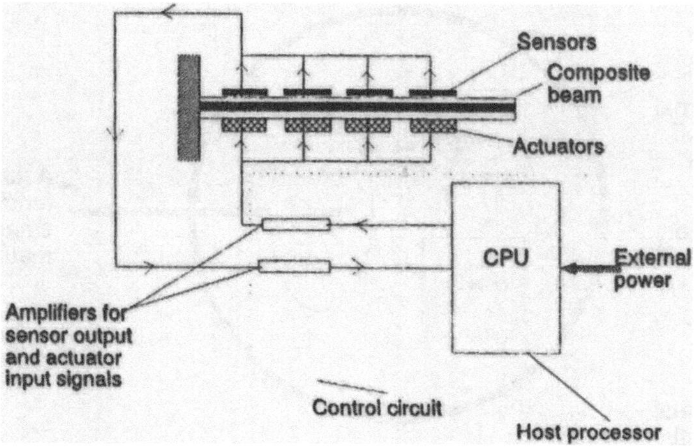

Figure 17.2. Incorporation of sensor, processor and actuator functions in an intelligent composite beam. Reprinted from Iyer, S. S. and Haddad, Y. M. (1994) Intelligent materials - An overview, *Int. J. Pres. Ves. & Piping* **58**, 335-44 with kind permission of Elsevier Science Limited.

restoring stress, strain or change in temperature, or stiffness, of the actuator mechanism that is coupled with the material. This change would be designed to neutralize the effect of the change in environment on the material, thereby adapting the material continuously to its environment. A typical intelligent composite cantilever beam which comprises of sensor, processor and actuator functions is illustrated in Fig. 17.2.

17.4. Artificial Intelligence in Materials

Different forms of substances that could be incorporated into the material as sensors and actuators are piezoelectric and piezoceramic devices. Optical fibres are used as sensors. Shape memory alloys, shape memory polymers and electrorheological fluids are employed as actuators. The following subsections describe the effectiveness of such materials as intelligent substances and their successful implementation in real time applications; e.g. Takagi (1990).

17.4.1. PIEZOELECTRIC AND PIEZOCERAMIC DEVICES

Piezoelectric and piezoceramic materials could be used as sensors and actuators in intelligent materials. These materials can convert a mechanical signal to an electrical voltage. Various researchers have developed models, through analytical and numerical simulation as well as experimental techniques, to verify the concept of piezoelectric materials as intelligent sensors and actuators. The reader is referred to, among others, Kraut (1969), Crawley and Luis (1985, 1987), Sung-Honein *et al.* (1991), Kyu Ha *et al.* (1991), and Lee *et al.* (1991).

A piezoelectric material is a crystal in which electricity or electric polarity is produced by pressure. Conversely, a piezoelectric material deforms when it is subjected to an electric field. The first characteristic expresses the so-called "*direct*" effect, while the second expresses the "*converse*" effect; e.g., Cady (1946), Gerber and Ballato (1985), and Ikeda (1990). Following the above characteristics of a piezoelectric crystal, if the pressure on the crystal is replaced by a stretch, the sign of the electric polarity would be reversed accordingly. This is determined by the crystal structural "*bias*" which establishes whether a given region on the surface is subjected to a positive or a negative mechanical effect. In the converse effect, the same unidirectional aspect determines the sign of deformation when the direction of an electric field is reversed in the crystal. It is this reversal of sign of mechanical strain with that of the electric field that distinguishes piezoelectricity from electrostriction; e.g. Cady (1946), Olson (1956) and Bailey and Hubbard (1985).

The basic quasi-static theoretical treatment of a piezoelectric material under loading is based on the definition of four parameters that describe the elastic and electric states of this material. These are the elastic stress (σ_{ij}), elastic strain (ϵ_{ij}), electric displacement (D_i) and electric field (ξ_j). Any two of these four parameters may be chosen to be the independent variables and the other two will accordingly be the dependent variables in the model. In addition to the above-mentioned parameters, the mechanical equilibrium equation ($\sigma_{ji,j} + \chi_i = 0$), Maxwell's equation ($D_{i,i} = 0$) and the electrical and mechanical boundary conditions would be specified for a complete description of the electromechanical state of the piezoelectric material.

Constitutive Relationships

The phenomenon of piezoelectricity is assumed to be linear, whereby the electric and elastic quantities are considered to be linearly related. Thus, the electric polarization (P_i) is seen to be related to the elastic stress σ_{ij} by the relation

$$P_i = d_{ijk}\sigma_{jk} \tag{17.1}$$

where the components of the third order tensor d_{ijk} are the piezoelectric strain coefficients.

The existence of such a polarization will result in an electric field ξ_i which would be linearly related to the polarization P_i through the relation $P_i = \epsilon_0\chi_{ij}\xi_j$ where ϵ_0 is the universal dielectric constant and χ_{ij} is the electric susceptibility coefficient of the material. Therefore, the complete equation for the direct piezoelectric effect is written as

$$P_i = d_{ikl}\sigma_{kl} + \epsilon_0\chi_{ij}\xi_j \tag{17.2}$$

or, in terms of the corresponding strains ϵ_{jk},

$$P_i = e_{ijk}\epsilon_{jk} + \epsilon_0\chi_{ij}\xi_j \tag{17.3}$$

where the components e_{ijk} are the piezoelectric stress coefficients.

An alternative form to equation (17.2) and (17.3) is expressed in terms of the electric displacement, i.e.,

$$D_i = d_{ikl}\sigma_{kl} + K_{ik}\xi_k \tag{17.4}$$

where K_{ik} is the permittivity tensor.

In the converse effect, strains (or stresses) are produced. They are assumed to be linearly related to the imposed electric field ξ_i, i.e.,

$$\epsilon_{jk} = d_{ijk}\xi_i \tag{17.5}$$

$$\sigma_{jk} = -e_{ijk}\xi_i \tag{17.6}$$

The occurrence of a stress (or strain) would further evoke a corresponding mechanical response in the crystal. The total stress contribution of a converse piezoelectric effect is, thus, expressed as:

$$\sigma_{ij} = E_{ijkl}\epsilon_{kl} - e_{kij}\xi_k \qquad (17.7)$$

The corresponding strain contribution is given by

$$\epsilon_{ij} = C_{ijkl}\sigma_{kl} + d_{kij}\xi_k \qquad (17.8)$$

where in (17.7) and (17.8), E_{ijkl} and C_{ijkl} are the elastic modulus and compliance tensors, respectively.

Table 17.1 describes the utilization of direct and converse effects as applied to the sensor and actuator functions of intelligent materials.

TABLE 17.1. Piezoelectric sensors and actuators

Type	Piezo-effect	Input	Output	Applications
Piezo ceramic (PZT)†	Direct	stress	voltage	sensors for mechanical loading
	Converse	voltage	strain	actuators for deformation control
Piezo electric polymer (PVDF)††	Direct	mechanical loading (static and dynamic)	voltage	sensors for static and dynamic loadings. Also, as passive vibration absorbers
	Converse	voltage	strain	strain rate control

† Lead zirconate titanate piezoelectric ceramics
†† Polyvinylidene fluoride

Piezoelectrics as Sensors and Actuators

As mentioned in the foregoing, mechanical displacement and electrical voltage are the varying parameters of the intelligent material when using piezoelectrics as sensors and actuators. Mechanical disturbance is converted into electrical voltage by a piezoelectric sensor. On the other hand, a piezoelectric actuator is activated by an electrical input to produce specific mechanical effect (e.g., strain or vibrations) through proper control algorithms; e.g. Sung-Kyu *et al* (1991). Such mechanical effect would then be used to compensate or control undesired effects such as deflections, excessive vibrations caused by the external stimuli on the engineering material or structure with which the intelligent material is incorporated. Sung-Kyu *et al* (1991) and Honein *et al* (1991) have successfully demonstrated that this active mechanical control could be effected on laminated composites by the use of distributed piezoelectric materials. Fundamental relationships have been derived from the basic principles, presented by equations (17.1) to (17.8) presented in the foregoing. A 3-D Finite Element procedure was adopted and supported by experimental results.

Piezoelectric-polymers as Intelligent Sensors and Actuators

Polyvinylidene fluoride (PVDF) is a piezoelectric polymer that can be used for sensor/actuator functions. The piezoelectric polymer may be embedded inside a structural member to actively control, for instance, the vibrations by dissipating the elastic energy imposed on the member (see, e.g., Ramachandran *et al.*, 1990). For this, a long bar of test specimen coupled with a layer of piezoelectric polymeric substance has been considered, with the lateral dimensions much smaller than the length. The polar direction is taken along the length of the specimen. The attenuation of mechanical vibrations in a passive absorbing element has been studied. This attenuation is achieved by converting a large fraction of elastic energy into electrical energy using the piezo-electric coupling effect and then dissipating the electrical energy using a simple resistive element. For efficient damping characteristics, the coupling coefficients must be large. In order to determine the damping factor (tan δ), constitutive equations of piezoelectric material coupled to the structural member were derived in a dynamic environment, where a harmonic plane wave propagating inside the material specimen has been considered. The results of the study indicate that it is possible to dissipate the mechanical vibratory energy imposed on the material through passive damping by piezoelectric polymers. It has also been proven through experimental work; e.g. Hagood *et al.* (1988), that it is possible to shift the peak damping to the frequency range of interest.

Active vibration control of a cantilever beam using distributed piezoelectric polymers and ceramics were studied by Honein *et al.* (1991), Lee *et al.* (1991) and Bailey and Hubbard (1985). All these studies included similar expressions derived from the fundamental principles of piezoelectricity, where piezoelectric sensors and actuators were used with a control algorithm to suppress the vibrational excitement.

Strain-rate Control Algorithm

Lee *et al.* (1991) used a "strain-rate control feedback mechanism" for the control algorithm. Based on the linear piezoelectric theory, the one-dimensional electrical displacement D in a piezoelectric material can be related to the mechanical strain ϵ in the same direction via the relationship:

$$D = dE\epsilon = e\epsilon \quad \textbf{(17.9)}$$

where d is the one-dimensional piezoelectric strain per charge constant, E is Young's modulus and e is the one-dimensional piezoelectric stress per charge constant. A piezoelectric polyvinylidene fluoride (PVF_2) film was used in this work as both sensor and actuator. Using a current amplifier to interface with the high impedance output of the piezoelectric material, piezoelectric strain rate sensors were created.

Bailey and Hubbard (1985), developed an active vibration damper for a cantilever beam using distributed parameter actuators on the basis of distributed parameter control theory. The distributed parameter actuator was the piezoelectric polymer (PVF_2). The control algorithm for the damper was based on the work done by Kalmann and Bartram (1960) on "Lyapunov's second method" for distributed parameter systems.

Numerous other articles have been published in the area of active vibration control of intelligent structures. Crawley and Luis (1987), for instance, have presented the use of piezoelectric actuators to suppress vibrational excitation in three different test specimens namely, aluminum, glass epoxy and graphite epoxy. Both analytical and experimental methods are presented and a scaling analysis has been performed to demonstrate the effectiveness in transmitting strain to the structure. Electronic damping of a large optical assembly has been studied by Forward *et al.* (1983). In this, piezoelectric ceramic strain transducers were used as sensors and actuators and the data taken during the study indicate the effectiveness of the devices even at high levels of acoustic and vibrational noise.

17.5. Optical Fibres as Sensors

Optical fibres have been used effectively as sensors in intelligent materials. Optical fibres may be classified, in general, into the following two types.

i) An extrinsic sensor which operates only as a transmitting medium for light but performs none of the sensing functions.

ii) An intrinsic fibre optic sensor which utilizes some intrinsic property of the fibre to detect a phenomenon or to quantify a measurement. A list of intrinsically measurable variables through the use of optical fibres is given in Table 17.2.

Glass and silica fibres form a basis for a broad range of sensors. The latter utilizes fibre properties to provide signals, indicative of external parameters such as force, temperature and deflection that are to be measured; e.g. Main (1985). The intrinsic properties of glass and silica qualify fibre optics as smart materials. Optical fibres are capable of performing as a sensor as well as a transmitter of the sensor's signal. Claus *et al.* (1988) developed an optical wave guide embedded in composites that can be used to determine the two-dimensional dynamic strain levels to which the material specimen is subjected to. This was carried out by using the change in the optical power transmitted in the fibre due to the induced strain in the structure and processing the resultant signal.

17.6. Shape Memory Alloys (SMA)

Shape Memory Alloys (SMA's) possess an interaction between the state of loading they are subjected to, the resulting strain and the thermal environment in which they are loaded. If these alloys are deformed at one temperature, they will completely recover their original shape when their thermal state is raised to a higher temperature. On the other hand, if the alloys are constrained during recovering, they can produce a mechanical effect (a recovery force) that is related to their temperature of transformation. Several alloy systems exhibit the phenomenon of shape memory (see, e.g., Wayman and Shimizu, 1972). A number of such alloy systems and their characteristics are given in Table 17.3.

TABLE 17.2. Applications of optical fibres

Variable	Methodology	Applications
Stress	Photoelastic effect	Fibre composites embedded with optical fibres can detect mechanical loading & vibrations
Strain	Change in optical power due to deformation	Strain could be sensed in structures embedded with optical fibres
Temperature	Thermal change in refractive index	Thermal state of fibre composites could be monitored during manufacturing by embedded optical fibres

Shape memory alloys have emerged as an alternative choice for situations involving dynamic control of large structures, which would often require vibration suppression and deflection control induced by adverse environment; e.g., Rogers *et al.* (1988). The mechanical deformation and thermal cycling of a shape memory alloy is illustrated by a stress-strain-temperature diagram in Figure 17.3.

TABLE 17.3. Alloy systems exhibiting shape memory effect

SME-alloy systems	Transportation temp.*	Recovery force for 2% strain in Kg/mm^2
Nitinol[1]	373 K	17
Cu-Zn-Al[2]	350 K	9
CANTIM 75[3]	480 K	14

[1] 49.93% nickel and 50.03% titanium
[2] 25.9% zinc, 4.04% aluminum and rest is copper
[3] 11.68% aluminum, 5.03% nickel, 2.00% Manganese, 0.96% titanium and rest is copper
* Temperature of transformation depends upon the composites of the alloy system.

As shown in Figure 17.3, the shape memory alloy is mechanically deformed to a plastic strain of 4% and the load is then removed (Curve OAB). To regain its original shape, the alloy is heated above its austenite end of transformation temperature A_f (Curve BCO'). The 4% strain is recovered between the temperatures of start and end of austenite transformation, A_s and A_f respectively.

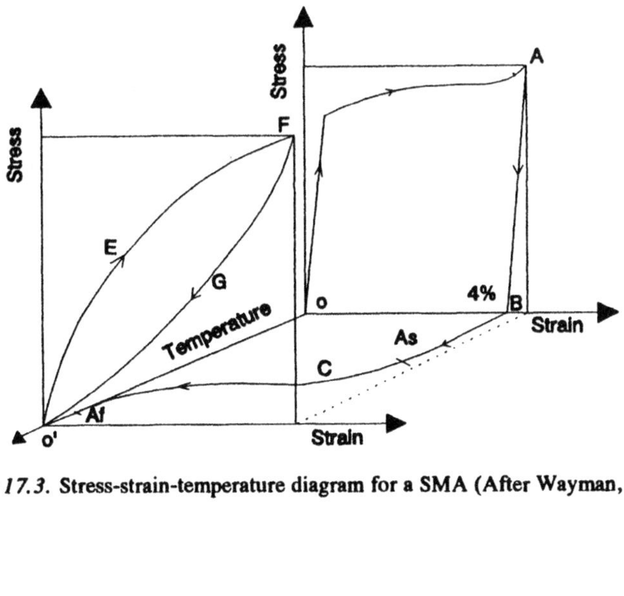

Figure 17.3. Stress-strain-temperature diagram for a SMA (After Wayman, 1989).

Martensite forms at a temperature below the temperature of martensite formation M_s, when the shape-memory alloy cools under no stress. Martensite can also form at temperatures above the temperature M_s if a stress is applied at such temperatures and the formed martensite is termed as 'Stress Induced Martensite' (SIM). If the alloy is stressed at a temperature above that of austenite end of transformation A_f, the alloy goes into a super elastic loop (O'EFGO') as illustrated in Fig. 17.3. This means that the strain of 4% is recovered completely on removal of load and the material behaves perfectly elastic.

The variation in the stress to produce SIM increases linearly with temperatures above M_s and obeys the modified Claussius-Clayperon relationship; e.g. Wayman (1989).

$$\frac{d\sigma_a}{dM_s} = -\frac{\Delta H}{\theta \epsilon_{Trans}} \tag{17.10}$$

In the equation above, σ_a is the applied stress above the Martensite formation temperature M_s to induce SIM.

θ is the ambient temperature
ΔH is the latent heat of phase transformation and
ϵ_{Trans} is the transformation strain of the super elastic loop

So far, shape memory effect has been considered only as one way effect, where an SMA wire, for instance, deformed below the temperature of Martensitic end of transformation M_f temperature can regain its original shape when heated to a temperature above that of A_f. But when cooled again to the temperature of Martensite start of transformation M_s, the wire's original shape remains and the material does not assume the 'deformed' shape. This is 'one way shape memory effect'. In the case of 'two way shape memory effect', however, a deformed SMA material below M_f regains its undeformed configuration when heated to a temperature above the temperature of Austenite end of transformation A_f (see, e.g., Delaey *et al.*, 1974). However, the undeformed configuration spontaneously attains its deformed shape when cooled below M_f. The specimen can, however, recover its undeformed configuration if heated to temperatures above A_f. Thus, it is possible to produce two geometric configurations of the material, by subjecting it to thermal cycling. The latter is termed as the "*trainability of two way shape memory effect*"; e.g. Wayman, 1989.

17.6.1. MATERIAL INTELLIGENCE USING SHAPE MEMORY ALLOYS

Thermomechanical environment subjects materials to cyclic thermal loadings, leading to fatigue and other undesirable mechanical effects. If the shape memory material is made to alter its mechanical properties with respect to a mechanical loading, many of the induced strains could be controlled. In this case, the thermal environment is sensed by an incorporated sensor and the SMA-material acts as an actuator by changing its mechanical response properties when heated (e.g., by passing an electric current through the SMA material).

In a multi-layered composite laminate with embedded SMA-fibres, excellent vibration suppression could be achieved when the laminate is subjected to dynamic loading; e.g., Rogers et al. (1988). Varying the mode shapes of induced vibration could be also achieved by varying the stiffness of SMA-fibres. This is accomplished by utilizing the large force created on constraining the micromechanical phase transformation from deformed state to undeformed state. Figure 17.4 illustrates, for instance, the effect of temperature on the variation of stiffness of nitinol fibres.

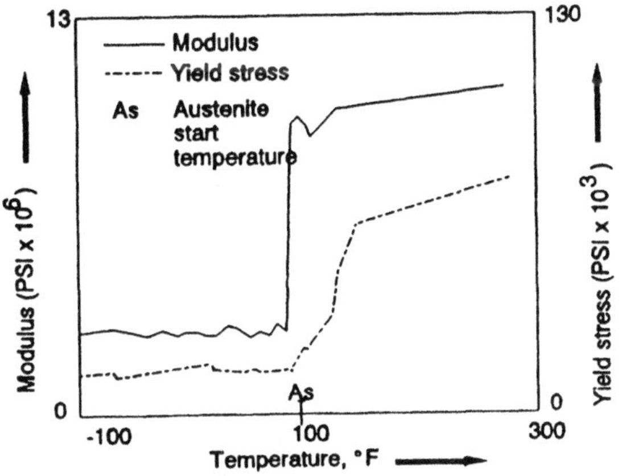

Figure 17.4. Approximate stiffness variation of Nitinol with temperature (After Jackson *et al.* , 1972).

It is also possible to use SMA-fibres as simple thermomechanical actuators rather than integrating them into a fibre-matrix system; e.g., Cross *et al.* (1969). This is achieved by coupling the thermomechanical actuator to the structural member externally. By ensuring proper coupling between the actuator and the structural member, the effects of the SMA actuator could be transferred to the parent material. Thus, shape memory alloys can be used effectively as actuators in intelligent materials when coupled with proper sensor and control algorithms.

17. 7. Shape Memory Polymers

Polymeric materials are generally viscoelastic in response behaviour and have the capability (e.g., Figure 17.5) of changing their dynamic properties (storage modulus (E'), loss modulus (E") and loss tangent (tan δ) with variations in environmental factors such as temperature, frequency and time; e.g. Ferry (1970), Murayama (1978), Nashif *et. al.* (1985), and Corsaro and Sperling (1989). Thus, polymeric materials have the capability of smart materials. This is accomplished by a sensor/actuator mechanism that could be incorporated in a structural member so that external stimuli such as mechanical vibrations could be sensed. Through a suitable control mechanism, the dynamic moduli of the polymeric material could be made to change (to adapt itself to the new environment) (see, for instance, Ganeriwala and Hartung, 1989). This could be achieved by shifting the loss factor (tan δ) towards the frequency spectrum that matches the imposed vibrational frequency, so that the absorption of the imposed vibrational frequency would be maximized. This shifting could be carried out by varying the loss modulus (E") or the loss factor (tan δ) of the polymer damper with respect to temperature or frequency.

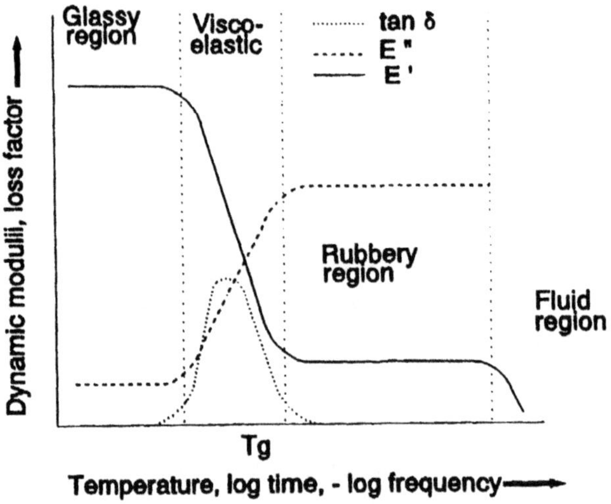

Figure 17.5. Schematic illustration of the variation of dynamic moduli of a polymer. Reprinted from Iyer, S. S. and Haddad, Y. M. (1994) Intelligent materials - An overview, *Int. J. Pres. Ves. & Piping* **58**, 335-44 with kind permission of Elsevier Science Limited.

17.7.1. MECHANISM OF SHAPE MEMORY IN A POLYMER

Shape memory polymers are unique polymeric materials which can recover their original shape before deformation at lower temperature (below the glass-transition temperature T_g), upon heating them to a temperature above T_g (see, e.g., Yoshiki and Shun-Ichi, 1988). This is an apparent advantage over ordinary polymers. An ordinary polymer when stressed, may not recover completely to its original undeformed configuration if the stress is released, thus, resulting in permanent deformation. In a shape memory polymer, however, the recovery loop is completed upon heating. Thus, a shape memory polymer is able to revert back to its original shape without undergoing any permanent deformation.

17.8. Electro-Rheological Fluids

The viscosity of certain fluids is influenced by the applied electric field. This phenomenon, termed "*Electroviscous Effect*", was reported around the turn of the century; e.g. Duff (1896). Researchers; Andrade *et al* (1946), have found an increase in the viscosity of conducting polar liquids of up to 100%, upon application of electric fields of the order of 1-10 KV/cm. For the electroviscous effect to occur, both polar molecules and conducting impurity ions are needed to be present. Large increases in viscosity, due to an applied electric field, for suspensions of finely divided solids in low viscosity oils was found as early as 1949. This effect termed as "*Winslow Effect*" is attributed to field induced fibre formation of the particles between the electrodes, thereby requiring additional shear stress for flow; e.g. Conrad and Sprecher, 1987.

The above said phenomenon has recently been termed as "*Electrorheology*" (see, e.g., Gandhi and Thomson, 1988), and has been applied in the development of actuator mechanisms in intelligent materials. When used with suitable sensors and control algorithms, electrorheological fluids can be made to change their properties by the application of electric field upon them.

The electrorheological behaviour of a suspension of fine silica particles in napthenic acid is governed by the Newtonian fluid flow principle (without an externally applied electric field). This principle is expressed as,

$$\tau = \eta\dot{\gamma}$$

where

τ is the applied shear stress
γ is the shear strain rate and
η is the Newtonian viscosity

When an electric field (ξ) is applied, the shear stress (τ) was found to increase to a critical value (τ_c) which must be overcome before any significant flow of the fluid occurs; e.g., Klass and Martinek (1967) and Uejima (1972). That is,

$$\tau = \tau_c(\xi) = \eta\dot{\gamma} \qquad \textbf{(17.11)}$$

where τ_c is independent of $\dot{\gamma}$, but increases with ξ.

Klass and Martinek (1967) used suspensions of silica particles in napthenic acid, and Uejima (1972) used cellulose in insulator oil to verify this phenomenon experimentally. The experimental verification indicates that τ_c is proportional to the square of the field, i.e., $\tau_c \propto \xi^2$. In the electrorheology phenomenon, the magnitude of electric field is the important parameter rather than, for instance, the spacing between the electrodes (see, for example, Conrad and Sprecher (1987).

17.8.1. MATERIAL INTELLIGENCE USING ELECTRO-RHEOLOGICAL FLUIDS

With reference to Fig. 17.6, a mechanical structural member which contains electrorheological fluid, when not activated, has a very low composite stiffness. This state represents the undisturbed configuration.

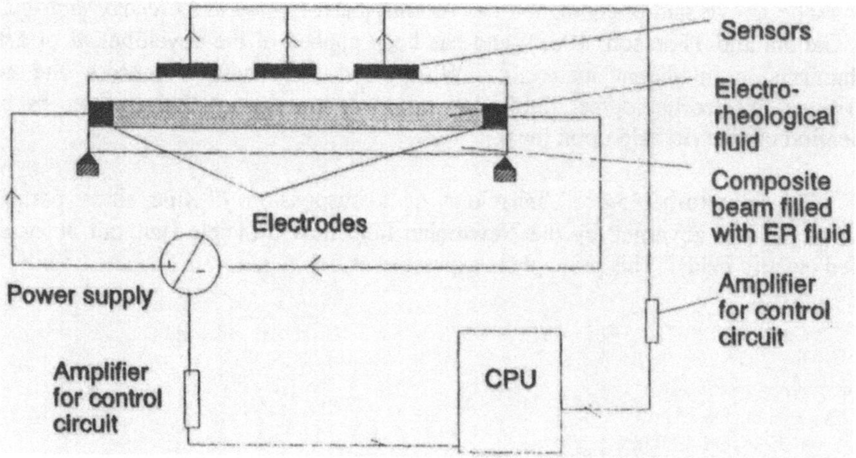

Figure 17.6. Electrorheological fluid as an actuator in a smart beam. Reprinted from Iyer, S. S. and Haddad, Y. M. (1994) Intelligent materials - An overview, *Int. J. Pres. Ves. & Piping* **58**, 335-44 with kind permission of Elsevier Science Limited.

When an environmental input (e.g. mechanical loading or a difference in thermal dient) causes, for instance, deflection in the structural member, it would be desirable to rease the stiffness to control the deflection. This is achieved by sensing the external chanical loading through incorporated sensors. The sensed signal is then processed in a croprocessor, which activates an auxiliary electric input to produce a desirable voltage. is voltage, when applied to the electrorheological fluid contained in the mechanical ictural member, increases the viscosity of the fluid, thus, practically converting it into a id. As a result, the overall stiffness of the specimen is increased, resisting the external ding and preventing deformation. The above said process could be made to take place 1/1000th of a second. Experimental investigations conducted, for instance, by Gandhi Thompson (1988) have verified the concept of electro-rheological fluids as intelligent terial actuators. These authors were able to illustrate that robot arms could be made ptable to external loading via changing its stiffness.

As demonstrated in this chapter, smart materials have the ability to improve mechanical ictures to be more advanced and reliable. Although the concepts of the techniques described this article were discovered decades ago, only recently that such techniques have emerged as ential constituents in intelligent materials methodology. The formulations for piezoelectrics icate the nature of direct and converse effects and their possible use in sensor and actuator hnologies. Discussions relating to shape memory alloys, shape memory polymers and ctrorheological fluids, illustrate the usage of these materials as actuators in smart material tems. The increase in stiffness of shape memory alloys and the change in the dynamic moduli shape memory polymers with temperature offers distinct advantages in controlling the static dynamic state of mechanical structures. Also the development of different feed back chanisms based on control algorithms and the increase in sophistication of microprocessor hnology and pattern recognition methodology will definitely play an important role in the 'ancement of processor function.

9. References

nad, I. (1988) 'Smart' structures and materials, *ARO Smart Materials, Structures & Mathematical Issues Workshop Proceedings*, Virginia Polytech. Inst. & State Univ., Blacksburg, VA, Sept. 15-16, pp. 13-16.

drade, C., Da. E.N. and Dodd, C. (1946) The effect of an electric field on the viscosity of liquids, *Proc. Roy. Soc. (London), Ser. A* **187**, 296-337.

ley, T. and Hubbard, J.E. (1985). Distributed piezoelectric polymer active vibration control of a cantilever beam, *J. Guidance* **8(5)**, pp. 605-11.

ly, W.G. (1946) *Piezoelectricity--An Introduction to the Theory & Applications of Electromechanical Phenomena in Crystals*, McGraw-Hill Inc., New York.

us, R.O., McKeeman, J.C., Mary, R.G. & Bennet, K.D. (1988) Optical fibre sensors and signal processing for smart materials and structures, *ARO Smart Materials, Structures, and Mathematical Issues Workshop Proceedings*, Virginia Polytech. Inst. & State Univ., Blacksburg, VA, Sept. 15-16, pp. 29-38.

rad, H. and Sprecher, A.F. (1987) Characteristics of ER fluids, *Advanced Mats. Conf.*, Colorado, Feb. 25-27, pp. 63-76.

Corsaro, R.D. & Sperling, L.H. (1989) Sound and vibration damping with polymers, in *Basic Viscoelas Definitions and Concepts*, ACS Symp., Series 424, Dallas, Texas, Apr. 9-14, pp. 5-22.

Crawley, E.F. and Luis, J.D. (1985). Use of piezoceramics as distributed actuators in large space structures, *A* Paper No. 85-0626, 126-32.

Crawley, E.F. and Luis, J.D. (1987). Use of piezoelectric actuators as elements of intelligent structures, *A* **25(10)**, 1373-85.

Cross, W.B., Kariotis, A.H. and Stimler, F.J. (1969) *Nitinol Characterization Study*, NASA Contractor Re NASA CR-1433.

Delaey, L., Krishnan, R.V., Tas, H. and Warlimont, H. (1974). Review - Thermoelasticity, pseudoelasticity & memory effects associated with martensitic transformations, *J. Mats. Sci.* **9**, 1521-35.

Duff, A.W. (1896). The viscosity of polarized dialectrics, *Phy. Rev.* **4**, 23-38.

Ferry, J.D. (1970) *Viscoelastic Properties of Polymers*, 2nd Edn., Wiley Interscience, New York.

Forward, R.L., Swigert, C.J. and Obal, M. (1983) *Electronic damping of a large optical bench, Shock Vibration Bulletin* **53**, 51-61.

Gandhi, M.V. and Thompson, B.S. (1988) A new generation of revolutionary ultra-aAdvanced intelli composite materials featuring electro-rheological fluids, *ARO Smart Materials, Structures Mathematical Issues Workshop Proceedings*, Virginia Polytech. Inst. & State Univ., Blacksburg, Sept. 15-16, 1988, pp. 63-8.

Ganeriwala, S.N. and Hartung, H.A. (1989). Fourier transform mechanical analysis and phenomenolo representation of viscoelastic material behaviour, *ACS Symp. Series* **424**, Dallas, Texas, Apr. 9-14, 1 pp. 92-110.

Gerber, E.A. & Ballato, A. (1985) *Precision Frequency Control*, Vol.1, Academic Press Inc., London.

Hagood, N.W., Crawley, E.F., Luis, J.D. and Anderson, E.H. (1988) Development of integrated component control of intelligent structures, *ARO Smart Materials, Structures & Mathematical Issues Work Proceedings*, Virginia Polytech. Inst. & State Univ., Blacksburg, VA, Sept. 15-16, pp. 80-104.

Honein, B., Braga, A.M.B. and Barbone, P. (1991) Wave propagation in piezoelectric layered media with s Applications, *J. of Intell. Mater. Syst. & Struct.* **2**, 542-57.

Ikeda, T. (1990) *Fundamentals of Piezoelectricity*, Tokohu Univ., Japan, Oxford Univ. Press.

Jackson, C.M., Wagner, H.J. and Wasilewski, R.J. (1972) *55-Nitinol - The Alloy with a Memory: Its Phy Metallurgy, Properties & Applications, A Report, NASA-SP* **5110**, Washington, D.C.

Kalmann, R.E. & Bartram, J.E. (1960) Control system analysis and design via the '*second*' method of Lyapu *J. Basic Engg., Trans. of ASME*, 371-400.

Klass, D.L. and Martinek, T.W. (1967) Electro viscous fluids, I. Rheological properties, *J. Appl. Phys.* **38**, 67

Kraut, E.A. (1969) New mathematical formulation for piezoelectric wave propagation, *Physical Review* **18** 1450-5.

Lee, C.K. O'Sullivan, T.C. and Chiang, W.W. (1991) *Piezoelectric strain rate sensor and actuator design active vibration control*, IBM Research Div. Yorktown Heights, New York, pp. 1-11.

Main, R.P. (1985) Fibre optic sensors-Future light, *Sensor Review (GB)* **5**, **3**, 133-9.

Murayama, T. (1978) *Dynamic Mechanical Analysis of Polymeric Materials*, Elsevier Scientific Pub. Oxford.

Nashif, A.D., Jones, D.I.G. and Henderson, J.P. (1985) *Vibration Damping*, John Wiley & Sons, Inc., New Yo

Olson, H.F. (1956) Electronic control of noise, vibration & reverberation, *J. Acoustical Soc. America* **28(5)**, 72.

Ramachandran, A.R., Xu, Q.C., Cross, L.E. and Newnham, R.E. (1990) Passive piezoelectric vibration damp *1st Joint U.S./Japan Conf. on Adaptive Structures*, Maui, Hawaii, Nov. 13-15, pp. 525-38.

Rogers, C.A. (1988) Workshop summary, *ARO Smart Materials, Structures & Mathematical Issues Work Proceedings*, Virginia Polytech. Inst. & State Univ., Blacksburg, VA, Sept. 15-16, pp. 1-9.

Rogers, C.A., Barker, D.K. and Jaeger, C.A. (1988) Introduction to smart materials and structures, *ARO S Materials, Structures & Mathematical Issues Workshop Proceedings*, Virginia Polytech. Inst. & S Univ., Blacksburg, VA, Sept. 15-16, 1988, pp. 17-28.

;ers, C.A., Liang, C. and Barker, D.K. (1988) Dynamic control concepts using SMA reinforced plates, *ARO Smart Materials, Structures & Mathematical Issues Workshop Proceedings,* Virginia Polytech. Inst. & State Univ., Blacksburg, VA, Sept. 15-16, 1988, pp. 39-62.

ıg-Kyu Ha, Charles K. and Fu-Kuo, Chang (1991) Analysis of laminated composites containing distributed piezoelectric ceramics, *J. of Intell. Mater. Syst. & Struct.* **2**, 59-71.

:agi, T. (1990) A Concept of intelligent material, *U.S.-Japan Workshop on Smart/Intelligent Materials and Systems*, Iqbal, A., Rogers, C.A., Andrew, C. & Masuo, A. (Eds.), Mar. 19-23, 1990, Honolulu, Hawaii, pp. 3-10.

ima, H. (1972) Dialectric mechanism and rheological properties of electro-fluids, *Jap. J. Appl. Phys.* **11(3)**, 319-26.

yman, C.M. (1989). The nature of the shape memory effect, *Proc. 1st Japan Int. SAMPE Symp.*, Nov. 28-Dec.1, 1989, pp. 189-94.

yman, C.M. and Shimizu, K. (1972) The shape memory 'Marmen' effect in alloys, *Metal Sci. Journal* **6**, 175-83.

ıhiki, S. and Shun-Ichi, H. (1988) *Development of Polymeric Shape Memory Material*, Mitsubishi Tech. Bulletin, Mitsubishi Heavy Ind. Ltd., New York, No. 184.

CHAPTER 18

PATTERN RECOGNITION AND CLASSIFICATION METHODOLOGY FOR THE CHARACTERIZATION OF MATERIAL RESPONSE STATES

18.1. Introduction

In the present Chapter, we discuss the design procedure of a computer-based expert system, in conjunction with a non-destructive quantitative examination technique, e.g., acousto-ultrasonics, for the identification of material response states.

Acousto-ultrasonics (AU) is a relatively new quantitative non-destructive examination technique that combines aspects of conventional "*Ultrasonic*" and "*Acoustic Emission*" practices. It has been proven to be a suitable approach to quantify microstructural and morphological states of materials and the related mechanical properties (e.g., Tanary *et al.*, 1992, and Haddad and Iyer, 1995&1996).

Acousto-ultrasonics may be interpreted as "*acoustic emission simulation with ultrasonic sources*" (Vary, 1987). In the acousto-ultrasonic technique, stress waves are "*simulated*" to resemble acoustic emission waves, but without disrupting the material, i.e., without the application of an external loading (Vary, 1988). The working hypothesis of the acousto-ultrasonic technique may be stated as:

> "*More efficient stress-energy transfer and strain redistribution, in the microstructure of the material specimen, during mechanical loading, would correspond to an enhanced mechanical strength of such material*".

In the AU practice, the multi-interactions of the ultrasonic-wave with the material microstructure usually result in complicated waveforms that are quite difficult to analyse. A relatively new approach to the analysis of AU signals is the use of "*Pattern-recognition and Classification Methodologies*". In this approach, acousto-ultrasonic waveforms are identified as belonging to a class, where each class represents one of different states of the tested material-property. For this purpose, each waveform is mathematically treated as a multi-parametric entity, which is called a "*pattern vector*".

Each component of such a pattern vector represents a value of a parameter, also called "*feature*", which is used for the identification of the AU signal. In the pattern-recognition (PR) practice, a computer-based pattern-recognition system, labelled "*Pattern-recognition Classifier*", is designed on the basis of AU signals pertaining to known material states.

Classification of unknown patterns is based on the so-called "*decision functions*". There are two main approaches to generate these decision functions, i.e., *deterministic* and *statistical*. The deterministic approach comprises decision rules that are established by the assumption that a minimum of a function of a so-called "*generalized distance*" between a pattern of values (outputs) of a particular event and various known classes of such an event indicates that the pattern belongs to the class indicated by the minimum of the said function (e.g., Tou and Gonzalez, 1974). The statistical approach, however, is based on the maximization of the probability of classifying a pattern as belonging to a particular class, when it appears to belong, at the same time, to another class (e.g., Fu, 1976).

Decision functions are usually determined by limited-size samples of pattern vectors that are selected for the design of the Pattern-recognition system. In this context, arbitrary decision functions are initially assumed and, then, through a sequence of iterative learning steps, these decision functions are made to approach satisfactory forms. This procedure is called "*Learning and Estimation*" of decision functions (e.g., Andrews, 1972, Chen, 1982 and Devijver, 1982). The technique has been recently used (e.g., Haddad and Iyer, 1995&1996, and Molina and Haddad, 1995&1996) to test solid polymers, which characteristically show high attenuation of ultrasonic waves.

18.2. The Acousto-Ultrasonics Technique

In the AU practice, *Figure 18.1*, a broadband transducer inputs a repetitive series of ultrasonic pulses into the test specimen. A receiving transducer, located at a specific distance from the sending transducer, captures the already transmitted wave. Both transducers are coupled at normal incidence to the surface of the specimen. The transmitted ultrasonic wave into the material specimen is considered to be affected by the microstructural and morphological properties of the material specimen that determine its mechanical performance. Accordingly, it is postulated that the captured AU signal would contain information concerning the overall mechanical response of the material specimen. Although a number of test configurations are possible, the most desired experimental configuration is the one in which the sending and receiving transducers are located on the same side of the specimen, as demonstrated in Fig. 18.1. This configuration is advantageous when inspecting, for instance, components of a large structure in service. Experimental work on solid polymers by, for instance, Lee and Williams (1991) and Iyer and Haddad (1993), on metals by Tanary (1988), and on different classes of composites by Vary (1982) and Williams and Lampert (1980) revealed the convenience of using the AU transducer configuration described above (Fig. 18.1).

Although the transmitting transducer injects longitudinal waves normal to the specimen surface, the sound waves radiated into the material produce oblique reflections and shear waves. The resultant stress waves, which consist of longitudinal and transverse components, propagate in the material specimen interacting with a significant portion of the microstructure along their path. In many situations, as discussed in the foregoing, it is possible to obtain information on the mechanical behaviour of the material from the AU wave propagation data. In this context, acousto-ultrasonic waveforms have been shown to be sensitive to interlaminar and adhesive bond-strength variations (Tanary, 1988). Acousto-ultrasonics have been proven to be useful in assessing micro-porosity and micro-cracking produced by fatigue cycling (Williams and Lampert, 1980). The technique has been also used in estimating the variation in strength of structural composites (Vary, 1987), as well as, in the evaluation of their residual strength and degradation due to cyclic fatigue and impact (Nayeb-Heshmi *et al.*,1986). The technique has been further used in determining the strength of wire ropes (Dos Reis and McFarland, 1986), the tensile strength of nylon ropes (Williams *et al.*,1984) and in the prediction of the filler content in wood and paper products (Dos Reis and McFarland, 1986). Strength of ceramic materials and the effect of hydrothermal aging on composites have been, also, evaluated successfully using the acousto-ultrasonic approach (Phani *et al.*,1986).

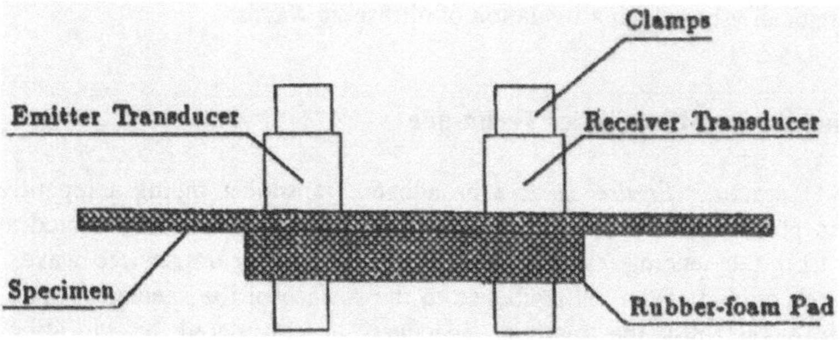

Figure 18.1. AU-transducer setup. "Reprinted from Molina, G. J. and Haddad, Y. M. (1995) On the identification of residual impact properties of materials by acousto-ultrasonics - A pattern recognition approach, *Acta Mechanica Sinica* , Vol 11, No. 1, February 1995, 34-43, with kind permission from Allerton Press, Inc."

As illustrated in Figure 18.2, the pulsar-sender is a printed circuit board capable of pulse generation. It emits electric voltage pulses to the piezoelectric transducer at a predetermined rate resulting in acoustic pressure waves that have frequencies in the ultra-sonic range of 1-20 MHz. As mentioned in the foregoing, the emitted wave is to interact with the microstructure of the material before being captured by the receiving transducer. As shown in Fig. 18.2, the pulsar-sender is connected to the pre-amplifier and, in turn, to an acoustic emission (AE) testing facility. The AE facility receives the captured wave-form, converts the signal from an analog to a digital form and, then, transfers the data to a digitization board. The latter possesses a signal processing capability. It digitizes the signal with the use of a specialized real-time data acquisition and signal processing soft-ware. As soon as a signal is digitized, the digitizer triggers the pulsar again to send an-other input-wave which interacts with the material and the sequence of operations mentioned above are repeated. After digitization, waveforms are stored electronically in separate data files for later analysis.

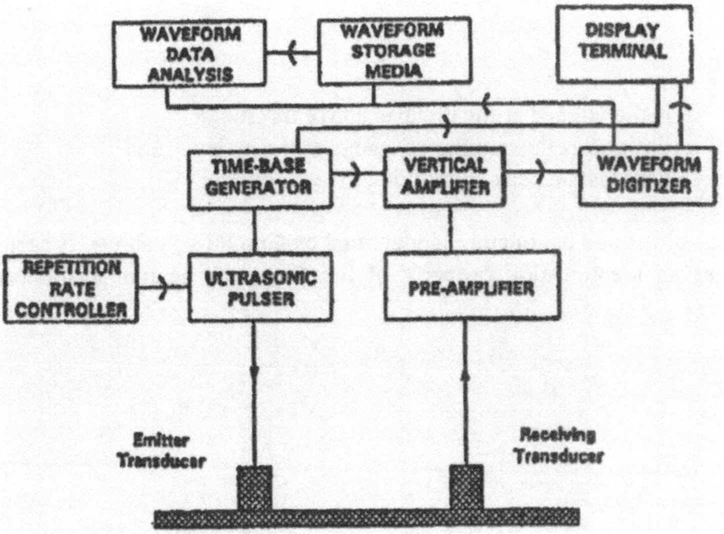

Figure 18.2. A schematic diagram of the acousto-ultrasonic test set-up. "Reprinted from *Int. J. Pres. Ves. & Piping* **67**, Molina, G. J. and Haddad, Y. M. , Acousto-ultrasonics approach to the characterization of impact properties of a class of engineering materials (1996), 307-15, with kind permission from Elsevier Science".

The rate of emission of input pulses is determined by a '*repetition rate utility*' which is operated by the controlling software. The latter initiates the waveform generation and reception. A properly designed software should, also, allow easy selection of threshold setting, gain, damping, frequency, etc.

A major concern with any quantitative non-destructive evaluation technique, such as Acousto-ultrasonics, is the reproducibility of measured data. A rational approach to the design of a suitable setup for AU data acquisition is a customized trial under the guide of previous research; *see*, for instance, Iyer (1993), Russell-Floyd and Phillips (1988), and Tanary (1988).

Acousto-Ultrasonic Parameter

A quantifying parameter to interpret the information contained in the received acousto-ultrasonic signal was originally proposed by Williams and Lampert (1980) and was adopted for the verification of the characterization of the mechanical performance of a class of materials by Tanary (1988), Tanary *et al.* (1992) and Iyer (1993). It is referred to as the '*Acousto-Ultrasonic Parameter* (AUP)' and is interpreted in the present work, with reference to Fig. 18.3, as follows

$$AUP(V_i) = \sum_{i=0}^{p} V_i(C_i - C_{i+1}) \tag{18.1}$$

where

V_i is the voltage at the ith level above threshold,
C_i designates the number of counts at the ith level, and
V_p denotes the peak amplitude of the waveform.

The acousto-ultrasonic parameter, as identified by Eqn. (18.1) above, is seen in the present work as an identification property of the wave propagation characteristics of the material.

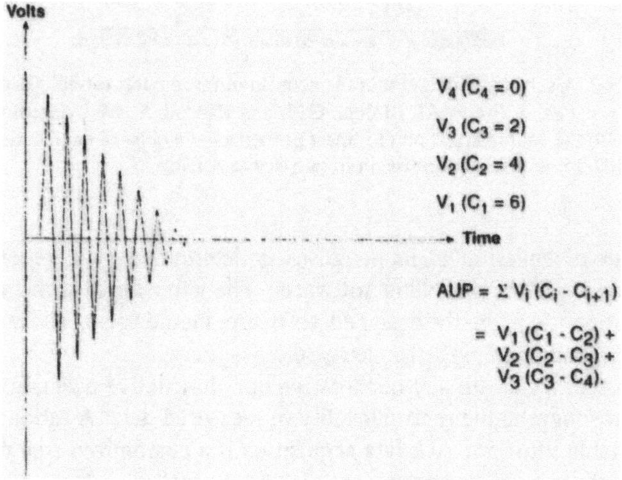

Figure 18.3. Calculation of the acousto-ultrasonic parameter (AUP). "Reprinted from Tanary, S., Haddad, Y. M., Fahr, A. and Lee, S. (1992) Nondestructive evaluation of adhesively bonded joints in graphite/epoxy composites using acousto-ultrasonics. *Journal of Pressure Vessel Technology, Transactions of the ASME*, August 1992, Vol. 114, 345-52, with kind permission of ASME"

Factors Affecting Acousto-ultrasonic Waveform Measurement
Other than material property variations, several important parameters relating to the experiment set-up would affect the waveform. These parameters may be grouped as "*external*" to the testing system, e.g., the pressure applied on the transducers, the type of couplant, between the transducers and the test specimen, and the distance between the transducers, and "*internal*" in the testing system, e.g., the frequency of the propagating wave, '*gain*' used to amplify the signal and the threshold voltage above which the signal is digitized (Russel-Floyd and Philips, 1988). These parameters and their effects on AU wave form are discussed, for instance, by Tanary !1988), Iyer (1993) and Molina (1994).

External Parameters. The type of couplant used to create continuity between the transducers and the test specimen is an important factor that affects the resultant waveform. Figure 18.4 presents experimental data (e.g., Haddad and Iyer, 1995) concerning the reproducibility of the acousto-ultrasonic results and the clarity of the transmitted signals for three different types of couplant, namely, acoustic emission gel SC-2, water and petroleum jelly. It is seen, from the latter figure, that, for the case of solid polyvinylchloride, tested at room temperature, the couplant SC-2 gives high and consistent readings of AUP. Such readings should also fall well within the calibration range of the instrumental setting. These results have also shown to be valid when testing other classes of materials including metals (e.g., Tanary, 1988) and polymeric base composite materials (e.g., Molina, 1994).

The change in distance between the sending and receiving transducers would also affect the test results quite significantly, as it corresponds to the extent of the material microstructure being examined by the travelling ultrasonic waveform. An illustrative example of the variation of AUP with the distance between the sender and receiver transducers is shown in Fig. 18.5 for the case of solid Polyvinylchloride tested at room temperature; e.g., Molina, 1994. Similar to the parameters mentioned above, the pressure applied on the transducers is an important factor that would affect the experimental data obtained. It was reported in the literature, e.g., Henneke (1983), that for repeated AU measurements, large loads (as much as 20 lbs.) could be applied to the transducers. In the experimental work of Iyer and Haddad (1993) and Molina and Haddad (1995, 1996), for instance, it is indicated that the load should be applied uniformly with a magnitude just sufficient to eliminate unwanted reverberations within the couplant (see, also, Tanary *et al.*, 1992).

Internal Parameters. The acousto-ultrasonic waveform is influenced by internal factors concerning the experimental hardware setting, such as gain and frequency of the injected wave. Following the discussions given by Tanary (1988) and Tanary *et al.*, (1992), an optimization of the instrumental setting should be carried out to find the best combination of transmitting frequency and amplitude gain. Thus, the system gain would be selected on the basis of sensitivity of the transmitting and receiving transducers. In the experimental work presented here, the set value is chosen, so that the signal to noise ratio is sufficiently high, while, the maximum signal amplitude is to be held below the saturation level of the receiving instruments. In this context, the frequency of the acousto-ultrasonic waveform should be set at a value which would result in optimum conditions for acousto-ultrasonic measurements. Further, this frequency should fall well within the band width of the broad band pulsar and receiving transducers. This consequently enables the sampling rate to be set at a reasonably high level and would still capture a significant number of waveforms without exceeding the memory space of the digitizer (e.g., Finkel, 1975, and Haddad and Iyer 1995&1996).

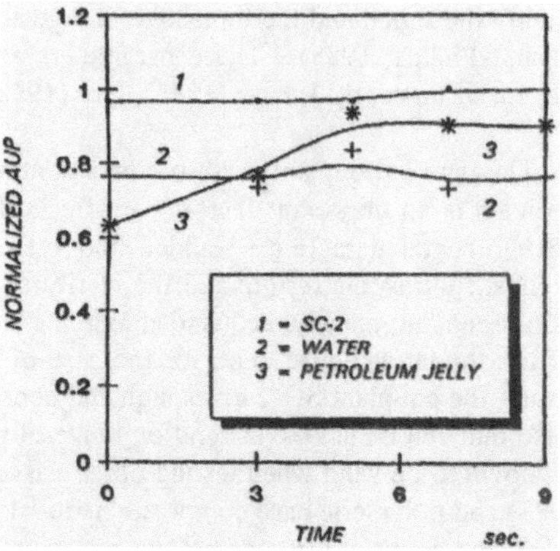

Figure 18.4. Variation of the acousto-ultrasonic parameter with couplant type. Material: solid Polyvinylchloride (PVC), tested at room temperature. "Reprinted from *Int. J. Pres. Ves. & Piping* **63**, Haddad, Y. M. and Iyer, S. S., An Acousto-ultrasonics pattern recognition approach for the mechanical characterization of engineering materials, 89-98 (1995), with kind permission from Elsevier Science".

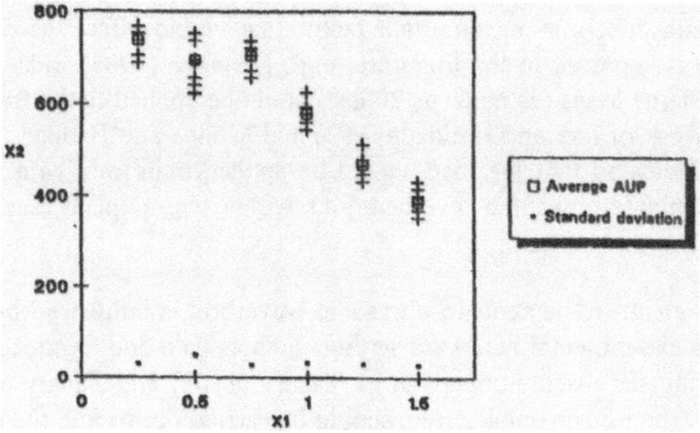

X1 : Distance between transducers in inches.
X2 : Acousto-Ultrasonics Parameter (AUP) in mVolts.

Figure 18.5. Sensitivity of AU measurements to distance between transducers. Material: solid Polyvinylchloride (PVC), tested at room temperature (Molina, 1994).

18.3. Fundamentals of the Design of Pattern-Recognition (PR) Systems

In statistical pattern recognition (e.g., Andrews, 1972 and Tou *et al.*, 1974), pattern values are classified into their respective classes by plotting their common feature values in a '*feature space*'. The latter is an Euclidean space where the coordinate axes represent the common features of interest to the problem being analysed. The design of a pattern recognition system generally involves the following three steps:

Step I
This step is concerned with the representation of input data which can be measured concerning the state of the material that is required to be recognized. The AU signal provides the raw data values that are necessary for the pattern classification process. Each point of the signal would be a characteristic feature in the time domain. However, the primitive measurements of the acousto-ultrasonic waveform could become very large. As stipulated in the foregoing, each of such primitive measurements would carry a 'small portion' of 'information' about the microstructure of the material which had been interrogated by the AU wave (e.g., Haddad and Iyer, 1993)..

Step II
Due to the large number of variables involved, it becomes necessary to '*extract*' important '*features*' from the primitive measurements. Each of the selected features would carry a small, but significant information for classification purposes and would be selected according to the physics of the problem. This is achieved through a process known as the '*feature extraction*' process. The latter constitutes the present *Step II*. In a typical study, the adopted software extracts in excess of 100 standard features that could be used for wave-form analysis.

Step III
The third step in pattern-recognition system design involves an optimum decision procedure associated with the classification process. On assuming, for instance, that a machine is to be designed to recognize 'M' different pattern classes denoted, for instance, by $\omega_1, \omega_2, ..., \omega_M$, then, the pattern space can be considered as consisting of M regions, each of which encloses the pattern points of a class ω_i ($i = 1,2, ..., M$). The recognition problem can, then, be viewed as that of generating the decision boundaries which separate the referred-to 'M' pattern-classes on the basis of the observed measurements. In Pattern Recognition, decision functions are established by the so-called "*learning and estimation*" procedures. They are implemented by means of a computer-based set of rules which constitutes the pertaining *"Pattern Recognition Classifier"*. Examples of decision func-

tions which may be employed in the design of corresponding PR-classifiers are:

(I) Empirical Bayesian,
(II) Linear Discriminant,
(III) K-Nearest Neighbour, and
(IV) Minimum Distance Classifier.

Figure 18.6 illustrates the concepts of the decision functions based on the classification schemes mentioned above (e.g., Haddad and Iyer, 1995&1996). Meanwhile, the reader is referred to Iyer (1993) and Molina (1994), among others, for details concerning these classification schemes.

For the purpose of establishing the set of sample patterns of known classification, the software is to be designed to split the input data values into two separate files after normalizing the data with respect, for instance, to its variance. The first file contains the values of the normalized feature values which are used to create the boundary between the pattern classes. Once the boundary is established, the classifier is evaluated using the data contained in the second file. The results of the "*training*" and "*evaluation*" processes are expressed as percentage of success in forming distinct clusters in the feature space from the known pattern values. The predictability of such a classifier is tested using a raw data set taken from an unknown sample of relevant input data whose values could be determined by an alternative technique.

Thus, the task of a pattern-recognition system may be defined as "*the categorization of input data into identifiable classes via the extraction of significant features or attributes from a background of relevant data*". Operationally speaking, a PR System would perform the following transformation

$$P_s \rightarrow F_s \rightarrow C_s \quad \textbf{(18.2)}$$

where the "Pattern-Space P_s" comprises the sets of feature values, also called primitive pattern vectors, which are extracted from either analogical or digitized descriptions of the material response states to be recognized; the "Feature Space F_s" comprises the pattern vectors formed by a selection of the features which carry the discriminatory power between classes for the given problem; and the "Classification Space C_s" is a frame of defined classes where a pattern of unknown classification is identified as belonging to a known class. The design of a "*Pattern Recognition System*" is the building of a series of procedures to perform this transformation.

18.3.1. FEATURE EXTRACTION AND NORMALIZATION

Figure 18.7 presents schematics of the procedure employed to perform the partial transformation from Pattern Space (P_s) to Feature Space (F_s). With reference to Figure 18.7,

AU signals pertaining to each class are divided in two sets; namely, "*Classifier*" and "*Testing Sets*". Classifier Sets are used for building the classifier, as well as for further testing of the "*Training*" and "Evaluation" performances of the designed classifier. Testing Sets are used for testing of the "Classification Performance".

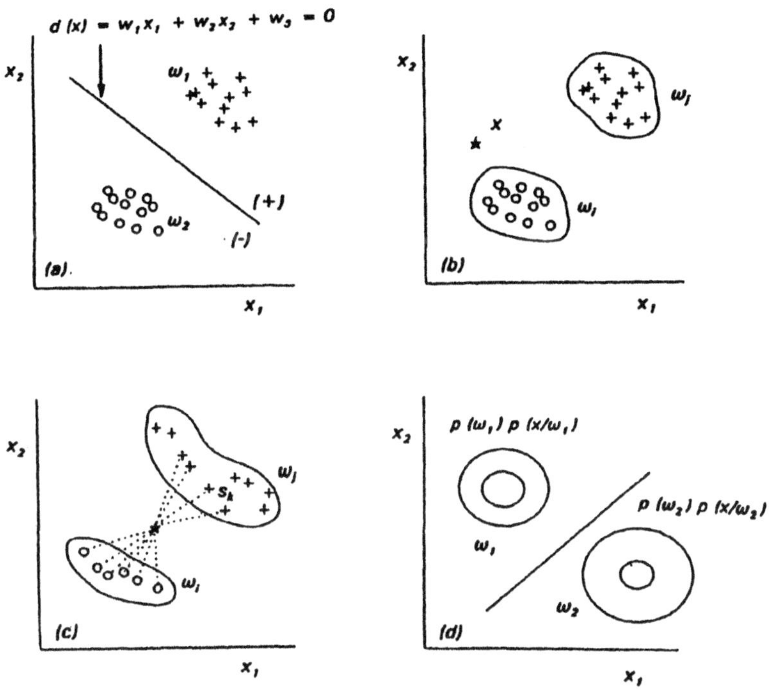

Figure 18.6. Two-parameter acousto-ultrasonic wave classification: (a) a simple decision function for two pattern classes; (b) the proximity concept as a classifier; (c) nearest neighbour classifier; (d) a schematic illustration of the empirical Bayesian classifier. (ω_1: a class of undamaged material states; ω_2: a class of damaged material states; x_1: peak amplitude of the acousto-ultrasonic waveform; x_2: inter-peak distance of the acousto-ultrasonic waveform). "Reprinted from Int. *J. Pres. Ves. & Piping* **63**, Haddad, Y. M. and Iyer, S. S., An Acousto-ultrasonics pattern recognition approach for the mechanical characterization of engineering materials, 89-98 (1995), with kind permission from Elsevier Science".

To carry out the feature extraction and normalization procedures, both Classifier and Testing sets have to be arranged as "*File-Trees*", respectively called "*Classifier Trees*" and "*Normalization Trees*", which correlate together the AU signals which are stored as computer files. A file tree is "*an arrangement of the records of digitized AU signals in related classes by means of a root Organization File*". An example of a Classifier Tree for the building of classifiers is presented in Figure 18.8. With reference to the latter figure, each class of the Classifier Tree comprises a limited number of specimens, each of them is represented by a number of AU-signal records taken successively at small time intervals, e.g., of one second.

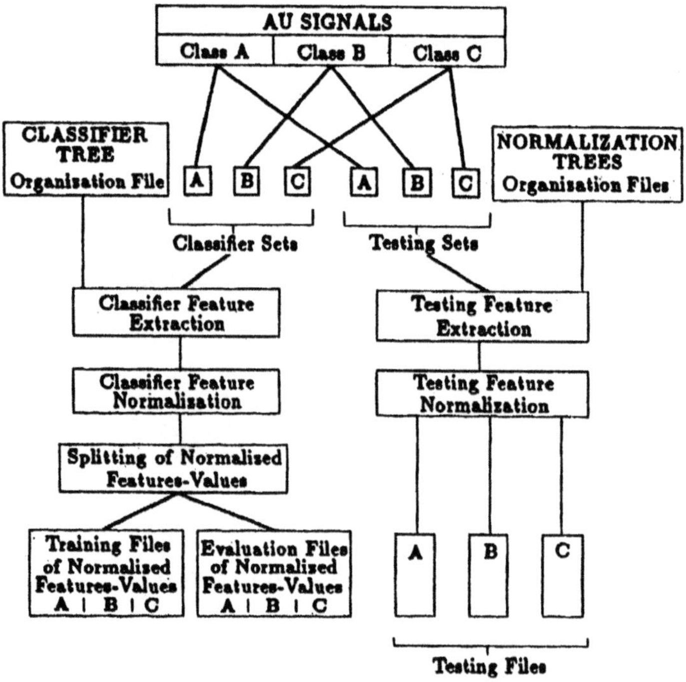

Figure 18.7. Schematics of the feature extraction and normalization procedures. "Reprinted from Molina, G. J. and Haddad, Y. M. (1995) On the identification of residual impact properties of materials by acousto-ultrasonics - A pattern recognition approach, *Acta Mechanica Sinica*, Vol 11, No. 1, February 1995, 34-43, with kind permission from Allerton Press, Inc.".

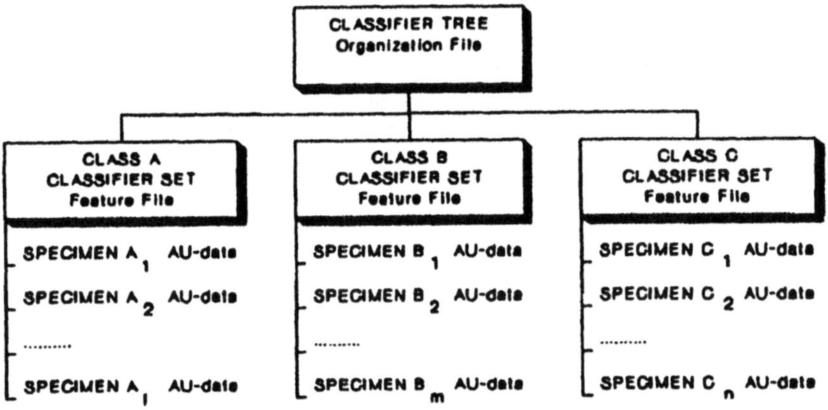

Figure 18.8. Classifier Tree. Arrangement of AU-records pertaining to classifier sets. "Reprinted from Molina, G. J. and Haddad, Y. M. (1995) On the identification of residual impact properties of materials by acousto-ultrasonics - A pattern recognition approach, *Acta Mechanica Sinica* , Vol 11, No. 1, February 1995, 34-43, with kind permission from Allerton Press, Inc.".

The software, designed for the purpose of this study, extracts 108 feature-values from each digitized record of an AU-signal. These values are of diverse order of magnitude in five domains of wave-description (i.e., Time, Frequency, Phase, Cepstral and Auto-correlation domains). Hence, Feature normalization is necessary so that the involved features would have comparable values in the dealt with "n-feature space". Feature normalization may be performed by means, for instance, of the Method of Zero-Mean Unit-Variance Mapping; e.g., Iyer (1993), and Haddad and Iyer, 1995 &1996. In the referred-to method, a normalized feature value ξ_j for feature j is obtained from the non-normalized value x_j of a pattern vector x as follows

$$\xi_j = \frac{x_j - \mu}{s} \tag{18.3}$$

where μ is the mean of the values of the j-feature for all pattern vectors being investigated; and s is the standard deviation of the values of the j-feature for all classes under consideration. This means that for a multiple class problem, the "global mean" of all values of the j-feature is used as the referrd-to mean μ, while the average of the standard deviation values for all the classes is taken as s. Normalized values will be loosely bounded to the range (-1,1), measured in standard deviation units.

As it is shown in Fig. 18.7, the "Classifier" and "Testing" sets are subjected to the successive operations of "*Feature Extraction*", "*Feature Normalization*" and, when it applies, "*Feature-value Splitting*". Feature-value Splitting has the purpose of obtaining two files of records representing the same AU signals for the Training and Evaluation of the classifiers, as it is explained in the following sub-section.

18.3.2. FEATURE SELECTION AND CLASSIFIER BUILDING

Figure 18.9 presents schematics of the procedure used in the present study for building and testing of classifiers by means of feature files. This procedure corresponds to the partial transformation from the "*Feature Space*" to the "*Classification Space*".

With the possible availability of a significantly large number of features for describing the AU signal, the problem of selecting an optimal set of best discriminating features could be very complex. To solve the problem of feature selection, a procedure may be established for the design of a classifier that would give the highest average of the classification performance for the classes being considered. For each particular classification problem, it is possible to establish a ranking of the feature's discriminating-ability between classes. In this context, for instance, the 'Fisher Distance', also called the 'Fisher Discriminatory Ratio', may be used for ranking the involved features; e.g., Iyer (1993).

For each type of classifier, the selection of features is guided by strictly following the adopted rule of ranking. Thus, no isolated features are chosen. Instead, only complete ranking-series may be taken. The procedure adopted for the design of one type of classifier is carried out in building a series of classifiers of this type by starting with the classifier designed by using the first-ranked feature only. Classifiers of the same type are, then, obtained by adding features from the succession already determined by the ranking series. The so-called "*performance-classification*" by means of the '*Testing Sets*' are then experimentally obtained for the designed series of classifiers. In the feature selection procedure, the number of features that gives the highest average of the obtained classification performance may be chosen to design the most appropriate classifier for a given classification problem.

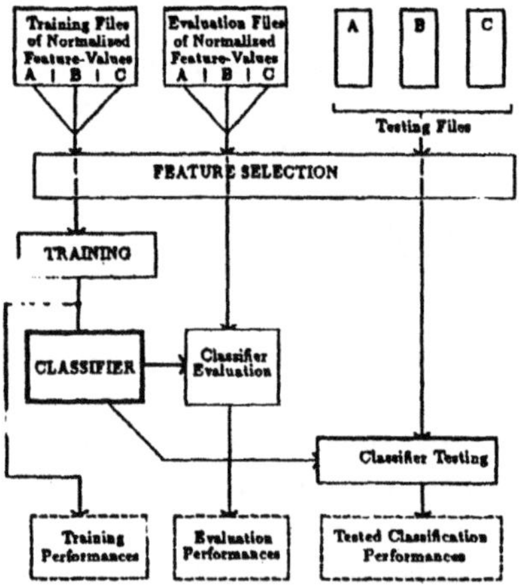

Figure 18.9. Procedures for building and testing of classifiers. "Reprinted from Molina, G. J. and Haddad, Y. M. (1995) On the identification of residual impact properties of materials by acousto-ultrasonics - A pattern recognition approach, *Acta Mechanica Sinica*, Vol 11, No. 1, February 1995, 34-43, with kind permission from Allerton Press, Inc.".

18.3.3. TRAINING AND PERFORMANCE EVALUATION OF THE CLASSIFIER

As discussed earlier in this chapter, the designed classifiers work on the basis of '*decision functions*'. The ability of these decision functions to distinguish between the same digitized records, used to build them, can be expressed as the percentage of correctly classified records. Since this percentage represents the effectiveness of the training of the classifier to recognize the AU signals used, it is referred to as the "*training*" performance.

A classifier may, also, be evaluated by the percentage of correct classification when classifying AU records obtained from the same specimens which were used in building the classifier. This percentage may be referred-to as "*Evaluation*" performance. The used procedure for testing the "*Training*" and "*Evaluation*" performances of a classifier is displayed in Figure 18.9.

18.4. Illustrative Applications

18.4.1. CHARACTERIZATION OF THE STRESS-RELAXATION RESPONSE

In this subsection, we present a case study concerning the use of the acousto-ultrasonic technique, in conjunction with statistical pattern recognition, to characterize the stress-relaxation response of a class of linear viscoelastic material; namely, solid Polycarbonate (PC), tested at room temperature. In this context, the value of the time-dependent "*relaxed*" stress is correlated with the "*acousto-ultrasonic parameter (AUP)*", introduced earlier by Eqn. (18.1), for different strain input levels. Statistical pattern recognition methodology, as described earlier in the previous section, is used to build a classifier for different "*relaxed*" stress states.

When employed for the case of a linear viscoelastic material, suitably at different times under a given level of strain input, the resultant acousto-ultrasonic waves would contain '*features*' pertaining to the time-dependent macro-mechanical property of stress-relaxation of such material. This poses a typical statistical pattern classification problem. The various material stress-relaxation states characterized by the particular features of the pertaining ultrasonic waveforms would form distinct clusters in an n-dimensional feature space. An unknown stress-relaxation response state of a specimen may then be matched to one of the clusters and classified as being the respective material stress state.

Experimental Procedure and Results

An uniaxially loaded test specimen configuration of Polycarbonate (PC) is adopted for performing the required stress-relaxation experiments. The material test specimens were prepared as per ASTM D-638. Stress-relaxation tests were carried out under constant strain levels of 0.01, 0.02,..., 0.07. The corresponding experimental relaxation curves are shown in Fig. 18.10. AU measurements were taken from material specimens already undergone stress relaxation experiments corresponding to the three strain levels 4%, 5%

and 6%, for the successive time intervals of 1,10,100,1000,10000 and 100000 sec., for each of the mentioned strain levels. For each considered stress state, five material specimens were tested. The different parameters concerning AU measurements and the values of these parameters, as considered in the present work, are outlined in Table 18.1.

The acousto-ultrasonic parameter (AUP) was calculated for each AU measurement in accordance with Eqn. (18.1). It is, then, normalized for each testing time interval t with respect to the maximum of the average values at the same time interval t, for the five specimens tested at this state. The normalization of AUP is carried out, in accordance to the following expression in order to eliminate the effect of possible material variations between the five tested material specimens.

$$AUP_{norm}(t)\Big|_{\varepsilon} = \frac{AUP\ measured\ at\ a\ relaxed\ stress\ state\ at\ time\ t}{Maximum\ of\ the\ average\ values\ of\ AUP\ for\ specimens\ tested\ at\ time\ t} \tag{18.4}$$

where $AUP_{norm}(t)|_{\epsilon}$ is the normalized value of the acousto-ultrasonic parameter at time t under a particular level of strain input ε.

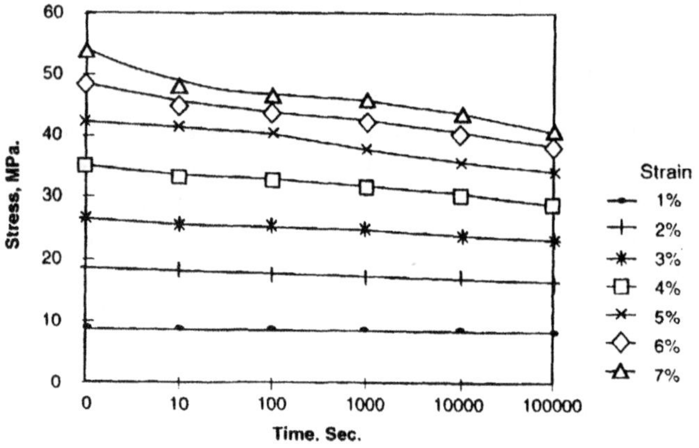

Figure 18.10. Stress-relaxation of solid polycarbonate (PC) at room temperature. "Reprinted from *Mechanics of Materials 24*, Haddad, Y. M. and Iyer, S. S., On the characterization of the stress-relaxation response of a class of linear viscoelastic material using acousto-ultrasonics: A pattern recognition approach, 199-211(1996), with kind permission from Elsevier Science".

TABLE 18.1. Factors affecting acousto-ultrasonic waveform measurement. Material: Solid Polycarbonate (PC); tested at room temperature (Haddad and Iyer, 1996)

Factor	Method adopted in the present research
Pressure on the transducer	Constant force clamps were used to obtain reproducible results; *see* Tanary (1988).
Couplant	Acoustic emission gel (SC-2) was chosen to obtain high, but consistent readings of AUP; *see* Iyer (1993).
Distance between the transducers	25.4 mm was chosen to obtain high readings of AUP that would still fall within the calibration level of the instrumentation, *see* Iyer (1993).
Internal parameters of experimental set-up	A frequency of 750 KHz and a sampling rate of 3.125 MHz with a gain of 60 decibels were chosen.

The magnitude of the "*relaxed*" stress in the material, for each testing time interval t, is also normalized with respect to the maximum of the average of its values obtained, at the same testing time interval t, for the five specimens tested at the particular strain level considered. It is expressed as

$$\sigma_{norm}(t)\big|_{\varepsilon} = \frac{\textit{Value of the "relaxed" stress in the specimen at time } t}{\textit{Maximum of the average of "relaxed" stresses in the specimens tested at time } t} \qquad \mathbf{(18.5)}$$

where $\sigma_{norm}(t)|_{\varepsilon}$ is the normalized value of the "*relaxed*" stress at time t under a particular level of strain input ε

Fig. 18.11 illustrates the correlations between the '*normalized acousto-ultrasonic parameter*', Eqn. (18.4), and the corresponding 'normalized relaxed stress', Eqn. (18.5), for the same "relaxed" stress state, i.e., at the same time for each of the strain levels considered. As shown in Fig. 18.11, for all strain levels considered, the value of the normalized AUP(t) increases linearly as the stress in the material relaxes at a constant strain level.

For the purpose of distinguishing between material states, three pattern classes were chosen for designing a pattern classifier. They are represented by the resultant acousto-ultrasonic waveforms belonging to three different stress states corresponding to time intervals: 1, 10000, and 100000 sec. These stress states are referred to, respectively as

Class 1, Class 2 and *Class 3*. The acousto-ultrasonic waveform characteristics obtained at these three stress states represent the raw data values for the pattern recognition analysis.

As discussed in the foregoing, important features that can readily distinguish a given waveform from other waveforms are ranked in order of their discriminatory power according to the problem being analysed. These selected features represent values in various domains whose units of measurement are different. Thus, a normalization process is initiated that normalizes the feature values of each of the class with respect to its mean and variance; e.g., Iyer (1993) and Haddad and Iyer (1995, 1996).

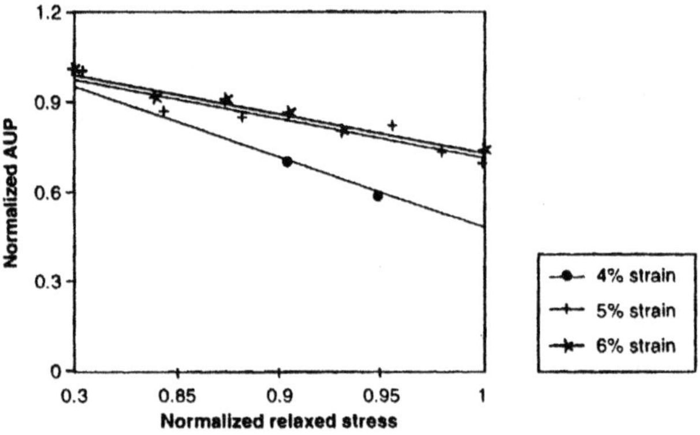

Figure 18.11. Correlation between AUP and '*relaxed*' stress at different times for strain levels of 4%, 5% and 6%. Material: Solid Polycarbonate (PC); tested at room temperature. "Reprinted from *Mechanics of Materials 24*, Haddad, Y. M. and Iyer, S. S., On the characterization of the stress-relaxation response of a class of linear viscoelastic material using acousto-ultrasonics: A pattern recognition approach, 199-211(1996), with kind permission from Elsevier Science".

The pattern classification approach that may be undertaken would depend on a prior knowledge of the decision boundaries between the different classes under consideration. For the purpose of designing a classifier, the normalized feature vectors are further split into training files and evaluation files. The normalized feature values of acousto-ultrasonic waveform obtained from two of the five specimens, tested at a particular relaxed stress state, were treated as '*unknown specimens*' and the acousto-ultrasonic waveforms obtained from them were used for testing the designed classifier. Each of Figures 18.12, 18.13 and 18.14 illustrates a two-dimensional feature space. The latter is an Euclidean feature space where the x-axis represents a selected feature and the y-axis represents another feature for the purpose of classification. The adopted features are selected through different '*iterations*' based on the extent of separation of '*clusters of pattern vectors*' belonging to the three pattern classes. Table 18.2 presents a list of features selected for the purpose of the present case study.

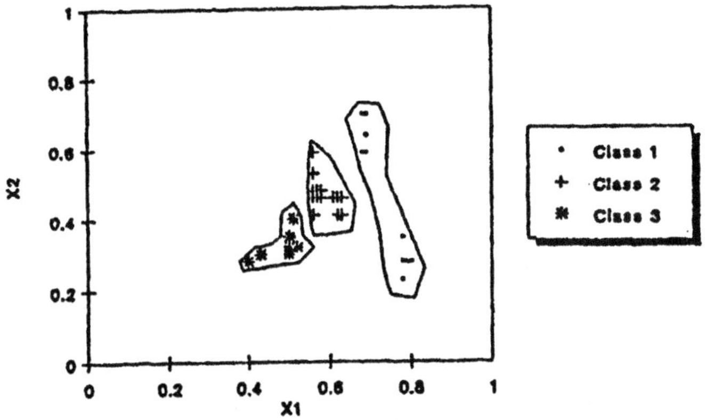

x_1 : Normalized 2nd greatest peak amplitude in power domain.
x_2 : Normalized % of partial power in 3rd octant in power domain.

Figure 18.12. Feature-Feature projection for two of the selected features for 4% strain level Material: Solid Polycarbonate (PC); tested at room temperature. "Reprinted from *Mechanics of Materials 24*, Haddad, Y. M. and Iyer, S. S., On the characterization of the stress-relaxation response of a class of linear viscoelastic material using acousto-ultrasonics: A pattern recognition approach, 199-211(1996), with kind permission from Elsevier Science".

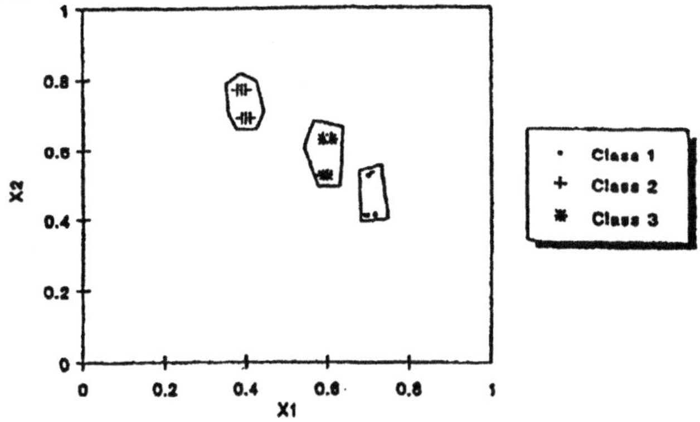

x_1 : Normalized 2nd greatest peak amplitude in power domain.
x_2 : Normalized % of partial power in 3rd octant in power domain.

Figure 18.13. Feature-Feature projection for two of the selected features for 5% strain level. Material: Solid Polycarbonate (PC); tested at room temperature. "Reprinted from *Mechanics of Materials 24*, Haddad, Y. M. and Iyer, S. S., On the characterization of the stress-relaxation response of a class of linear viscoelastic material using acousto-ultrasonics: A pattern recognition approach, 199-211(1996), with kind permission from Elsevier Science".

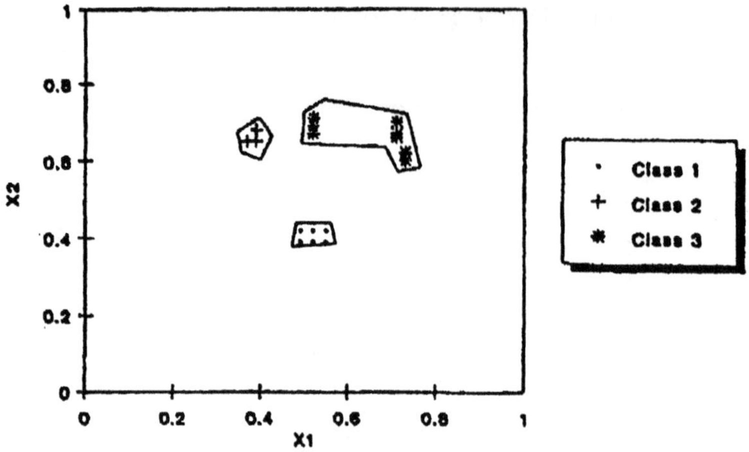

x_1 : Normalized 2nd greatest peak amplitude in power domain.
x_2 : Normalized % of partial power in 3rd octant in power domain.

Figure 18.14. Feature-Feature projection for two of the selected features for 6% strain level. Material: Solid Polycarbonate (PC); tested at room temperature. "Reprinted from *Mechanics of Materials 24*, Haddad, Y. M. and Iyer, S. S., On the characterization of the stress-relaxation response of a class of linear viscoelastic material using acousto-ultrasonics: A pattern recognition approach, 199-211(1996), with kind permission from Elsevier Science".

With reference to Figures 18.12, 18.13 and 18.14, for each considered strain level, Pattern class 1 corresponds to the 'original' or 'unrelaxed' stress state. Pattern class 2 represents the 'relaxed' stress state corresponding to the time interval of 10000 sec, Pattern class 3 represents the 'relaxed' stress state corresponding to the time interval of 100000 sec. The results for training, evaluation and testing of the classifiers designed at the strain levels of 4%, 5% and 6% are listed in Table 18.3. As seen from Table 18.3, for the training and evaluation process of the designed classifier, 100% of the pattern values belonging to each of the three pattern classes forms three separate clusters for each of the strain levels considered.

For the case of 4% strain level, the classification of unknown specimens to their respective class is almost 100%. For the 5% strain level, however, the classification rate is in the range of 71-91%. For 6% strain level, the classification rate is in the range of 89-98%. This indicates a satisfactory level of classification.

A review of literature revealed that no research work has been carried out yet, with the exception of the work presented here, on correlating the acousto-ultrasonic wave propagation data with the time-dependent behaviour of material systems, e.g., stress-relaxation of viscoelastic materials. Vary (1988), however, obtained correlations between the so-called "stress wave factor (SWF)", a quantifying parameter similar to the acousto-ultrasonic parameter (AUP), and the reduction in tensile strength of fibre-reinforced composites. In evaluating adhesive bonded joints under tensile loading, Tanary (1988), obtained a straight line fit for normalized AUP that decreased with the increase in stress level. Williams et al.(1984), also, reported a change in AU measurement of nylon ropes as the tension of the rope was increased. In this, the stress wave factor was found to decrease with the increase in the stress level.

TABLE 18.2. Normalized features selected for the identification of the stress-relaxation response of solid Polycarbonate (PC) specimens, tested at room temperature (Haddad and Iyer, 1996)

Acousto-Ultrasonic feature description	**Domain of the waveform**	**Feature number***
Inter peak distance from 1st to 2nd greatest.	Time	13
Number of peaks above the signal base line.	Power	37
2nd greatest peak amplitude	Power	43
% of partial power in the 3rd octant.	Power	49**
Greatest peak position	Phase	58
Greatest peak amplitude.	Time	5**
2nd greatest peak position	Cross-correlation	96*

* Corresponding to number identification in the employed software.
** For 5% strain level, features #5 and #96 were used instead of #49.

18.4.2. IDENTIFICATION OF RESIDUAL IMPACT PROPERTIES

Low-energy repeated-impact constitutes an important degrading factor in the residual ability of solid polymers to withstand static and/or dynamic loadings. Since this type of polymer degradation is likely to affect in-service structural components, quantitative non-destructive examination techniques are often considered to assess repeated-impact damage in polymeric material systems.

This case study is concerned with the application of acousto-ultrasonics, in conjunction with Pattern Recognition and Classification techniques, to the identification of residual impact properties of a class of polymeric material, namely, solid Polyvinylchloride (PVC), at room temperature. PVC specimens of different low-energy repeated impact damage states are processed by Acousto-ultrasonics (AU) to retrieve AU signals in the form of digitized records. These AU signals are grouped as distinct classes, each pertaining to a known level of repeated impact damage. Describing features of these AU signals are used to build pattern recognition classifiers. These classifiers are used to identify unknown damage states in other PVC specimens by classifying the retrieved AU signals as belonging to one of the already defined damage states (classes). Again, the obtained results indicate that Acousto-ultrasonics in combination with Pattern Recognition and Classification methodology can be used for the quantitative non-destructive identification of damage states in PVC specimens of unknown low-energy repeated impact conditions.

TABLE 18.3. Training, evaluation and testing of the classifier for the identification of the stress-relaxation response of solid Polycarbonate (PC) specimens, tested at room temperature. Classification scheme: K-Nearest Neighbour Classifier (Haddad and Iyer, 1996)

Strain level	Time, sec.	Corresponding pattern class	Training (%)	Evaluation (%)	Identification of unknown specimens (%)
4%	1	1	100.0	100.0	100.0
	10000	2	100.0	91.04	97.10
	100000	3	100.0	98.75	100.0
5%	1	1	100.0	100.0	91.40
	10000	2	100.0	100.0	71.23
	100000	3	100.0	100.0	79.27
6%	1	1	100.0	100.0	98.80
	10000	2	100.0	100.0	89.0
	100000	3	100.0	100.0	96.25

Controlled Parameters for AU Data-acquisition

The values of relevant parameters pertaining to the transducer configuration and instrumentation setup used for the acousto-ultrasonic measurements in the present case study are presented in Table 18.4.

Material, Test Specimens and Classifiers

A solid polymer, Polyvinylchloride (PVC) is chosen for this study. The material test specimens, of rectangular cross-section (i.e., 25.4 mm by 4.76 mm) and length of 210 mm are cut of a constant-thickness sheet (4.76 mm). They were subjected to different levels of controlled low-energy repeated-impact by repeatedly dropping a weight of 0.9 Kg from a height of 1.2 m a number of times.

TABLE 18.4. Controlled parameters for AU data-acquisition. Material: Solid Polyvinylchloride (PVC); tested at room temperature (Molina and Haddad, 1995)

Parameter	**Set Value**
Voltage Range	0 to +12 Volts
Gain	30 decibels
Wave Frequency	750 KHz
Sampling Rate	3.125 MHz
Distance Between Transducers	19 mm (0.75")
Couplant Medium	Ultragel II™

Three levels of repeated-impact damage are obtained by varying the number of impacts applied to each specimen. Accordingly, three classes of impact level are defined:

Class A : No impact,
Class B : Five impacts, and
Class C : Twenty impacts.

The specimens belonging to the above-mentioned classes of impact were tested using acousto-ultrasonics under identical experimental conditions as outlined in Table 18.4. Based on the retrieved AU signals, the following four classifiers are designed for each of the three input states (classes) using the procedure discussed earlier:

i) Linear Discriminant Classifier designed for the first best-ranked feature.
ii) Empirical Bayesian Classifier designed for the three best-ranked features.
iii) K-Nearest Neighbour Classifier designed for the four best-ranked features.
iv) Minimum Distance Classifier designed for the four best-ranked features.

Experimental Results of Pattern Recognition and Classification

Figure 18.15 displays the averages of "*Training*" and "*Evaluation*" performances for the three classes of impact as pertaining to each of the four designed classifiers mentioned above. As shown in the figure, good performances are obtained for the four designed classifiers. The maximum of the averages of both Training and Evaluation Performances are reached for the case of the K-nearest Neighbour Classifier.

Figure 18.16 presents the "*Classification*" performance for the four designed classifiers as the percentages of correctly classified AU signals that were not used in the classifier design. The experimental classification performance is obtained for each class by classifier testing with the respective Testing Set. Averages of the pertaining performances for the three classes of impact are also presented in Figure 18.16. Good performances are obtained for the four types of built classifiers. The maximum of Classification performance is reached for the case of Empirical Bayesian Classifier for the three impact classes under consideration.

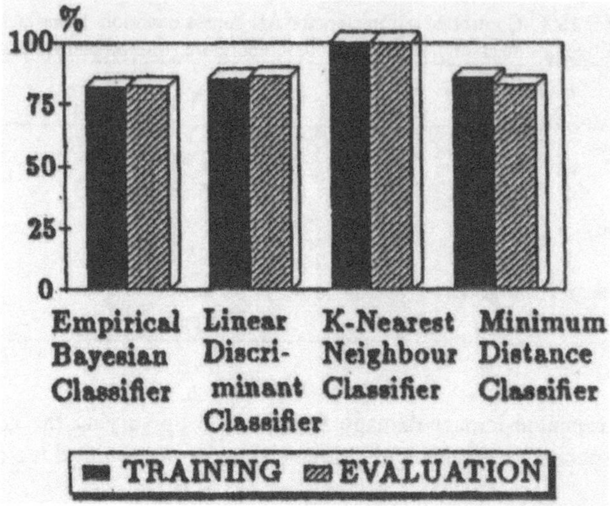

Figure 18.15. Training and evaluation performances in percentage of correctly AU-signals for the four designed classifiers as pertaining to repeated-impact damage classes of solid Polyvinylchloride (PVC) specimens; tested at room temperature. "Reprinted from Molina, G. J. and Haddad, Y. M. (1995) On the identification of residual impact properties of materials by acousto-ultrasonics - A pattern recognition approach, *Acta Mechanica Sinica* , Vol 11, No. 1, February 1995, 34-43, with kind permission from Allerton Press, Inc.".

Discussion of Classifier's Performance

With reference to Fig. 18.15, good results are experimentally obtained concerning the "*Training*" and "*Evaluation*" performances of the four designed classifiers. The fact that training and evaluation performances are as high as shown in Fig. 18.15 is an indication of the good repeatability of the AU-signal for the same specimen as well as between specimens of the same level of impact damage. On carrying out the present study, it was observed that classifiers that do not show good training and evaluation performances give subsequently poor "*classification*" performance. Thus, if a classifier is to be used in further identification of unknown-classification signals, excellent training and evaluation performances should be a minimum requirement for such classifier.

???? With reference to Fig. 18.16, better classifier performance corresponds, in general, to the identification of class A comprising undamaged specimens. However, on carrying out the present study, it was observed that AU effective signal-discrimination between different levels of impact damage may be more difficult than that between damaged and undamaged-specimen classes. This may be explained by a more overlap of the probability density functions pertaining to the two classes of effective damage (B & C) than that occurring between the probability density functions pertaining to the class A (of undamaged specimens) and any of the other mentioned two classes (B & C) of damaged specimens; Figure 18.17. An alternative explanation of the different discriminating ability of the classifier for classes of undamaged and damaged specimens might be given by a dispersion of testing-set values for the damaged specimens that is more extensive than that for the undamaged ones. However, such a phenomenon was not observed in the present study. As an example, Figure 18.18 presents a graph of the "*cluster boundaries*" for the Classifier Sets compared with the feature values for the three pertaining Testing-Sets. This two-feature space plot is illustrated for the case of the two best-ranked features. Similarly, more dispersion is observed among feature values for damaged Testing-Sets B and C, than that for values of the same features in the undamaged Testing-Set of class A. Figure 18.18 shows, in addition, that values of class A may be easily identified in the correct class, even if they show some dispersion with respect to the original Classifier-Set values. On the contrary, classes B and C may present a more difficult identification situation because of their overlapping. Bartos (1993), for instance, studied the feasibility of performing pattern analysis of impact-damage due to single impacts of different levels of energy on graphite-epoxy composite panels. Although using a different AU procedure from the one described here, Bartos (1993) obtained Bayesian Classifier's performances that are in good agreement with the performances obtained in the present study for this type of classifier.

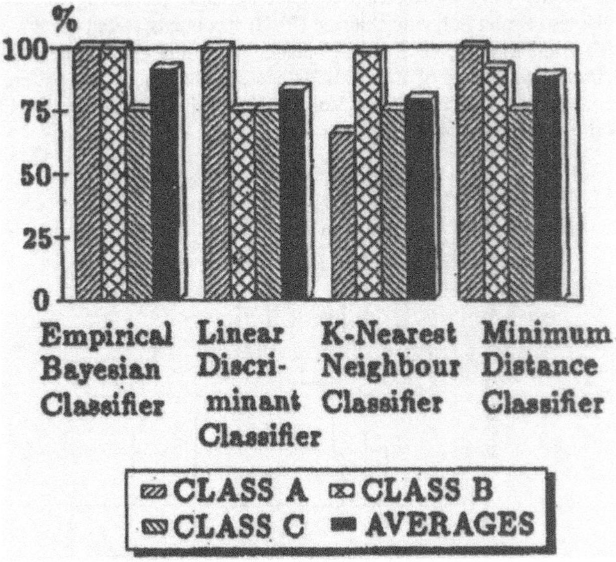

Figure 18.16. Classification performance in percentage of correctly classified AU-signals for the four designed classifiers of AU-signals pertaining to repeated-impact damage classes of solid Polyvinylchloride specimens; tested at room temperature. "Reprinted from Molina, G. J. and Haddad, Y. M. (1995) On the identification of residual impact properties of materials by acousto-ultrasonics - A pattern recognition approach, *Acta Mechanica Sinica* , Vol 11, No. 1, February 1995, 34-43, with kind permission from Allerton Press, Inc.".

With the above-discussed limitations, the presented application shows the viability of characterizing low-energy repeated impact damage in PVC specimens by AU methodology combined with Pattern Recognition and Classification (*see* Molina, 1994).

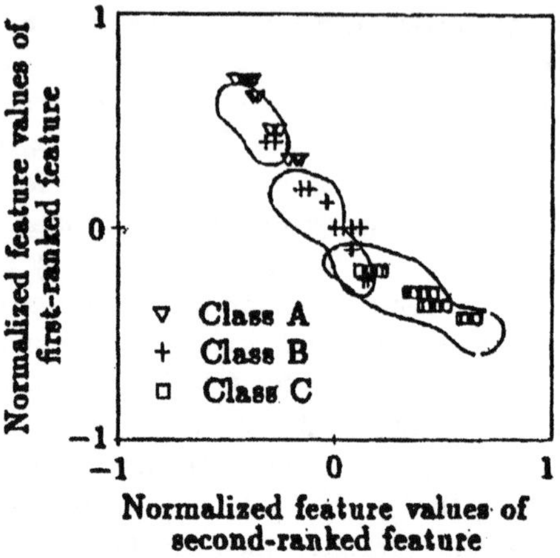

Figure 18.17. Probability density function for the first ranked feature (Greatest peak amplitude in the wave-autocorrelation domain). Data pertaining to repeated-impact damage classes of solid Polyvinylchloride (PVC) specimens; tested at room temperature. "Reprinted from Molina, G. J. and Haddad, Y. M. (1995) On the identification of residual impact properties of materials by acousto-ultrasonics - A pattern recognition approach, *Acta Mechanica Sinica* , Vol 11, No. 1, February 1995, 34-43, with kind permission from Allerton Press, Inc.".

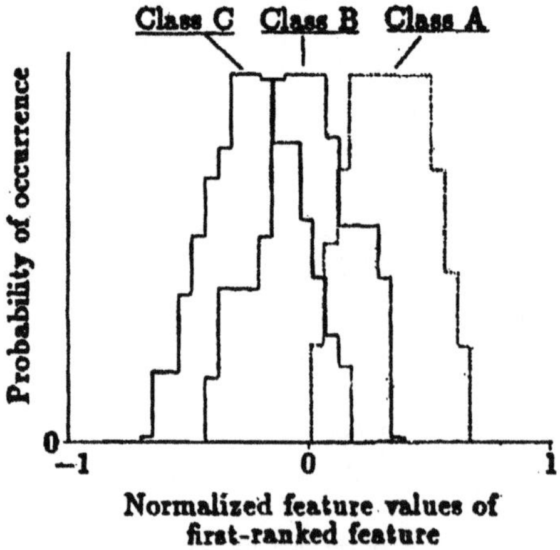

Figure 18.18. Cluster boundaries of classifier sets for the two best ranked features: First greatest peak amplitude in the wave-autocorrelation domain and second greatest peak amplitude in the wave-frequency domain. "Reprinted from Molina, G. J. and Haddad, Y. M. (1995) On the identification of residual impact properties of materials by acousto-ultrasonics - A pattern recognition approach, *Acta Mechanica Sinica* , Vol 11, No. 1, February 1995, 34-43, with kind permission from Allerton Press, Inc."

18.5 Design and Testing of a Pattern Recognition System

Following our discussion in the previous section, the choice of different sets of reference patterns may imply, according to Equation (18.3), different mean and/or standard deviation values for the normalization procedure and corresponding normalization outcomes. Various *Normalization Trees* can be proposed for the normalization of the involved Testing Sets. In the following, we discuss the suitability of a number of *Normalization Trees* for an adequate testing of PR-classifiers. In this context, we confine our attention to the impact case study presented above, whereby the following Normalization Trees are employed (e.g., Haddad and Molina, 1998):

NT(i): Normalization Tree (i), Figure 18.19, comprises the three Testing Sets.

NT(ii): Normalization Tree (ii), Figure 18.20, constitutes the pattern-vectors pertaining to a single specimen from one of the classes and the entire "Testing Sets" for the two other classes.

NT(iii): Normalization Tree (iii), Figure 18.21, includes the entire "Testing Set" for one of the classes.

NT(iv): Normalization Tree (iv), Figure 18.22, pertains to the "Testing Set" for one of the classes plus the three Classifier Sets previously used for building the PR-classifier.

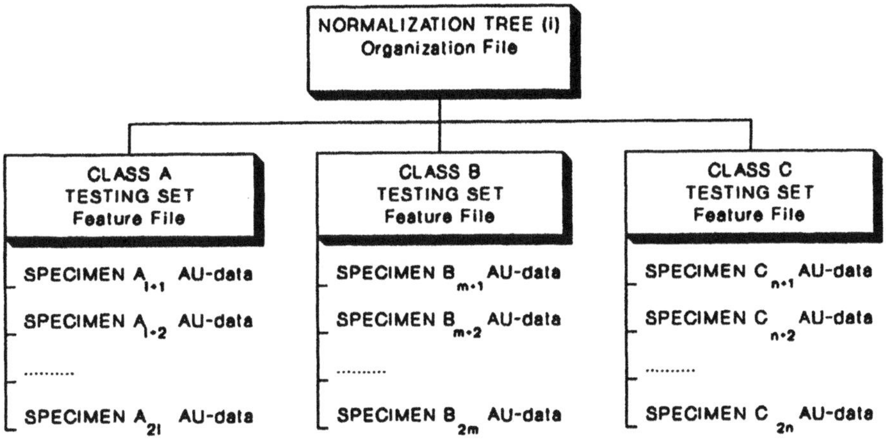

Figure 18.19. Normalization tree (i). Arrangement of AU-data for the normalization of "Testing Sets". "Reprinted from Haddad, Y. M. and Molina, G. J. (1998) On the design of acousto-ultrasonics - pattern recognition classifiers for the identification of material response states, Energy Sources Technology Conference & Exhibition, ETCE98-4572, Houston, Texas, February, 1998, with kind permission of ASME".

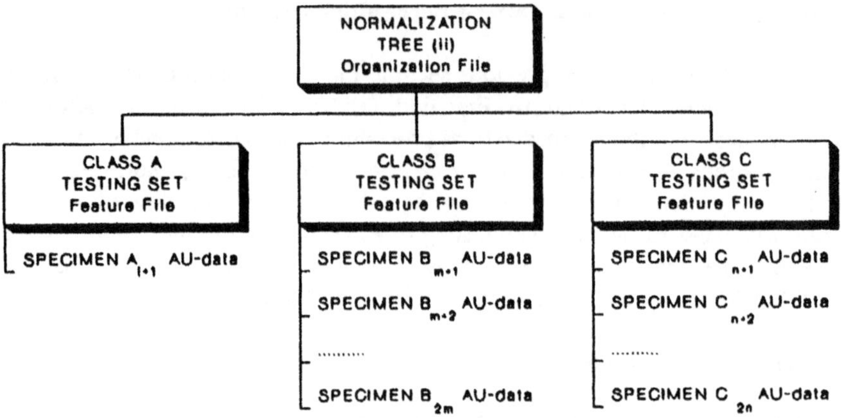

Figure 18.20. Normalization tree (ii). Arrangement of AU-data for the normalization of "Testing Sets". "Reprinted from Haddad, Y. M. and Molina, G. J. (1998) On the design of acousto-ultrasonics - pattern recognition classifiers for the identification of material response states, Energy Sources Technology Conference & Exhibition, ETCE98-4572, Houston, Texas, February, 1998, with kind permission of ASME".

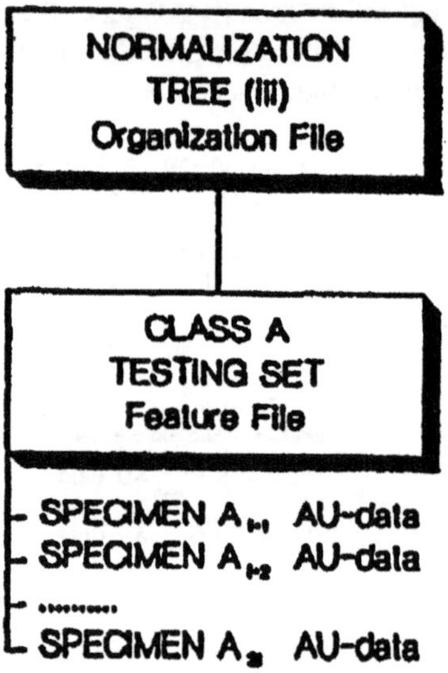

Figure 18.21. Normalization tree (iii). Arrangement of AU-data for the normalization of "Testing Sets". "Reprinted from Haddad, Y. M. and Molina, G. J. (1998) On the design of acousto-ultrasonics - pattern recognition classifiers for the identification of material response states, Energy Sources Technology Conference & Exhibition, ETCE98-4572, Houston, Texas, February, 1998, with kind permission of ASME".

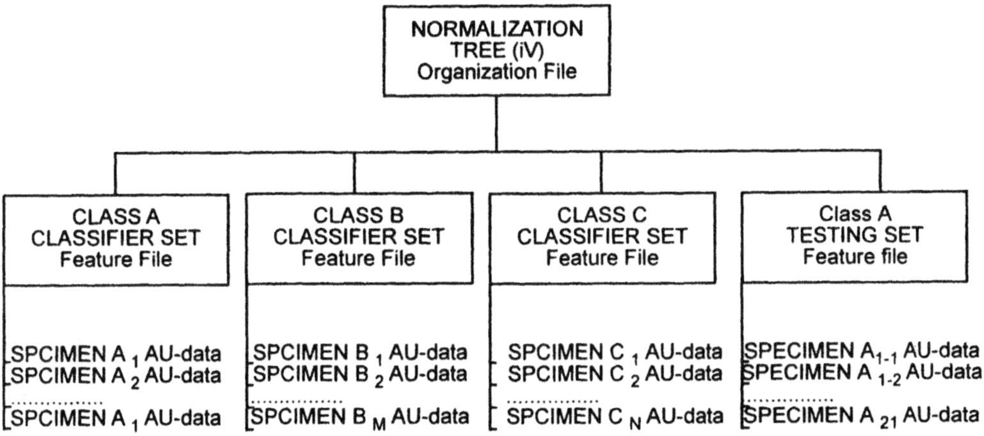

Figure 18.22. Normalization tree (iv). Arrangement of AU-data for the normalization of "Testing Sets". "Reprinted from Haddad, Y. M. and Molina, G. J. (1998) On the design of acousto-ultrasonics - pattern recognition classifiers for the identification of material response states, Energy Sources Technology Conference & Exhibition, ETCE98-4572, Houston, Texas, February, 1998, with kind permission of ASME".

The above types of Normalization Trees are examples of the normalization arrangements that may be proposed for the testing of PR-classifiers:

- *NT(i)* should be optimal for the employed normalization procedure and it may be a reference case for comparison with other Normalization Trees. Design of *NT (i)* requires, however, *a priori* complete knowledge of classification for the Testing Sets. It demands, before any normalization or classification to be carried out, that patterns be correctly identified into pertaining Testing Sets. Thus, building of *NT(i)* would not be possible if we were dealing with pattern vectors of unknown classification. Given this, *NT(i)* may be inadequate for testing the PR-classifiers.

- *NT (ii)* is a special case of *NT(i)*. It presents the situation that may arise in practice when normalizing by *NT(i)* a set of either known or unknown-classification patterns that comprises fewer patterns for one class than for the other two classes.

- *NT(iii)* is a special case of *NT(i)*. It presents the situation that may evolve when normalizing by *NT(i)* a set comprising only unknown-classification patterns which actually belong to a single class.

- *NT(iv)* is the proposed arrangement to solve the problem of adequate normalization in this study. The case represented in Fig. 18.22 pertains to the use of *NT(iv)* for testing the PR-classifiers by means of known-classification Testing Sets. If *NT*(iv) is employed for the normalization of unknown-classification patterns, they should be arranged as displayed in Fig. 18.23.

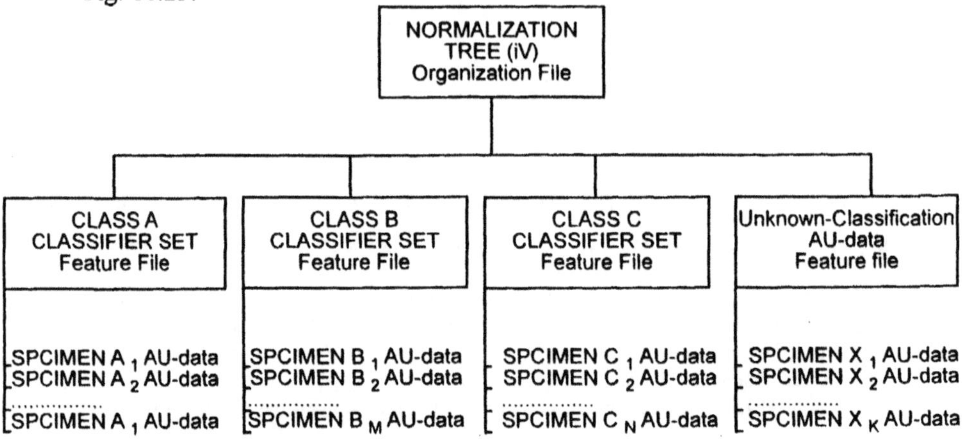

Figure 18.23. Normalization tree (iv). Arrangement for the normalization of AU-data with unknown classification.

Table 18.5 displays the experimental results for the four types of PR-classifiers when employed for testing the recognition performance for the considered three impact states of the material. The results of classification are presented, for each situation, as percentage of the correctly identified pattern vectors, as normalized by each of the above proposed types of Normalization Trees.

The results of Table 18.5 show, for the same type of PR-classifier, when identifying the same material state, significant differences of classification performance if the pertaining Testing Set was normalized by different Normalization Trees. In general, normalization by *NT(ii)* shows the highest classification performance, while *NT(iii)* shows the lowest classification performances. For *NT(iv)*, however, classification performances are quite close to those pertaining to *NT(i)*, specially for high rates of recognition.

It is evident from the results above that the normalization process can strongly influence the testing performance. Previous work showed that pattern vector discrimination is influenced by the overlapping of the probability density functions, of occurrence of a given feature-value in a class, between classes; e.g., Haddad and Iyer (1995), and Molina and Haddad (1995&1996). Figure 18.24 shows, for instance, the probability density function for the first-ranked feature (greatest peak amplitude in the wave-autocorrelation domain of description) in the three considered impact classes, as experimentally obtained for the *Classifier Sets* dealt with earlier. In general, the more the referred to probability density functions are separated for any two classes, the higher the corresponding classification performances for these classes.

Case *NT(ii)* can be analysed on the basis of the normalization expression (18.3). If the hypothesis is made that the global mean $\mu_{(ii)}$ of *NT(ii)* is not larger than the global mean $\mu_{(i)}$ of *NT(i)*, i.e., when,

$$\mu_{(ii)} \leq \mu_{(i)} \tag{18.6}$$

and

$$x_j - \mu_{(ii)} \geq x_j - \mu_{(i)} \tag{18.7}$$

where x_j is the non-normalized value of feature j in a pattern vector x.

It can be reasonably assumed, for the purpose of this analysis, that the standard deviation $s_{(ii)}$ for *NT(ii)*, as an average of corresponding values for all classes, where one of them is represented by a sample of pattern vectors pertaining to a single specimen, is not larger than $s_{(i)}$, obtained for the three classes considered in *NT(i)*, i.e.,

$$s_{(ii)} \leq s_{(i)} \tag{18.8}$$

By applying the above inequalities, (18.6) to (18.8), to the normalization equation (18.3), it follows that:

$$\frac{x_j - \mu_{(ii)}}{x_{(ii)}} \geq \frac{x_j - \mu_{(i)}}{x_{(i)}} \tag{18.9}$$

$$\xi_{j\,(ii)} \geq \xi_{j\,(i)} \tag{18.10}$$

where in the above equations (18.9) and (18.10), ξ_j is the normalized value of feature j for the pattern vector x, and subscripts *(i)* and *(ii)* refer, respectively, to *NT(i)* and *NT(ii)*.

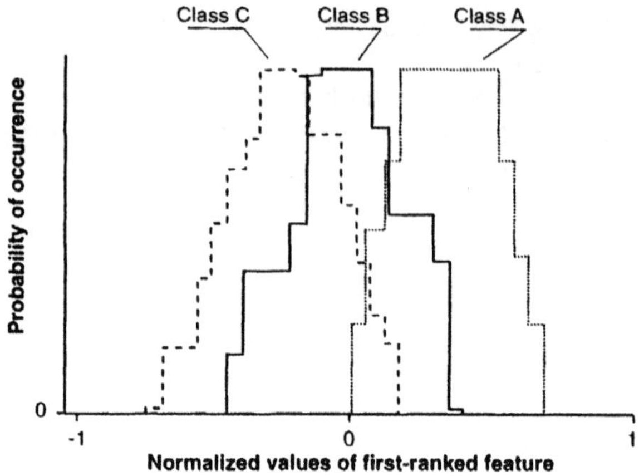

Figure 18.24. Probability density functions for the best ranked feature (Greatest peak amplitude in the wave autocorrelation domain).

Equations (18.6) to (18.10) above, indicate that *NT(ii)* may give higher values for normalized features than those correspondingly obtained by *NT (i)*. When, on the contrary, the mean of the feature values for the class is higher than the global mean of the Classifier, no such relation can be predicted for the normalized values. The analysis of equations (18.6) through (18.10) is schematically shown in Fig. 18.24, where the probabilities of occurrence p_j of feature j are assumed as Gaussian distributions.

Figure 18.25 displays the probability density functions of unnormalized feature j for the concerned classes in *NT(ii)*. This figure shows, also, the distributions of feature j as normalized by *NT(i)*. Figure 18.25 suggests that normalization by *NT(ii)* may reduce the overlapping of the probability density function distributions for the class under consideration. Thus, it may lead to a higher classification performance as confirmed by the corresponding results in Table 18.5.

TABLE 18.5. Testing performance for PR-classifiers of low-energy repeated-impact states in PVC specimens, tested at room temperature, when using different "Normalization Trees" for the testing sets

Type of Classifier	Class	Normalization Tree "NT"			
		(i)	*(ii)*	*(iii)*	*(iv)*
(I) Empirical Bayesian Classifier	Class A	93.33%	96.00%	20.67%	88.00%
	Class B	74.00%	75.00%	54.00%	79.00%
	Class C	75.00%	79.00%	6.00%	74.00%
	Average	80.78%	83.33%	26.89%	80.33%
(II) Linear Discriminant Classifier	Class A	99.33%	100.00%	38.67%	90.00%
	Class B	27.00%	30.00%	21.00%	34.00%
	Class C	57.00%	60.00%	39.00%	44.00%
	Average	61.11%	63.33%	32.89%	56.00%
(III) K-Nearest Neighbours Classifier	Class A	98.67%	100.00%	44.67%	91.33%
	Class B	66.00%	65.00%	27.00%	63.00%
	Class C	52.00%	62.00%	2.00%	47.00%
	Average	72.22%	75.67%	24.56%	67.11%
(IV) Minimum Distance Classifier	Class A	93.33%	100.00%	18.67%	88.00%
	Class B	75.00%	74.00%	54.00%	75.00%
	Class C	70.00%	84.00%	10.00%	65.00%
	Average	79.44%	86.00%	27.56%	76.00%

In general, Classifier performances for a given class, as tested by *NT(ii)*, depend on the relative position of the mean of the unnormalized class with respect to the global mean, for each of the concerned features. Since *NT(ii)* is a particular case of *NT(i)*, the latter may not be adequate for the normalization of Testing Sets wich are required for testing the Classifier.

Analysis of the case NT (iii) on the basis of the normalization expression (18.3) is trivial. Since global mean $\mu_{(iii)}$ and standard deviation $s_{(iii)}$ are those of the single class in the *NT(iii)*, the normalized distribution will be centred around zero-mean. Figure 18.26 presents schematics that can be used to compare the results pertaining to *NT(iii)* with those of *NT(i)*, where the probability of occurrence p_j of feature *j* is assumed to have a Gaussian distribution. It shows that *NT(iii)* gives, for the class under consideration, a normalized density function which may significantly overlap with the other two classes. This may explain why the classification performances are the lowest when experimentally tested by *NT(iii)*. Since *NT(iii)* is a particular case of *NT(i)*, the latter may not be adequate for the normalization of the involved Testing Sets in the testing of the pertaining Classifier.

Classification performances for *NT(iv)* are very close to those of *NT(i)*, specially for high percentages of recognition. Thus, *NT(iv)* may be adequate for testing the PR-classifier, given that the latter will be employed only when showing high recognition rates for the considered classes. Further, an important advantage of *NT(iv)* is that such normalization arrangement can always be built for both known and unknown classification patterns, as respectively, displayed in Figures 18.25 and 18.26. Thus, when using a PR Classifier, the operator must be provided with files containing the Classifier Sets, by which an *NT(iv)* can always be determined. Given this condition, testing of Classifiers by *NT(iv)* may give an adequate estimate of the probability of correct recognition for pattern vectors of *a priori* unknown classification.

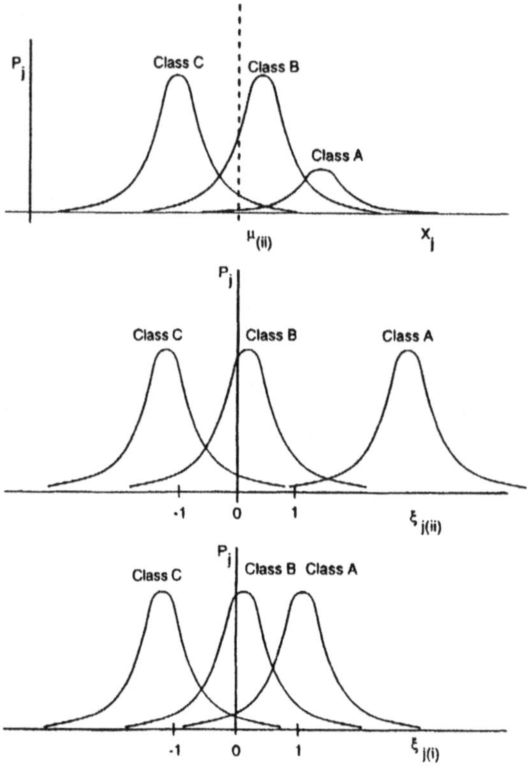

Figure 18.25. Schematics of the effects of Normalization Tree (ii) on the probability density function of the dealt with classes.

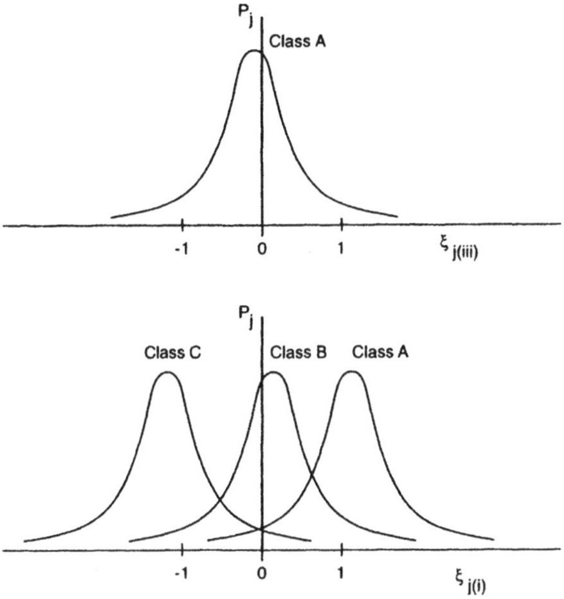

Figure 18.26. **Schematics of the effects of Normalization Tree (iii) on the probability density function of the dealt with classes.**

The presentation in the foregoing shows the influence of the "*normalization procedure*" on the testing of computer-based Classifier performance. There are a number of Normalization Trees that can be built in a multiple-class problem. However, a critical examination of the most intuitively obvious "*Normalization Tree (i)*" for the normalization of pattern vectors shows that such a normalization arrangement may be inadequate for the testing the PR-classifier.

Design of appropriate testing procedures requires, however, special consideration to the fact that testing a PR-classifier is performed in practice by pattern vectors whose classifications are *a priori* known. The intended purpose of such testing is to give an estimate of the Classifier success in the recognition of pattern vectors of unknown classification.

Testing of the Classification Performance requires that the Normalization Tree employed in the testing process should be available for the further normalization of pattern vectors of unknown classification. This study proposes a Normalization Tree comprising pattern-vectors from Testing and Classifier sets, which can always be built, irrespective of whether known or unknown classification was used for the concerned patterns. Experimental results show that testing of a PR-classifier by such proposed normalization arrangement may be adequate for the estimation of the PR-classifier performance.

18.6. References

Andrews, H.C. (1972) *Introduction to Mathematical Techniques in Pattern Recognition*, John Wiley & Sons Inc., Toronto, Canada.

Bartos, A.L. (1993) Acousto-ultrasonic pattern analysis of impact-damaged composite panels, *Topical Conference Proceedings of the 2nd Int. Conf. on Acousto-ultrasonics*, Vary A. (Editor), The American Society for Nondestructive Testing, Inc., Columbus, OH, pp. 133-42.

Chen, C.H. (1982) Statistical pattern recognition, in *Digital Waveform Processing and Recognition*, Chen, C.H. (Editor), CRC Press Inc., Boca Raton, Florida, pp. 59-74.

Devijver, P.A. (1982) Statistical pattern recognition, in *Applications of Pattern Recognition*, Fu, K.S. (Editor), CRC Press Inc., Boca Raton, pp. 15-36.

Dos Reis, H.L.M., and McFarland, D.M. (1986) On the acousto-ultrasonic evaluation of wire rope using the stress wave factor technique, *British Journal of Nondestructive Testing* **28(3)**, 155-56.

Finkel, J. (1975) Computer Aided Experimentation--Interfacing to Mini Computers, John Wiley and Sons, New York.

Fu, K.S. (1976) Introduction, in *Digital Pattern Recognition*, Fu, K.S., Keidel, W.D. and Walter, H. (Editors), Springer Verlag, New York, pp. 1-14.

Haddad, Y. M. and Iyer, S. S. (1995) An acousto-ultrasonics pattern recognition approach for the mechanical characterization of engineering materials, *Int. J. Pres. Ves. & Piping* **63**, 89-98.

Haddad, Y. M. and Iyer, S. S. (1996) On the characterization of the stress-relaxation response of a class of linear viscoelastic material using acousto-ultrasonics: A pattern recognition approach, *Mechanics of Materials* **24**, 199-211.

Haddad, Y. M. and Molina, G. J. (1998) On the design of acousto-ultrasonics - pattern recognition classifiers for the identification of material response states, Energy Sources Technology Conference & Exhibition, ASME, Houston, Texas, February 1998, ETCE98-4572.

Henneke II, E.G. (1983) NASA Contractor Report 3670, Feb. 1983.

Iyer, S.S. (1993) *On the Characterization of the Viscoelastic Response of a Class of Materials Using Acousto-ultrasonics - A Pattern Recognition Approach*, Master's Thesis, Univ. of Ottawa, Canada.

Iyer, S.S., and Haddad, Y.M. (1993) On the characterization of the linear viscoelastic response of a class of materials by acousto-ultrasonics - A pattern recognition approach, Proceedings of the 2nd Canadian Int. Composites Conf. and Exhibition (*CANCOM '93*), Ottawa, Canada, Wallace, W., Gauvin, R., and Hoa, S.V. (Editors), Canadian Assoc. for Composite Structures and Materials, pp. 479-89.

Lee, O.S. and Williams, J.H., Jr. (1991) Ultrasonic wave characterization of polymers, *Materials Evaluation* **49 (3)**, March 1991, pp. 68-72.

Molina, G. J. (1994) *On the Characterization of Impact Properties of Engineering Materials. An Acousto-Ultrasonics Pattern Recognition Approach*, Master's Thesis, University of Ottawa, Canada.

Molina, G. J. and Haddad, Y. M. (1995) On the identification of residual impact properties of materials by acousto-ultrasonics - A pattern recognition approach, *Acta Mechanica Sinica* **11(1)**, 34-43.

Molina, G. J. and Haddad, Y. M. (1996) Acousto-ultrasonics approach to the characterization of impact properties of a class of engineering materials, *Int. J. Pres. Ves. & Piping* **67**, 307-15.

Nayeb-Heshmi, N., Cohen, M.D., Zotos, J. and Poormand, R. (1986) Ultrasonic Characterization of Graphite/Epoxy Composite Material Subjected to Fatigue and Impact, *Journal of Nondestructive Evaluation* 5 (3/4), 119-31.

Phani, K.K., Niyogi, S.K., Maitra, A.K. and Roychaudhury, M. (1986) Strength and elastic modulus of a porous brittle solid: An acousto-ultrasonic study, *Journal of Material Science* **21(2)**, 4335-41, Dec. 1986.

Russell-Floyd, R. and Phillips, M.G. (1988) A critical assessment of acousto-ultrasonics as a method of nondestructuve examination for carbon fibre-reinforced thermoplastic laminates, *NDT International* **21(4)**, 247-57.

Tanary, S. (1988) *Characterization of Adhesively Bonded Joints Using Acousto-ultrasonics*, Master's Thesis, Univ. of Ottawa, Canada.

Tanary, S., Haddad, Y.M., Fahr, A. and Lee, S. (1992) Nondestructive evaluation of adhesively bonded joints in graphite/epoxy composites using acousto-ultrasonics, *J. Pressure Vessel Technology, Transactions of the ASME* **114**, Aug. 1992, 344-52.

Tou, J.T., and Gonzalez, R.C. (1974) *Pattern Recognition Principles*, AddisonWesley, Massachusetts.

Vary, A. (1982) Acousto-ultrasonic characterization of fiber reinforced composites, *Materials Evaluation.***4(6)**, 650-62.

Vary, A. (1987) *The Acousto-ultrasonics Approach*, NASA Technical Memorandum 89843.

Vary, A. (1988) The Acousto-Ultrasonic Approach, in *Acousto-Ultrasonics: Theory and Applications*, Edited by J.C. Duke, Jr., Plenum Press, New York, 1-21.

Williams, J.H., Jr., Hainsworth, J. and Lee, S.S. (1984) Acousto-ultrasonic nondestructive evaluation of double braided nylon ropes using the stress wave factor, *Fibre Science and Technology* **21(3)**, 169-80.

Williams, J.H., Jr., and Lampert, N.R. (1980) Ultrasonic evaluation of impact damaged graphite fiber composite, Material Evaluation **38(12)**, 68-72.

18.7. Further Reading

ASTM 3039-90 (1990) Standard Test methods for Impact Resistance of Rigid Plastic, *Annual Book of ASTM Standards* **08.04**, American Society for Testing and Materials, pp. 524-35.

Haddad, Y.M. (1995) *Viscoelasticity of Engineering Materials*, Kluwer, Dordrecht.

Huntsberger, D.V. and Billingsley, P. (1973) *Elements of Statistical Inference*, Allen and Bacon, New York, PP. 97-199.

APPENDIX D

THE Z-TRANSFORM

D.1. Introduction

The subject of this Appendix is of assistance particularly to the reader of Chapter 13, where the *z-transform* is used frequently in the presented analytical treatment. In this context, this appendix serves as a brief introduction of the concept of the *z-transform* and a number of its basic properties. For more comprehensive treatment of the subject matter, the reader is encouraged to consult the references provided at the end of the appendix.

The z-transform is the discrete-time counter-part of the Laplace transform. Meantime, it is the corresponding generalization of the discrete-time Fourier transform. Both Laplace and Fourier transforms have been dealt with in Appendix B. The properties of the z-transform parallel, in essence, those of Laplace transform, but, with some apparent variances or distinctions. These distinctions, as will be discussed in this Appendix, result essentially from fundamental differences between continuous- and discrete-time signals and systems. Similar to Laplace transform in the continuous-time case, and to Fourier transform in analysing both continuous-time and discrete-time data (signals), the z-transform is an important tool in performing transformations related to sequences in general.

The z-transformation of a sequence x[n] is denoted here by X(z), and is expressed as

$$X(z) = \sum_{n=-\infty}^{+\infty} x[n] Z^{-n} \qquad (1)$$

where z is a *complex variable*. Alternatively, the z-transform of the sequence x[n] may be written as Z{x[n]}. In this case, the relationship between x[n] and its z-transform is interpreted as

$$x[n] \overset{Z}{\leftrightarrow} X[z] \qquad (2)$$

As already mentioned, similar to both Laplace and Fourier transforms (*Appendix B*), the z-transform possesses basic properties which make it a valuable tool in the analysis of "***discrete-time***" signals and systems. Some of these properties are presented below.

D.2. Properties of the z-Transform

Linearity

If $\quad x_1[n] \overset{Z}{\leftrightarrow} X_1(z)$, ROC (Region of Convergence) = R_1

and $\quad x_2[n] \overset{Z}{\leftrightarrow} X_2(z)$, ROC = R_2

then, $\quad A x_1[n] + B x_2[2] \overset{Z}{\leftrightarrow} AX_1(z) + B X_2(z)$, ROC contains $R_1 \cap R_2$ **(3)**

i.e., the region of convergence (ROC) of the indicated linear combination is (at least) the intersection of the individual regions of convergence R_1 and R_2.

Time-shifting

If $\quad x[n] \overset{Z}{\leftrightarrow} X(z)$, ROC = R_x

then, $\quad x[n-n_o] \overset{Z}{\leftrightarrow} z^{-n_o}X(z)$, ROC = R_x (except for the possible addition or deletion of the origin or infinity). **(4)**

Frequency of Shifting

If $\quad x[n] \overset{Z}{\leftrightarrow} X(z)$, ROC = R_x

then,

$$e^{j\Omega_o n}x[n] \overset{Z}{\leftrightarrow} X(e^{-j\Omega_o}z), \quad \text{ROC} = R_x \tag{5}$$

Time-reversal

If $\quad x[n] \overset{Z}{\leftrightarrow} X(z)$, ROC = R_x

then,

$$x[-n] \overset{Z}{\leftrightarrow} X\left(\frac{1}{z}\right), \quad \text{ROC} = \frac{1}{R_x} \tag{6}$$

i.e., the region of convergence of x[-n] is an inversion of R_x. In other words, if z is in the region of convergence of X [n], then $\frac{1}{z_0}$ is in the region of convergence for x[-n].

Convolution Property

$$\text{If} \quad x_1[n] \overset{Z}{\leftrightarrow} X_1(z), \quad \text{ROC} = R_1$$

$$x_2[n] \overset{Z}{\leftrightarrow} X_2(z), \quad \text{ROC} = R_2$$

then,

$$x_1[n] * x_2[n] \overset{Z}{\leftrightarrow} X_1(z)X_2(z), \quad \text{ROC contains} \quad R_1 \cap R_2 \tag{7}$$

i.e., when two polynomials or power series $X_1(z)$ and $X_2(z)$ are multiplied, the coefficients in the polynomial representing the product are the convolution of the coefficients in the polynomials $X_1(z)$ and $X_2(z)$.

Differentiation (in the Z-domain)

$$\text{If} \quad x[n] \overset{Z}{\leftrightarrow} X(z), \quad \text{ROC} = R_x$$

then, by differentiating both sides of the z-transform expression (1), it follows that

$$nx[n] \overset{Z}{\leftrightarrow} -z\frac{dX(z)}{dz}, \quad \text{ROC} = R_x \tag{8}$$

D.3. Relations Between the z-Transform and Fourier Transform

There are a number of important relations between the z-transform and Fourier transform. In order to illustrate such relationships, one may express the complex variable z in the following polar form

$$z = r\,e^{j\omega} \tag{9}$$

in which r is the magnitude of z and ω is its angle.

In view of (1) and (9), one writes

$$X(r\,e^{j\omega}) = \sum_{n=-\infty}^{\infty} x[n](r\,e^{j\omega})^{-n}$$

or,

$$X(r\,e^{j\omega}) = \sum_{n=-\infty}^{\infty} \left\{x[n]r^{-n}\right\}e^{-j\omega n} \tag{10}$$

In view of the definition of Fourier transform (*Appendix B*), it is apparent that the above expression, reads as $X(r\,e^{j\omega})$, is the Fourier transform of the sequence x[n] multiplied by r^{-n}. That is

$$\begin{aligned} X(re^{j\omega}) &= S\left\{x[n]\,r^{-n}\right\} \\ &= X(z) \end{aligned} \tag{11}$$

for r =1, i.e. $|Z| = 1$, the z-transform (11) reduces to the Fourier transform of the sequence x[n], i.e.,

$$X(z)\,\big|_{z=e^{j\omega}} = S\left\{x[n]\right\} \tag{12}$$

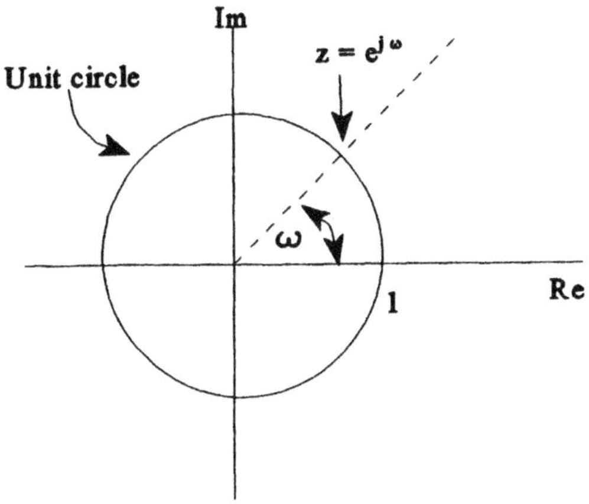

Figure D.1. The "*unit circle*" on the complex z-plane. (The z-transform reduces to the Fourier transform for values of *z* on the unit circle.)

Thus, the z-transform reduces to the Fourier transform when the magnitude of the transform variable is equal to unity (i.e., for $|Z| = 1$ or $z = e^{j\omega}$). In other words, and with reference to Figure D.1, the z-transform reduces to the Fourier transform, of the considered sequence, on the contour, in the complex z-plane, of a circle with a radius of unity ($|Z| = 1$). This circle in the z-plane is referred to as the "***unit circle***" and plays an important role in the discussions concerning the properties of the z-transform.

The convergence of the z-transform following immediately from relation (12). Referring to this relation, it is apparent that, for the convergence of z-transform, one requires that the Fourier transform of $x[n]\, r^{-n}$ converges. Thus, for a consequence $x[n]$, one may expect this convergence to occur for some values of $|Z| = r$ and not for others. In other words, there is associated with the z-transform, of a sequence $x[n]$, a rank of values of the complex variable z for which the transform X(z) converges. This rank of values is referred to as "***the region of convergence (ROC)***".

Example D.1

Consider the signal $x[n] = a^n u[n]$, where $u[n]$ is the unit-step time series. Then,

$$X(z) \triangleq \sum_{n=-\infty}^{+\infty} x[n] z^{-n} \qquad \textbf{(13)}$$

and,

$$X(z) = \sum_{n=-\infty}^{\infty} a^n u[n] z^{-n}$$

$$= \sum_{n=0}^{\infty} (a z^{-1})^n$$

For convergence of $X^{n=0}(z)$, we require that $\sum_{n=0}^{\infty} |a z^{-1}|^n < \infty$. Thus, the region of conveyance is the range of values of z for which $|a z| < 1$, or, equivalently, $|z| > |a|$. Then,

$$X(z) = \sum_{n=0}^{\infty} (a z^{-1})^n = \frac{1}{1 - a z^{-1}} = \frac{z}{z - a}, \quad |z| > |a|$$

Consequently, the z-transform converges for any finite value of *a*.

Example D.2

Consider $x[n] = -a^n u[-n-1]$, where u [n] is the unit-step time series. Then,

$$\begin{aligned} X(z) &= -\sum_{n=-\infty}^{+\infty} a^n u[-n-1] z^{-n} \\ &= -\sum_{n=-\infty}^{-1} a^n z^{-n} \\ &= -\sum_{n=1}^{\infty} a^{-n} z^n = -1 - \sum_{n=0}^{\infty} (a^{-n} z)^n \end{aligned} \tag{14}$$

If $|a^{-1} z| < 1$, or, equivalently, $|z| < |a|$, the sum in (14) converges and

$$X(z) = 1 - \frac{1}{1 - a^{-1} z} = \frac{1}{1 - a z^{-1}} = \frac{z}{z - a} \tag{15}$$

As with the Laplace transform, the determination of the z-transform requires both the algebraic expression and the region of convergence.

Example D.3

Consider the following signal which is the sum of two real exponential functions, and where u[n] is the unit-step time series.

$$x[n] = \left(\frac{1}{2}\right)^n u[n] + \left(\frac{1}{3}\right)^n u[n] \tag{16}$$

The z-transform is then

$$X(z) = \sum_{n=-\infty}^{+\infty} \left\{\left(\frac{1}{2}\right)^n u[n] + \left(\frac{1}{3}\right)^n u[n]\right\} z^{-n}$$

$$= \sum_{n=-\infty}^{+\infty} \left(\frac{1}{2}\right)^n u[n] z^{-n} + \sum_{n=-\infty}^{+\infty} \left(\frac{1}{3}\right)^n u[n] z^{-n}$$

$$= \sum_{n=0}^{\infty} \left(\frac{1}{2} z^{-1}\right)^n + \sum_{n=0}^{\infty} \left(\frac{1}{3} z^{-1}\right)^n$$

(17)

$$= \frac{1}{1 - \frac{1}{2} z^{-1}} + \frac{1}{1 - \frac{1}{3} z^{-1}} = \frac{2 - \left(\frac{5}{6}\right) z^{-1}}{\left(1 - \frac{1}{2} z^{-1}\right)\left(1 - \frac{1}{3} z^{-1}\right)}$$

$$= \frac{z\left(2z - \frac{5}{6}\right)}{\left(z - \frac{1}{2}\right)\left(z - \frac{1}{3}\right)}$$

(18)

For the convergence of X(z), both sums in equation (17) must converge, which requires that both

$$, \left|\frac{1}{2} z^{-1}\right| < 1 \quad \text{and} \quad \left|\frac{1}{3} z^{-1}\right| < 1$$

or, equivalently, $|z| > \frac{1}{2}$ and $|z| > \frac{1}{3}$. Thus, the region of convergence is $|z| > \frac{1}{2}$.

Example D.4

The z-transform for the Example D.3 above can also be obtained using the results of Example D.1. From the linearity property of the z-transform, that is, if x [n] is the sum of two terms, then X (z) will be the sum of the z-transforms of the individual terms and will converge when both z-transforms converge (*Section D.2 above*).

From Example D.3, one concludes that:

$$\left(\frac{1}{2}\right)^{n} u[n] \overset{Z}{\leftrightarrow} \frac{1}{1-\frac{1}{2}z^{-1}}, \quad |z| > \frac{1}{2} \tag{19}$$

$$\left(\frac{1}{3}\right)^{n} u[n] \overset{Z}{\leftrightarrow} \frac{1}{1-\frac{1}{3}z^{-1}}, \quad |z| > \frac{1}{3} \tag{20}$$

and consequently,

$$\left(\frac{1}{2}\right)^{n} u[n] + \left(\frac{1}{3}\right)^{n} u[n] \overset{Z}{\leftrightarrow} \frac{1}{1-\frac{1}{2}z^{-1}} + \frac{1}{1-\frac{1}{3}z^{-1}}, \quad |z| > \frac{1}{2} \tag{21}$$

D.4. Regions of Convergence for the z-Transform

In this section, a number of properties of the "***regions of convergence***" for the z-transform are presented (e.g., Oppenheim et al., 1983):

P.1: The ROC of X(z) consists of a ring in the z-plane centered about the origin.
This property is illustrated in Fig. D.2, below, and follows from the fact that the ROC consists of those values of $z = r\,e^{j\omega}$ for which x [n] r^{-n} has a Fourier transform that converges. Thus, convergence is dependent only on r = | z | and not on ω. Consequently, if a specific value of z is in the ROC, then all values of z on the same circle (i.e., with the same magnitude) will be in the ROC. This by itself guarantees that the ROC will consist of concentric rings. As a result of Property 6, below, the ROC must, in effect, consist of only a single ring. <u>In some</u>

cases, the inner boundary of the ROC may extend inward to the origin, thus reducing the RO. For other specific cases, however, the other boundary can extend outward to infinity (e.g., Oppenheim et al., 1983).

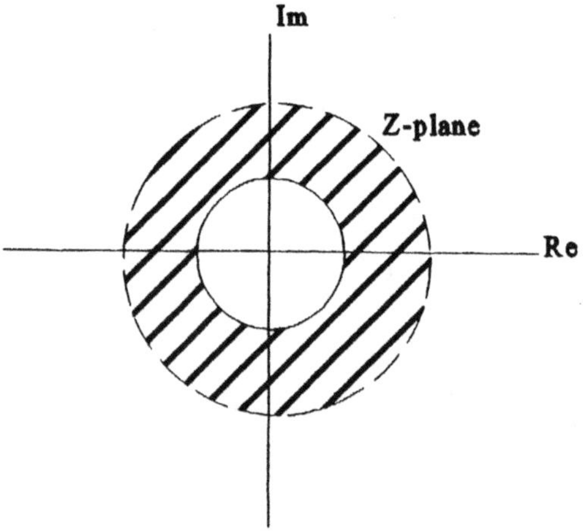

Figure D.2. ROC as a ring in the z-plane. For specific cases the inner boundary can extend inward to the origin in which case the ROC becomes a disc.

P.2: The ROC does not contain any poles.
As with the Laplace transform, this property is simply a consequence of the fact that a pole, X (z) is infinite and therefore by definition does not converge.

P.3: If x[n] is of finite duration, thus the ROC is the entire z-plane, except possibly z=0 and/or z =∞.
A finite-duration sequence has only a finite number of nonzero values, extending, say from $n = N_1$, to $n = N_2$, where N_1 and N_2 are finite. Thus, the z-transform X(z) is the sum of a finite number of terms, specifically,

$$X(z) = \sum_{n=N_1}^{N_2} x[n] z^{-n} \qquad (22)$$

In this context, the following remarks may be made:

- For z not equal to zero or infinity, each term in the sum (22) will be finite and consequently x (z) will converge.
- If N_1 is negative and N_2 positive so that x[n] has nonzero values both for n <0 and n>0, then the summation in (22) includes terms with positive powers of z and negative powers of z. As $|z| \to 0$, terms involving negative powers of z become unbounded, and as $|z| \to \infty$, those involving positive powers of z become unbounded. Consequently, for N_1 negative and N_2 positive the ROC does not include z = 0 or $z=\infty$.
- If N_1 is zero or positive, there are only negative powers of z in expression (22), and consequently the ROC includes $z = \infty$.
- If N_2 is zero or negative, there would be only positive powers of z in equation (22) and consequently the ROC includes z = 0.

P. 4: If x[n] is a right sided sequence, and if the circle $|z| = r_0$ is in the ROC, then all finite values of z for which $|z| > r_0$ will also be in the ROC.

In the context of this property, the following remarks may be made:

- A right-sided sequence is zero prior to some value of n, say N_0. If the circle $|z| = r_0$ is in the ROC, then, $x[n]\, r_0^{-n}$ can be (absolutely) summed up or, equivalently, the Fourier transform of $x[n]\, r_0^{-n}$ converges.
- Since x [n] is right-sided, the term x [n] multiplied by any real exponential sequence which, with increasing n, decays faster than r_0^{-n} can also be (absolutely) summed up. This (more) rapid exponential decay will further alternate sequence values of n to become unbounded since $x[n]\, z^{-n} = 0$ for $n < N_1$.
- For right-sided sequences in general, the z-transform takes the form

$$X(z) = \sum_{n=N_1}^{\infty} x[n]\, z^{-n} \tag{23}$$

where N_1 is finite and may be positive or negative:

- If N_1 is negative, then the summation in (23) includes terms with positive sources of z which become unbounded as $|z| \to \infty$. Consequently, for the right-sided sequences in general, the ROC will not include infinity. For the particular class of "***casual*** *" sequences, however, N_1 will be non negative, and, consequently, the ROC will extend to infinity.

*A signal is often referred to as "***casual***" if it corresponds to the impulse response of a casual system, i.e., is zero for t < 0 (continuous time), or, n < 0 (discrete time).

P. 5: If x[n] is a left-sided sequence and if the circle | z | = r_o is in the ROC, then, all values of z for which 0 < | z | < r_o will also be in the ROC.

- In general, for left-sided sequences, from the definition of the z-transform, the summation for the z-transform can be written in the form:

$$X(z) = \sum_{n=-\infty}^{N_2} x[n]\, z^{-n} \tag{24}$$

where N_2 may be positive or negative.

- If N_2 is positive, then eqn. (24) includes negative powers of z, which become unbounded as $|z| \to 0$. Thus, for left-sided sequences in general, the ROC will not contain z=0. For the particular class of left-sided sequences which are anti-casual [i.e., $x[n] = 0;\ n \geq 0$, so that N_2 in (24) is less than or equal to zero], the ROC will contain $z = 0$.

P. 6: If x[n] is two-sided and if the circle | z | = r_o is in the ROC, then the ROC will consist of a ring in the z-plane which includes the circle | z | = r_o.

- The ROC for a two-sided signal can be examined by expressing x [n] as the sum of a right-sided and left-sided signals:
- The ROC for the right-sided component is a region bounded on the inside by a circle and extending outward to (and possibly including) infinity.
- The ROC for the left-sided component is a region bounded on the outside by a circle and extending inward to, and possibly including, the origin.
- The ROC for the composite signal includes the intersection of the above two zones. As illustrated in Fig. D.3, the overlap (assuming it exists) is a ring in the z-plane.

D.5. The Inverse z-Transform

With the "***inverse z-transform***", we seek to determine a sequence when its z-transform is known. Expressing, as dealt with earlier, the z-transform as the Fourier transform of an exponentially weighted sequence, i. e.,

$$X(r\, e^{j\omega}) = S\{x[n]\, r^{-n}\} \tag{25}$$

where $|z| = r$ is in the ROC.

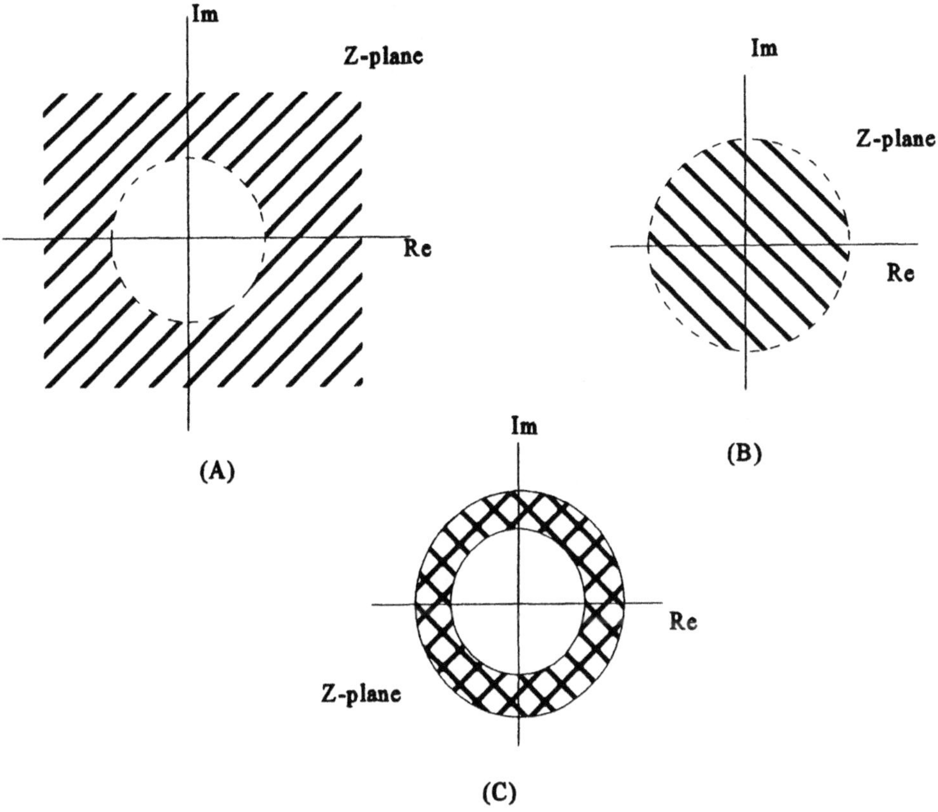

Figure D.3. (A) ROC for a right-sided sequence; (B) ROC for left-sided sequence; (C) The intersection of the ROC's in (A) and (B) represents the ROC for a two-sided sequence that is the sum of a right-sided and left-sided sequences.
(Adapted after Oppenheim et al., 1983)

Meantime, applying the inverse Fourier transform to both sides of (25),

$$x[n]\,r^{-n} = S^{-1}\left\{X(r\,e^{j\omega})\right\}$$

or

$$x[n] = r^{n}\,S^{-1}\left\{X(r\,e^{j\omega})\right\} \tag{26}$$

Using the inverse Fourier transform expression, it follows that

$$x[n] = r^{n} \frac{1}{2\pi} \int_{2\Pi} X(r e^{j\omega}) e^{j\omega n} d\omega$$

or

$$x[n] = \frac{1}{2\pi} \int_{2\Pi} X(r e^{j\omega}) (r e^{j\omega n}) d\omega \tag{27}$$

Changing the variable of integration from ω to *z*. With $z = r e^{j\omega}$ and r fixed, then $d z = j r e^{j\omega} = j \omega d\omega$ or $d\omega = (1/j) z^{-1} dz$. The integration in (27) is over a 2 π interval in ω which, in terms of z, corresponds to one interval around the circle $| z | = r$. Consequently, in terms of an integration in the z-plane, equation (27) can be written as

$$x[n] = \frac{1}{2\pi j} \oint X(z) z^{n-1} dz \tag{28}$$

where the $\oint$ denotes an integral about counter clockwise closed circular contour centered at the origin and with radius *r*. The value of *r* can be chosen as any value for which X(z) converges.

Equation (28) is the formal expression for the inverse z-transform and is the discrete-time expression for the inverse Laplace transform. Formal evaluation of the inverse transform integral Eqn. (28) requires the use of contour integration in the complex plane.

There are however, a number of alternative procedures for obtaining a sequence from its z-transform. As with Laplace transforms, one particular useful procedure for rational z-transforms consists of expanding the algebraic expression into a partial fraction expansion and recognizing the sequence associated with the individual terms.

TABLE D.1. Some common z-transform pairs**

x (n)	X (z)	Region of convergence, "ROC"
$\delta(n) = \begin{cases} 1 & n=0 \\ 0 & n \neq 0 \end{cases}$	1	All z
$\delta(n-m)$	z^{-m}	All z (except 0 if m > 0) or All z (except ∞ if m < 0)
$u(n) = \begin{cases} 1 & n \geq 0 \\ 0 & n<0 \end{cases}$	$\frac{z}{z-1}$	$\lvert z \rvert > 1$
$u(-n-1)$	$\frac{z}{1-z^{-1}}$	$\lvert z \rvert < 1$
$a^n u(n)$	$\frac{z}{z-a}$	$\lvert z \rvert > \lvert a \rvert$
$-a^n u(-n-1)$	$\frac{z}{z-a}$	$\lvert z \rvert < \lvert a \rvert$
$na^n u(n)$	$\frac{az}{(z-a)^2}$	$\lvert z \rvert > \lvert a \rvert$
$-na^n u(-n)$	$\frac{az}{(z-a)^2}$	$\lvert z \rvert < \lvert a \rvert$
$-na^n u(-n-1)$	$\frac{az}{(z-a)^2}$	$\lvert z \rvert < \lvert a \rvert$

x (n)	X (z)	Region of convergence, "ROC"
$\sin(n\omega)\,u(n)$	$\dfrac{z \sin \omega}{z^2 - 2z\cos\omega + 1}$	$\lvert z \rvert > 1$
$a^n \sin(n\omega)\,u(n)$	$\dfrac{az \sin \omega}{z^2 - 2az\cos\omega + a^2}$	$\lvert z \rvert > \lvert a \rvert$
$a^n \cos(n\omega)\,u(n)$	$\dfrac{z(z - a\cos\omega)}{z^2 - 2az\cos\omega + a^2}$	$\lvert z \rvert > \lvert a \rvert$

** where u(n) is the unit-step time series.

TABLE D.2. Fundamental properties of the z-transform

Property	x (n) y (n)	X (z) Y (z)	Region of convergence, "ROC"
1. Linearity	$\alpha x(n) + \beta y(n)$	$\alpha X(z) + \beta Y(z)$	
2. Time-shift	$x(n - m)$	$z^{-m} X(z)$	
3. Convolution	$\sum_{k=-\infty}^{\infty} x(k)y(n-k)$	$X(z)\,Y(z)$	
4. Product	$x(n)\,y(n)$	$\dfrac{1}{2\pi i} \oint_c X(\omega)\, Y\left(\dfrac{z}{\omega}\right) \omega^{-1} d\omega$	

Property	x (n) y (n)	X (z) Y (z)	Region of convergence, "ROC"
5. Time-transpose	x (-n)	$X(z^{-1})$	
6. Time-multiplication	nx (n)	$-z\dfrac{dX(z)}{dz}$	
7. Correlation	$\sum_{k=-\infty}^{\infty} x(k)y(n+k)$	$X(z^{-1})\,Y(z)$	
8. Exponential multiplication	$a^{n}x(n)$	$X(a^{-1}z)$	

D.6. Problems

1. Find the regions of convergence concerning the z-transform properties given in Table D.2 above.

2. Find X(z) and the corresponding regions of convergence for the following time series:

(i) $$X(n)=\frac{(n+k-1)!}{n!(k-1)}\,a^{n}u(n)$$

where u(n) is the unit-step time series.

(ii) $$X(n)=a^{|n|}$$

(iii) $$x(n)=t$$

(iv)

$$x(n) = e^{-at}$$

$$x(n) = 1 - e^{-at}$$

(v)

$$x(n) = \sin \omega t$$

$$x(n) = \cos \omega t$$

(vi)

$$x(n) = e^{-at} \sin \omega t$$

$$x(n) = e^{-at} \cos \omega t$$

D.7. References

Oppenheim, A. V., Willsky, A. S. and Young, I. T. (1983) *Signals and Systems*, Prentice-Hall, Englewood Cliffs, N. J.

D.8. Further Reading

Cadzow, J. A. (1987) *Foundations of Digital Signal Processing and Data Analysis*, Macmillan, New York.
Chen, C. T. (1979) *One-Dimensional Digital Signal Processing*, Marcel Dekker, New York.
Dorf, R. C. (1980) *Modern Control Systems*, Third Edition, Addison-Wesley, London.
Franks, L. E. (1969) *Signal Theory*, Prentice-Hall, Englewood Cliffs, N. J.
Hamming, R. W. (1977) *Digital Filters*, Prentice-Hall, Englewood Cliffs, N. J.
Jury, E. I. (1964) *Theory and Application of the z-Transform Method*, John Wiley, New York.
Oppenheim, A. V. and Schafer, R. W. (1975) *Digital Signal Processing*, Prentice-Hall, Englewood Cliffs, N. J.
Papoulis, A. (1977) *Signal Processing*, McGraw-Hill, New York.

SUBJECT INDEX

CUMULATIVE SUBJECT INDEX